P. D. Panagiotopoulos

Inequality Problems in Mechanics and Applications

Convex and Nonconvex Energy Functions

1985 Birkhäuser
Boston · Basel · Stuttgart

Author's Address:
P. D. Panagiotopoulos

Aristotle University	and	RWTH Aachen
Department of		Lehrstuhl und Institut
Civil Engineering		für Technische Mathematik
Thessaloniki		Aachen
Greece		F.R.G.

Library of Congress Cataloging in Publication Data

Panagiotopoulos, P. D., 1950
Inequality problems in mechanics and applications.
Includes bibliographical references.
1. Mechanics, Analytic. 2. Calculus of variations. 3. Inequalities (Mathematics)
I. Title.
QA808.P36 1984 531'.01'51526 82-9519
AACR2

CIP-Kurztitelaufnahme der Deutschen Bibliothek

Panagiotopoulos, Panagiotis D.:
Inequality problems in mechanics and applications : convex and nonconvex energy functions / P. D. Panagiotopoulos. – Boston ; Basel ; Stuttgart : Birkhäuser, 1985.

ISBN-13: 978-0-8176-3094-2 e-ISBN-13: 978-1-4612-5152-1
DOI: 10.1007/978-1-4612-5152-1

Softcover reprint of the hardcover 1st edition 1985

To my parents

Contents

Preface

In a remarkably short time, the field of inequality problems has seen considerable development in mathematics and theoretical mechanics. Applied mechanics and the engineering sciences have also benefitted from these developments in that open problems have been treated and entirely new classes of problems have been formulated and solved.

This book is an outgrowth of seven years of seminars and courses on inequality problems in mechanics for a variety of audiences in the Technical University of Aachen, the Aristotle University of Thessaloniki, the University of Hamburg and the Technical University of Milan. The book is intended for a variety of readers, mathematicians and engineers alike, as is detailed in the Guidelines for the Reader.

It goes without saying that the work of G. Fichera, J. L. Lions, G. Maier, J. J. Moreau in originating and developing the theory of inequality problems has considerably influenced the present book.

I also wish to acknowledge the helpful comments received from C. Bisbos, J. Haslinger, B. Kawohl, H. Matthies, H. O. May, D. Talaslidis and B. Werner. Credit is also due to G. Kyriakopoulos and T. Mandopoulou for their exceptionally diligent work in the preparation of the final figures. Many thanks are also due to T. Finnegan and J. Gateley for their friendly assistance from the linguistic standpoint. I would also like to thank my editors in Birkhäuser Verlag for their cooperation, and all those who helped in the preparation of the manuscript.

I must apologize to those whose work was inadvertently neglected in compiling the literature for this book. I shall welcome all comments and corrections from readers.

P. D. Panagiotopoulos
Aachen, January 1983

Introduction

The aim of the present book is the study of problems in mechanics whose variational forms are inequalities, expressing the principle of virtual power in its inequality form. The treatment of such problems, here called inequality or unilateral problems, differs fundamentally from that of problems whose variational forms are equalities. We shall call the latter equality or bilateral problems.

The inequality form of the principle of virtual power is attributed to Fourier; since that time little work has been done, until recently, on inequality problems, primarily because there existed no sufficiently developed mathematical theory. The mathematical study of inequality problems began in the early sixties with the works of G. Fichera, J. L. Lions and G. Stampacchia. J. J. Moreau connected the theory of variational inequalities to convex analysis and especially to subdifferentiability by introducing the notion of superpotential; the theory of maximal monotone operators was then related to inequality problems. Independently of the above authors, who mainly treated mathematical aspects of the theory of inequality problems, G. Maier formulated and studied such problems in applied mechanics using the methods of optimization theory. Until the present decade, all the inequality problems studied were related to convex energy functions. Since the gradient (or more generally the subdifferential) of a convex function is a monotone mapping, it is apparent that the inequality problems studied were firmly tied to monotonicity, e.g. only monotone stress-strain laws and boundary conditions could be treated. In order to overcome this limitation, an approach no longer requiring convexity was initiated by the author, making use of two new notions: the generalized gradient of F. H. Clarke and the derivate container of J. Warga. In the case of nonconvex energy functions, we obtain variational expressions which we call hemivariational inequalities. While in the convex case static variational inequalities generally give rise to minimization problems of potential and complementary energy, in the nonconvex case substationarity problems result. We would like to point out here that all inequality problems may also be expressed in terms of multivalued operator equations.

It is worth noting that while bilateral problems are connected with differentiable energy functions, inequality problems involve functions which are nondifferentiable. In the former case, the generalized forces are derived from a potential (by forming the gradient), and in the latter case from a convex superpotential (by forming the subdifferential) or a nonconvex superpotential (by forming Clarke's generalized gradient or Warga's derivate container). Because of the multivalued character and lack of differentiability of superpotentials, one could speak of "multivalued mechanics" or of "nonsmooth mechanics".

Concerning numerical treatment, we must distinguish between dynamic and static inequality problems. For the former, we may employ a time discretization procedure to reduce the problem to a sequence of static problems. Alternatively, the nondifferentiable energy functions may be smoothed by a regularization procedure, transforming the problem into a sequence of dynamic equality problems which give rise to nonlinear differential equations. For static problems there are two main possibilities. If the problem can be formulated as a minimization problem, then optimization algorithms are used; otherwise, a smoothing procedure can be employed to obtain a sequence of nonlinear algebraic equations.

The present book is divided into Introductory Topics (Chapters 1 and 2), Inequality Problems (Chapters 3 to 9), and Numerical Applications (Chapters 10 and 11).

After beginning with the necessary mathematical background in the first two Chapters, we then deal in Chapters 3 and 4 with convex and nonconvex superpotentials in various areas of nonrelativistic mechanics. These two Chapters emphasize the relation of inequality problems to multivalued differential or integral equations, to minimization theory and to substationarity theory. The rest of Part 2 concerns the mathematical study of unilateral boundary value problems (B.V.P.s). Starting with friction problems (Chapter 5), we examine B.V.P.s arising from subdifferential boundary conditions and material laws for deformable bodies (Chapter 6). We then formulate and study inequality problems in the theory of von Kármán plates, in which context special attention is paid to eigenvalue problems for variational inequalities (Chapter 7). Chapters 8 and 9 concern inequality problems in thermoelasticity, plasticity and viscoplasticity. The emphasis in this part in favor of convex problems is caused by the fact that the mathematical theory of hemivariational inequalities started very recently and is still in full development. The last two Chapters are written in a different spirit; they are more engineering oriented, although the use of complicated techniques of mechanics is avoided. In these Chapters, actual applications from the engineering sciences illustrate the numerical treatment of inequality problems. The numerical calculations use well-known optimization algorithms, which are available in the form of computer programs, in the belief that this constitutes a substantial help for the engineer.

Guidelines for the Reader

The choice of material in Part 1 was governed by the requirements of the subsequent Chapters. Every definition or proposition in this Part is applied in the rest of the book. We expect the reader to have some knowledge of basic functional analysis. The reader who is familiar with the theory of Sobolev spaces need not read the first Chapter. Each of Chapters 5 to 9 could be read separately.

In spite of a tendency toward mathematical exactness, the book is written as well for a reader unfamiliar with functional analysis who is interested only in mechanics and in applications. In this case, Chapter 1 could be omitted, and in Chapter 2 proofs could be disregarded and the reader should assume that X is simply n-dimensional space and that $\langle \cdot, \cdot \rangle$ denotes its usual inner product. Chapter 3 can then be read up to Sec. 3.3.4, and Secs. 4.1.1 and 4.1.2 can be passed over. In the remaining Chapters, the existence propositions need not be read. Each variational expression can be understood in the "engineering sense" (cf., the book by K. Washizu [333]). In Chapters 10 and 11 the problems are discretized, and therefore these two Chapters are addressed equally to readers familiar with functional analysis or not.

One further remark on terminology. The term "pointwise" refers generally (as, e.g., on pp. 111, 112) to a condition that holds "at almost every point"; however, in all classical formulations of boundary conditions (as, e.g., in Sec. 3.3.1) or B.V.P.'s, it has the meaning "at every point". In any case of ambiguity we have made clear the distinction.

Part 1
Introductory Topics

Chapter 1

Essential Notions and Propositions of Functional Analysis

The aim of Chapter 1 is to provide some notions and propositions of functional analysis which will be necessary in the next chapters for the study of inequality problems in mechanics. Commencing with the notion of topological vector spaces and the corresponding notion of duality, we give some properties of certain function spaces. Particular attention is paid to Sobolev spaces and spaces of functions of bounded deformation for which the trace theorems and some imbedding properties are given. Korn's inequalities and the Green–Gauss theorem are also presented. Elements of the theory of vector-valued functions and distributions as well as of differential calculus close this chapter.

In preparing this chapter a number of well-known works have been followed, in particular those of:

R. A. Adams [2], N. Bourbaki [24] [25] [26], P. G. Ciarlet [41], J. Dieudonné [70], N. Dunford and J. T. Schwarz [79], G. Fichera [98], I. M. Gelfand and G. E. Schilow [109], G. Köthe [164], J. L. Lions [179] [182] [183], J. L. Lions and E. Magenes [181], J. Nečas [231], A. P. Robertson and W. Robertson [284], L. Schwartz [297], W. I. Smirnov [302], F. Treves [320], M. M. Vainberg [324], N. Ya. Vilenkin [325], B. von Querenburg [277], and K. Yosida [337].

1.1 Topological Vector Spaces and Related Subjects

1.1.1 Topological Spaces and Continuous Mappings

A topological vector space is a set which carries two structures, one of topological space (T.S.) and one of vector space (V.S.). These two structures comply with a compatibility requirement in that the algebraic operations are continuous. First, let us review briefly the definitions and some elementary properties of T.S.s and V.S.s.

A T.S. is a set $\mathscr{X}$ endowed with a structure that makes possible definitions of convergence and continuity. In order to define such a structure it suffices to specify the open subsets of $\mathscr{X}$. A set S of subsets of $\mathscr{X}$ which contains

(i) every union of its elements,
(ii) every finite intersection of its elements and
(iii) $\varnothing$ and $\mathscr{X}$

is a topology on $\mathscr{X}$.

The subsets of S are called open sets, $X = \{\mathscr{X}, S\}$ is a topological space and its elements x are the points of X. A subset A of X is said to be closed if its complement $X - A$ is open. The sets $\varnothing$ and $\mathscr{X}$ are both closed. A set $F \subset X$ is called a neighborhood of a point x if there exists an open set $O \in S$ such that $x \in O \subset F$ ($F \equiv O$ possible). A point $x \in A \subset X$ is called an interior point of A if there exists a neighborhood F of x such that $F \subset A$. The set of all the interior points of A is the interior of A (int A). The closure of A, denoted $\bar{A}$, is the set of those points of X (closure points of A) such that every neighborhood of x has a nonempty intersection with A. As is obvious, A is open (resp. closed) if and only if $A = \text{int}\, A$ (resp. $A = \bar{A}$). The boundary of A is the set $\bar{A} - \text{int}\, A$. We say that a set B is dense in another set A if $A \subset \bar{B}$. In particular, B is said to be dense in $\mathscr{X}$, or everywhere dense, if $\bar{B} = \mathscr{X}$. A T.S. X is separable if $\mathscr{X}$ contains a finite or denumerable dense subset. A point $x \in A$ is a limit point of A if every neighborhood of x contains at least one point of A different from x. A point x is a limit of the sequence $\{x_i\}$ and we write $x_i \to x$ if every neighborhood of x contains all the elements of the sequence up to a finite number. Then we say that the sequence $\{x_i\}$ is convergent to x.

By means of the notion of neighborhood, an equivalent definition for the topology can be given. The set F_x of all neighborhoods of a point x, called also the filter of neighborhoods of the point x, has the following properties:

(i) $x \in F, \forall F \in F_x$;
(ii) if $F_1 \in F_x$ and $F_2 \in F_x$, then $F_1 \cap F_2 \in F_x$;
(iii) if $F_1 \in F_x$ and $F_1 \subset F_2 \subset \mathscr{X}$, then $F_2 \in F_x$; and
(iv) if $F_1 \in F_x$, there exists a neighborhood $F_2 \in F_x$ such that $F_1 \in F_y, \forall y \in F_2$.

Conversely, if for every point x of a set $\mathscr{X}$ a set F_x of subsets of $\mathscr{X}$ is defined such that the foregoing conditions are satisfied, then there exists exactly one topology on $\mathscr{X}$ such that for each $x \in \mathscr{X}$ F_x is the set of neighborhoods of x. A subset G_x of F_x, with the property that for $F \in F_x$ there exists $G \in G_x$ such that $G \subset F$, is called a base of neighborhoods of x.

On a set $\mathscr{X}$, several different topologies can be defined. We say that the topology S_1 is finer or stronger than the topology S_2, and that S_2 is coarser or weaker than S_1, if $S_2 \subset S_1$. In this case, every subset of $\mathscr{X}$ which is open in S_2 is also open in S_1. It may also occur that, given two topologies on $\mathscr{X}$, neither is finer than the other, i.e., they are not comparable. The discrete topology on $\mathscr{X}$ (in this case F_x consists of every subset of $\mathscr{X}$ containing x) is the strongest topology on $\mathscr{X}$, and the trivial topology ($F_x = \{\mathscr{X}\}$) is the weakest. If X is a T.S. and $\mathscr{X}_1 \subset \mathscr{X}$, then $\mathscr{X}_1$ can be topologized by defining the open sets of $\mathscr{X}_1$ as the intersections of the open sets of $\mathscr{X}$ with $\mathscr{X}_1$. This topology is called the induced topology on $\mathscr{X}_1$. Let $\{X_\alpha\}$, $\alpha \in \Gamma$, be a family of T.S.s $X_\alpha = \{\mathscr{X}_a, S_a\}$, $\alpha \in \Gamma$, and let $\Pi_{\alpha \in \Gamma} X_a$ be their Cartesian product. Then the product topology is the least fine topology containing sets of the form $\prod_{\alpha \in \Gamma} O_a$, where $O_a \in S_a$ and $O_a = \mathscr{X}_a$, up to a finite number of O_a.

A T.S. X is called a Hausdorff or separated topological space (H.T.S.) if for any two distinct points x and y of X disjoint neighborhoods F and G of x and y respectively can be determined. A H.T.S. has the property that a sequence converges to one point at most. In such a space a set consisting of a single point is closed.

An important notion is the continuity of a mapping and the semi-continuity of a functional. Let $X = \{\mathscr{X}, S\}$ and $Y = \{\mathscr{Y}, T\}$ be two T.S.s. The function $f: X \to Y$ is said to be continuous at x if for every neighborhood G of $f(x) \in Y$ there is a neighborhood F of x such that $f(F) \subset G$. If f is continuous at every $x \in X$, f is said to be continuous on X. If both f and f^{-1} are continuous, f is called bicontinuous. A mapping $f: X \to Y$ which is bijective and bicontinuous is a homeomorphism. A functional $f: X \to \bar{\mathbb{R}} = [-\infty, +\infty]$ is said to be lower (resp. upper) semicontinuous at x (written in abbreviated form l.s.c. (resp. u.s.c.)), if the following holds: for every $a \in \mathbb{R}$ with $f(x) > a$ (resp. $f(x) < a$) a neighborhood F of x can be determined such that $f(y) > a$ (resp. $f(y) < a$) for every $y \in F$. We say that f is l.s.c. or u.s.c. on X if this property holds for every $x \in X$.

An important class of T.S.'s is that of metric spaces. A function $d: \mathscr{X} \times \mathscr{X} \to \mathbb{R}$ is called a distance on the set $\mathscr{X}$ if

(i) $d(x, y) \geq 0$, and $d(x, y) = 0$ if and only if $x = y$,
(ii) $d(x, y) = d(y, x)$, $\forall x, y \in \mathscr{X}$ and
(iii) $d(x, y) \leq d(x, z) + d(z, y)$, $\forall x, y, z \in \mathscr{X}$.

The set $B(x, a) = \{y \mid y \in \mathscr{X}, d(x, y) < (\text{resp.} \leq) a\}$ is called the open (resp. closed) ball with center x and radius a. An open or closed set $F \subset \mathscr{X}$ with $x \in F$ is a neighborhood of x if it contains for some $a > 0$ an open ball $B(x, a)$. The set F_x of all neighborhoods of x as previously defined, generates a metric topology on X. A T.S. $X = \{\mathscr{X}, S\}$ is metrizable if a distance function d can be defined introducing a metric topology that coincides with S. A metrizable space is separated. In a metric space X, a set A is bounded if there exists $a \in \mathbb{R}$ such that $d(x, y) \leq a$ for every $x, y \in A$. A subset A of a metric space X is said to be complete if every Cauchy sequence in A has a limit in A. We recall that a sequence $\{x_i\} \in X$ is a Cauchy sequence if for every $\varepsilon > 0$ there exists a natural number $N(\varepsilon)$ such that for $i, j \geq N(\varepsilon)$, $d(x_i, y_j) \leq \varepsilon$. In a metric space X the continuity of a mapping $f: X \to Y$ at a point x, where Y is a T.S., is equivalent to the condition that $f(x_j) \to f(x)$ whenever $x_j \to x$.

1.1.2 Locally Convex Topological Vector Spaces, Normed Spaces and Linear Mappings

We recall the notion of a vector space (V.S.), or linear space, over the field of real numbers. A vector space (V.S.) over the field of reals is a set $\mathscr{X}$ for which two functions A_V: $\mathscr{X} \times \mathscr{X} \to \mathscr{X}$ and M_S: $\mathbb{R} \times \mathscr{X} \to \mathscr{X}$ are defined, written

$$A_V(x, y) = x + y, \quad \forall x, y \in \mathscr{X}$$

and
$$M_S(a, x) = ax, \quad \forall a \in \mathbb{R}, \quad x \in \mathscr{X}$$

respectively, which have the following properties:

(i) $x + y = y + x, \forall x, y \in \mathscr{X}$;
(ii) $x + (y + z) = (x + y) + z, \forall x, y, z \in \mathscr{X}$;
(iii) there exists a zero element, denoted 0 and called origin in $\mathscr{X}$ such that $x + 0 = x, \forall x \in \mathscr{X}$;
(iv) for every $x \in \mathscr{X}$ there exists a unique element of $\mathscr{X}$ denoted $-x$ such that $x + (-x) = 0$;
(v) $\alpha(\beta x) = (\alpha\beta)x, \forall x \in \mathscr{X}, \forall \alpha, \beta \in \mathbb{R}$;
(vi) $(\alpha + \beta)x = \alpha x + \beta x, \forall x \in \mathscr{X}, \forall \alpha, \beta \in \mathbb{R}$;
(vii) $1x = x, \forall x \in \mathscr{X}$; and
(viii) $\alpha(x + y) = \alpha x + \alpha y, \forall x, y \in \mathscr{X}, \forall \alpha \in \mathbb{R}$.

The functions A_V and M_S are called vector addition, or simply addition and scalar multiplication respectively.

A subset $\mathscr{X}_1$ of a V.S. $\mathscr{X}$ is called a vector subspace of $\mathscr{X}$ if $x + y \in \mathscr{X}_1$ $\forall x, y \in \mathscr{X}_1$ and $\alpha x \in \mathscr{X}_1$ $\forall x \in \mathscr{X}_1$, $\forall \alpha \in \mathbb{R}$. If A and B are subsets of a V.S. $\mathscr{X}$, then we define the set

$$\lambda A + \mu B = \{\lambda x + \mu y \mid x \in A, y \in B, \lambda, \mu \in \mathbb{R}\}.$$

If $\mathscr{A}$ is a set of subsets A of a V.S. $\mathscr{X}$, then $\mathscr{A} + x = \{x + A \mid A \in \mathscr{A}\}$. A subset A of a V.S. $\mathscr{X}$ is convex if and only if $\lambda x + \mu y \in A$, $\forall x, y \in A$ when $\lambda > 0$, $\mu > 0$ and $\lambda + \mu = 1$. A subset A of a V.S. $\mathscr{X}$ is said to be linearly independent if $\Sigma_{i=1}^n \lambda_i x_i = 0$ implies $\lambda_1 = \lambda_2 = \cdots = \lambda_n = 0$ for every $n \in N$ and for every choice of $x_1, \ldots, x_n$ in A. A linearly independent set A constitutes a base (or Hamel basis) of a V.S. $\mathscr{X}$ if every $x \in \mathscr{X}$ can be expressed as a linear combination of elements of A. Every V.S. has a base. If a base of a V.S. contains a finite number, say n, of elements, then every base has n elements and $\mathscr{X}$ is an n-dimensional V.S.

The notion of quotient space, we recall further plays an important role in many proofs given in this book. Let M be a linear subspace of the V.S. $\mathscr{X}$. For $x, y \in M$ the property $x - y \in M$ introduces an equivalence relation. The quotient set $\mathscr{X}/M$ is the set of the equivalence classes for the relation $x - y \in M$. To each $x \in \mathscr{X}$ there corresponds the class *modulo* the relation $x - y \in M$ (i.e., the y's such that $x - y \in M$), through the canonical mapping from $\mathscr{X}$ onto $\mathscr{X}/M$. If the class of x is denoted by $[x]$, we can turn

$\mathscr{X}/M$ into a V.S. by defining the vector addition and the scalar multiplication by the relations

$$[x] + [y] = [x + y] \text{ and } \lambda[x] = [\lambda x].$$

The mapping $f\colon \mathscr{X} \to \mathscr{Y}$ between the V.S.s $\mathscr{X}, \mathscr{Y}$ is said to be linear if

$$\begin{aligned} &f(x + y) = f(x) + f(y), \\ &f(ax) = af(x), \qquad \forall x, y \in \mathscr{X}, \qquad \forall a \in \mathbb{R}. \end{aligned} \tag{1.1.1}$$

The set $\{x \mid f(x) = 0\}$ is called the kernel of f (ker f). This set is a linear subspace of $\mathscr{X}$. The set $\mathscr{L}\{\mathscr{X}, \mathscr{Y}\}$ of all linear mappings from $\mathscr{X}$ into $\mathscr{Y}$ can be equipped with a vector space structure if addition and scalar multiplication are defined on $\mathscr{L}(\mathscr{X}, \mathscr{Y})$ by means of

$$(f + g)(x) = f(x) + g(x) \quad \text{and} \quad \lambda f(x) = f(\lambda x), \qquad \lambda \in \mathbb{R}. \tag{1.1.2}$$

Let us consider further a topology S on the vector space $\mathscr{X}$. This topology is said to be compatible with the linear structure of $\mathscr{X}$ if the mappings A_V and M_S are continuous when $\mathscr{X} \times \mathscr{X}$ and $\mathbb{R} \times \mathscr{X}$ are equipped with the corresponding product topologies. A vector space with a compatible topology is called a topological vector space (T.V.S.). In a T.V.S. X the translation $f\colon x \to x + x_0$ is a homeomorphism of X into itself, as can be easily verified and thus if G_0 is a base of neighborhoods of the origin, then $G_0 + x$ is a base of neighborhoods of x. A subset A of a T.V.S. X is bounded if for every neighborhood of zero F, there exists a number $\alpha \geq 0$ such that $A \subset \alpha F$. A linear mapping $f\colon X \to Y$, where X, Y are T.V.S.s, is continuous on X if and only if it is continuous at the origin. If $f \in \mathscr{L}(X, Y)$ and is bijective, $f^{-1}\colon Y \to X$ is a linear surjective mapping. If moreover, f is bicontinuous, then it is a linear homeomorphism (called an isomorphism) between the two T.V.S.s., with the two spaces termed isomorphic. A T.V.S. with a Hausdorff topology is called a separated or Hausdorff topological vector space (H.T.V.S.). T.V.S.s with a base of convex neighborhoods of the origin are particularly interesting; such spaces are called locally convex topological vector spaces (L.C.T.V.S.). If, furthermore, the topology of a L.C.T.V.S. X is Hausdorff, then X is called a locally convex separated topological vector space. We shall use the abbreviation L.C.H.T.V.S.

Let $\mathscr{X}$ be a V.S. Then a function $p\colon \mathscr{X} \to \mathbb{R}$ is a seminorm on $\mathscr{X}$ if and only if it has the properties

(i) $p(x) \geq 0$, $\forall x \in \mathscr{X}$,
(ii) $p(\lambda x) = |\lambda| p(x)$, $\forall x \in \mathscr{X}$, $\forall \lambda \in \mathbb{R}$ and
(iii) $p(x + y) \leq p(x) + p(y)$, $\forall x, y \in \mathscr{X}$.

If, furthermore, $p(x) = 0$ implies that $x = 0$, the seminorm is said to be a norm and is usually denoted by $\|x\|$. Let $\mathscr{P}$ be a set of seminorms on a V.S. $\mathscr{X}$. Then the sets $\{x \mid \sup p_i(x) \leq \varepsilon,\ \varepsilon > 0,\ p_i \in \mathscr{P}\}$ form a base of closed neighborhoods of the origin and determine a locally convex topology on $\mathscr{X}$.

This topology is the coarsest topology compatible with the linear structure of $\mathscr{X}$ for which every seminorm of $\mathscr{P}$ is continuous.

A T.V.S. X is normable if its topology is metric, defined by the distance $d(x, y) = \|x - y\|$. This topology is the norm topology on X and is a separated topology. A pair consisting of a V.S. $\mathscr{X}$ and a norm $\|\cdot\|$ on $\mathscr{X}$ defines a normed space. $\mathscr{X}$ is topologized by the norm topology. Two norms $\|\cdot\|$ and $|\cdot|$ on a V.S. $\mathscr{X}$ are equivalent if and only if there exist constants $c_1, c_2 > 0$ such that

$$c_1 \|x\| \le |x| \le c_2 \|x\|, \qquad \forall x \in \mathscr{X}. \tag{1.1.3}$$

Any two norms on a finite dimensional V.S. are equivalent. If X and Y are two normed spaces and $f \in \mathscr{L}(X, Y)$, f is bounded, i.e., it maps bounded sets into bounded sets, if and only if

$$\|f(x)\|_Y \le c\|x\|_X, \quad c \text{ const} > 0, \quad \forall x \in X \tag{1.1.4}$$

A linear operator $f: X \to Y$, where X, Y are given normed spaces, is continuous if and only if it is bounded. The space of such operators is denoted by $L(X, Y)$. Two normed spaces X and Y are called isometric if there exists an injection $f: X \to Y$ such that $\|x\|_X = \|f(x)\|_Y$ for all $x \in X$. A complete normed space X is called a Banach space (B-space). To a normed space X which is not complete, we may associate a B-space $\hat{X}$ such that there exists a linear isometry from X onto a dense linear subspace of $\hat{X}$. The space $\hat{X}$ is uniquely determined up to a linear isometry and is called the completion of X. A closed linear subspace of a B-space is itself a B-space. We say that a B-space Y is imbedded into a B-space X if $Y \subset X$ algebraically and if

$$\|x\|_X \le c\|x\|_Y \quad c \text{ const} > 0, \quad \forall x \in Y \tag{1.1.5}$$

i.e., if the injection of Y into X is continuous. The imbedding is dense if $\bar{Y} = X$, and compact if it maps bounded sets into relatively compact sets. We recall that in a B-space a set A is compact (resp. relatively compact) if and only if every sequence in A (resp. in $\bar{A}$) contains a convergent (in the norm-topology) subsequence to an element of A (resp. $\bar{A}$).

Let X be a B-space and M a closed linear subspace of X. The quotient space X/M equipped with the quotient norm

$$\|[x]\|_{X/M} = \inf_{y \in M} \|x - y\| \tag{1.1.6}$$

is also a B-space. Some examples of T.V.S.s will be given in Sec. 1.3.

1.2 Duality in Topological Vector Spaces

1.2.1 Duality. Weak and Strong Topologies

If $\mathscr{X}$ is a V.S., a linear mapping $f: \mathscr{X} \to \mathbb{R}$ is called a linear functional or linear form on $\mathscr{X}$. The space $\mathscr{L}(\mathscr{X}, \mathbb{R})$ of all linear functionals on $\mathscr{X}$ is called the algebraic dual of $\mathscr{X}$ and is denoted by $\mathscr{X}^*$. If f is a nonzero linear functional on a T.V.S. X then the equation $f(x) = 0$ (resp. $f(x) = a$, $a \in \mathbb{R}$, defines for all x a hyperplane (resp. an affine hyperplane) H (resp. H_a) in X. It is worth noting that a hyperplane is either closed (i.e., $H = \bar{H}$) or everywhere dense (i.e., $\bar{H} = X$). The subsets $\{x \in X \mid f(x) < a\}$ and $\{x \in X \mid f(x) > a\}$ are called open halfspaces bounded by H_a and the subsets $\{x \in X \mid f(x) \leq a\}$ and $\{x \in X \mid f(x) \geq a\}$ are the respective closed halfspaces. It is said that an affine hyperplane separates (resp. separates strictly) two subsets of X, A, and B, if

$$\forall x \in A \quad f(x) \leq a \qquad (\text{resp. } f(x) < a)$$

and

$$\forall y \in B \quad f(y) \geq a \qquad (\text{resp. } f(y) > a)$$

Among the elements of $\mathscr{X}^*$, the continuous linear functionals play an important role. They constitute a subspace of $\mathscr{X}^*$ called the (continuous) dual of $\mathscr{X}$, denoted by $\mathscr{X}'$. The existence of nontrivial (i.e., nonzero) continuous linear functionals on a L.C.H.T.V.S. is ensured by the Hahn–Banach theorem. There are many forms of this theorem, which is one of the fundamental theorems of functional analysis; here we shall give only an analytic and a geometric form.

(a) Analytic form: Let p be a seminorm on a V.S. $\mathscr{X}$. Let $\mathscr{X}_1$ be a linear subspace of $\mathscr{X}$ and $f \in \mathscr{X}_1^*$ such that $|f(x)| \leq p(x)$, $\forall x \in \mathscr{X}_1$. Then there exists $\tilde{f} \in X^*$ extending f (i.e. $\tilde{f}(x) = f(x)$, $\forall x \in X_1$) such that $|\tilde{f}(x)| \leq p(x)$, $\forall x \in \mathscr{X}$. (b) Geometric form: Let A be an open, convex, nonempty subset of a T.V.S. X, and M a linear subspace of X which does not meet A. Then there exists a closed affine hyperplane H_a which contains M and which does not meet A.

A corollary of this theorem is that two disjoint, convex, nonempty subsets of a T.V.S., one of which is open, are separated by a closed affine hyperplane; if both are open, the separation is strict. If x_1 and x_2 are two points of a L.C.H.T.V.S. X, then there exists a closed affine hyperplane separating them, which defines a continuous linear functional on X. Another corollary is that every closed, convex subset of a L.C.H.T.V.S. X is the intersection of all the closed halfspaces which contain it. Let X be a L.C.T.V.S. and $\mathscr{X}'$ its dual. Henceforth we denote by x' the elements of $\mathscr{X}'$ and by $\langle x', x\rangle$ the value of x' at the point x. As will be seen further on, we can denote this value by $\langle x, x'\rangle$ as well. Using the foregoing notation, we can exhibit a kind of symmetry existing between X and $\mathscr{X}'$. To each element $x \in X$ there corresponds a linear functional on $\mathscr{X}'$ defined by $\tilde{x}(x') = x'(x)$, $x' \in \mathscr{X}'$. This is a linear mapping from $\mathscr{X}$ into $\mathscr{X}'^*$. If, furthermore, X is provided with a Hausdorff topology, this mapping is injective. Thus X is identified (up to an isomorphism) with a subspace $\tilde{\mathscr{X}}$ of $\mathscr{X}'^*$. Accordingly, there is a kind of symmetry between X and $\mathscr{X}'$ in the sense that each space is isomorphic to a subspace of the algebraic dual of the other. From the definitions of the spaces X and $\mathscr{X}'$, $\langle x', x\rangle$ is a bilinear form on $\mathscr{X}' \times \mathscr{X}$ and has the following properties, as results from the application of the Hahn–Banach theorem:

$$\text{(i) } \forall x \neq 0 \text{ in } \mathscr{X}, \text{ there exists } \quad x' \in \mathscr{X}' \text{ with } \langle x', x\rangle \neq 0 \text{ and;} \tag{1.2.1}$$

$$\text{(ii) } \forall x' \neq 0 \text{ in } \mathscr{X}', \text{ there exists } \quad x \in \mathscr{X} \text{ with } \langle x', x\rangle \neq 0. \tag{1.2.2}$$

Let us consider further the V.S.s $\mathscr{X}$ and $\mathscr{X}'$ and a bilinear form $\langle x', x\rangle$ on $\mathscr{X}' \times \mathscr{X}$ with the properties (1.2.1) and (1.2.2). A consequence of these properties is that there exists a linear injective mapping of $\mathscr{X}'$ into $\mathscr{X}^*$ (resp. of $\mathscr{X}$ into $\mathscr{X}'^*$) in which the image of $x' \in \mathscr{X}'$ (resp. of $x \in \mathscr{X}$) is the linear form $\langle x', x\rangle$ on $\mathscr{X}$ (resp. on $\mathscr{X}'$). Thus $\mathscr{X}'$ (resp. $\mathscr{X}$) is isomorphic to a subspace of $\mathscr{X}^*$ (resp. on $\mathscr{X}'^*$). The pair $\{\mathscr{X}, \mathscr{X}'\}$ of the vector spaces $\mathscr{X}$ and $\mathscr{X}'$ is called[1] a dual pair and the form $\langle x', x\rangle$ a duality pairing. Until now, we have considered $\mathscr{X}'$ independently of its topology; henceforth, the algebraic symmetry between $\mathscr{X}$ and $\mathscr{X}'$ will be extended to a topological symmetry. There is a class of topologies on $\mathscr{X}'$ for which $\mathscr{X}'$ is a L.C.H.T.V.S. having $\mathscr{X}$ as a dual space. Let A be a subset of a T.V.S. X. The subset of $\mathscr{X}'$

$$A^0 = \{x' \mid x' \in \mathscr{X}', \sup_A |\langle x', x\rangle| \leq 1\} \tag{1.2.3}$$

is called the polar set of A. Let $\mathscr{G}$ be a set of bounded subsets of X, with the properties:

(i) if $A_1, A_2 \in \mathscr{G}$, there exists $A_3 \in \mathscr{G}$ such that $A_1 \cup A_2 \subset A_3$, and
(ii) if $A_1 \in \mathscr{G}$ and $\lambda \in \mathbb{R}$, there exists $A_2 \in \mathscr{G}$ such that $\lambda A_1 \subset A_2$.

The locally convex topology generated on $\mathscr{X}'$ by considering as a basis of neighborhoods of zero the set $\mathscr{G}^0$ consisting of the polars of the elements of $\mathscr{G}$, is called the $\mathscr{G}$-topology.

Some examples of $\mathscr{G}$-topologies follow. If $\mathscr{G}$ is the family of all finite sets of X, the $\mathscr{G}$-topology on $\mathscr{X}'$ is called the weak topology determined by $\mathscr{X}$ and is denoted by $\Sigma(\mathscr{X}', \mathscr{X})$. We denote the weak dual space $\{\mathscr{X}', \Sigma(\mathscr{X}', \mathscr{X})\}$ of X by X'_w. It is also possible for the topology $\Sigma(\mathscr{X}', \mathscr{X})$ to be generated on $\mathscr{X}'$ by a set of seminorms of the form $p_i(x') = |\langle x', x_i\rangle|$. Then a base of (closed) neighborhoods of zero is formed by the sets[2]

$$\{x' \mid x' \in \mathscr{X}' \mid |\langle x', x_i\rangle| \leq \varepsilon \qquad i = 1, 2, \ldots, n\} \tag{1.2.4}$$

where ε varies over the set of positive reals and $\{x_1, \ldots, x_n\}$ over all finite subsets of X.

Further, we define the strong topology $\bar{S}$ on $\mathscr{X}'$. It is generated by taking as a family $\mathscr{G}$ the family of all bounded subsets of X. This topology is also called the topology of bounded convergence, since a sequence of $X'_{\bar{S}} = \{\mathscr{X}',$

[1] Recall here the duality in projective geometry.
[2] Cf. (1.2.3). $\langle x', x\rangle \leq \varepsilon$ if and only if $\langle x', x/\varepsilon\rangle \leq 1$.

$\tilde{S}\}$ converges to zero if it converges uniformly to zero on every bounded subset of X. If X is a normed V.S. the strong topology on X' is defined by the norm

$$\| x' \| = \sup_{\substack{x \in X \\ x \neq 0}} \frac{|\langle x', x \rangle|}{\| x \|} = \sup_{\|x\| \leq 1} |\langle x', x \rangle| \tag{1.2.5}$$

We can define the weak topology $\Sigma(\mathscr{X}, \mathscr{X}')$ on $\mathscr{X}$ analogously. The base of neighborhoods of zero generating $\Sigma(\mathscr{X}, \mathscr{X}')$ on $\mathscr{X}$ is formed by the sets

$$\{x \mid x \in \mathscr{X}, |\langle x_i', x \rangle| \leq \varepsilon, \qquad i = 1, 2, \ldots, n\} \tag{1.2.6}$$

Here $\{x_1', \ldots, x_n'\}$ varies over the finite subsets of $\mathscr{X}'$, as does ε over the positive reals. We denote $\{\mathscr{X}, \Sigma(\mathscr{X}, \mathscr{X}')\}$ by X_w. Note that a finite-dimensional V.S. $\mathscr{X}$ has only one topology.

1.2.2 Topologically Dual Pairs of Vector Spaces

Let $\mathscr{X}$ and $\mathscr{X}'$ be V.S.s and $\langle x', x \rangle$ a bilinear form obeying (1.2.1) and (1.2.2). Then $\{\mathscr{X}, \mathscr{X}'\}$ is a dual pair. Suppose that $\mathscr{X}$ is equipped with the weak topology $\Sigma(\mathscr{X}, \mathscr{X}')$. This topology is the weakest locally convex Hausdorff topology on $\mathscr{X}$ under which the dual space of X_w is $\mathscr{X}'$. Suppose that $\mathscr{X}'$ is equipped with the weak topology $\Sigma(\mathscr{X}', \mathscr{X})$. This topology is the weakest locally convex Hausdorff topology on $\mathscr{X}'$ under which the dual space of X_w' is X. It results that we can write $\langle x', x \rangle$ or $\langle x, x' \rangle$ for the value of x (resp. x') at x' (resp. x), and that the roles of X and X' in a dual pair are interchangeable.

Let X be a L.C.H.T.V.S. with the topology T, and $\mathscr{X}'$ the dual space. There now arises the question whether it is possible to determine all the locally convex Hausdorff topologies T on $\mathscr{X}$ under which the dual of $X = \{\mathscr{X}, T\}$ is exactly $\mathscr{X}'$. Such a topology is said to be compatible with the duality between $\mathscr{X}$ and $\mathscr{X}'$. According to the Mackey–Arens theorem, all the topologies T on $\mathscr{X}$ which are compatible with the duality are stronger than the weak topology $\Sigma(\mathscr{X}, \mathscr{X}')$ and weaker than a certain topology Σ_M, called the Mackey topology,[3] i.e.,

$$\Sigma(\mathscr{X}, \mathscr{X}') \subset T \subset \Sigma_M \quad \text{on} \quad \mathscr{X} \tag{1.2.7}$$

Similarly all the topologies T' on $\mathscr{X}'$ which are compatible with the duality between $\mathscr{X}$ and $\mathscr{X}'$ are stronger than $\Sigma(\mathscr{X}', \mathscr{X})$ and weaker than the Mackey topology Σ_M', i.e.,

$$\Sigma(\mathscr{X}', \mathscr{X}) \subset T' \subset \Sigma_M' \quad \text{on} \quad \mathscr{X}'. \tag{1.2.8}$$

[3] Σ_M is the topology of uniform convergence on the set of all absolutely convex, compact (with respect to the topology $\Sigma(\mathscr{X}', \mathscr{X})$), subsets of $\mathscr{X}'$ [284].

A dual pair $\{\mathscr{X}, \mathscr{X}'\}$ is a topologically dual pair, and is denoted by $\{X, X'\}$ if $\mathscr{X}$ and $\mathscr{X}'$ are equipped with topologies compatible with the duality between $\mathscr{X}$ and $\mathscr{X}'$.

We close this section by defining reflexivity. A L.C.H.T.V.S. X is called reflexive if

$$X = (X'_{\bar{S}})'_{\bar{S}}. \tag{1.2.9}$$

1.2.3 Duality in Normed and Hilbert Spaces

Let X be a normed space carrying the usual norm topology and let $\mathscr{X}'$ be its dual, i.e., the space of all continuous linear functionals on X. Many topologies can be considered on $\mathscr{X}$, but when dealing with a normed space we shall consider the natural norm topology on $\mathscr{X}$ (i.e., the space $X_{\bar{S}}$), unless otherwise stated. $X'_{\bar{S}}$, denoted simply X', is a B-space with the norm $\| x' \|$ given by (1.2.5). It is called the strong dual of X.

As in the general case of a T.V.S., we introduce the natural duality pairing $\langle x', x\rangle = x'(x)$ for every $x \in X$ and every $x' \in X'$. But since X and X' are normed, they do not exhibit the topological and algebraic symmetry of the general L.C.H.T.V.S.s, i.e., X is not the dual of X' unless X is reflexive in the sense of (1.2.9). The dual of X' is the space X'', called the bidual of X. Then, as in the case of L.C.H.T.V.S.s [see (1.2.4, 6)], the topologies $\Sigma(\mathscr{X}, \mathscr{X}')$ on $\mathscr{X}$ and $\Sigma(\mathscr{X}', \mathscr{X})$ on $\mathscr{X}'$ may be defined. We call the first the weak topology on $\mathscr{X}$ and the second the weak$_*$ (weak star) topology on $\mathscr{X}'$ in order to distinguish it from the weak topology $\Sigma(\mathscr{X}', \mathscr{X}'')$ on $\mathscr{X}'$. Both topologies are locally convex and separated and are weaker than the norm topologies.

A sequence $\{x_n\}$ in a normed space X is said to be weakly convergent to $x_0 \in X$ if

$$\lim_{n\to\infty} x'(x_n) = x'(x_0), \qquad \forall x' \in X'. \tag{1.2.10}$$

A sequence $\{x'_n\}$ in the dual space X' of a normed space X is weakly$_*$ convergent to an element $x' \in X'$ if and only if

$$\lim_{n\to\infty} x'_n(x) = x'_0(x), \qquad \forall x \in X. \tag{1.2.11}$$

Weak limits are unique if they exist. The strong convergence always implies the weak (resp. the weak$_*$) convergence in X (resp. in X'). The following results are of importance. If a B-space is reflexive (i.e. $X = (X'_{\bar{S}})'_{\bar{S}}$), then every strongly bounded sequence in that space contains a subsequence weakly convergent to an element of X (this proposition will be used often in this book). If X is a separable normed-space, then every bounded sequence in $X'_{\bar{S}}$ contains a subsequence weakly$_*$ convergent to an element of $X'_{\bar{S}}$. A B-space X is reflexive if and only if every strongly bounded sequence of X contains a subsequence weakly convergent to an element of X (Eberlein–

Shmulyan theorem). Let M be a closed linear subspace of a B-space X (M is a B-space as well), and let M' be its dual space. For $x'|_M \in M'$, the relation

$$\langle x', x\rangle = \langle x'|_M, x\rangle \quad \forall x \in M, \quad \forall x' \in X' \tag{1.2.12}$$

holds by definition.

The duality theory of B-spaces is applied also to Hilbert spaces. To recall the definition of a Hilbert space, let $\mathscr{X}$ be a V.S. and (x, y) a symmetric bilinear form (inner product) on $\mathscr{X} \times \mathscr{X}$ which is positive definite, i.e.,

$$(x, x) > 0, \quad \forall x \in \mathscr{X}, \quad x \neq 0. \tag{1.2.13}$$

The V.S. $\mathscr{X}$, together with this bilinear form, is called a real Hausdorff pre-Hilbert space. Then $(x, x)^{1/2}$ is a norm on $\mathscr{X}$. If the bilinear form is nonnegative, i.e.,

$$(x, x) \geq 0, \qquad \forall x \in \mathscr{X}, \tag{1.2.14}$$

the resulting space is not Hausdorff and $(x, x)^{1/2}$ is a seminorm on $\mathscr{X}$. A Hausdorff pre-Hilbert space which is complete is called a Hilbert space (abbreviated as H-space). Furthermore, a norm $\| x \|$ on a normed space $\mathscr{X}$ is a Hilbert norm, i.e., there exists a positive definite, symmetric bilinear form (x, y) such that $\| x \| = (x, x)^{1/2}$, if and only if the equality

$$\| x \|^2 + \| y \|^2 = \tfrac{1}{2}(\| x + y \|^2 + \| x - y \|^2), \qquad \forall x, y \in X \tag{1.2.15}$$

is satisfied. If M is a closed vector subspace of a H-space then X/M with the norm defined in (1.1.6) is also a H-space. Two vectors x, y of a H-space X are orthogonal to each other if $(x, y) = 0$. The orthogonal complement $M^\perp$ of a subset M of X is $M^\perp = \{y \mid y \in X, (x, y) = 0, \forall x \in M\}$. If X_1 is a closed linear subspace of X, then $X = X_1 \oplus X_1^\perp$. Any vector $x \in X$ can be decomposed uniquely into the form $x = x_1 + x_1^\perp$, $x_1 \in X_1$, $x_1^\perp \in X_1^\perp$. In order to study the duality of H-spaces, we consider a H-space X and its dual X', and we denote by

$$\langle x', x\rangle = x'(x), \qquad x' \in X', \qquad x \in X. \tag{1.2.16}$$

the value of functional x' at x. The dual space X' is equipped with the norm (1.2.5). We define the duality mapping $\mathfrak{J} \in L(X, X')$ by the relation

$$\mathfrak{J}x : y \to \mathfrak{J}x(y) = \langle \mathfrak{J}x, y\rangle = (x, y), \qquad \forall x, y \in X. \tag{1.2.17}$$

By the Riesz theorem we know that $\mathfrak{J}$ is a surjective isometry from X onto X', called the canonical isometry between X and X'. The space X' endowed with the inner product

$$(x', y') = (\mathfrak{J}^{-1}x', \mathfrak{J}^{-1}y') = \langle x', \mathfrak{J}^{-1}y\rangle, \qquad \forall x', y' \in X' \tag{1.2.18}$$

is a H-space as well. When a H-space X is identified with its dual space X', the canonical isometry $\mathfrak{J}$ is the identity mapping. In this case, the duality pairing $\langle x, y\rangle$ is equal to the inner product (x, y), and X is called a pivot Hilbert space.

1.2.4 Transpose of a Continuous Linear Mapping. Scales of Hilbert Spaces. The Lax–Milgram Theorem

Let X and Y be two H-spaces, X' and Y' their duals, $\langle x', x\rangle_X$ and $\langle y', y\rangle_Y$ the respective duality pairings, and $A: X \to Y$ a continuous linear operator, i.e., $A \in L(X,Y)$. The transpose operator of A is a mapping $A': Y' \to X'$ defined by

$$\langle A'y', x\rangle_X = \langle y', Ax\rangle_Y, \quad \forall y' \in Y', \quad \forall x \in X. \tag{1.2.19}$$

If X and Y are pivot spaces, A' is identical to the adjoint operator A^* of A. A^* is defined by the relation

$$(A^*y, x)_X = (y, Ax)_Y, \quad \forall y \in Y, \quad \forall x \in X. \tag{1.2.20}$$

where $(\cdot, \cdot)_X$ and $(\cdot, \cdot)_Y$ denote the inner products on X and Y respectively. From (1.2.17), there exists between A' and A^* the relation

$$A' = \mathfrak{J}A^*\mathfrak{K}^{-1}, \tag{1.2.21}$$

where $\mathfrak{J}$ and $\mathfrak{K}$ are the canonical isometries between X and X' and between Y and Y'.

We shall encounter quite often the following situation: two H-spaces X and Y are given such that $X \subset Y$, where the injection is continuous and dense. Then the dual space Y' is densely imbedded into X'. The duality pairing $\langle x', x\rangle_X$ on $X' \times X$ is identified with the unique extension of the duality pairing $\langle y', y\rangle_Y$ on $Y' \times X$. If Y is a pivot space, then

$$X \subset Y = Y' \subset X', \tag{1.2.22}$$

where the injections are continuous and dense, and the unique extension of the inner product of Y is identified with the duality pairing on $X' \times X$. If M is a closed linear subspace of a H-space X, we denote by $\tilde{M}^{\perp}$ the closed linear subspace of the dual space X':

$$\tilde{M}^{\perp} = \{x' \,|\, x' \in X', \langle x', x\rangle = 0, \forall x \in M\} \tag{1.2.23}$$

The dual of M is isometric to the space $X'/\tilde{M}^{\perp}$, and the dual of X/M is isometric to $\tilde{M}^{\perp}$.

In many problems we shall consider a family of H-spaces $\{X_{\alpha_1}\}$ depending on a parameter α which varies over the reals with the property that for $\alpha_2 < \alpha_1$, X_{α_1} is imbedded into X_{α_2}. Such a family is called a scale of H-spaces. For our purposes, scales of H-spaces with the following properties will prove useful:

(i) if $\alpha_1 < \alpha_2$, $\alpha_i \in \mathbb{R}$, then $X_{\alpha_2} \subset X_{\alpha_1}$, where the imbeddings are dense;
(ii) if $\alpha_1 < \alpha_2 < \alpha_3$, then

$$\|x\|_{\alpha_2} \leq \|x\|_{\alpha_1}^{(\alpha_3-\alpha_2)/\alpha_3-\alpha_1)} \|x\|_{\alpha_3}^{(\alpha_2-\alpha_1)/(\alpha_3-\alpha_1)}, \qquad \forall x \in X_{\alpha_3}, \tag{1.2.24}$$

where $\| x \|_{\alpha_i}$ denotes the norm of X_{α_i}; and

(iii) the dual of X_α is the space $X_{-\alpha}$, the space X_0 is a pivot space and the duality pairing on $X_{-\alpha} \times X_\alpha$ is identified with the unique extension of the inner product of X_0.

With respect to an operator $A \in L(X, X')$, where X and X' are H-spaces, we define a continuous bilinear form on X by means of

$$a(x, y) = \langle Ax, y \rangle, \qquad \forall x, y \in X. \tag{1.2.25}$$

If, conversely, the continuous bilinear form $a(x, y)$, is given on X, it then defines an operator $A \in L(X, X')$ where $Ax \in X'$ and $Ax: y \to a(x, y)$, $\forall x \in X$. A is said to be coercive if a constant $c > 0$ can be determined such that

$$a(x, x) \geq c \| x \|^2, \qquad \forall x \in X \tag{1.2.26}$$

According to the Lax–Milgram theorem, any coercive operator $A: X \to X'$ is an isomorphism from X into X'. We shall apply this theorem under the equivalent form that the variational problem, "Find $x \in X$ such that

$$a(x, y) = \langle f, y \rangle, \quad \forall y \in X, \quad f \in X' \text{"} \tag{1.2.27}$$

admits a unique solution.

1.3 Certain Function Spaces and Their Properties

1.3.1 The Spaces $C^m(\Omega)$, $C^m(\bar{\Omega})$, $D(\Omega)$, $D(\bar{\Omega})$ and $L^p(\Omega)$

In this section, we shall briefly describe the properties of the function spaces which we shall use throughout this book. All the function spaces are defined, unless otherwise specified, on an open bounded connected subset Ω of $\mathbb{R}^n$ with boundary Γ. We shall, in all cases, assume that Γ possesses the smoothness necessary for the mathematical formulation and study of the problem being considered. A precise but very restrictive assumption is that Γ is an $(n-1)$-dimensional, infinitely smooth manifold such that Ω is locally on one side of Γ. Obviously, these hypotheses can be generalized, extended, and changed, and Ω and Γ can have a more general form depending on the nature of the problem being studied. The dimensions of the problem, the form of the operator equations, the boundary conditions, the functional framework, the theorems necessary for the treatment of the problem—all these factors determine the properties of Ω and Γ. To take an example, for the trace theorems of Sobolev spaces and for the Korn inequalities, which will often be used, a Lipschitz boundary Γ is sufficient. This means that (i) a neighborhood F of each point $x \in \Gamma$ may be represented by a Lipschitz continuous (cf., (1.3.3), with $n = 1$) hypersurface, and (ii) F is divided by Γ

into a set of exterior and a set of interior points. For further details, see [2], [98], [231]. A discussion of the most general admissible Ω and Γ for each problem to be studied is beyond the scope of the present book. Without insisting further on this point we shall assume hereafter that Γ is sufficiently regular.

Let us introduce the following function spaces.

(a) For any nonnegative integer $m < \infty$, let $C^m(\Omega)$ be the V.S. consisting of all real-valued functions on Ω which have continuous partial derivatives $D^\alpha\phi$ of order $|\alpha| \le m$. For $x = (x_1, \dots, x_n)$ and α_i, $i = 1, \dots, n$, nonnegative integers, we use the notations

$$\alpha = (\alpha_1, \dots, \alpha_n), \qquad |\alpha| = \alpha_1 + \cdots + \alpha_n,$$

$$D^\alpha\phi(x) = D^{\alpha_1}D^{\alpha_2} \cdots D^{\alpha_n}\phi(x) = \frac{\partial^{\alpha_1+\cdots+\alpha_n}\phi(x_1, \dots, x_n)}{\partial^{\alpha_1}_{x_n} \cdots \partial x_n^{\alpha_n}}, \; D^0\phi = \phi$$

We denote by $C_c^m(\Omega)$ the subspace of $C^m(\Omega)$ consisting of all functions $\phi \in C^m(\Omega)$ with compact support. By the support of ϕ (supp ϕ) we mean the smallest closed subset of Ω outside which ϕ is zero. Further, let $C^\infty(\Omega) = \bigcap_{m=0}^{\infty} C^m(\Omega)$. For any compact subset A of Ω, let $D_A(\Omega)$ be the set of all functions $\phi \in C_c^\infty(\Omega)$ such that the supp $\phi \subset A$. $D_A(\Omega)$ is a L.C.H.T.V.S. with a topology generated by the set of seminorms

$$P_{A,m}(\phi) = \sup_{|\alpha| \le m,\, x \in A} |D^\alpha\phi(x)|, \qquad m < \infty.$$

The strict inductive limit[4] of the spaces $D_A(\Omega)$ as A ranges over all compact subsets of Ω is a L.C.H.T.V.S. $C_c^\infty(\Omega)$ topologized in this way is denoted by $D(\Omega)$. It can be proved that

$$p(\phi) = \sup_{x \in \Omega} |\phi(x)| \tag{1.3.1}$$

is a seminorm defining this topology. In $D(\Omega)$ a sequence $\{\phi_n\}$ tends to zero if

(i) a compact subset A of Ω can be determined, such that supp $\phi_n \subset A$, $n = 1, 2, \dots$, and

(ii) all the sequences $\{D^\alpha\phi_n(x)\}$, $\alpha \ge 0$, converge on A uniformly to zero.

(b) Let $\bar{\Omega}$ be the closure of Ω. $C^m(\bar{\Omega})$ is the V.S. of those functions $\phi \in C^m(\Omega)$ such that for each $\alpha = (\alpha_1, \dots, \alpha_n)$ with $|\alpha| \le m$ the derivative $D^\alpha\phi$ possesses a unique continuous extension to $\bar{\Omega}$ (e.g., if $D^\alpha\phi$ is bounded and uniformly continuous on Ω for $|\alpha| \le m$). Note that $C^m(\Omega) \neq C^m(\bar{\Omega})$. We can turn $C^m(\bar{\Omega})$ into a B-space by the norm

$$\|\phi\| = \max_{0 \le |\alpha| \le m} \sup_{x \in \Omega} |D^\alpha\phi(x)| \tag{1.3.2}$$

The space of C_c^∞-functions on $\bar{\Omega}$ is denoted by $C_c^\infty(\bar{\Omega})$. The space $D(\bar{\Omega})$ can be defined analogously to $D(\Omega)$.

[4] Cf. [337], p. 28.

For $0 < n \leq 1$, we consider the space $C^{m,n}(\bar{\Omega})$ consisting of those functions $\phi \in C^m(\bar{\Omega})$ for which $D^\alpha(\phi)$, with $0 \leq |\alpha| \leq m$, satisfies the inequality

$$|D^\alpha\phi(x) - D^\alpha\phi(y)| \leq c|x - y|^n, \qquad \text{c const.} > 0 \qquad \forall x, y \in \Omega \tag{1.3.3}$$

called the Hölder condition of exponent n. If $n = 1$, (1.3.3) is a Lipschitz condition.[5] The V.S. $C^{m,n}(\bar{\Omega})$ is a B-space with respect to the norm

$$\|\phi\|_{C^{m,n}} = \|\phi\|_{C^m} + \max_{0 \leq |\alpha| \leq m} \sup_{\substack{x, y \in \Omega \\ x \neq y}} \frac{|D^\alpha\phi(x) - D^\alpha\phi(y)|}{|x - y|^n}. \tag{1.3.4}$$

(c) We denote by $L^p(\Omega)$, $p \in \mathbb{R}_+$ the set of all real-valued measurable functions f on Ω such that $\int_\Omega |f(x)|^p \, d\Omega < \infty$. If we identify elements of $L^p(\Omega)$ which are almost everywhere (a.e.) equal in Ω, then $L^p(\Omega)$ actually consists of the equivalence classes of measurable functions; two functions f_1 and f_2 are equivalent if $f_1 - f_2 = 0$ a.e. in Ω. If $1 \leq p < \infty$,

$$|f|_p = \left(\int_\Omega |f(x)|^p \, d\Omega\right)^{1/p} \tag{1.3.5}$$

is a norm on L^p. Equipped with this norm, $L^p(\Omega)$ is a B-space. $L^2(\Omega)$ is a H-space in which the inner product is given by

$$(f_1, f_2) = \int_\Omega f_1 f_2 \, d\Omega. \tag{1.3.6}$$

A measurable function f on Ω for which $|f(x)| \leq c$ a.e. on Ω is said to be essentially bounded. The infimum of the constants c is denoted by $\operatorname{ess\,sup}_{x \in \Omega} |f(x)|$. The V.S. of all equivalence classes of essentially-bounded functions on Ω is denoted by $L^\infty(\Omega)$, two functions f_1, f_2 being identified if $f_1 - f_2 = 0$ a.e. on Ω. $L^\infty(\Omega)$ is a B-space with respect to the norm

$$|f|_\infty = \operatorname*{ess\,sup}_{x \in \Omega} |f(x)| \tag{1.3.7}$$

Here we have assumed the well-known Lebesgue measure on Ω. Similarly, we can define the spaces L^p, $1 \leq p \leq \infty$, with respect to a measure space $(\Omega, \mathscr{B}, \mu)$ (cf., [330] [337]).

Next, let us recall the Minkowski inequality

$$|f_1 + f_2|_p \leq |f_1|_p + |f_2|_p, \qquad 1 \leq p < \infty \tag{1.3.8}$$

and the general Hölder inequality stating that if $f_i \in L^{p_i}(\Omega)$ and $\sum_{i=1}^m 1/p_i = 1/r$, where $p_i, r \in [1, \infty]$, $i = 1, 2, \ldots m$, then $f_1 f_2 \ldots f_m \in L^r(\Omega)$ and

$$|f_1 f_2 \cdots f_m|_r \leq |f_1|_{p_1} \cdots |f_m|_{p_m}. \tag{1.3.9}$$

If $1 \leq p \leq q \leq \infty$ and mes $\Omega < \infty$, $L^q(\Omega)$ is imbedded into $L^p(\Omega)$. Moreover, $D(\Omega)$ is dense in $L^p(\Omega)$ if $1 \leq p < \infty$. We say that p and p' are conjugate exponents if $p, p' \in [1, \infty]$ and $1/p + 1/p' = 1$. Let $1 \leq p < \infty$; then every

[5] A Lipschitz boundary is also called a $C^{0,1}$–boundary. The $C^{m,n}$–boundary is defined likewise.

continuous linear functional $\tilde{g}$ on $L^p(\Omega)$ has the form

$$\langle \tilde{g}, f \rangle = \tilde{g}(f) = \int_\Omega g(x) f(x)\, d\Omega \tag{1.3.10}$$

where $g \in L^{p'}(\Omega)$; $[L^p(\Omega)]'$ is for $1 \leq p < \infty$ isometrically isomorphic to $[L^{p'}(\Omega)]$, and we may write $[L^p(\Omega)]' = L^{p'}(\Omega)$. It is worth noting that $[L^\infty(\Omega)]' \neq L^1(\Omega)$. For $1 < p < \infty$ $L^p(\Omega)$ is reflexive, and for $1 \leq p < \infty$ $L^p(\Omega)$ is separable. We shall often encounter bounded sequences in $L^\infty(\Omega) = [L^1(\Omega)]'$. Due to the separability of $L^1(\Omega)$, there exists a subsequence converging weakly$_*$ (cf., Sec. 1.2.3) to an element of $L^\infty(\Omega)$.

1.3.2 Spaces of Distributions

A linear continuous functional f defined on the space $D(\Omega)$ is called a generalized function or a distribution on Ω. f belongs to the dual space of $D(\Omega)$, denoted by $\mathscr{D}'(\Omega)$. The value $f(\phi)$, also written $\langle f, \phi \rangle$, is called the value of f at the test function ϕ. A linear form on $D(\Omega)$ is a distribution if and only if it has one of the following two properties:

(i) if the sequence $\{\phi_n\} \in D(\Omega)$ converges to zero in the topology of $D(\Omega)$, the sequence $\{\langle f, \phi_n \rangle\}$ converges to zero as well; or
(ii) for every compact set $\Omega_1 \subset \Omega$ there exists a nonnegative integer m such that for every $\phi \in D_{\Omega_1}(\Omega) \, |\langle f, \phi \rangle| \leq c \|\phi\|, c$ const > 0 where $\|\phi\|$ is given by (1.3.2).

Two distributions f_1 and f_2 are said to be equal on Ω if

$$\langle f_1, \phi \rangle = \langle f_2, \phi \rangle, \qquad \forall \phi \in D(\Omega). \tag{1.3.11}$$

The weak topology is introduced on $\mathscr{D}'(\Omega)$ and this space becomes a L.C.H.T.V.S. Then $f_n \to f$ in $\mathscr{D}'(\Omega)$ if and only if

$$\langle f_n, \phi \rangle \to \langle f, \phi \rangle, \qquad \forall \phi \in D(\Omega). \tag{1.3.12}$$

Note that if $f_n \to f$ in the weak topology of $\mathscr{D}'(\Omega)$, then $f_n \to f$ in the strong topology of $\mathscr{D}'(\Omega)$ as well [337]. A well-known example in the theory of distributions is the Dirac distribution δ_{x_0}, defined for $x_0 \in \Omega$ by the relation $\langle \delta_{x_0}, \phi \rangle = \phi(x_0)$ for every $\phi \in D(\Omega)$. A function f defined a.e. on Ω is said to be locally integrable ($f \in L^1_{\text{loc}}(\Omega)$) if $f \in L^1(\Omega_1)$ for every measurable set Ω_1 with $\bar{\Omega}_1$ a subset of Ω. To such a function f (or similarly to a continuous function f) there corresponds a distribution $\tilde{f}$ defined by means of

$$\langle \tilde{f}, \phi \rangle = \int_\Omega f\phi \, d\Omega, \qquad \forall \phi \in D(\Omega). \tag{1.3.13}$$

$\tilde{f}$ is called the distribution equivalent to the function f (we may speak simply of the distribution f by ignoring the tilde ($\tilde{}$)). If a distribution is equivalent to a locally integrable function, it is called a regular distribution; otherwise it is called singular (e.g., the Dirac distribution). Singular distributions are often written as if they were ordinary functions. Thus if $f \in \mathscr{D}'(\Omega)$ and $\phi \in D(\Omega)$, we may write

$$\langle f, \phi \rangle = \int_\Omega f\phi \, d\Omega \tag{1.3.14}$$

Note that generally we cannot speak of the value of a distribution at a point x.

Measures constitute an important class of distributions. A measure f on Ω is a linear continuous functional on the space $C_c^0(\Omega)$ of the continuous functions with compact support in Ω. If $\phi \in C_c^0(\Omega)$, we denote by $\langle f, \phi \rangle$ the value of the measure f at the function ϕ, also called the integral of ϕ with respect to f. Besides $\langle f, \phi \rangle$ and $f(\phi)$, the notations $\int_\Omega \phi \, df$ and $\int \phi(x) \, df(x)$ are also in use. The space of the measures on Ω is denoted by $\mathscr{M}(\Omega)$. The δ-distribution is an example of a measure. Another measure on Ω is the well-known Lebesgue measure. We say that f is a bounded measure on Ω if a real $c > 0$ exists such that

$$|\langle f, \phi \rangle| \leq c \sup_{x \in \Omega} |\phi(x)|, \qquad \forall \phi \in C_c^0(\Omega). \tag{1.3.15}$$

The space of bounded measures on Ω is denoted by $\mathscr{M}^1(\Omega)$ and is the dual of $C_c^0(\Omega)$, when the latter is normed by the norm $\sup_{x \in \Omega} |\phi(x)|$. The space $\mathscr{M}^1(\Omega)$ provided with the norm

$$\|f\|_{\mathscr{M}^1(\Omega)} = \sup_{\phi \in C_c^0(\Omega)} \frac{|\langle f, \phi \rangle|}{\|\phi\|}, \qquad \|\phi\| = \sup_{x \in \Omega} |\phi(x)| \tag{1.3.16}$$

is a nonreflexive B-space.

A distribution f is by definition nonnegative if

$$\langle f, \phi \rangle \geq 0, \quad \forall \phi \geq 0, \quad \phi \in D(\Omega). \tag{1.3.17}$$

It is proved [297], in that case, that f is a measure.

The product of a distribution $f \in \mathscr{D}'(\Omega)$ and a function $\bar{\phi} \in D(\Omega)$ is defined by

$$\langle \bar{\phi} f, \phi \rangle = \langle f, \bar{\phi}\phi \rangle. \tag{1.3.18}$$

In general, the product of two distributions is not defined. The derivatives $D^\alpha f$ of a distribution f are defined by

$$\langle D^\alpha f, \phi \rangle = (-1)^{|\alpha|} \langle f, D^\alpha \phi \rangle. \tag{1.3.19}$$

$D^\alpha f$ is a continuous linear functional on $D(\Omega)$ and hence a distribution on Ω. The derivative of a distribution is a linear continuous mapping from $\mathscr{D}'(\Omega)$ into $\mathscr{D}'(\Omega)$.

Note that the space $L^p(\Omega)$ is imbedded into $\mathscr{D}'(\Omega)$. Suppose that p and p'

are conjugate exponents. It follows that if $f \in \mathscr{D}'(\Omega)$ and

$$|\langle f, \phi\rangle| \leq c \|\phi\|_{L^{p'}(\Omega)} \quad c \text{ const} > 0, \quad \forall \phi \in D(\Omega) \tag{1.3.20}$$

then $f \in L^p(\Omega)$.

Analogously to $\mathscr{D}'(\Omega)$, the space of distributions $\mathscr{D}'(\mathbb{R}^n)$ can be defined. An important subspace of $\mathscr{D}'(\mathbb{R}^n)$ is the space of tempered distributions $\mathscr{S}'(\mathbb{R}^n)$. We denote by $S(\mathbb{R}^n)$ the subspace of $C^\infty(\mathbb{R}^n)$ such that for every $\alpha = (\alpha_1, \ldots \alpha_n)$ and $\beta = (\beta_1, \ldots, \beta_n)$

$$\sup_{x \in \mathbb{R}^n} |x_1^{\beta_1} \cdots x_n^{\beta_n} D^\alpha f(x)| < \infty.$$

For the topology which is defined by the family of seminorms

$$p(f) = \sup_{x \in \mathbb{R}^n} |P(x) D^\alpha f(x)|, \tag{1.3.21}$$

where $P(x)$ is a polynomial, $S(\mathbb{R}^n)$ is a L.C.T.V.S. having $C_c^\infty(\mathbb{R}^n)$ as a dense subspace. $S(\mathbb{R}^n)$ is the space of the C^∞-functions on $\mathbb{R}^n$ rapidly decreasing at infinity. For any $f \in S(\mathbb{R}^n)$, the Fourier transform $\hat{f}$ of f is defined by

$$\hat{f}(y) = \int_{\mathbb{R}^n} e^{-2\pi i\langle x, y\rangle} f(x)\, dx, \qquad dx = dx_1 \cdots dx_n, \tag{1.3.22}$$

where $\langle x, y\rangle = \sum_{i=1}^n x_i y_i$. The transformation $f \to \hat{f}$ is an isomorphism of $S(\mathbb{R}^n)$ onto itself. The inverse Fourier transform of $\hat{f}$ is given by

$$f(x) = \int_{\mathbb{R}^n} e^{2\pi i\langle x, y\rangle} \hat{f}(y)\, dy, \qquad dy = dy_1 \cdots dy_n. \tag{1.3.23}$$

The dual space $\mathscr{S}'(\mathbb{R}^n)$ is a subspace of $\mathscr{D}'(\mathbb{R}^n)$ and is called the space of tempered distributions. A tempered distribution is a distribution which "does not increase too rapidly at ∞." For a tempered distribution f we can define its Fourier transform $\hat{f}$ by setting

$$\langle \hat{f}, \phi\rangle = \langle f, \hat{\phi}\rangle, \qquad \forall \phi \in S(\mathbb{R}^n). \tag{1.3.24}$$

1.3.3 Sobolev Spaces

A very important family of B-spaces is that of the Sobolev spaces, which have proved quite a suitable tool in the mathematical study of boundary value problems (B.V.P.s). Let $1 \leq p < \infty$ and l be an integer. We denote by $W^{l,p}(\Omega)$ the space of functions $f \in L^p(\Omega)$ whose distributional (or weak) derivatives $D^\alpha f$, up to order l, are also in $L^p(\Omega)$, i.e.,

$$W^{l,p}(\Omega) = \{f \mid f \in L^p(\Omega),\ D^\alpha f \in L^p(\Omega),\ |\alpha| \leq l\}. \tag{1.3.25}$$

Equipped with the norm

$$\|f\|_{l,p} = \left\{\sum_{0 \leq |\alpha| \leq l} |D^\alpha f|_p^p\right\}^{1/p}, \tag{1.3.26}$$

$W^{l,p}(\Omega)$ is a separable B-space (and reflexive if $p > 1$). Obviously, $W^{0,p} = L^p$. The injection of $W^{l,p}(\Omega)$ into $L^p(\Omega)$ is continuous, and for $l_1 \geq l$ and $p_1 \geq p$, $W^{l_1,p_1} \subset W^{l,p}$. In general, $\bar{D}(\Omega) \neq W^{l,p}(\Omega)$ unless $\Omega = \mathbb{R}^n$ or $l = 0$. Thus we may introduce the closure of $D(\Omega)$ in $W^{l,p}(\Omega)$ with respect to the $W^{l,p}$–norm. Denoted by $\mathring{W}^{l,p}(\Omega)$ it is for the norm defined in (1.3.26), a closed vector subspace of $W^{l,p}(\Omega)$ and thereby a B-space. The space $W^{l,2}(\Omega)$ (resp. $\mathring{W}^{l,2}(\Omega)$) is usually denoted by $H^l(\Omega)$ (resp. $\mathring{H}^l(\Omega)$) and the norm $\|\cdot\|_{l,2}$ is denoted simply by $\|\cdot\|_l$. Note that $D(\bar{\Omega})$ is dense in $H^l(\Omega)$. The norm $\|\cdot\|_l$ turns $H^l(\Omega)$ (resp. $\mathring{H}^l(\Omega)$) into a Hilbert space with the inner product

$$((f_1, f_2))_l = \sum_{0 \leq |\alpha| \leq l} (D^\alpha f_1, D^\alpha f_2), \tag{1.3.27}$$

where $(f_1, f_2) = \int_\Omega f_1 f_2 \, d\Omega$ is the inner product in $L^2(\Omega)$. On the space $\mathring{H}^l(\Omega)$ the inner product

$$(f_1, f_2)_l = \sum_{|\alpha| = l} (D^\alpha f_1, D^\alpha f_2) \tag{1.3.28}$$

defines a norm $|\cdot|_l$ which is equivalent to the norm (1.3.26) (Friedrich's Lemma). On $H^l(\Omega)$

$$|f|_l = \left(\sum_{l = |\alpha|} |D^\alpha f|_2^2 \right)^{1/2} \tag{1.3.29}$$

is simply a seminorm. A function $f \in H^l(\Omega)$ with the property $|f|_l = 0$ is a.e. equal to a polynomial of x of degree $\leq l - 1$. All such functions together constitute a closed linear subspace $\mathcal{R}_{l-1}(\Omega)$, of $H^l(\Omega)$. Then

$$|||f||| = (|f|_l^2 + |f|_2^2)^{1/2} \tag{1.3.30}$$

is a norm on $H^l(\Omega)$ ($l \geq 1$) equivalent to the usual norm. We consider further the quotient space $Q^l(\Omega) = H^l(\Omega)/\mathcal{R}_{l-1}(\Omega)$ which has the norm [cf., (1.1.6)]

$$\|[f]\|_l = \inf \|f - g\|_l, \qquad g \in \mathcal{R}_{l-1}(\Omega). \tag{1.3.31}$$

It may be proved [41] that the seminorm $|f|_l$ is a norm on the space $Q^l(\Omega)$ equivalent to the norm $\|[f]\|_l$.

The Sobolev spaces $W^{l,p}(\Omega)$ can also be defined for l negative (negative orders of differentiation!). By $W^{-l,p'}(\Omega)$, where $1 \leq p' \leq \infty$, $1 \leq l$ and $1/p + 1/p' = 1$ we denote the space of distributions f on Ω such that,

$$f = \sum_{|\alpha| \leq l} D^\alpha g_\alpha, \qquad g_\alpha \in L^{p'}(\Omega). \tag{1.3.32}$$

The space $W^{-l,p'}(\Omega)$ is for $p' < \infty$ a B-space with respect to the norm

$$\|f\|_{-l,p'} \doteq \sup_{\phi \in \mathring{W}^{l,p}(\Omega)} \frac{|\langle f, \phi \rangle|}{\|\phi\|_{l,p}} \tag{1.3.33}$$

where $\langle f, \phi \rangle = \sum_{|\alpha| \le l} \int_\Omega (-1)^{|\alpha|} g_\alpha D^\alpha \phi \, d\Omega$. It may be proved that $(W^{l,p}(\Omega))'$ is for $p > 1$ the completion of $L^{p'}(\Omega)$ with respect to the norm in (1.3.33) (now with $\phi \in W^{l,p}$, not $\phi \in \mathring{W}^{l,p}$), and that for $p > 1$

$$(\mathring{W}^{l,p}(\Omega))' = W^{-l,p'}(\Omega). \tag{1.3.34}$$

For the case $p = 2$, the dense and compact imbeddings

$$\mathring{H}^l(\Omega) \subset L^2(\Omega) \subset H^{-l}(\Omega). \tag{1.3.35}$$

are often used.

We now define the spaces $H^s(\mathbb{R}^n)$ and $H^s(\Omega)$, where s is generally a real number. To this end, the notion of tempered distribution is used. The Sobolev space $H^s(\mathbb{R}^n)$, $s \in \mathbb{R}$, is the space of tempered distributions f such that their Fourier transforms $\hat{f}$ when multiplied by $(1 + x^2)^{s/2}$ are elements of $L^2(\mathbb{R}^n)$. Thus we may define on $H^s(\mathbb{R}^n)$ the norm

$$\| f \|_{H^s(\mathbb{R}^n)} = |(1 + |x|^2)^{s/2} \hat{f}(x)|_2 . \tag{1.3.36}$$

This norm turns $H^s(\mathbb{R}^n)$ into a H-space. If s is a positive integer, the norm $\| \cdot \|_{H^s(\mathbb{R}^n)}$ is equivalent to the usual norm (1.3.26). The dual space of $H^s(\mathbb{R}^n)$ is $H^{-s}(\mathbb{R}^n)$, and the topology on $H^{-s}(\mathbb{R}^n)$ defined by the norm in (1.3.36) is equivalent to the topology generated by the norm in (1.3.33) if s is a negative integer. Additionally, let us define the spaces $H^s(\Omega)$ for s arbitrary real. We say that $f \in H^s(\Omega)$ for $s \in \mathbb{R}_+$, if f is the restriction to Ω of an element $\tilde{f}$ of $H^s(\mathbb{R}^n)$. The norm

$$\| f \|_s = \inf \| \tilde{f} \|_{H^s(\mathbb{R}^n)}, \qquad \tilde{f} = f \quad \text{a.e. on } \Omega \tag{1.3.37}$$

turns $H^s(\Omega)$ into a H-space. If $s > n/2 + \alpha$ ($\alpha > 0$ integer) $H^s(\Omega) \subset C^m(\bar{\Omega})$ with continuous injection. We denote as well the closure of $D(\Omega)$ in $H^s(\Omega)$ $(s > 0)$ by $\mathring{H}^s(\Omega)$. For $s \le 1/2$,

$$\mathring{H}^s(\Omega) = H^s(\Omega), \quad \text{and for} \quad s > 1/2, \ \mathring{H}^s(\Omega) \subset H^s(\Omega).$$

It can be proved that for $s > 1/2$ and if the boundary is appropriately regular,

$$\mathring{H}^s(\Omega) = \left\{ f \mid f \in H^s(\Omega), \frac{\partial^j f}{\partial n^j} = 0 \text{ on } \Gamma \text{ for } 0 \le j < s - \frac{1}{2} \right\}, \tag{1.3.38}$$

where $\partial^j f / \partial n^j$ denotes the jth normal distributional derivative of f defined on Γ.

Next, the spaces $H^{-s}(\Omega)$, where $s \in \mathbb{R}_+$, are defined by duality, i.e.,

$$H^{-s}(\Omega) = (\mathring{H}^s(\Omega))'. \tag{1.3.39}$$

The family of spaces $H^s(\mathbb{R}^n)$ for $s \in \mathbb{R}$ is an example of a Hilbert scale of spaces and has the properties mentioned in Sec. 1.2.4. We take as $H^\circ(\mathbb{R}^n)$ the space $L^2(\mathbb{R}^n)$. For the spaces $H^s(\Omega)$ and for every positive integer N, a Hilbert scale of spaces $\{H^s_{(N)}(\Omega)\}$ can be constructed, which for $0 \le s \le N$ contains all the spaces $H^s(\Omega)$. The norm of the space $H^s(\Omega)$ can be expressed for $s = l + \mu$, where l is a positive integer and $0 < \mu < 1$, by the formula

$$\| f \|_s = \left(\| f \|_l^2 + \sum_{|\alpha| = l} \int_\Omega \int_\Omega \frac{|D^\alpha f(x_1) - D^\alpha f(x_2)|^2}{|x_1 - x_2|^{n + 2\mu}} d\Omega \, d\Omega \right)^{1/2} \tag{1.3.40}$$

The norm on $H^{-s}(\Omega)$, $s \geq 0$, is defined by (1.2.5). For an arbitrary set of bounded indices the spaces $H^s(\Omega)$ share the properties (i), (ii), and (iii) of a Hilbert scale of spaces (cf., Sec. 1.2.4).

1.3.4 Trace Theorem. Imbedding Properties of Sobolev Spaces

Within the framework of a Hilbertian theory, the study of nonhomogeneous B.V.P.s for differential operators on Ω compels us to consider the traces of a function $f \in H^l(\Omega)$ and of its derivatives on the boundary Γ. Hence a class of Sobolev spaces on Γ (denoted by $H^s(\Gamma)$) containing the traces of f and of its derivatives on Γ will now be defined.

We shall only outline the definition of the spaces $H^s(\Gamma)$ for $s \in \mathbb{R}$ (cf., [181], [231]). Consider first an open covering of Γ by a finite number of open bounded subsets $\{U_j\}$, $j = 1, 2, 3, \ldots \nu$, second an associated set of appropriately smooth[6] mappings ψ_j: $U_j \to B = \{y \mid y = \{y', y_n\} \in \mathbb{R}^n,\ y' \in \mathbb{R}^{n-1}, |y'| < 1, |y_n| < 1\}$ such that sufficiently smooth ψ_j^{-1} exist, and finally a partition of unity $\{g_j\}$ subordinate to this covering. If f is a function defined on Γ, we define $w_j(f)$ on $\mathbb{R}^{n-1}$ by

$$w_j(f)(y', 0) = (g_j f)(\psi_j^{-1}(y', 0)). \tag{1.3.41}$$

We denote by $H^s(\Gamma)$ the set of functions f such that $w_i(f) \in H^s(\mathbb{R}^{n-1})$ for $1 \leq i \leq \nu$. This definition is independent of the choice of the open covering, the mappings ψ, and the partition of unity. The norm

$$\| f \|_{s,\Gamma} = \left(\sum_{j=1}^{\nu} \| w_j(f) \|^2_{H^s(\mathbb{R}^{n-1})} \right)^{1/2} \tag{1.3.42}$$

turns $H^s(\Gamma)$ into a H-space. If $H^0(\Gamma) = L^2(\Gamma)$ is a pivot space, then

$$H^{-s}(\Gamma) = (H^s(\Gamma))' \tag{1.3.43}$$

for every $s \in \mathbb{R}$. It can be proved that $C_c^\infty(\Gamma)$ is dense in $H^s(\Gamma)$ for $s \geq 0$, and that the imbeddings

$$H^s(\Gamma) \subset L^2(\Gamma) \subset H^{-s}(\Gamma). \tag{1.3.44}$$

are dense and compact.

The following trace theorem for $H^l(\Omega)$ spaces is of importance: Let γ be the linear mapping

$$f \to \gamma f = \{\gamma_i f\}, \qquad \gamma_i f = \left.\frac{\partial^i f}{\partial n^i}\right|_\Gamma \quad \gamma_i : D(\bar{\Omega}) \to D(\Gamma), \tag{1.3.45}$$

where $i = 0, 1, \ldots, l-1$. Then

(i) γ_i has a unique extension (by continuity) to a continuous linear mapping, still denoted by γ_i, from $H^l(\Omega)$ onto $H^{l-i-1/2}(\Gamma)$ $(0 \leq i \leq l-1)$.

[6] If Γ is uniformly C^m-regular, the mappings belong to C^m[2].

We retain the notation $\gamma f = \{\gamma_i f\}$, $i = 0, 1, \dots l-1$;

(ii) γ is a continuous linear operator from $H^l(\Omega)$ onto $\prod_{i=0}^{l-1} H^{l-i-1/2}(\Gamma)$;

(iii) for a $g = \{g_i\} \in \prod_{i=0}^{l-1} H^{l-i-1/2}(\Gamma)$ there exists $\bar{g} \in H^l(\Omega)$ and a linear continuous mapping $g \to \bar{g}$ from $\prod_{i=0}^{l-1} H^{l-i-1/2}(\Gamma)$ into $H^l(\Omega)$ such that

$$\gamma_i \bar{g} = g_i, \qquad 0 \le i \le l-1; \quad \text{and} \tag{1.3.46}$$

(iv) the kernel of the operator γ is the space $\mathring{H}^l(\Omega)$.

Thus we may write

$$\mathring{H}^l(\Omega) = \{f \mid f \in H^l(\Omega), \gamma^i f = 0 \quad \text{for} \quad 0 \le i \le l-1\}. \tag{1.3.47}$$

For the sake of simplicity, instead of $\gamma_i f$, we shall often write $\partial^i f/\partial n^i|_\Gamma$ in the sense of the foregoing trace theorem. It should be noted that analogous trace theorems also hold for other classes of Sobolev spaces (cf., [2]); we cite here the Gagliardo trace theorem [108], according to which there exists a linear continuous trace application γ_0 from $W^{1,1}(\Omega)$ onto $L^1(\Gamma)$.

Finally, we shall mention certain important properties of Sobolev spaces. The following imbeddings hold for $\Omega \subset \mathbb{R}^n$ "regular"[7] if $j \ge 0$, $m \ge 0$ and $1 \le p < \infty$ (Sobolev's imbedding theorem):

(i) if $mp < n$, then

$$W^{j+m,p}(\Omega) \subset W^{j,q}(\Omega), \quad \text{for} \quad 1 \le q \le \frac{np}{n-mp}; \tag{1.3.48}$$

(ii) if $mp = n$, then (1.3.48) holds for $1 \le q < \infty$; and

(iii) if $mp > n > (m-1)p$ (resp. $n = (m-1)p$)

$$W^{j+m,p}(\bar{\Omega}) \subset C^{j,\lambda}(\bar{\Omega}) \quad \text{for} \quad 0 < \lambda \le m - \frac{n}{p} \qquad (\text{resp. } 0 < \lambda < 1) \tag{1.3.49}$$

At this point we also mention the useful imbedding

$$H^s(\Omega) \subset L^p(\Omega) \quad \text{for} \quad \frac{1}{p} = \frac{1}{2} - \frac{s}{n} > 0, \; s \ge 0. \tag{1.3.50}$$

By the Rellich–Kondrachov theorem, the following imbeddings are compact for $j \ge 0$, $m \ge 1$ and $1 \le p < \infty$:

(i′) Suppose that $mp \le n$. Then the imbedding

$$W^{j+m,p}(\Omega) \subset W^{j,q}(\Omega) \tag{1.3.51}$$

is compact if $n > mp$ and $1 \le q < np/(n-mp)$ or if $n = mp$ and $1 \le q < \infty$;

[7] More precisely, (i), (i′), (ii) and (ii′) hold if Ω has the cone property. (iii) and (iii′) hold if Ω has the strong local Lipschitz property [2].

(ii′) if $mp > n$, then the imbedding in (1.3.51) is compact for $1 \leq q \leq \infty$; and

(iii′) if $mp > n$, the imbedding

$$W^{j+m,p}(\Omega) \subset C^j(\bar{\Omega}), \tag{1.3.52}$$

is compact, and if $mp > n \geq (m-1)p$ and $0 < \lambda < m - n/p$ then the imbedding in (1.3.49) is compact.

Both the imbedding and the compact imbedding theorems hold, with an arbitrary Ω, for the respective $\mathring{W}$ spaces.

We mention, finally the compact imbeddings

$$H^s(\Omega) \subset H^{s-\alpha}(\Omega), \qquad \forall \alpha > 0, \qquad s \in \mathbb{R} \tag{1.3.53}$$

and

$$H^1(\Omega) \subset L^2(\Gamma). \tag{1.3.54}$$

For a systematic study of Sobolev spaces, the reader is referred to [181] [2] [231].

1.3.5 The Space of Functions of Bounded Deformation

A recently defined function space [304] [200] [314] which has proved important in the theory of plasticity is the space $BD(\Omega)$ of the functions of bounded deformation on $\Omega \subset \mathbb{R}^n$. By definition,

$$BD(\Omega) = \{f \mid f = \{f_i\}, f_i \in L^1(\Omega),$$
$$\varepsilon_{ij}(f) = \tfrac{1}{2}(f_{i,j} + f_{j,i}) \in \mathcal{M}^1(\Omega) \quad i, j = 1, \ldots, n\}. \tag{1.3.55}$$

$BD(\Omega)$ is a nonreflexive B-space for the norm

$$\| f \|_{BD(\Omega)} = \sum_{i=1}^{n} | f_i |_1 + \sum_{i,j=1}^{n} \| \varepsilon_{ij}(f) \|_{\mathcal{M}^1(\Omega)}. \tag{1.3.56}$$

Obviously, $W^{1,1}(\Omega) \subset BD(\Omega)$. A distribution f such that $\varepsilon_{ij}(f) \in \mathcal{M}^1(\Omega)$, $i, j = 1, \ldots, n$, is a function of $[L^{n/n-1}(\Omega)]^n$. Suppose that Γ is a C^1-boundary. The following trace theorem holds on $BD(\Omega)$: there exists a surjective, continuous linear mapping $\gamma_0 \colon BD(\Omega) \to [L^1(\Gamma)]^n$ which is uniquely determined and is such that

$$\gamma_0(f) = f|_\Gamma, \qquad \forall f \in BD(\Omega) \cap [C^0(\bar{\Omega})]^n. \tag{1.3.57}$$

The injection

$$BD(\Omega) \subset [L^p(\Omega)]^n \tag{1.3.58}$$

is a compact imbedding, if $1 \leq p < n/(n-1)$. The same injection is continuous if $p = n/(n-1)$. Therefore, if $\Omega \subset \mathbb{R}^3$, every bounded sequence in $BD(\Omega)$ has a subsequence which converges strongly in $[L^1(\Omega)]^n$.

1.4 Additional Topics

1.4.1 Elements of the Theory of Vector-Valued Functions and Distributions

We are concerned here with vector-valued functions defined on a real interval, $[0, T]$, $T < \infty$. Let X be a B-space. A function $f\colon [0, T] \to X$ is said to be absolutely continuous on $[0, T]$ if for every $\varepsilon > 0$ there exists $\delta(\varepsilon) > 0$ and n natural such that $\sum_{j=1}^{n} \| f(t'_j) - f(t_j) \| \le \varepsilon$, whenever $\sum_{j=1}^{n} | t'_j - t_j | \le \delta(\varepsilon)$ for every sequence of mutually disjoint intervals (t'_j, t_j). The following differentiability theorem holds. If X is a reflexive B-space, then every X-valued absolutely continuous function $f\colon [0, T] \to X$ is a.e. differentiable on $(0, T)$ and

$$f(t) = f(0) + \int_0^t \frac{df}{dt}(t)\, dt \qquad 0 \le t \le T. \tag{1.4.1}$$

The derivative $(df/dt)(t)$, also called the strong or ordinary derivative of the function $f\colon [0, T] \to X$, is by definition the strong limit of the difference quotient

$$\frac{f(t + \delta t) - f(t)}{\delta t}. \tag{1.4.2}$$

We denote by $\mathscr{D}'(0T, X)$ the space of all continuous linear mappings from $D(0, T)$ into X. An element $f \in \mathscr{D}'(0T, X)$ is called an X-valued, or vectorial, distribution on $(0, T)$. We define for $f \in \mathscr{D}'(0T, X)$ the derivative $D^\alpha f$ of order α of $f (D^\alpha f \in \mathscr{D}'(0T, X))$ by means of

$$\langle D^\alpha f, \phi \rangle = (-1)^\alpha \left\langle f, \frac{d^\alpha \phi}{dt^\alpha} \right\rangle, \qquad \forall \phi \in D(0T). \tag{1.4.3}$$

The space of all (equivalence classes of) X-valued measurable functions from $[0, T]$ to X such that

$$\| f(t) \|_X \in L^p(0T), \qquad 1 \le p \le \infty \quad \text{is denoted by} \quad L^p(0T, X).$$

It is a B-space with respect to the norm

$$\| f \|_{L^p(0T, X)} = \left(\int_0^T \| f(t) \|_X^p \, dt \right)^{1/p} \tag{1.4.4}$$

if $1 \le p < \infty$, or to the norm

$$\| f \|_{L^\infty(0T, X)} = \operatorname*{ess\,sup}_{(0, T)} \| f(t) \|_X \tag{1.4.5}$$

if $p = \infty$. Note that $L^p(0T, L^p(\Omega)) = L^p((0T) \times \Omega)$.

When X is a H-space, $L^2(0T, X)$ is a H-space as well, with the scalar product

$$(f_1, f_2)_{L^2(0T, X)} = \int_0^T (f_1(t), f_2(t))_X \, dt. \tag{1.4.6}$$

If X is a reflexive B-space, $L^p(0T, X)$ with $1 < p < \infty$ is also reflexive, and its dual space can be identified with $L^{p'}(0T, X')$ where p' is the conjugate exponent of p. If X is nonreflexive, the dual space is $L^{p'}_w(0T, X')$. Here the index w indicates that $L^{p'}_w(0T, X')$ contains weakly measurable[8] functions f such that $\| f \|_{L^{p'}(0T, X')} < \infty$. We shall often deal with bounded sequences $\{f_n\}$ in $L^\infty(0T, X)$ where X is a H-space. Such a sequence contains a weakly$_*$ convergent subsequence to an element $f_0 \in L^\infty(0T, X)$, i.e.,

$$\int_0^T (f_n, \phi)_X \, dt \to \int_0^T (f_0, \phi)_X \, dt, \qquad \forall \phi \in L^1(0T, X'). \tag{1.4.7}$$

To every function $f \in L^p(0T, X)$, where X is a B-space and $p \geq 1$, there corresponds a uniquely determined distribution, again denoted by f, which is defined by

$$\langle f, \phi \rangle = \int_0^T f(t)\phi(t) \, dt, \qquad \forall \phi \in \mathscr{D}(0T). \tag{1.4.8}$$

If $f \in L^1(0T, X)$ then the function $F(t) = \int_0^t f(t) \, dt$ is differentiable a.e. on $(0, T)$ and

$$\frac{dF}{dt}(t) = f(t) \qquad \text{a.e on } (0, T). \tag{1.4.9}$$

Moreover, Lebesgue's theorem states that for $\delta t \neq 0$

$$\lim_{\delta t \to 0} \frac{1}{\delta t} \int_t^{t+\delta t} f(t) \, dt = f(t) \qquad \text{a.e on } (0, T). \tag{1.4.10}$$

Suppose that X is a B-space, m is a positive integer and $1 \leq p \leq \infty$. We introduce the space

$$W^{m,p}(0T, X) = \{ f \mid D^i f \in L^p(0T, X), 0 \leq i \leq m \}. \tag{1.4.11}$$

We denote by $AC^{m,p}(0T, X)$ the space of all X-valued absolutely continuous functions, whose derivatives $d^i f/dt^i$ defined a.e. on $[0, T]$ are absolutely continuous for $1 \leq i \leq m - 1$, and for $0 \leq i \leq m$, they are elements of $L^p(0T, X)$. If $f \in W^{m,p}(0T, X)$, there exists an $\tilde{f} \in AC^{m,p}(0T, X)$ such that

$$f(t) = \tilde{f}(t) \qquad \text{a.e on } (0, T) \tag{1.4.12}$$

[8] In the sense that the real-valued function $t \to \langle f(t), x \rangle_X$ is measurable for every $x \in X$.

and conversely. Let X and Y be H-spaces such that (1.2.22) holds with dense imbeddings. Then the space

$$\tilde{W}^{1,2}(0T, X) = \{f \mid f \in L^2(0T, X), Df \in L^2(0T, X')\} \tag{1.4.13}$$

is considered. It can be proved that a function $\tilde{f} \in AC^{1,2}(0T, X')$ exists such that (1.4.12) holds. Accordingly, Df can be considered as the derivative df/dt of f. The following continuity theorem holds: any function $f \in \tilde{W}^{1,2}(0T, X)$ is a.e. continuous from $[0, T]$ into Y, i.e., $f \in C^0(0T, Y)$. If $f_1, f_2 \in \tilde{W}^{1,2}(0T, X)$, then the function $t \to \langle f_1(t), f_2(t)\rangle_X$ is on $[0, T]$ absolutely continuous and

$$\frac{d}{dt}\langle f_1(t), f_2(t)\rangle_X = \left\langle f_1(t), \frac{df_2(t)}{dt}\right\rangle_X + \left\langle \frac{df_1(t)}{dt}, f_2(t)\right\rangle_X \quad \text{a.e on } (0, T). \tag{1.4.14}$$

Because of the continuity theorem, the initial condition $f = f_0 \in Y$ at $t = 0$ makes sense. This theorem therefore is applied in the study of dynamic B.V.P.s.

1.4.2 Elements of Differential Calculus

Let $F: X \to Y$ be a nonlinear operator, and X and Y two B-spaces. The set of operators F is denoted by $N(X, Y)$. We shall use the notion of the graph of F, defined as the set

$$\mathcal{G}(F) = \{(x, y) \mid x \in X, y \in Y, y = F(x)\}. \tag{1.4.15}$$

Let $\mathcal{O}$ be an open subset of X. We can say that the graphs of the operators $F: \mathcal{O} \to Y$, $\bar{F}: \mathcal{O} \to Y$ are tangential at $x_0 \in \mathcal{O}$ if

$$\lim_{\substack{x \to x_0 \\ x \neq x_0}} \frac{\| F(x) - \bar{F}(x) \|_Y}{\| x - x_0 \|_X} = 0. \tag{1.4.16}$$

We are looking for the linear functions whose graphs are tangential to the graph of a nonlinear operator at $x_0 \in \mathcal{O}$. There exists at most one linear function $\bar{F}: X \to Y$ having the form $\bar{F}(x) = F(x_0) + L(x - x_0)$, where L is a linear mapping, whose graph is tangent to the graph of F at x_0. The mapping $F: \mathcal{O} \to Y$ is said to be Fréchet-differentiable, (or simply differentiable) at x_0, if there exists a linear continuous operator from X into Y, denoted by $(dF/dx)(x_0)$, such that the graph $\mathcal{G}(F)$ is tangent at x_0 to the graph $\mathcal{G}(\bar{F})$ of an operator $\bar{F}: X \to Y$ defined by

$$\bar{F}(x) = F(x_0) + \frac{dF}{dx}(x_0)(x - x_0). \tag{1.4.17}$$

F is said to be Fréchet-differentiable in $\mathcal{O}$ if it is so at every $x_0 \in \mathcal{O}\cdot$ $(dF/dx)(x_0)(\cdot - x_0)$ is called the Fréchet-differential of F at x_0. $(dF/dx)(x_0)$: $X \to Y$ is called the Fréchet-derivative at x_0 and belongs to the space $L(X, Y)$. It can be easily verified that the Fréchet-derivative of a constant mapping $f(x) = c$ is zero; that of a linear continuous mapping is the mapping itself.

Further, let X be a L.C.H.T.V.S., Y a B-space and $\mathcal{O}$ an open subset of X. For $F: \mathcal{O} \to Y$, $x_0 \in \mathcal{O}$ and $h \in X$ we can determine a number $c > 0$ such that for $|\mu| < c$ $x_0 + \mu h \in \mathcal{O}$. Let

$$\lim_{\mu \to 0} \left\| \frac{F(x_0 + \mu h) - F(x_0)}{\mu} - F'(x_0, h) \right\|_Y = 0. \tag{1.4.18}$$

$h \to F'(x_0, h)$ is called the Gâteaux-differential of F at x_0 in direction h. If the limit in (1.4.18) exists for every $h \in X$ then F is said to be Gâteaux-differentiable at x_0, and $h \to F'(x_0, h)$ is the Gâteaux-differential at x_0. It is homogeneous in h, i.e.,

$$F'(x_0, \alpha h) = \alpha F'(x_0, h), \qquad \forall \alpha \in \mathbb{R}. \tag{1.4.19}$$

If furthermore, $h \to F'(x_0, h)$ is linear and continuous in h, we can write it as $F'(x_0)h$ and we call $F'(x_0): X \to Y$ the Gâteaux-derivative of F at x_0.

If F is Fréchet-differentiable at x_0, it is Gâteaux-differentiable as well. If F' exists and is continuous in a neighborhood of x_0, then F is Fréchet-differentiable at x_0. In both cases

$$\frac{dF}{dx}(x_0) = F'(x_0). \tag{1.4.20}$$

Assume that there exists the Gâteaux-derivative $f'(x_0) \in \mathscr{X}'$ of a functional $f: \mathcal{O} \to \mathbb{R}$. $f'(x_0)$ is called the gradient of f at x_0 and is denoted by grad $f(x_0)$. Obviously

$$\langle \operatorname{grad} f(x), h \rangle = \frac{d}{d\mu} f(x + \mu h) \Big|_{\mu = 0}, \qquad \forall h \in \mathcal{O}\ \mu \in \mathbb{R}. \tag{1.4.21}$$

An operator $\mathscr{K}: \mathcal{O} \to \mathscr{X}'$ such that for every $x \in \mathcal{O}$

$$\mathscr{K}(x) = \operatorname{grad} f(x) \tag{1.4.22}$$

is called a potential operator, and the functional f is called the potential of the operator $\mathscr{K}$ (see also [324])

1.4.3 Supplementary Notions and Propositions

Several additional definitions and propositions which are to be used in subsequent chapters follow.

(a) Suppose that X and Y are two B-spaces, and let F be a nonlinear operator from X into Y. We say that F is bounded if it maps bounded sets into bounded sets. If F maps weakly convergent sequences in X into strongly convergent sequences in Y, then F is said to be completely continuous. An operator $F: X_1 \to Y$, where $X_1 \subset X$, is said to be compact if F is continuous and maps bounded subsets of X_1 into relatively compact subsets of Y. If X is a reflexive B-space and F is completely continuous, then F is compact. The complete continuous linear operators $f: X \to Y$ where X and Y are H-spaces, are always compact.

An operator $F: X \to X'$, where X' is the dual space of a reflexive B-space X is called pseudomonotone if the following hold:

(i) F is bounded; and
(ii) for any sequence $\{x_n\} \in X$ such that $x_n \to x_0$ weakly in X and $\limsup \langle F(x_n), x_n - x_0 \rangle \leq 0$, the inequality

$$\liminf \langle F(x_n), x_n - x \rangle \geq \langle F(x_0), x_0 - x \rangle, \qquad \forall x \in X \tag{1.4.23}$$

holds.

(b) If I is the identity operator and A is a continuous linear operator $A: X \to X$, where X is a B-space, then the set of all real numbers λ such that $A - \lambda I$ has no bounded inverse is called the spectrum $\sigma(A)$ of A. If $\ker(A - \lambda I) \neq \emptyset$ and $\lambda \in \sigma(A)$, λ is called an eigenvalue of A. A nonzero element $x \in X$ such that $x \in \ker(A - \lambda I)$ is an eigenvector corresponding to the eigenvalue λ. Let us consider more generally the equation

$$f(x, \lambda) = 0, \quad x \in X, \quad \lambda \in \mathbb{R}. \tag{1.4.24}$$

A point (x_0, λ_0) is a bifurcation point of (1.4.24) if, first, (x_0, λ_0) lies on a curve of solutions $x = x(s)$, $\lambda = \lambda(s)$ (s is a parameter) through (x_0, λ_0), and, second if in every neighborhood of (x_0, λ_0) a solution of (1.4.24) exists, distinct from $(x(s), \lambda(s))$.

(c) We shall close this section by giving some additional results of the theory of Sobolev spaces which are of considerable importance in the study of inequality problems. We assume that Ω is an open, bounded, connected subset of $\mathbb{R}^3$ with a Lipschitz boundary Γ. We consider on Ω the H-spaces $[L^2(\Omega)]^3$ and $[H^1(\Omega)]^3$ equipped with the usual norms. Let $\varepsilon = \{\varepsilon_{ij}\}$, $i, j = 1, 2, 3$, be a symmetric tensor defined on $[L^2(\Omega)]^6$ by

$$\begin{gathered} \varepsilon_{ij} = \varepsilon_{ij}(u) = \tfrac{1}{2}(u_{i,j} + u_{j,i}), \\ \forall u = \{u_i\} \in [H^1(\Omega)]^3. \end{gathered} \tag{1.4.25}$$

Further the finite-dimensional subspace

$$\mathcal{R} = \left\{ v \mid v \in [H^1(\Omega)]^3, \int_\Omega \varepsilon_{ij}(v)\varepsilon_{ij}(v)\, d\Omega = 0 \right\} \tag{1.4.26}$$

is considered. The functions $v \in \mathcal{R}$ are denoted by r and can be put in the form

$$r = r(x) = a + b \wedge x, \quad a, b \in \mathbb{R}^3, \quad x \in \Omega. \tag{1.4.27}$$

Having in mind a deformable body (small deformation theory) occupying Ω when it is undeformed, we consider u as representing the displacement vector, ε the strain tensor and r the vector of rigid-body displacements.

The following inequalities are very important.

Korn's first inequality: suppose that V_0 is a closed linear subspace of $[H^1(\Omega)]^3$ with $V_0 \cap \mathscr{R} = \{0\}$. Then there exists a constant $c > 0$ such that

$$\int_\Omega \varepsilon_{ij}(u)\varepsilon_{ij}(u)\, d\Omega \geq c\|u\|_1^2, \qquad \forall u \in V_0, \tag{1.4.28}$$

where $\|\cdot\|_1$ denotes the norm of $[H^1(\Omega)]^3(\|u\|_1^2 = \sum_{i=1}^3 \|u_i\|_1^2)$.

Korn's second inequality: there exists a constant $c > 0$ such that

$$\int_\Omega \varepsilon_{ij}(u)\varepsilon_{ij}(u)\, d\Omega + |u|_2^2 \geq c\|u\|_1^2, \qquad \forall u \in [H^1(\Omega)]^3, \tag{1.4.29}$$

where $|u|_2^2 = \sum_{i=1}^3 |u_i|_2^2$ is the norm of $[L^2(\Omega)]^3$. The proofs of these inequalities can be found in [98].

Further, the spaces $[L^2(\Gamma)]^3$ and $[H^{1/2}(\Gamma)]^3$ are defined on Γ. The imbedding $[L^2(\Gamma)]^3 \supset [H^{1/2}(\Gamma)]^3$ is compact and dense. According to the trace theorem of Sobolev spaces, for each $U \in [H^{1/2}(\Gamma)]^3$ a $u \in [H^1(\Omega)]^3$ can be determined (not uniquely) with the properties[9]

$$\gamma(u) = U \quad \text{and} \quad \|u\|_1 \leq c\|U\|_{[H^{1/2}(\Gamma)]^3} \quad c \text{ const} > 0. \tag{1.4.30}$$

Moreover,[10]

$$\gamma(u) = u|_\Gamma, \qquad \forall u \in [D(\bar{\Omega})]^3. \tag{1.4.31}$$

Let $\sigma = \{\sigma_{ij}\}$, $i, j = 1, 2, 3$, be another tensor (e.g., a stress tensor). The V.S.s

$$H = \{\sigma \mid \sigma = \{\sigma_{ij}\},\ \sigma_{ij} = \sigma_{ji},\ \sigma_{ij} \in L^2(\Omega)\} \tag{1.4.32}$$

and

$$\Sigma = \{\sigma \mid \sigma \in H,\ \sigma_{ij,j} \in L^2(\Omega)\}, \tag{1.4.33}$$

equipped respectively with the norms

$$\|\sigma\|_H = \left[\int_\Omega \sigma_{ij}\sigma_{ij}\, d\Omega\right]^{1/2} \tag{1.4.34}$$

and

$$\|\sigma\|_\Sigma = \left[\int_\Omega \sigma_{ij}\sigma_{ij}\, d\Omega + \int_\Omega \sigma_{ij,j}\sigma_{ik,k}\, d\Omega\right]^{1/2}, \tag{1.4.35}$$

are H-spaces. By $n = \{n_i\}$ we denote the outward unit normal vector to Γ.

[9] $\|U\|_{[H^{1/2}(\Gamma)]^3}^2 = \sum_{i=1}^3 \|U_i\|_{H^{1/2}(\Gamma)}^2$, and analogously for $\|F\|_{[H^{-1/2}(\Gamma)]^3}$.

[10] In this sense we shall often write simply $u|_\Gamma$ instead of γu.

A "trace" mapping $\bar{\gamma} \in L(\Sigma, [H^{-\frac{1}{2}}(\Gamma)]^3)$ exists with the property that for an $F \in [H^{-\frac{1}{2}}(\Gamma)]^3$ a $\sigma \in \Sigma$ can be determined such that

$$\begin{aligned} \bar{\gamma}(\sigma) &= \sigma_{ij}|_\Gamma n_j, \qquad \forall \sigma \in [D(\bar{\Omega})]^6, \\ \bar{\gamma}(\sigma) &= F, \qquad \|\sigma\|_\Sigma \leq c \|F\|_{[H^{-\frac{1}{2}}(\Gamma)]^3}. \end{aligned} \tag{1.4.36}$$

Here c is a constant > 0 and $\bar{\gamma}$ denotes the trace mapping. For any $\sigma \in \Sigma$ and $v \in [H^1(\Omega)]^3$, the Green–Gauss formula

$$\int_\Omega \sigma_{ij} \varepsilon_{ij}(v)\, d\Omega + \int_\Omega \sigma_{ij,j} v_i \, d\Omega = \langle \bar{\gamma}(\sigma), \gamma(v) \rangle \tag{1.4.37}$$

holds. Here $\langle \cdot, \cdot \rangle$ denotes the duality pairing on $[H^{-1/2}(\Gamma)]^3 \times [H^{1/2}(\Gamma)]^3$. (1.4.37), also called the generalized Green–Gauss formula results easily through extending by density the standard Green–Gauss formula as formulated for $\tau \in [D(\bar{\Omega})]^6$. To this end it suffices to consider the density of $[D(\bar{\Omega})]^6$ in Σ (see e.g., [13]). From (1.4.37), it results that if $\sigma \in \Sigma$, then $\bar{\gamma}\sigma$ is uniquely defined as an element of $[H^{-\frac{1}{2}}(\Gamma)]^3$.

For the sake of simplicity, we shall often omit in this book γ and $\bar{\gamma}$ in the notation of the traces, if no ambiguity occurs.

The vectors U and F defined on the boundary Γ can be "decomposed" into the normal and tangential components U_N, U_T and F_N, F_T with respect to Γ. Obviously, if $a \in [L^2(\Gamma)]^3$, $a_N = a_i n_i$ and $a_T = a - a_N n$. The case $a \in [H^{1/2}(\Gamma)]$ is more involved. Suppose that Γ is regular,[11] and let

$$H_T = \{v \mid v \in [H^{1/2}(\Gamma)]^3,\ v_i n_i = 0 \quad \text{a.e. on } \Gamma\}. \tag{1.4.38}$$

If $a \in [H^{1/2}(\Gamma)]^3$, then the mapping $a \to \{a_N, a_T\}$ is an isomorphism from $[H^{1/2}(\Gamma)]^3$ onto $H^{1/2}(\Gamma) \times H_T$. In the dual spaces, a'_N and a'_T are uniquely determined by the relation

$$\langle a', a \rangle = \langle a'_N, a_N \rangle_{1/2} + \langle a'_T, a_T \rangle_{H_T}, \quad \forall a \in [H^{1/2}(\Gamma)]^3, \tag{1.4.39}$$

where $\langle \cdot, \cdot \rangle_{1/2}$ and $\langle \cdot, \cdot \rangle_{H_T}$ denote the duality pairings on $H^{-1/2}(\Gamma) \times H^{1/2}(\Gamma)$ and $H'_T \times H_T$ respectively. The mapping $a' \to \{a'_N, a'_T\}$ is an isomorphism from $([H^{1/2}(\Gamma)]^3)'$ onto $H^{-\frac{1}{2}}(\Gamma) \times H'_T$. There exists a linear continuous mapping γ_N (resp. γ_T) from $[H^1(\Omega)]^3$ onto $H^{1/2}(\Gamma)$ (resp. H_T) such that

$$\gamma_N(u) = u_i|_\Gamma n_i, \qquad \gamma_T u = u|_\Gamma - \gamma_N(u) n, \qquad \forall u \in [D(\bar{\Omega})]^3 \tag{1.4.40}$$

and

$$\gamma(u) = \gamma_N(u) + \gamma_T(u), \qquad \forall u \in [H^1(\Omega)]^3. \tag{1.4.41}$$

Similarly, there exists a linear continuous mapping $\bar{\gamma}_N$ (resp. $\bar{\gamma}_T$) from Σ onto

[11] It suffices that Γ be a $C^{1,1}$-boundary [150].

$H^{-1/2}(\Gamma)$ (resp. H'_T) such that

$$\bar{\gamma}_N(\sigma) = \sigma_{ij}|_\Gamma n_j n_i, \qquad \bar{\gamma}_T(\sigma) = \{\sigma_{ij}|_\Gamma n_j\} - \bar{\gamma}_N(\sigma)n, \qquad \forall \sigma \in [D(\bar{\Omega})]^6 \tag{1.4.42}$$

and

$$\langle \bar{\gamma}(\sigma), U \rangle = \langle \bar{\gamma}_N(\sigma), U_N \rangle_{1/2} + \langle \bar{\gamma}_T(\sigma), U_T \rangle_{H_T}, \qquad \forall \sigma \in \Sigma, \qquad \forall U \in [H^{1/2}(\Gamma)]^3. \tag{1.4.43}$$

Suppose that $\{U_N, U_T\} \in H^{1/2}(\Gamma) \times H_T$ (resp. $\{F_N, F_T\} \in H^{-1/2}(\Gamma) \times H'_T$) is given. Then a function $u \in [H^1(\Omega)]^3$ (resp. $\sigma \in \Sigma$) can be determined (non-uniquely) such that

$$\gamma_N(u) = U_N, \quad \gamma_T(u) = U_T \quad (\text{resp. } \bar{\gamma}_N(\sigma) = F_N, \bar{\gamma}_T(\sigma) = F_T). \tag{1.4.44}$$

The mappings $\{U_N, U_T\} \to u$ and $\{F_N, F_T\} \to \sigma$ are continuous and linear. Then the Green–Gauss formula

$$\int_\Omega \tau_{ij}\varepsilon_{ij}(v)\, d\Omega + \int_\Omega \tau_{ij,j} v_i\, d\Omega = \langle \bar{\gamma}_N(\tau), \gamma_N(v) \rangle_{1/2} + \langle \bar{\gamma}_T(\tau), \gamma_T(v) \rangle_{H_T} \tag{1.4.45}$$

holds for every $\tau \in \Sigma$ and $v \in [H^1(\Omega)]^3$. In the above formulas we shall often write simply $\langle \cdot,\cdot \rangle$ instead of $\langle \cdot,\cdot \rangle_{1/2}$ or $\langle \cdot,\cdot \rangle_{H_T}$, if no ambiguity occurs. Note that the previous results can be extended for a $C^{0,1}$-boundary, in which case a slight modification of the spaces of U_N and U_T is required [132].

For $v \in BD(\Omega)$ the Green–Gauss formula holds in the following form [304] [314]. Let $\tilde{D}(\bar{\Omega}) = \{\tau \mid \tau = \{\tau_{ij}\}\ \tau_{ij} = \tau_{ji},\ \tau_{ij} \in D(\bar{\Omega}),\ i, j = 1, 2, 3\}$. Then the relation

$$\int_\Omega \tau_{ij}\, d\varepsilon_{ij}(v) + \int_\Omega \tau_{ij,j} v_i\, d\Omega = \int_\Gamma (\gamma_0 v)_i \tau_{ij} n_j\, d\Gamma, \qquad \forall \tau \in \tilde{D}(\bar{\Omega}) \qquad \forall v \in BD(\Omega) \tag{1.4.46}$$

holds. If S is a surface dividing Ω into the nonempty sets Ω_+ and Ω_-, then there exist two linear continuous mappings γ_+ and γ_- from $BD(\Omega)$ onto $[L^1(S)]^3$ such that for every $\tau \in \tilde{D}(\bar{\Omega})$ and for supp $v \subset \Omega$

$$\int_{\Omega_-} \tau_{ij,j} v_i\, d\Omega + \int_{\Omega_-} \tau_{ij}\, d\varepsilon_{ij}(v) = \int_S (\gamma_- v)_i \tau_{ij} n_j\, ds, \qquad \forall v \in BD(\Omega) \tag{1.4.47}$$

and

$$\int_{\bar{\Omega}_-} \tau_{ij,j} v_i\, d\Omega + \int_{\bar{\Omega}_-} \tau_{ij}\, d\varepsilon_{ij}(v) = \int_S (\gamma_+ v)_i \tau_{ij} n_j\, ds, \qquad \forall v \in BD(\Omega) \tag{1.4.48}$$

hold. $\gamma_+ v$ and $\gamma_- v$ are called the external and the internal traces of v on S

with respect to Ω. If $v \in C^0(\bar{\Omega})$, then $\gamma_+ v = \gamma_- v = u|_s$, but generally the two traces are not equal. If p is a continuous seminorm on $BD(\Omega)$ which at the same time is a norm on $\mathscr{R}$, then the norm

$$|||u||| = p(u) + \sum_{i,j=1}^{n} \|\varepsilon_{ij}(u)\|_{\mathscr{M}^1(\Omega)} \tag{1.4.49}$$

is equivalent to the original norm (1.3.56) [314].

Chapter 2

Elements of Convex Analysis

The purpose of this chapter is to provide some notions and fundamental results of convex analysis which will be used throughout this book. Starting with the notion of convexity, some propositions on convex and lower semicontinuous functionals as well as on the minimization of functionals on convex sets are given. The notion of subdifferential is introduced and its relation to one-sided Gâteaux-differentiability is illustrated. There follows the definition of the conjugate functional and some propositions on the conjugacy operation. The chapter closes with some elements of the theory of maximal monotone operators. Our attention is concentrated on the maximal monotone operators on $\mathbb{R}$ as they allow a compact formulation of general classes of variational inequalities. In the present chapter, the relation between convex analysis and the theory of variational inequalities becomes clear.

We rely primarily on the papers of U. Dieter [69], W. Fenchel [95], J. J. Moreau [209], and U. Mosco [215], [216], and on the monographs of A. Auslender [15], V. Barbu and T. Precupanu [17], H. Brézis [29], I. Ekeland and R. Temam [91], R. T. Rockafellar [288], [289], and L. C. Young [338].

2.1 Convex Sets and Functionals

2.1.1 Definitions

Let X be a L.C.H.T.V.S. space and K a subset of X. We recall that K is said to be convex if

$$\lambda x_1 + (1 - \lambda)x_2 \in K \tag{2.1.1}$$

whenever $x_1 \in K$, $x_2 \in K$ and $0 < \lambda < 1$ (Fig. 2.1.1). All linear subspaces of X (including X) are convex. By convention, the empty set $\varnothing$ is convex. The general characteristic of convex sets is that they contain, along with any two distinct points x_1 and x_2, the line segment between x_1 and x_2 (Fig.

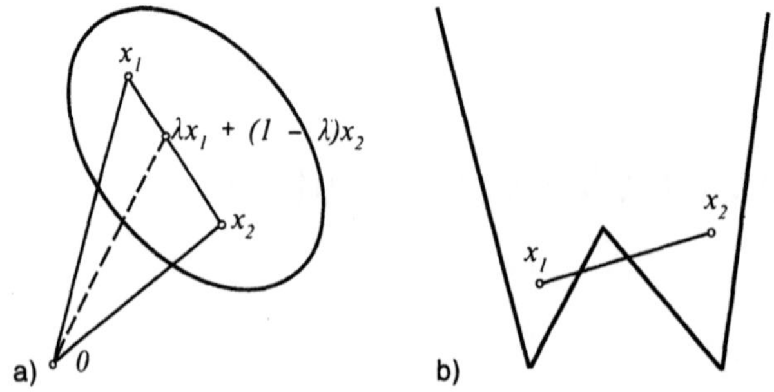

Fig. 2.1.1. A convex and a nonconvex set in $\mathbb{R}^2$.

2.1.1). Any intersection of convex sets is convex; the same does not hold for an arbitrary union of convex sets. Given a set $K_1 \subset X$, then the necessarily convex set of all finite linear combinations $\sum_i \lambda_i x_i$, $x_i \in K_1$, with $\sum_i \lambda_i = 1$, $i = 1, 2, \ldots, n$, is a set called the affine hull of K_1. If, additionally, $\lambda_i \geq 0$, $i = 1, 2, \ldots, n$, the set is called the convex hull of K_1 and is denoted by co K_1; it is the smallest convex subset of X which contains K_1. A point $x \in K_1$ is said to be a relatively interior point of K_1 if it is an interior point of K_1 when K_1 is regarded (topologically) as a subset of its affine hull. The set of all relative interior points of K_1 is called the relative interior of K_1 and is denoted by relint K. Of special interest are convex cones. A set $K \subset X$ is a cone if, for $x \in K$, $\lambda x \in K$ for every $\lambda \geq 0$. Moreover K is a convex cone if $x_1 + x_2 \in K$ for $x_1 \in K$ and $x_2 \in K$. A real-valued functional $f: K \to \mathbb{R}$ is convex (resp. strictly convex) on K if for each $x_1 \in K$, $x_2 \in K$ and $0 < \lambda < 1$

$$f(\lambda x_1 + (1 - \lambda)x_2) \leq (\text{resp. } <)\ \lambda f(x_1) + (1 - \lambda)f(x_2). \tag{2.1.2}$$

A geometrical interpretation of this definition is given in Fig. 2.1.2. A functional f is said to be concave (resp. strictly concave) if and only if $-f$ is convex (resp. strictly convex). A linear functional is at the same time convex and concave, but not strictly. The convexity definition given in (2.1.2) can

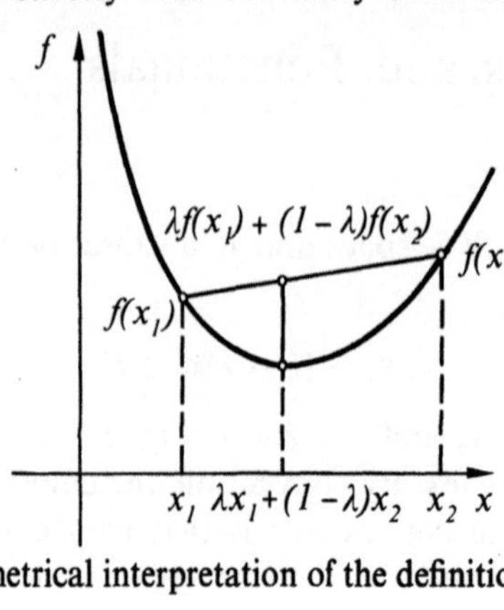

Fig. 2.1.2. Geometrical interpretation of the definition of convexity.

be extended to include the case of convex functionals with possibly infinite values: a functional $f: K \to \bar{\mathbb{R}}$ is defined to be convex on K if for every $x_1 \in K$ and $x_2 \in K$ (2.1.2) holds whenever the right-hand side makes sense. It is obvious that the right-hand side cannot be defined if $f(x_1) = -f(x_2) = \pm\infty$. Because of the fact that a convex functional may have infinite values, we can limit our attention to functionals defined on all of X. Indeed, if $f: K \to \mathbb{R}$ is convex on K, then we can define an extension $\bar{f}$ of f to all of X by setting $\bar{f}(x) = f(x)$ for $x \in K$ and $\bar{f}(x) = \infty$ for $x \notin K$. To every convex set K we can associate a convex functional $I_K: X \to \bar{\mathbb{R}}$, defined by

$$I_K(x) = \begin{cases} 0 & \text{for } x \in K \\ \infty & \text{for } x \notin K \end{cases} \tag{2.1.3}$$

and called the indicator of K. With respect to a convex functional $f: X \to \bar{\mathbb{R}}$ the set

$$\{(x, \lambda) \mid f(x) \leq \lambda, \lambda \in \mathbb{R}, x \in X\} \tag{2.1.4}$$

is defined, called the epigraph of f and denoted by epi f. An equivalent general definition of convexity results, as may be easily established, if we define $f: X \to \bar{\mathbb{R}}$ as being convex, whenever epi f is a convex subset of $X \times \mathbb{R}$. The effective domain $D(f)$ of a convex functional f on X is defined by

$$D(f) = \{x \mid x \in X, f(x) < \infty\}. \tag{2.1.5}$$

A functional f is said to be proper if $f: X \to (-\infty, +\infty]$ and $f \not\equiv \infty$. If f is convex, λf $(\lambda \geq 0)$ is convex. For f_1 and f_2 convex, $f_1 + f_2$ is convex as well (defining[1] $(f_1 + f_2)(x) = +\infty$ for $f_1(x) = -f_2(x) = \pm\infty$).

2.1.2 Lower Semicontinuous Convex Functionals

Next we give the definition of lower semicontinuous (resp. upper semicontinuous) functionals (cf., Sec. 1.1.1). A functional $f: X \to \bar{\mathbb{R}}$, $X = \{\mathscr{X}, T\}$, is called lower semicontinuous (l.s.c.) on X if for every $\lambda \in \mathbb{R}$ the set

$$\{x \mid x \in X, f(x) \leq \lambda\} \tag{2.1.6}$$

is closed in X. For f l.s.c., $-f$ is upper semicontinuous (u.s.c.), and conversely. From this definition it follows that a functional $f: X \to \bar{\mathbb{R}}$ is l.s.c. (resp. u.s.c.) on X if and only if

$$\liminf_{x_n \to x} f(x_n) \geq f(x) \qquad \forall x \in X, \tag{2.1.7}$$

$$\text{(resp. } \limsup_{x_n \to x} f(x_n) \leq f(x), \qquad \forall x \in X, \qquad (2.1.8))$$

[1] cf., [91] p. 67.

for every sequence $\{x_n\}$ converging to x in the T-topology of X.[2] If (2.1.7) (resp. (2.1.8)) holds at a point x, f is called l.s.c. (resp. u.s.c.) at x. The property of lower semicontinuity is closely related to the closedness of epi f. The following proposition holds.

Proposition 2.1.1. A functional $f: X \to \bar{\mathbb{R}}$ is l.s.c. if and only if epi f is a closed subset of $X \times \mathbb{R}$.

This proposition is easily proved by noting that f is l.s.c. on X if and only if $F(x, a) = f(x) - a$ is l.s.c. on $X \times \mathbb{R}$. We leave the proof as an exercise for the reader.

Let X be a L.C.H.T.V.S. and T a topology on X compatible with the duality between $\mathscr{X}$ and $\mathscr{X}'$ (cf., (1.2.7)). The following important proposition is used in many existence proofs.

Proposition 2.1.2. A convex subset of X, closed with respect to the topology T, is closed with respect to every other topology compatible with the duality between $\mathscr{X}$ and $\mathscr{X}'$.

Proof. By the Hahn-Banach theorem, a closed convex subset of X is the intersection of all the closed halfspaces containing it. But all the closed affine hyperplanes and all the closed halfspaces are the same for every topology T on X compatible with the duality, q.e.d.

In a normed V.S. a convex set is norm-closed if and only if it is weakly closed. This proposition, combined with Prop. 2.1.1, gives the following interesting result for a convex, l.s.c. functional $f: X \to \bar{\mathbb{R}}$.

Proposition 2.1.3. If a convex functional is l.s.c. with respect to the topology T, it is l.s.c. with respect to every topology compatible with the duality between $\mathscr{X}$ and $\mathscr{X}'$.

If X is a normed V.S., a functional $f: X \to \bar{\mathbb{R}}$ is l.s.c. on X if and only if it is weakly l.s.c. As we associate with a functional f its epigraph epi f, so we may associate with a set K the indicator I_K. It is easily shown that K is closed if and only if its indicator is l.s.c.

The following two propositions on the continuity properties of convex functionals deserve mention (for the proofs, see [17], [91]).

Proposition 2.1.4. Let $f: X \to \bar{\mathbb{R}}$ be convex. f is proper and continuous on int $D(f)$, where int $D(f) \neq \emptyset$, if and only if a neighborhood F of a point $x_0 \in X$ exists such that $f(x) \not\equiv -\infty$ on F and $f(x) < a < \infty$, $\forall x \in F$, $a \in \mathbb{R}$.

[2] With respect to a sequence $\{a_n\} \in \mathbb{R}$, we define $\liminf_{n\to\infty} a_n = \lim_{n\to\infty} (\inf_{p\geq 0} a_{n+p})$, $\limsup a_n = \lim_{n\to\infty} (\sup_{p\geq 0} a_{n+p})$, and similarly for $\liminf_{x_n\to x} f(x_n) = \lim_{n\to\infty} (\inf_{p\geq 0} f(x_{n+p}))$, $\limsup_{x_n\to x} f(x_n) = \lim_{n\to\infty} (\sup_{p\geq 0} f(x_{n+p}))$. Note that $\liminf f(x_n) = -\limsup -f(x_n)$.

Proposition 2.1.5. In a *B*-space X, a convex, l.s.c. functional $f: X \to \bar{\mathbb{R}}$ is continuous on int $D(f)$.

2.2 Minimization of Convex Functionals

2.2.1 Existence of a Minimum

Convexity is of great importance in the study of optimization problems, as will become clear with the following propositions.

We consider the minimization problem of a convex functional on a convex set $K \subset X$, where X is a L.C.H.T.V.S. We seek a point x_0 which is a solution to the problem

$$f(x_0) = \inf \{f(x) \mid x \in K\} \quad \text{or} \quad f(x_0) = \inf_K f(x). \tag{2.2.1}$$

If f achieves its infimum on K for $x = x_0 \in K$, we can write

$$f(x_0) = \min \{f(x) \mid x \in K\} \quad \text{or} \quad f(x_0) = \min_K f(x). \tag{2.2.2}$$

Every x_0 satisfying (2.2.2) is a solution of the optimization problem.

Let K be a nonempty, convex, closed subset of X, and f a convex, l.s.c., proper functional $f: K \to \bar{\mathbb{R}}$. It is obvious that f can be extended to all of X, and hence the solution of (2.2.2) is sought in X. The following proposition, which is fundamental in optimization theory, establishes certain conditions which ensure that f attains its minimum on K.

Proposition 2.2.1. Let X be a reflexive Banach space with the norm $\|\cdot\|$ and

$$\lim f(x) = \infty \quad \text{when} \quad \|x\| \to \infty, \qquad x \in K \subset X, \tag{2.2.3}$$

or let

$$K \text{ be bounded.} \tag{2.2.4}$$

Then problem (2.2.2) admits at least one solution. If, furthermore, f is strictly convex, the solution is unique.

Proof. Let $\{x_n\} \in K$ be a minimizing sequence such that

$$f(x_n) \to \inf_{x \in K} f(x) = I. \tag{2.2.5}$$

Clearly, $-\infty \leq I < \infty$. Moreover, from (2.2.3) it results that $\|x_n\| \leq c$, where c is a constant; indeed, if this is not true, then a subsequence $\{x_{n'}\}$ exists such that $\|x_{n'}\| \to \infty$ for $n' \to \infty$. Then $f(x_{n'}) \to \infty$, and thus $I = \infty$. Similarly, if instead of (2.2.3) (2.2.4) holds, the boundedness of $\{x_n\}$ results. But in a reflexive Banach space we may extract from a bounded sequence a subsequence, again denoted by $\{x_{n'}\}$, such that $x_{n'} \to \bar{x}$ weakly in X (cf., Sec.

1.2.3). Since K is closed and convex, it is weakly closed as well (cf., Sec. 2.1.2), and thus $\bar{x} \in K$. f is weakly lower semicontinuous, since it is convex and l.s.c. on K, and hence

$$f(\bar{x}) \leq \liminf_{n' \to \infty} f(x_{n'}). \tag{2.2.6}$$

Therefore, due to (2.2.5)

$$f(\bar{x}) \leq I, \tag{2.2.7}$$

which implies that $I \neq -\infty$ and $\bar{x} = x_0$. Let x_1 and x_2 be two solutions of (2.2.1), where f is now assumed to be strictly convex. Then $\frac{1}{2}(x_1 + x_2) \in K$, and because of the strict convexity of f, $f(\frac{1}{2}(x_1 + x_2)) < \frac{1}{2}[f(x_1) + f(x_2)] = \inf_{x \in K} f(x)$. Accordingly, $x_1 = x_2$, q.e.d.

The solutions of problem (2.2.2) constitute a convex closed subset of X, as can be easily shown by considering the convexity and the lower semicontinuity of f. This subset is nonempty if Prop. 2.2.1 holds.

2.2.2 Variational Inequalities

Some variational inequalities equivalent to problem (2.2.1) will now be obtained. X is a L.C.H.T.V.S. and $\langle \cdot, \cdot \rangle$ denotes the duality pairing.

Proposition 2.2.2. Let $f = f_1 + f_2$ be a proper functional, where f_1 and f_2 are convex, l.s.c. functionals on K and suppose, that grad f_1 exists on X. For $x_0 \in K$, the following conditions are equivalent to each other:

$$f(x_0) = \inf_K f(x); \tag{2.2.8}$$

$$\langle \text{grad } f_1(x_0), x - x_0 \rangle + f_2(x) - f_2(x_0) \geq 0, \qquad \forall x \in K; \text{ and} \tag{2.2.9}$$

$$\langle \text{grad } f_1(x), x - x_0 \rangle + f_2(x) - f_2(x_0) \geq 0, \qquad \forall x \in K. \tag{2.2.10}$$

Proof. From (2.2.8), by using the convexity of f_2 it follows for $x_1 = (1 - \lambda)x_0 + \lambda x$, where $\lambda \in (0, 1)$, that

$$f(x_0) \leq f(x_1) \leq f_1(x_1) + (1 - \lambda) f_2(x_0) + \lambda f_2(x), \qquad \forall x \in K, \tag{2.2.11}$$

or

$$\frac{1}{\lambda}[f_1((1 - \lambda)x_0 + \lambda x) - f_1(x_0)] + f_2(x) - f_2(x_0) \geq 0, \qquad \forall x \in K. \tag{2.2.12}$$

If we let λ tend to 0, (2.2.9) results. Conversely, from (2.2.9), by means of the well-known inequality[3]

$$f_1(x) - f_1(x_0) \geq \langle \text{grad } f_1(x), x - x_0 \rangle, \qquad \forall x \in K, \tag{2.2.13}$$

[3] (2.2.13) and (2.2.15) are necessary and sufficient for the convexity of $f: K \to \mathbb{R}$, if grad f exists on K [91].

the relation

$$f(x) - f(x_0) \geq 0, \qquad \forall x \in K, \tag{2.2.14}$$

i.e., (2.2.8), results. By adding to (2.2.9) the monotonicity inequality

$$\langle \text{grad } f_1(x) - \text{grad } f_1(x_0), x - x_0 \rangle \geq 0, \tag{2.2.15}$$

which is valid because of the convexity of f_1, (2.2.10) results. To prove the converse, we set $x = (1 - \lambda)x_0 + \lambda x_1$, $x_1 \in K$, $\lambda \in (0, 1)$ into (2.2.10). Because of the convexity of f_2, the relation

$$\langle \text{grad } f_1((1 - \lambda)x_0 + \lambda x_1), x_1 - x_0 \rangle + f_2(x_1) - f_2(x_0) \geq 0, \qquad \forall x_1 \in K, \tag{2.2.16}$$

results. Let

$$h(\lambda) = f_1((1 - \lambda)x_0 + \lambda x_1) + \lambda(f_2(x_1) - f_2(x_0)). \tag{2.2.17}$$

$\lambda \to h(\lambda)$ is convex and

$$h'(\lambda) = \langle \text{grad } f_1((1 - \lambda)x_0 + \lambda x_1), x_1 - x_0 \rangle + f_2(x_1) - f_2(x_0) \geq 0 \qquad \forall \lambda \in (0,1]. \tag{2.2.18}$$

Therefore, the inequality $h(1) \geq h(0)$ holds. It implies (2.2.14) and thus the variational inequality (2.2.9), q.e.d.

It is shown in [300] that if grad $f_1(x)$ does not exist, two variational inequalities equivalent to (2.2.8) can again be derived (see also [66]).

2.3 Subdifferentiability

2.3.1 Definitions and Related Propositions

The differential properties of a functional f can be described, if f is differentiable, by means of the tangent hyperplanes to the graph of f. If f is convex but not necessarily differentiable, then the supporting hyperplanes to the epigraph of f describe the differential properties of f, as is made obvious by considering a convex functional $f: \mathbb{R} \to \bar{\mathbb{R}}$. This leads to the definition of the subdifferential, a notion which generalizes the classical differential.

In this section, we assume that X is a L.C.H.T.V.S., X' its dual, $\langle \cdot, \cdot \rangle$ the duality pairing between X and X', and $f: X \to \bar{\mathbb{R}}$. The vector $x' \in X'$, for which

$$f(x_1) - f(x) \geq \langle x', x_1 - x \rangle, \qquad \forall x_1 \in X \tag{2.3.1}$$

holds, where $f(x)$ is finite at $x \in X$, is called the subgradient of f at x. The set of all $x' \in X'$ satisfying (2.3.1) is called the subdifferential of f at x and is denoted by $\partial f(x)$. We then write

$$x' \in \partial f(x). \tag{2.3.2}$$

The set $\{x \mid \partial f(x) \neq \emptyset\}$ is denoted by $D(\partial f)$, and is called the domain of ∂f.

The mapping $\partial f\colon X \to X'$ is multivalued and is called the subdifferential of f. The affine function L defined by $L(x_1) = f(x) + \langle x', x_1 - x\rangle$, $\forall x_1 \in X$ is called the supporting hyperplane of epi f at the point $(x, f(x))$. Accordingly, (2.3.1) states for $f(x)$ finite that the supporting hyperplane to epi f at $(x, f(x)) \in X \times \mathbb{R}$ is nonvertical. If $\partial f(x) \neq \emptyset$, f is said to be subdifferentiable at x, $\partial f(x) = \emptyset$ for $x \notin D(f)$ and $f \not\equiv \infty$. From (2.3.1), it follows readily that a necessary and sufficient condition in order that x_0 minimize f on X is that

$$0 \in \partial f(x_0). \tag{2.3.3}$$

This condition indicates the role of subdifferentials in optimization theory. Further, several propositions on ∂f are proved.

Proposition 2.3.1. The set $\partial f(x)$ is convex and closed with respect to the topology $\Sigma(\mathscr{X}', \mathscr{X})$.

Proof. The cases $\partial f(x) \neq \emptyset$ and $\partial f(x) = \{x'\}$, $x' \in X'$, are trivial. Suppose that x'_1, $x'_2 \in \partial f(x)$. It follows from (2.3.1) that $\lambda x'_1 + (1 - \lambda)x'_2 \in \partial f(x)$, $\forall \lambda \in (0, 1)$, and thus $\partial f(x)$ is convex. Assume that $\partial f(x)$ is not $\Sigma(\mathscr{X}', \mathscr{X})$-closed, and let x'_0 be a closure point of $\partial f(x)$. If $x'_0 \notin \partial f(x)$, then we can determine a vector $x_1 \in X$ such that

$$f(x_1) - f(x) < \langle x'_0, x_1 - x\rangle. \tag{2.3.4}$$

The $\Sigma(\mathscr{X}', \mathscr{X})$-continuity of $\langle \cdot, x_1 - x\rangle$ implies that there exists a neighborhood F of x'_0 such that (2.3.4) is valid for every $x' \in F$. It follows that $F \cap \partial f(x) = \emptyset$, and this leads to a contradiction, since x'_0 is a closure point of $\partial f(x)$, q.e.d.

The existence of subdifferentials is ensured in the case of convex functionals by means of the following two propositions.

Proposition 2.3.2. Let $f\colon X \to \bar{\mathbb{R}}$ be convex, and suppose that f is finite and continuous at $x_0 \in X$. Then $\partial f(x_0) \neq \emptyset$.

Proof. Since f is continuous at x_0, the point $(x_0, f(x_0) + \varepsilon) \in$ intepi f for every $\varepsilon > 0$. $(x_0, f(x_0))$ belongs to the boundary of the convex set epi f. By the Hahn–Banach theorem, there exists a closed affine hyperplane $\mathscr{H}$ through $(x_0, f(x_0))$ which separates $(x_0, f(x_0))$ from intepi f. Assume that $\mathscr{H}$ is described by the equation

$$\langle x', x\rangle + a'a = b, \quad x \in X, \quad x' \in X', \quad a, a', b \in \mathbb{R}, \tag{2.3.5}$$

where

$$\langle x', x\rangle + a'a \geq b, \qquad \forall (x, a) \in \text{epi } f \tag{2.3.6}$$

and

$$\langle x', x_0\rangle + a' f(x_0) = b. \tag{2.3.7}$$

$a' \neq 0$, i.e., the hyperplane is not vertical. Indeed, if $a' = 0$, then $\langle x - x_0, x' \rangle \geq 0$, $\forall x \in D(f)$ and therefore $x' = 0$, which is not possible. Thus $a' > 0$, and from (2.3.5–7), dividing by a' we obtain the relation

$$f(x) - f(x_0) \geq \left\langle -\frac{x'}{a'}, x - x_0 \right\rangle, \qquad \forall x \in X, \tag{2.3.8}$$

i.e.,

$$-\frac{x'}{a'} \in \partial f(x_0). \tag{2.3.9}$$

Therefore, $\partial f(x_0) \neq \emptyset$, q.e.d.

We may generalize this result.

Proposition 2.3.3. On the assumptions of Prop. 2.3.2, $\partial f(x)$ is nonempty for every $x \in \operatorname{int} D(f)$.

Proof. By Prop. 2.1.4, since f is finite and continuous at x_0, there exists a neighborhood of x_0 on which f is not everywhere equal to $-\infty$ and is bounded above by a constant $a < \infty$. Thus f is finite and continuous on $\operatorname{int} D(f)$. Prop. 2.3.2 completes the proof, q.e.d.

The special case $f = I_K$ (see (2.1.3)), where K is a nonempty convex subset of a L.C.H.T.V.S. X, is very important. Then

$$\partial I_K(x) = \{x' \mid I_K(x_1) - I_K(x) \geq \langle x', x_1 - x \rangle, \forall x_1 \in X\}, \tag{2.3.10}$$

or, equivalently,

$$\partial I_K(x) = \{x' \mid \langle x', x_1 - x \rangle \leq 0, \forall x_1 \in K, \text{ for } x \in K\}. \tag{2.3.11}$$

The geometrical meaning of the variational inequality

$$\langle x', x_1 - x \rangle \leq 0, \qquad \forall x_1 \in K, \qquad x \in K \tag{2.3.12}$$

is that x' is an outward normal vector to K at x. In general, the set of all vectors x' satisfying (2.3.12) forms an outward normal cone to K at x. This cone (a) is empty provided $x \notin K$ (b) has at least the zero element if $x \in K$, and (c) has only the zero element if $x \in \operatorname{relint} K$ [213], [288].

Subdifferentiability is closely related to the notion of one-sided Gâteaux-differentiability. This provides a method for the "construction" of the subdifferential for a given functional.

2.3.2 One-Sided Directional Gâteaux-Differential

A functional $f: X \to \bar{\mathbb{R}}$, where X is a L.C.H.T.V.S., is said to be one-sided directional Gâteaux-differentiable at x_0 if there exists $\tilde{f}'(x_0, h)$ such that

$$\lim_{\mu \to 0+} \frac{f(x_0 + \mu h) - f(x_0)}{\mu} = \tilde{f}'(x_0, h), \qquad \forall h \in X. \tag{2.3.13}$$

It should be noted that $+\infty$ and $-\infty$ are allowed as limits in (2.3.13). Functional $h \to \tilde{f}'(x_0, h)$ is the one-sided directional Gâteaux-differential of f at x_0 with respect to the direction h. If $\tilde{f}'(x_0, h) = -\tilde{f}'(x_0, -h)\ \forall h \in X$, then f is Gâteaux-differentiable at x_0. It can be readily shown that $\tilde{f}'(x_0, \cdot)$ is a convex, positively homogeneous function of h. One of the most important properties of convex functionals is their one-sided directional Gâteaux-differentiability [122] [288].

Proposition 2.3.4. Assume that $f: X \to \bar{\mathbb{R}}$ is convex. Then f is one-sided directional Gâteaux-differentiable at every $x \in X$ with $f(x) \neq \pm\infty$.

Proof. It is apparent that for every h the difference quotient in (2.3.13) is nondecreasing if $0 < \mu < \infty$. Accordingly, the limit in (2.3.13) exists and furthermore

$$\tilde{f}'(x, h) = \inf_{\mu > 0} \frac{f(x + \mu h) - f(x)}{\mu}, \tag{2.3.14}$$

q.e.d.

We shall prove the following interesting properties.

Proposition 2.3.5. For $f: X \to \bar{\mathbb{R}}$ convex and $f(x) \neq \pm\infty$,

$$f(x_1) - f(x) \geq \tilde{f}'(x, x_1 - x), \qquad \forall x_1 \in X \tag{2.3.15}$$

and

$$\tilde{f}'(x, x_1 - x) \geq -\tilde{f}'(x, -(x_1 - x)), \qquad \forall x_1 \in X. \tag{2.3.16}$$

Proof. (2.3.15) follows immediately from (2.3.13) by setting $h = x_1 - x$ and $0 < \mu \leq 1$. Because of the convexity of f,

$$\tilde{f}'(x, x_1 - x) = \lim_{\mu \to 0_+} \frac{f(x + \mu(x_1 - x)) - f(x)}{\mu} \leq f(x_1) - f(x). \tag{2.3.17}$$

To prove (2.3.16), it suffices to show that

$$\frac{f(x + \mu h) - f(x)}{\mu} + \frac{f(x - \mu h) - f(x)}{\mu} \geq 0, \qquad \forall h \in X \tag{2.3.18}$$

for $\mu > 0$. (2.3.16) results for $\mu \to 0$ and for $h = x_1 - x$. (2.3.18) is derived from (2.1.2) by setting $x_1 = x + \mu h$, $x_2 = x - \mu h$ and $\lambda = \frac{1}{2}$, q.e.d.

In the one-dimensional case $f: \mathbb{R} \to \bar{\mathbb{R}}$, a simple geometric meaning can be given to (2.3.15) and (2.3.16) (cf., Fig. 2.3.1).

The following propositions illustrate the relation between subdifferential and one-sided directional Gâteaux-differential.

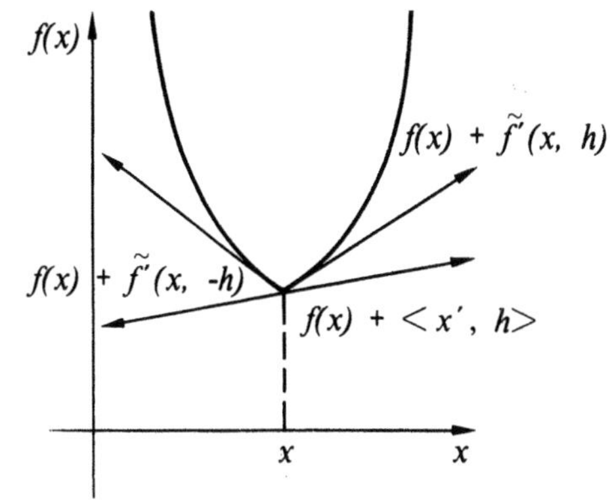

Fig. 2.3.1. The geometrical meaning of (2.3.15) and (2.3.16).

Proposition 2.3.6. Let f be convex and $f(x_0) \neq \pm\infty$. If $x' \in \partial f(x_0)$, then

$$\tilde{f}'(x_0, h) \geq \langle x', h \rangle, \qquad \forall h \in X. \tag{2.3.19}$$

Proof. The subdifferential inequality (2.3.1) yields (2.3.19) by setting $x_1 = x + \lambda h$, $\lambda > 0$, and letting $\lambda \to 0_+$, q.e.d.

This result can be improved by the following proposition.

Proposition 2.3.7. Let f: $X \to \bar{\mathbb{R}}$ be a convex functional and suppose that it is bounded on a neighborhood of $x_0 \in X$. Then

$$\tilde{f}'(x_0, h) = \max\{\langle x', h \rangle \mid x' \in \partial f(x_0)\}, \qquad \forall h \in X. \tag{2.3.20}$$

Proof. From Prop. 2.1.4, f is continuous at x_0. We can show that $(x_0 + \lambda h,\ f(x_0) + \lambda \tilde{f}'(x_0, h)) \notin$ intepi f for every $\lambda \geq 0$. For fixed $\lambda_0 \geq 0$ there exists a closed affine hyperplane $\mathscr{H}$ through $(\tilde{x}_0, \tilde{a}) = (x_0 + \lambda_0 h, f(x_0) + \lambda_0 \tilde{f}'(x_0, h))$, which separates intepi f from this point. For $\lambda_0 = 0$, (2.3.9) results. Suppose that $\lambda_0 > 0$. The inequality

$$\langle x', \tilde{x}_0 - x_0 \rangle + a'(\tilde{a} - a) \leq 0, \tag{2.3.21}$$

which holds for every $(x_0, a) \in$ epi f yields, by setting $a = f(x_0)$ and since $a' > 0$,

$$\left\langle -\frac{x'}{a'}, h \right\rangle \geq \tilde{f}'(x_0, h). \tag{2.3.22}$$

(2.3.22) along with (2.3.19) implies (2.3.20), q.e.d.

The two last propositions easily permit a simple construction of the set $\partial f(x)$. As we can see, if f maps $\mathbb{R}$ into $\mathbb{R}$ then the subgradients x' are the slopes of the nonvertical lines through $(x, f(x))$, which have no point in common with intepi f. From $\tilde{f}'(x, 1) = f'_+(x)$ and $\tilde{f}'(x, -1) = -f'_-(x)$ (right and left derivatives), it follows that $f'_-(x) \leq x' \leq f'_+(x)$. Assume that f is a convex, l.s.c., proper functional on $\mathbb{R}$. In this case the right and left derivatives f'_+ and f'_- may be extended, when $x \notin D(f)$, by setting $f'_+ = f'_- = \infty$

(resp. $f'_+ = f'_- = -\infty$) for points lying to the right (resp. to the left) of $D(f)$. We may then write for $f: \mathbb{R} \to \bar{\mathbb{R}}$

$$\partial f(x) = \{x' \in \mathbb{R} \mid f'_-(x) \leq x' \leq f'_+(x)\}. \tag{2.3.23}$$

For the left and right derivatives of a convex functional on $\mathbb{R}$, the following two propositions hold.

Proposition 2.3.8. Let f be a convex, l.s.c. proper functional on $\mathbb{R}$. Then f'_+ and f'_- are nondecreasing functions on $\mathbb{R}$ and have finite values on the int $D(f)$, and for $x_1 < x < x_2$

$$f'_+(x_1) \leq f'_-(x) \leq f'_+(x) \leq f'_-(x_2). \tag{2.3.24}$$

Proof. Due to Prop. 2.3.4, $f'_+(x)$ and $f'_-(x)$ exist for $x \in D(f)$ and $f'_-(x) \leq f'_+(x)$. The monotonicity of the difference quotients defining the right and left derivatives f'_+ and f'_- implies that these derivatives are finite in int $D(f)$. If $x_1 < x_2$ and $x_1, x_2 \in D(f)$ then the inequality $f'_+(x_1) \leq f'_-(x_2)$ is readily proved. If $x_1, x_2 \notin D(f)$, the same inequality holds by definition, q.e.d.

Proposition 2.3.9. Let f be a convex, l.s.c. and proper functional on $\mathbb{R}$. Then for every x, $\lim_{z\to x+} f'_\pm(z) = f'_+(x)$ and $\lim_{z\to x-} f'_\pm(z) = f'_-(x)$.

Proof. From (2.3.24) it results that f'_+ and f'_- are nondecreasing and that

$$f'_+(x) \leq \lim_{z\to x+} f'_-(z) \leq \lim_{z\to x+} f'_+(z). \tag{2.3.25}$$

But for $z \to x_+$, $\lim f(z) = f(x)$ (Prop. 2.1.5). If $x, x_1 \in D(f)$, $x < x_1$ (the case $x, x_1 \notin D(f)$ is trivial),

$$\frac{f(x_1) - f(x)}{x_1 - x} = \lim_{z\to x+} \frac{f(x_1) - f(z)}{x_1 - z} \geq \lim_{z\to x+} f'_+(z), \tag{2.3.26}$$

and thus

$$\lim_{z\to x+} f'_+(z) \leq \lim_{x_1\to x+} \frac{f(x_1) - f(x)}{x_1 - x} = f'_+(x). \tag{2.3.27}$$

Accordingly, $\lim_{z\to x+} f'_\pm(z) = f'_+(x)$. Similarly, the other equality results, q.e.d.

Suppose further that β is a function on $\mathbb{R}$ such that

$$f'_-(x) \leq \beta(x) \leq f'_+(x) \qquad \forall x \in \mathbb{R}. \tag{2.3.28}$$

Due to (2.3.24), β is a nondecreasing function, and moreover (Prop. 2.3.9)

$$\beta_+(x) = \lim_{z\to x+} \beta(z) = f'_+(x) \quad \text{and} \quad \beta_-(x) = \lim_{z\to x-} \beta(z) = f'_-(x). \tag{2.3.29}$$

From these relations and from (2.3.23), it is concluded that β (resp. ∂f) determines ∂f (resp. β) completely. We can thus write

$$\partial f(x) = \{x' \in \mathbb{R} \mid \beta_-(x) \leq x' \leq \beta_+(x)\}. \tag{2.3.30}$$

Finally, the case in which ∂f has only one element is discussed.

Proposition 2.3.10. Let $f: X \to \bar{\mathbb{R}}$ be convex, and suppose that $\operatorname{grad} f(x)$ exists at x. Then $\partial f(x) = \{\operatorname{grad} f(x)\}$. Conversely, if f is finite and continuous at x and if $\partial f(x)$ has only one element, then $\operatorname{grad} f$ exists at x and $\partial f(x) = \{\operatorname{grad} f(x)\}$.

Proof. From the well-known inequality

$$f(x_1) - f(x) \geq \langle \operatorname{grad} f(x), x_1 - x \rangle, \qquad \forall x_1 \in X, \tag{2.3.31}$$

it results that $\operatorname{grad} f(x) \in \partial f(x)$. Suppose that x' is another element of $\partial f(x)$. Since $\operatorname{grad} f(x)$ exists, $\tilde{f}'(x, h) = \langle \operatorname{grad} f(x), h \rangle$, and thus by means of (2.3.19)

$$\langle x' - \operatorname{grad} f(x), h \rangle \leq 0, \qquad \forall h \in X, \tag{2.3.32}$$

which implies for $h = \pm h_1$ that $x' = \operatorname{grad} f(x)$. Conversely, from Prop. 2.3.7

$$\tilde{f}'(x, h) = \langle x', h \rangle, \qquad \forall h \in X, \tag{2.3.33}$$

where x' is the unique element of $\partial f(x)$. Accordingly, $x' = \operatorname{grad} f(x)$, q.e.d.

The notion of the subdifferential will be illustrated by means of two simple examples.

(i) Let $f: \mathbb{R} \to \bar{\mathbb{R}}$ be given by

$$f(x) = \begin{cases} -g(x - a_1) & \text{if } x \leq a_1 \\ 0 & \text{if } a_1 \leq x \leq a_2 \\ \infty & \text{if } x > a_2 \end{cases}. \tag{2.3.34}$$

Then (Fig. 2.3.2)

$$\partial f(x) = \begin{cases} -g & \text{if } x < a_1 \\ [-g, 0] & \text{if } x = a_1 \\ 0 & \text{if } a_1 < x < a_2 \\ [0, \infty) & \text{if } x = a_2 \\ \varnothing & \text{if } x > a_2 \end{cases}. \tag{2.3.35}$$

(ii) Let $f: \mathbb{R} \to \bar{\mathbb{R}}$ be given by

$$f(x) = \begin{cases} -(a^2 - x^2)^{1/2} & \text{if } |x| \leq a \\ \infty & \text{otherwise} \end{cases}. \tag{2.3.36}$$

Then

$$\partial f(x) = \begin{cases} \dfrac{x}{\sqrt{a^2 - x^2}} & \text{if } |x| < a \\ \varnothing & \text{otherwise} \end{cases}. \tag{2.3.37}$$

Note that $\partial f(\pm a) = \varnothing$, although $\pm a \in D(f)$. [288]

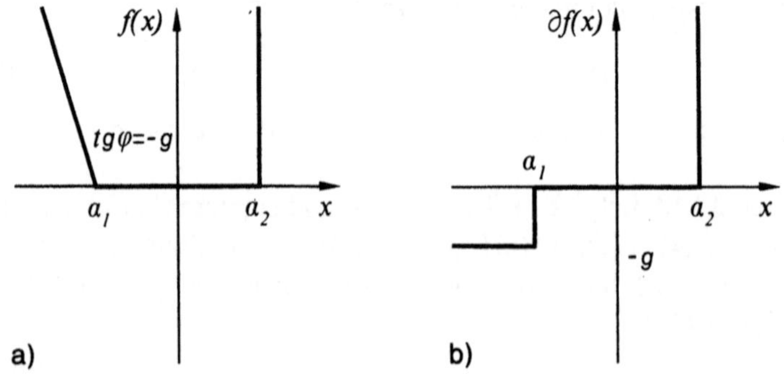

Fig. 2.3.2. The graphs of (2.3.34) and (2.3.35).

2.4 Subdifferential Calculus

2.4.1 The Subdifferential of a Sum of Functionals and of a Composite Functional

In this section some propositions concerning operations with subdifferentials will be proved. Again, X is a L.C.H.T.V.S.

Proposition 2.4.1 Let $f: X \to \mathbb{R}$ and $\lambda > 0$. Then for every $x \in D(\partial f)$

$$\partial(\lambda f)(x) = \lambda \partial f(x). \tag{2.4.1}$$

Proof. Results immediately from (2.3.1), q.e.d.

Proposition 2.4.2. Let $f_1: X \to \bar{\mathbb{R}}$ and $f_2: X \to \bar{\mathbb{R}}$. Then for every $x \in D(\partial f_1) \cap D(\partial f_2)$

$$\partial f_1(x) + \partial f_2(x) \subset \partial(f_1 + f_2)(x). \tag{2.4.2}$$

Proof. If $x' = x'_1 + x'_2$ with $x'_1 \in \partial f_1(x)$ and $x'_2 \in \partial f_2(x)$, we have

$$(f_1 + f_2)(z) = f_1(z) + f_2(z) \geq f_1(x) + f_2(x) + \langle x'_1, z - x \rangle$$
$$+ \langle x'_2, z - x \rangle = (f_1 + f_2)(x) + \langle x', z - x \rangle, \qquad \forall z \in X, \tag{2.4.3}$$

and hence $x' \in \partial(f_1 + f_2)(x)$. This proves the inclusion (2.4.2), q.e.d.

This inclusion holds in the form of equality of sets, if certain additional conditions given in the following propositions hold.

Proposition 2.4.3. Assume that $f_1: X \to \bar{\mathbb{R}}$ and $f_2: X \to \bar{\mathbb{R}}$ are convex and that grad f_2 exists at the point x. Then if $x \in D(\partial f_1)$,

$$\partial(f_1 + f_2)(x) = \partial f_1(x) + \partial f_2(x). \tag{2.4.4}$$

Proof. It is sufficient to show that every $x' \in \partial(f_1 + f_2)(x)$ can be written in the form $x'_1 + \operatorname{grad} f_2(x)$, where $x'_1 \in \partial f_1(x)$ and $\partial f_2(x) = \{\operatorname{grad} f_2(x)\}$. If we replace z in the inequality

$$f_1(z) + f_2(z) - f_1(x) - f_2(x) \geq \langle x', z - x \rangle, \qquad \forall z \in X \tag{2.4.5}$$

by $x + \mu(z - x)$, $\mu > 0$, divide by μ and let $\mu \to 0_+$, then the relation

$$\tilde{f}'_1(x, z - x) \geq \langle x' - \operatorname{grad} f_2(x), z - x \rangle, \qquad \forall z \in X \tag{2.4.6}$$

results, which, combined with (2.3.15), yields

$$x' - \operatorname{grad} f_2(x) \in \partial f_1(x), \tag{2.4.7}$$

q.e.d.

A more general result can be proved as well [91].

Proposition 2.4.4. Suppose that $f_1: X \to (-\infty, +\infty]$ and $f_2: X \to (-\infty, +\infty]$ are convex and l.s.c., and that a point $x_0 \in D(f_1) \cap D(f_2)$ exists at which f_1 is continuous. Then

$$\partial(f_1 + f_2)(x) = \partial f_1(x) + \partial f_2(x), \qquad \forall x \in X. \tag{2.4.8}$$

Proof. It suffices to show that every $x' \in \partial(f_1 + f_2)(x)$ can be written in the form $x' = x'_1 + x'_2$, where $x'_1 \in \partial f_1(x)$ and $x'_2 \in \partial f_2(x)$, i.e., we must prove the existence of x'_1 and x'_2. We consider the convex subsets of $X \times \mathbb{R}$.

$$\begin{aligned} \mathscr{K}_1 &= \{(z, a) \mid f_1(z) - f_1(x) - \langle x', z - x \rangle \leq a\} \\ \mathscr{K}_2 &= \{(z, a) \mid f_2(z) - f_2(x) \leq -a\} \end{aligned} \tag{2.4.9}$$

The inequality

$$f_1(z) + f_2(z) - f_1(x) - f_2(x) \geq \langle x', z - x \rangle, \qquad \forall z \in X \tag{2.4.10}$$

implies that $\mathscr{K}_1$ and $\mathscr{K}_2$ may have common points only on their boundaries. Moreover $\operatorname{int} \mathscr{K}_1 \neq \varnothing$, since $\mathscr{K}_1$ is the epigraph of the convex functional

$$g(z) = f_1(z) - f_1(x) - \langle x', z - x \rangle, \tag{2.4.11}$$

which is continuous at x_0. Then $\operatorname{int} \mathscr{K}_1$ and $\mathscr{K}_2$ may be separated by a closed affine hyperplane $\mathscr{H}$ which is nonvertical (cf., the proof of Prop. 2.3.2). Accordingly, a functional $z' \in X'$ and a number $b \in \mathbb{R}$ exist such that

$$f_2(x) - f_2(z) \leq \langle z', z \rangle + b \leq f_1(z) - f_1(x) - \langle x', z - x \rangle, \qquad \forall z \in X, \quad b \in \mathbb{R}. \tag{2.4.12}$$

It follows for $z = x$ that $b = -\langle z', x \rangle$, and for this reason (2.4.12) implies that $-z' \in \partial f_2(x)$ and $z' + x' \in \partial f_1(x)$. Thus $x'_1 = z' + x'$ and $x'_2 = -z'$, q.e.d.

We shall now define the subdifferential of a composite functional. Let X (resp. Y) be a L.C.H.T.V.S. and X' (resp. Y') its dual space. We denote by $l: X \to Y$ a linear continuous function and by $l': Y' \to X'$ its transpose.

Proposition 2.4.5. Assume that $f: Y \to (-\infty, +\infty]$ is convex and l.s.c. and that a point $y_0 = lx_0$ exists at which f is finite and continuous. Then

$$\partial(f \circ l)(x) = l' \, \partial f(lx), \qquad \forall x \in X. \tag{2.4.13}$$

Proof. Assume that $y' \in \partial f(lx)$. We see that the composite functional $f \circ l: x \to f(lx)$ has the same properties as f. Then

$$f(y) - f(lx) \geq \langle y', y - lx \rangle_Y, \qquad \forall y \in Y, \tag{2.4.14}$$

and thus

$$(f \circ l)(z) - (f \circ l)(x) \geq \langle y', l(z - x) \rangle_Y = \langle l'y', z - x \rangle_X, \qquad \forall z \in X. \tag{2.4.15}$$

Accordingly,

$$l' \, \partial f(lx) \subset \partial(f \circ l)(x), \qquad \forall x \in X. \tag{2.4.16}$$

We will now show the inverse inclusion. Suppose that $x' \in \partial(f \circ l)(x)$; then let us consider the affine subspace of $Y \times R$

$$\mathscr{L} = \{(lz, \langle x', z - x \rangle_X + (f \circ l)(x)) \text{ for } z \in X\}. \tag{2.4.17}$$

Because of the subdifferential inequality resulting from $x' \in \partial(f \circ l)(x)$, $\mathscr{L}$ and epi f may have common points only on their boundaries. Moreover, the assumptions about f imply that intepi $f \neq \varnothing$ and thus intepi f and $\mathscr{L}$ may be separated by a nonvertical, closed affine hyperplane $\mathscr{H}$ which contains $\mathscr{L}$. Hence a functional $y' \in Y'$ exists such that

$$\langle y', lz \rangle_Y + b = \langle x', z - x \rangle_X + (f \circ l)(x), \qquad \forall z \in X. \tag{2.4.18}$$

For $z = 0$, $b = (f \circ l)(x) - \langle x', x \rangle$, and thus (2.4.18) implies that $x' = l'y'$. Moreover, since $\mathscr{H}$ does not meet intepi f,

$$\langle y', y \rangle_Y + (f \circ l)(x) - \langle l'(y'), x \rangle_X \leq f(y), \qquad \forall y \in Y, \tag{2.4.19}$$

or

$$\langle y', y - lx \rangle_Y \leq f(y) - f(lx), \qquad \forall y \in Y. \tag{2.4.20}$$

This last inequality implies that $y' \in \partial f(lx)$, or that $x' = l'y' \in l' \, \partial f(lx)$, q.e.d.

The following example illustrates the addition rule. Let f be a convex functional on X such that grad f exists everywhere, and let $K \neq \varnothing$ be a convex subset of X. The minimization problem of f over K is equivalent to the minimization of $f + I_K$ over all of X. For x_0 to be a solution of this problem, it is necessary and sufficient (cf., (2.3.3)) that

$$0 \in \partial(f(x_0) + I_K(x_0)). \tag{2.4.21}$$

(2.4.21) is equivalent (Prop. 2.4.3) to the relation

$$-\text{grad} \, f(x_0) \in \partial I_K(x_0). \tag{2.4.22}$$

2.4.2 The Relative Interior of $R(\partial f)$

We denote by $R(\partial f)$ the range of ∂f, i.e., the set $\{x' \mid x' \in \partial f(x), \, x' \in X', \, x \in X\}$. Further, we shall give a proposition on the relative interior of $R(\partial f)$, which will be useful for the study of "nonlinear" inequality problems.

For the proof of this proposition, which uses a nonclassical separation theorem for convex sets, refer to [268] and [296].

Proposition 2.4.6. (a) Suppose that $\mathcal{Q}$ is a finite dimensional space and f is a convex, l.s.c., proper functional on $\mathcal{Q}$. Then in order that

$$0 \in \text{relint } R(\partial f), \tag{2.4.23}$$

the following conditions are necessary and sufficient: that $\mathcal{Q}$ be the direct sum of the vector spaces $\mathcal{Q}_1$ and $\mathcal{Q}_2$, that f be invariant on $\mathcal{Q}_1$ and that there exist constants $c_1 > 0$ and c_2 such that

$$f(q) \geq c_1 |q|_{\mathcal{Q}} + c_2, \qquad \forall q \in \mathcal{Q}_2. \tag{2.4.24}$$

(b) Let X be a B-space, $\mathcal{Q}$ a finite dimensional subspace of X, $\mathcal{Q}'$ its dual, $x' \in X'$, and $x'|_{\mathcal{Q}} \in \mathcal{Q}'$ (c.f., (1.2.12)). Further, let Φ be a convex, l.s.c. and proper functional on X. Then in order that

$$x'|_{\mathcal{Q}} \in \text{relint } R(\partial \Phi_{\bar{x}}), \tag{2.4.25}$$

where $\Phi_{\bar{x}}$ is a convex, l.s.c. and proper functional on $\mathcal{Q}$ defined by

$$\Phi_{\bar{x}}(q) = \Phi(\bar{x} + q) \quad \text{for} \quad \bar{x} \in D(\Phi), \tag{2.4.26}$$

it is necessary and sufficient that $\mathcal{Q}$ be the direct sum of the vector spaces $\mathcal{Q}_1$ and $\mathcal{Q}_2$ as in a), such that the function $F: x \to \Phi(x) - \langle x', x \rangle$ is invariant on $\mathcal{Q}_1$, and that there exist constants $c_3 > 0$ and c_1, c_2 such that for any $\tilde{x} \in \tilde{\mathcal{Q}}$, where $\tilde{\mathcal{Q}}$ is closed and $X = \mathcal{Q} \oplus \tilde{\mathcal{Q}}$, the inequality

$$\Phi(\tilde{x} + q) - \langle x', q \rangle_X \geq c_1 - c_2 \|\tilde{x}\|_X + c_3 \|q\|_X, \qquad \forall q \in \mathcal{Q}_2 \tag{2.4.27}$$

holds.

Note that a consequence of this proposition is that the sets relint $R(\partial \Phi_{\bar{x}})$ and $\overline{R(\partial \Phi_{\bar{x}})}$ are independent of $\bar{x}$, if $\bar{x} \in D(\Phi)$. Indeed, the necessary and sufficient condition characterizing relint $R(\partial \Phi_{\bar{x}})$ is independent of the choice of $\bar{x} \in D(\Phi)$. Moreover, by taking into account that the closure of $R(\partial \Phi)$ is identical to the closure of relint $R(\partial \Phi)$, as may be easily verified (cf., also [288]), we find that the same is true for $\overline{R(\partial \Phi_{\bar{x}})}$.

2.5 Conjugates of Convex Functionals

2.5.1 The Classes $\Gamma(X)$ and $\Gamma_0(X)$

Let $\{X, X'\}$ be a dual pair of L.C.H.T.V.S.s and $\langle x', x \rangle$ the duality pairing. For a functional $f: X \to \bar{\mathbb{R}}$ there arises the question as to whether an affine continuous function $\langle x', \cdot \rangle - \mu$, $\mu \in \mathbb{R}$, exists which is a minorant of f, i.e.,

$$f(x) \geq \langle x', x \rangle - \mu, \qquad \forall x \in X. \tag{2.5.1}$$

A necessary and sufficient condition is that

$$\mu \geq \sup_{x \in X} (\langle x', x \rangle - f(x)). \tag{2.5.2}$$

This relation introduces the conjugate functional f^c, which is defined on X' by the relation

$$f^c(x') = \sup_{x \in X} (\langle x', x \rangle - f(x)). \tag{2.5.3}$$

Clearly, the supremum in (2.5.3) may be taken only over $D(f)$. f^c can be regarded as the pointwise supremum of the family of affine continuous functionals $g(\cdot) = \langle \cdot, x \rangle - \mu$ with $(x, \mu) \in \text{epi} f$. Let us denote by $\Gamma(X)$ the set of functions f: $X \to \bar{\mathbb{R}}$, which are the pointwise suprema of a family of affine continuous functionals $\langle x', \cdot \rangle + a$, $a \in \mathbb{R}$ on X. The following propositions can be proved.

Proposition 2.5.1. Assume that f: $X \to \bar{\mathbb{R}}$ is convex and l.s.c. Then if an x_0 exists such that $f(x_0) = -\infty$, f cannot take any finite value.

Proof. Arguing by contradiction, we assume that x_1 exists such that $f(x_1) \in \mathbb{R}$. Then, because of the convexity of f, $f(\lambda x_0 + (1 - \lambda)x_1) = -\infty$ for every $0 < \lambda < 1$. The lower semicontinuity of f implies as $\lambda \to 0$ that $f(x_1) = -\infty$, which contradicts the initial assumption, q.e.d.

Proposition 2.5.2. The class $\Gamma(X)$ consists exactly of the convex, l.s.c., proper functionals f: $X \to \bar{\mathbb{R}}$ and of the constants $+\infty$ and $-\infty$.

Proof. Suppose that $f \in \Gamma(X)$. Clearly, f is convex and l.s.c. If $\Gamma(X) = \emptyset$, the pointwise supremum is $-\infty$. This supremum cannot be $-\infty$ if $\Gamma(X) \neq \emptyset$. Conversely, let $f \equiv \pm\infty$. Clearly $f \in \Gamma(X)$. Taking f as convex, l.s.c. and proper, let us choose a point $(x_0, a_0) \in X \times \mathbb{R}$ such that $f(x_0) > a_0$. Obviously, $(x_0, a_0) \notin \text{epi} f$ which is closed in $X \times \mathbb{R}$, and therefore (x_0, a_0) can be separated strictly from epi f by a closed affine hyperplane $\mathscr{H}$ described by the equation

$$\langle x', x \rangle + a'a = b, \tag{2.5.4}$$

where $x \in X$, $x' \in X$, and $a, a', b \in \mathbb{R}$. Then

$$\langle x', x_0 \rangle + a'a_0 < b \tag{2.5.5}$$

and

$$\langle x', x \rangle + a'a > b, \quad \forall (x, a) \in \text{epi} f. \tag{2.5.6}$$

If $x_0 \in D(f)$, then it results for $x = x_0$ and $a = f(x_0)$ that $a' > 0$, and therefore

$$a_0 < \frac{b}{a'} - \frac{\langle x', x_0 \rangle}{a'} < f(x_0). \tag{2.5.7}$$

We have thus constructed a continuous affine functional from X into $\mathbb{R}$, whose values lie between $f(x_0)$ and a_0. Therefore $f \in \Gamma(X)$. A similar functional can be constructed if $x_0 \notin D(f)$. If $a' \neq 0$ we can proceed as before; if $a' = 0$, (2.5.5) states that $\langle x', \cdot \rangle < b$, at x_0. On the other hand (2.5.6) implies that $\langle x', \cdot \rangle > b$ on $D(f)$. Thus we may determine $x'_1 \in X$, $b_1 \in \mathbb{R}$, and $c > 0$ such that

$$a_0 < b_1 - \langle x'_1, x_0 \rangle + c(b - \langle x', x_0 \rangle) < f(x_0), \tag{2.5.8}$$

q.e.d.

Further, we denote by $\Gamma_0(X)$ the set of functionals $f \in \Gamma(X)$ such that $f \not\equiv \pm\infty$. It is obvious that $\Gamma_0(X)$ consists precisely of the convex, l.s.c. and proper functionals on X.

Let F_1 and F_2 be two functions from X into $\bar{\mathbb{R}}$. If F_2 is the pointwise supremum of the continuous affine functions which are for every x less than F_1, then F_2 is the largest minorant of F_1 in $\Gamma(X)$, and conversely. F_2 is called the Γ-regularization of F_1. Obviously, if $F_1 \in \Gamma(X)$, then F_1 is identical to its Γ-regularization. Similarly we can define the largest l.s.c. minorant of $F_1 : X \to \mathbb{R}$, which is called the l.s.c.-regularization of F_1 and is denoted by $\bar{F}_1$. If F is convex and has a continuous affine minorant, $\bar{F}_1 = F_2$. It may also be proved [91] that

$$F_2 \leq \bar{F}_1 \leq F_1. \tag{2.5.9}$$

We are now going to prove some important propositions on the relation between f and f^c, where f is convex.

Proposition 2.5.3. Let f be a convex functional on X. Then f^c is a convex, l.s.c. functional on X'. If, moreover, f is proper, then f^c is proper as well, and conversely.

Proof. Since $f^c \in \Gamma(X')$, we can apply Prop. 2.5.2, and thus f^c is a convex and l.s.c. functional on X'. The proof of the second part of the proposition is trivial, q.e.d.

We now define the functional f^{cc} on X by the relation

$$f^{cc}(x) = (f^c)^c(x) = \sup_{x' \in X'} (\langle x', x \rangle - f^c(x')), \qquad \forall x \in X. \tag{2.5.10}$$

f^{cc} is the conjugate of $f^c \cdot f^{cc}$ is clearly the Γ-regularization of f.

Proposition 2.5.4. For $f \in \Gamma(X)$

$$f^{cc} = f. \tag{2.5.11}$$

Proof. By Prop. 2.5.2 we see that f is the pointwise supremum of the affine functions $\langle x', \cdot \rangle - \mu$ such that $\mu = f^c(x')$. Thus from (2.5.10) we obtain (2.5.11), q.e.d.

It is evident that if $f_1 \leq f_2$ on X, then $f_1^c \geq f_2^c$ on X'. Since f^{cc} is the Γ-regularization of f, the inequality $f \geq f^{cc}$ holds on X, and thus $f^c \leq f^{ccc}$

on X. On the other hand,

$$f^{ccc}(x') = \sup_{x \in X} (\langle x', x\rangle - f^{cc}(x)) \le f^c(x'), \tag{2.5.12}$$

and therefore

$$f^{ccc} = f^c. \tag{2.5.13}$$

If f is any function on X taking values in $\bar{\mathbb{R}}$, where X is a normed space, the functions f^c and f^{cc} are convex. Moreover, f^c is weakly$_*$ l.s.c. in X' and f^{cc} is weakly l.s.c. in X [17].

2.5.2 The Conjugacy Operation

From the propositions of the foregoing section, it results that if $f \in \Gamma(X)$ and $g \in \Gamma(X')$, then the relation $f = g^c$ implies that $g = f^c$, and conversely. Then the functionals f and g are called mutually conjugate, or dual. Suppose that the constant functions $f = \pm\infty$ and $g = \mp\infty$, which are conjugate to each other, are excluded, i.e. let us consider the classes $\Gamma_0(X)$ and $\Gamma_0(X')$. The conjugacy operation $f \to f^c$ can be considered therefore as a one-to-one correspondence between the elements of the classes $\Gamma_0(X)$ and $\Gamma_0(X')$. The operation $f \to f^c$ is called the Fenchel transformation and is closely related to the Legendre transformation, if f is a convex differentiable functional[4] (see [288], [338]). The definition of f^c in (2.5.3) yields the inequality

$$f(x) + f^c(x') \ge \langle x', x\rangle, \quad \forall x \in X, \quad \forall x' \in X' \tag{2.5.14}$$

which holds for any convex, proper functional f and its conjugate f^c. This inequality is called Fenchel's inequality [95]. Note that the theory of conjugacy can be based on the following elementary problem: let W be the set of the proper functionals $f : X \to (-\infty, +\infty]$ and $g : X' \to (-\infty, +\infty]$ such that

$$f(x) + g(x') \ge \langle x', x\rangle, \quad \forall x \in X, \quad \forall x' \in X'. \tag{2.5.15}$$

We wish to determine functionals $(f_0, g_0) \in W$ such that if $(f, g) \in W$ and $f < f_0$, $g < g_0$, then $f = f_0$ and $g = g_0$. It results easily (cf., (2.5.3)) that $f_0 = g_0^c$ and $g_0 = f_0^c$, i.e., the pairs (f_0, g_0) are precisely the pairs of dual functionals $f_0 \in \Gamma_0(X)$ and $g_0 \in \Gamma_0(X')$.

The following proposition exhibits the relation between ∂f and ∂f^c.

Proposition 2.5.5. Assume that f is a convex, proper functional on X. The following conditions are equivalent to one another.

(i) $x' \in \partial f(x)$; (2.5.16)

(ii) $\sup_{z \in X}(\langle x', z\rangle - f(z))$ is achieved at $z = x$; (2.5.17)

[4] The names Fenchel–Young transformation, Legendre–Fenchel transformation and polarity transformation are encountered in the literature.

(iii) $f(x) + f^c(x') \le \langle x', x\rangle$; and (2.5.18)

(iv) $f(x) + f^c(x') = \langle x', x\rangle$. (2.5.19)

If, additionally, f is l.s.c., these conditions are equivalent to:

(v) $x \in \partial f^c(x')$; and (2.5.20)

(vi) $\sup_{z' \in X'}(\langle z', x\rangle - f^c(z'))$ is achieved at $z' = x'$. (2.5.21)

Proof. The subgradient inequality resulting from (2.5.16) implies (2.5.17), and conversely. Because of (2.5.3), (2.5.17) is equivalent to (2.5.18) or to (2.5.19). Reasoning similarly we see that (2.5.20) is equivalent to (2.5.21), which is again equivalent to

$$f^{cc}(x) + f^c(x') = \langle x', x\rangle. \tag{2.5.22}$$

But this last condition is equivalent to (2.5.19), because of (2.5.11), q.e.d.

From this proposition we note that the subdifferential ∂f^c coincides with the inverse mapping $(\partial f)^{-1}$. The same happens if X is a reflexive B-space. The conjugacy operation will be illustrated further by two examples:

(i) Suppose that $f = I_K$, where K is a nonempty, convex subset of X. Then I_K^c is given on X' by

$$I_K^c(x') = \sup_{x \in K}\langle x', x\rangle. \tag{2.5.23}$$

I_K^c is called the support function of K. Now let K be a linear subspace M of $\mathbb{R}^n$. Then the supremum in (2.5.23) is ∞, unless $\langle x', x\rangle = 0$, $\forall x \in M$. Accordingly, $I_K^c = I_{M^\perp}$, where $M^\perp$ denotes the orthogonal complement of M.

(ii) Assume that $X = X' = H$, where H is a Hilbert space. The functional $f_0 : x \to \frac{1}{2}\|x\|_H^2$ is the unique solution of the equation $f = f^c$. Indeed, we can show immediately that $f_0 = f_0^c$. Moreover, if f is another convex functional such that $f = f^c$, then it is proper as well, and thus, from (2.5.14), $f(x) \ge \frac{1}{2}\|x\|_H^2 = f_0(x)$. But the conjugacy operation reverses the inequalities. Accordingly, $f^c \le f_0^c$ and thus $f_0 = f$.

2.6 Maximal Monotone Operators

2.6.1 Definitions and Fundamental Results

Let $\mathscr{X}$ and $\mathscr{X}'$ be two vector spaces and A a mapping from $\mathscr{X}$ into the power set $\mathscr{P}(\mathscr{X}')$ of $\mathscr{X}'$. The mapping A is called a multivalued operator or multivalued mapping or multifunction. In this case, from $A(x_1) = x_1'$ and $A(x_1) = y_1'$ it does not follow that $x_1' = y_1'$, as happens with single-valued functions. Considering A as a subset of $\mathscr{X} \times \mathscr{X}'$, we can write $A(x) = \{y \in \mathscr{X}' \mid (x, y) \in A\}$. The set $D(A) = \{x \mid x \in \mathscr{X}, A(x) \ne \varnothing\}$ is called the domain of A and the set $R(A) = \bigcup_x A(x)$, $x \in \mathscr{X}$, the range of A. Because A is multivalued, we will write $y \in A(x)$, where $x \in D(A)$ and $y \in \mathscr{X}'$. If

A and B are two multivalued operators on $\mathscr{X}$, then $\lambda A + \mu B$, $\lambda, \mu \in \mathbb{R}$ is a multivalued operator mapping x into $\lambda A(x) + \mu B(x) = \{\lambda y + \mu z \mid y \in A(x), z \in B(x)\}$. Moreover, $D(\lambda A + \mu B) = D(A) \cap D(B)$. Suppose further that $\mathscr{X}$ and $\mathscr{X}'$ are placed in separating duality by means of the duality pairing $\langle x', x \rangle$ for $x \in \mathscr{X}$, $x' \in \mathscr{X}'$. The multivalued mapping $A : \mathscr{X} \to \mathscr{P}(\mathscr{X}')$ is said to be monotone if

$$\langle y_1 - y_2, x_1 - x_2 \rangle \geq 0, \tag{2.6.1}$$

$$\forall x_1, x_2 \in D(A), \qquad \forall y_1 \in A(x_1), \qquad \forall y_2 \in A(x_2).$$

If $\geq$ is replaced by $>$, then A is said to be strictly monotone. Hereafter by the term monotone operator we mean a multivalued monotone operator, unless otherwise stated.

Proposition 2.6.1. Let f be a convex proper functional on $\mathscr{X}$. Then ∂f is a monotone multivalued function from $\mathscr{X}$ into $\mathscr{P}(\mathscr{X}')$.

Proof. Assume that $y_1 \in \partial f(x_1)$ and $y_2 \in \partial f(x_2)$. If we add the respective variational inequalities defining $\partial f(x_1)$ and $\partial f(x_2)$, (2.6.1) immediately results, q.e.d.

The graph of the multivalued operator $A: \mathscr{X} \to \mathscr{P}(\mathscr{X}')$ is the set $\mathcal{G}(A) = \{(x, y) \mid (x, y) \in D(A) \times \mathscr{X}', \ y \in A(x)\}$. Then $\mathcal{G}(A_1) \subset \mathcal{G}(A_2)$, if and only if $A_1(x) \subset A_2(x)$ $\forall x \in \mathscr{X}$. The set $\mathscr{A}$ of the monotone operators from $\mathscr{X}$ into $\mathscr{P}(\mathscr{X}')$ can be partially ordered by graph inclusion. It can be shown, furthermore, that every totally ordered subset of $\mathscr{A}$ has an upper bound. Then by means of the well-known Zorn Lemma [277] [337] $\mathscr{A}$ contains at least one maximal element, which is called a maximal monotone operator. Accordingly, a monotone operator $A: \mathscr{X} \to \mathscr{P}(\mathscr{X}')$ is called maximal monotone if and only if $\mathcal{G}(A) \subset \mathcal{G}(B)$ implies that $A = B$, where $B: \mathscr{X} \to \mathscr{P}(\mathscr{X}')$ is an arbitrary monotone operator, i.e., if and only if $\mathcal{G}(A)$ is not properly contained in any other monotone subset of $\mathscr{X} \times \mathscr{X}'$. Another definition equivalent to the previous one may also be given: an operator $A: \mathscr{X} \to \mathscr{P}(\mathscr{X}')$ is called maximal monotone if and only if i) A is monotone and ii) for every $x \in \mathscr{X}$ and $y \in \mathscr{X}'$ such that

$$\langle y - y_1, x - x_1 \rangle \geq 0, \quad \forall x_1 \in D(A), \quad \forall y_1 \in A(x_1) \tag{2.6.2}$$

the relation

$$y \in A(x) \tag{2.6.3}$$

holds.

Suppose that $A: H \to H$, where H is a Hilbert space, and let $I : H \to H$ be the identity operator.

Proposition 2.6.2. If $A + I$ is surjective, i.e., $R(A + I) = H$, and A is monotone, then A is maximal monotone.

Proof. Arguing by contradiction, we suppose that there exists a monotone operator $B\colon H \to H$ such that $\mathcal{G}(B) \supset \mathcal{G}(A)$. Thus, if $(x, y) \in \mathcal{G}(B)$, then there exists $x_1 \in D(A)$ such that $x + y \in (I + A)x_1$. It results that $x + y \in x + Bx$ and $x + y \in x_1 + Bx_1$. Since B is monotone,

$$(z_1 - z, x_1 - x)_H \geq 0, \qquad \forall z_1 \in Bx_1, \qquad \forall z \in Bx, \tag{2.6.4}$$

and thus, for $z_1 = x + y - x_1$ and for $z = y$, we obtain the inequality

$$(x - x_1, x_1 - x) \geq 0, \tag{2.6.5}$$

which implies that $x_1 = x$ and $y \in Ax$, q.e.d.

The converse is also true [29]. For the generalization of this result, see [17].

The following proposition relates the theory of maximal monotone operators to subdifferentiation. For the sake of brevity the proposition is given without proof (see e.g. [17], [29]).

Proposition 2.6.3. The subdifferential ∂f of a convex, l.s.c., proper functional f on X, where X is a B-space, is a maximal monotone operator.

2.6.2 Maximal Monotone Graphs in $\mathbb{R}^2$

The class of the monotone operators $\beta\colon \mathbb{R} \to \mathscr{P}(\mathbb{R})$ is subsequently considered. The set $\mathscr{A}$ of such monotone mappings is ordered with respect to the coordinatewise partial ordering. A subset of $\mathbb{R}^2$ is totally ordered if and only if for any two pairs (x_1, x_1') and (x_2, x_2') of this subset either $x_1 \leq x_2$ and $x_1' \leq x_2'$ or $x_1 \geq x_2$ and $x_1' \geq x_2'$. A complete nondecreasing curve in $\mathbb{R}^2$ is the graph $\mathcal{G}(\beta)$ of a maximal monotone mapping $\beta\colon \mathbb{R} \to \mathscr{P}(\mathbb{R})$. In a Cartesian coordinate system such a graph is similar to the graph of a continuous nondecreasing function, with the difference that it may contain vertical segments as well. The maximal monotone graphs in $\mathbb{R}^2$ will be widely used in later chapters for the formulation of unilateral boundary conditions. A proposition now follows relating the complete nondecreasing curves in $\mathbb{R}^2$ and the subdifferentials ∂f of convex, l.s.c. and proper functionals on $\mathbb{R}$.

Proposition 2.6.4. Let $\beta\colon \mathbb{R} \to \mathscr{P}(\mathbb{R})$ be a maximal monotone mapping. A convex, l.s.c., proper functional $f\colon \mathbb{R} \to \bar{\mathbb{R}}$ can be determined up to an additive constant such that

$$\beta = \partial f. \tag{2.6.6}$$

Proof. Assume that for $a \in \mathbb{R}$, $\beta(a)$ is finite. Let

$$f(x) = \int_a^x \beta(t)\, dt. \tag{2.6.7}$$

This integral is well-defined and finite in the interval r on which β is finite, since β is nondecreasing. Furthermore, it is defined[5] on the closure $\bar{r}$ of r. For $x \notin \bar{r}$, clearly $f(x) = \infty$. Let $x_1, x_2 \in r$, $x_3 = \lambda x_1 + (1-\lambda)x_2$, $\lambda \in (0, 1)$ and $x_1 < x_2$. Then

$$f(x_3) - f(x_1) = \int_{x_1}^{x_3} \beta(t)\, dt \leq (x_3 - x_1)\beta(x_3) \tag{2.6.8}$$

and

$$f(x_2) - f(x_3) = \int_{x_3}^{x_2} \beta(t)\, dt \leq (x_2 - x_3)\beta(x_2). \tag{2.6.9}$$

The convexity of f on r follows from (2.6.8) and (2.6.9). Then because of the continuity of f on $\bar{r}$, f is a convex, l.c.s., proper functional on $\mathbb{R}$. For any $x \in r$ and any $z > x$, the relation

$$\frac{f(z) - f(x)}{z - x} = \frac{1}{z - x} \int_x^z \beta(t)\, dt \geq \beta(x) \tag{2.6.10}$$

holds. Thus for $z \to x_+$, $f'_+(x) \geq \beta(x)$. Arguing as before, the inequality $f'_-(x) \leq \beta(x)$ for $z \to x_-$ results. If $x \notin r$, these two inequalities hold by definition. From (2.3.29) we conclude that $f'_+ = \beta_+$ and $f'_- = \beta_-$, $\forall x \in \mathbb{R}$. Thus

$$f'_- = \beta_- \leq \beta \leq \beta_+ = f'_+, \qquad \forall x \in \mathbb{R} \tag{2.6.11}$$

and because of (2.3.30) it follows that $\beta = \partial f$. Assume that there exists another functional $f^* \in \Gamma_0(\mathbb{R})$ such that $\beta = \partial f^*$. Then (2.3.28) yields the relation

$$f^{*\prime}_- \leq \beta \leq f^{*\prime}_+, \tag{2.6.12}$$

and thus

$$f^{*\prime}_- = \beta_- = f'_- \quad \text{and} \quad f^{*\prime}_+ = \beta_+ = f'_+. \tag{2.6.13}$$

But $\operatorname{int} D(f^*) = \operatorname{int} D(f) = \operatorname{int} r$, as is easily seen (cf., Prop. 2.3.8) since $f'_\pm$ (resp. $f^{*\prime}_\pm$) are finite on $\operatorname{int} D(f)$ (resp. $\operatorname{int} D(f^*)$) and due to $r \subset D(f) \subset \bar{r}$. The lower semicontinuity of f and f^* implies that f and f^* are determined on $\mathbb{R}$ by their values in $\operatorname{int} r$. Obviously from (2.6.13), we find that if[6] $\operatorname{int} r \neq \emptyset$, then $f'_+(x) - f^{*\prime}_+(x) = 0$ and $f'_-(x) - f^{*\prime}_-(x) = 0$, $\forall x \in \operatorname{int} r$. Thus $f = f^* + a$, in $\operatorname{int} r$, where a is a constant, q.e.d.

Proposition 2.6.5. The graphs of the subdifferentials ∂f, where $f \in \Gamma_0(\mathbb{R})$, are precisely the complete nondecreasing curves of $\mathbb{R}^2$.

Proof. Follows immediately from Prop. (2.3.8), (2.3.9) and (2.6.4), q.e.d.

Note that the proof of Prop. 2.6.4 provides a method for the determination of f, when β is given. In the following simple examples, yet another method is illustrated.

[5] As the limit of Riemann integrals or as a Lebesgue integral.

[6] The case of r consisting of a single element is trivial.

(i) Let β be given for $x \in \mathbb{R}$ by

$$\beta(x) = \begin{cases} \varnothing & \text{if} \quad x < -1 \\ (-\infty, 0] & \text{if} \quad x = -1 \\ 0 & \text{if} \quad -1 < x < 1. \\ [0, \infty) & \text{if} \quad x = 1 \\ \varnothing & \text{if} \quad x > 1 \end{cases} \tag{2.6.14}$$

Then β is a maximal nonotone operator (Fig. 2.6.1a). For the determination of f, we define a continuous and single-valued function β_ε (called a regularized operator), which depends on $\varepsilon > 0$ (Fig. 2.6.1b). One possible form of β_ε is

$$\beta_\varepsilon(x) = \begin{cases} \dfrac{x+1}{\varepsilon} & \text{if} \quad x \leq -1 \\ 0 & \text{if} \quad -1 \leq x \leq 1. \\ \dfrac{x-1}{\varepsilon} & \text{if} \quad x \geq 1 \end{cases} \tag{2.6.15}$$

i.e., $\beta_\varepsilon \in C^0(\mathbb{R})$. We can also apply another regularization procedure such that $\beta_\varepsilon \in C^1(\mathbb{R})$, etc. For $\varepsilon \rightarrow 0_+$ the graph $\mathcal{G}(\beta_\varepsilon)$ "coincides" with the graph

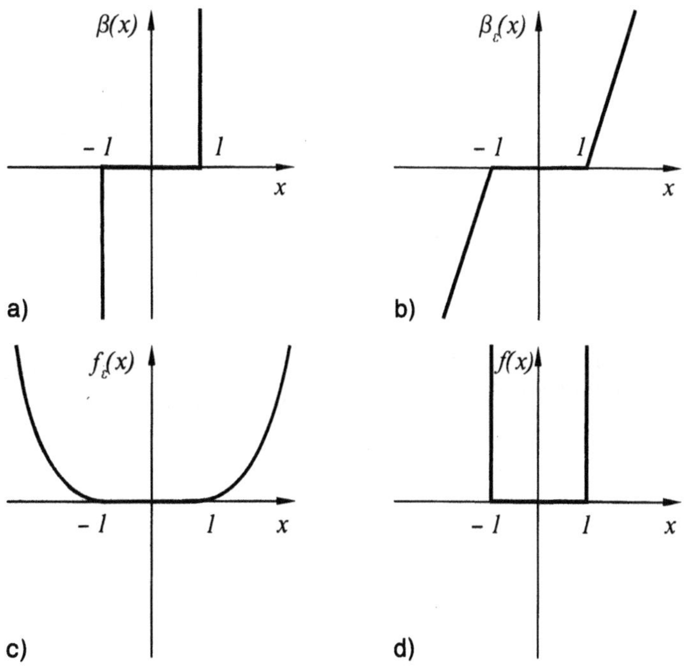

Fig. 2.6.1. Regularization procedure.

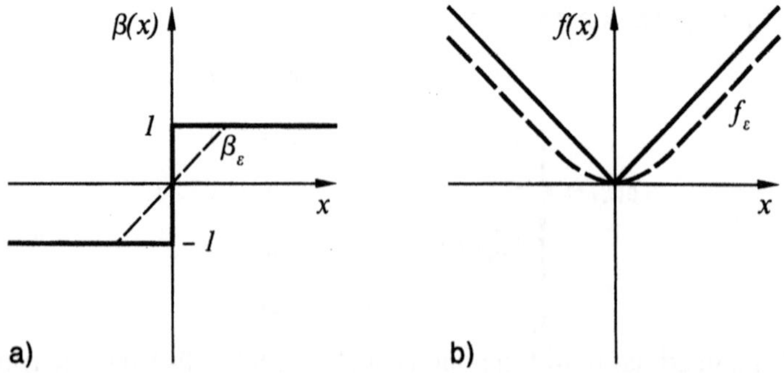

Fig. 2.6.2. Regularization of $f(x) = |x|$, $x \in \mathbb{R}$. The functions β and β_ε.

$\mathcal{G}(\beta)$. f_ε results from β_ε through integration (Fig. 2.6.1c):

$$f_\varepsilon(x) = \begin{cases} \dfrac{(x+1)^2}{2\varepsilon} & \text{if} \quad x \leq -1 \\ 0 & \text{if} \quad -1 \leq x \leq 1. \\ \dfrac{(x-1)^2}{2\varepsilon} & \text{if} \quad x \geq 1 \end{cases} \tag{2.6.16}$$

From f_ε we obtain f for $\varepsilon \to 0_+$. Thus (Fig. 2.6.1d)

$$f(x) = \begin{cases} 0 & \text{if} \quad |x| \leq 1 \\ \infty & \text{if} \quad |x| > 1 \end{cases}. \tag{2.6.17}$$

The method of regularization is closely related to Yosidá's approximation of maximal monotone operators (see Sec. 7.2.2 and [29]).

(ii) The mapping β: $\mathbb{R} \to \mathscr{P}(\mathbb{R})$ is defined by means of

$$\beta(x) = \begin{cases} -1 & \text{if} \quad x < 0 \\ [-1, +1] & \text{if} \quad x = 0. \\ +1 & \text{if} \quad x > 0 \end{cases} \tag{2.6.18}$$

It is obvious that β is maximal monotone. It follows, either directly or by regularization, that $\beta = \partial f$, where (Fig. 2.6.2) $f(x) = |x|$.

Part 2
Inequality Problems

Chapter 3

Variational Inequalities and Superpotentials

The aim of this chapter is to explain the origins of the inequality problems encountered in mechanics. To do this we introduce certain notions of convex analysis into mechanics; more precisely, we consider material laws and boundary conditions involving subdifferentials of convex functionals.

In the first section, we introduce certain classes of multivalued laws in mechanics which give rise to variational inequalities and convex minimization problems. In the second section, the notion of superpotential is defined and the duality of variational principles is discussed. The third section deals with examples of subdifferential material laws and boundary conditions, first in a Euclidean space and then, in extended form, in a function space.

3.1 Mechanical Laws and Constraints

3.1.1 Generalized Forces and the Principle of Virtual Power

The virtual power method was historically the first approach to the description of forces acting on a mechanical system. The first vague reference to this method can be found in Aristotle,[1] who tried to relate forces and velocities. In more modern times, d'Alembert applied this method for the first time [3] [130]. Compared to the classical vector-analysis approach to mechanics, the method of virtual power is commonly considered more abstract and more difficult to apply. However, the great advantage of the virtual power method lies in the fact that it is inherent in the nature of the concept of force.

Let us consider a mechanical system Σ. In order to determine the forces acting on Σ, we consider with respect to Σ the space $\mathcal{U}$ of all fields of possible velocities, or possible displacements which take place in a time

[1] See, e.g., [10] Arist. *Metaph.* I 1,1052β28.

interval. Depending on the nature of each problem, $\mathscr{U}$ is understood to consist of velocities or displacements, denoted hereafter in this Chapter by v and u respectively. Further we denote by $\mathscr{F}$ a real V.S., whose elements f are called forces. Let $\langle v, f\rangle$, $v \in \mathscr{U}$, $f \in \mathscr{F}$ be a bilinear form with the properties ($\mathscr{U}$ is a linear space)

(i) for each $v \neq 0$ in $\mathscr{U}$ there exists $f \in \mathscr{F}$ such that $\langle v, f\rangle \neq 0$, (3.1.1)

and

(ii) for each $f \neq 0$ in $\mathscr{F}$ there exists $v \in \mathscr{U}$ such that $\langle v, f\rangle \neq 0$. (3.1.2)

If $\mathscr{U}$ is the space of displacements, v is replaced in (3.1.1, 2) by u. The bilinear form $\langle\cdot,\cdot\rangle$ is the power (resp. work) of the force f for the velocity v (resp. displacement u). Due to (i) and (ii), the bilinear form $\langle\cdot,\cdot\rangle$ defines locally convex Hausdorff topologies on $\mathscr{U}$ and $\mathscr{F}$, and $\{\mathscr{U}, \mathscr{F}\}$ becomes a dual pair of L.C.H.T.V.S.s (cf., Sec. 1.2.1). Accordingly, we may define the force f as a linear continuous functional on the space $\mathcal{U}$, after the latter has been topologized by a topology satisfying (1.2.7). $\mathcal{U}$ is also called the space of virtual velocities (resp. displacements) and $\langle\cdot,\cdot\rangle$ the virtual power (resp. work). Then we can say that a force f acting on Σ is given if the virtual power (resp. work) $\langle v, f\rangle$ (resp. $\langle u, f\rangle$), is given on $\mathscr{U}$ so as to satisfy (i) and (ii), or if on $\mathcal{U}$ a topology not stronger than the Mackey-topology and not weaker than the weak topology is given. The triplet $\Sigma = \{\mathscr{U}, \langle v, f\rangle, \mathscr{F}\}$, consisting of the vector spaces $\mathscr{U}$ and $\mathscr{F}$ and the bilinear form $\langle v, f\rangle$ with the properties (i) and (ii), is called a mechanical element [229]. If the system Σ_0 consists of the subsystems $\Sigma_1, \ldots, \Sigma_n$, and if for every Σ_i, $i = 1, 2, \ldots, n$ the $\Sigma_i = \{\mathscr{U}_i, \langle v_i, f_i\rangle, \mathscr{F}_i\}$ is known, then we define $\mathscr{U} = \prod_{i=1}^n (\mathscr{U})_i$, $\mathscr{F} = \prod_{i=1}^n \mathscr{F}_i$ and

$$\langle v, f\rangle = \sum_{i=1}^n \langle v_i, f_i\rangle \qquad v \in \mathscr{U}, \quad f \in \mathscr{F}. \tag{3.1.3}$$

Obviously, $\Sigma = \{\mathscr{U}, \langle v, f\rangle, \mathscr{F}\}$. As two examples, let us consider a material point $\mathfrak{m}$ and a rigid body Ω in $\mathbb{R}^3$ [197]. For the material point, $\mathcal{U} = \mathbb{R}^3$ and the virtual power of a force $f \in \mathbb{R}^3$ acting on $\mathfrak{m}$ is $\langle v, f\rangle = v_i f_i$. For the rigid body, $\mathcal{U} = \mathbb{R}^6$. Indeed, the velocity v at a point $P \in \Omega$ takes the well-known form

$$v_P = v_A + \omega \wedge \overrightarrow{AP}, \tag{3.1.4}$$

where v_A is the velocity of a reference point $A \in \mathbb{R}^3$ and ω is the angular velocity of Ω. The space $\mathscr{F}$ consists of the forces f and the moments M_P such that

$$M_p = M_A + \overrightarrow{PA} \wedge f. \tag{3.1.5}$$

It is also the space $\mathbb{R}^6$. The duality pairing $\langle v, f\rangle$ expressing the virtual power is then

$$\langle v, f \rangle = v_{P_i} f_i + \omega_i M_{P_i}, \quad i = 1, 2, 3. \tag{3.1.6}$$

We shall now consider a continuous mechanical system occupying at a time t an open bounded subset Ω of $\mathbb{R}^3$. Let $\mathcal{U}$ be the V.S. of the virtual velocities of the system. The motion of the body is governed by the following principle of virtual power.

(H_1) At any time t and for any field of virtual velocities, the virtual power of all the internal and external forces impressed on the system is equal to the virtual power of the inertial forces.

The system is considered in an inertial frame of reference. Among the virtual velocity fields of Ω, of importance are the rigidifying velocity fields (see App. II). The V.S. of rigidifying velocity fields is denoted by $\mathcal{U}_0$. We postulate the following [110]:

(H_2) The virtual power Π_i of the internal forces of the body is zero for any rigidifying velocity field at any time.

Due to the linearity of Π_i, the value $\Pi_i(v)$ at v of the virtual power remains unchanged if v is replaced by $v + v_0$, $v_0 \in \mathcal{U}_0$. Further, we write $\Pi_i(v)$ in the form $\langle v, f \rangle$, and (H_2) is equivalent to the statement that $\Pi_i(v) = \langle v, f \rangle = 0$, $\forall v \in \mathcal{U}_0$. We are thus led to consider the quotient space $\bar{\mathcal{U}} = \mathcal{U}/\mathcal{U}_0$, which is called the space of objective virtual velocities (see App. II). The internal forces may be defined as the continuous linear functionals[2] on $\bar{\mathcal{U}}$. We denote their space by $\bar{\mathcal{F}}$. Obviously, the elements of both $\bar{\mathcal{U}}$ and $\bar{\mathcal{F}}$ are objective quantities, as is easily seen by considering the invariance of the duality mapping. For the further formulation of a continuum theory, we have to choose in addition to the space $\mathcal{U}$ the precise form of the linear mapping $v \to \langle v, f \rangle$ expressing the virtual power of the system. Hence in the framework of a local theory we may consider that the successive gradients of the vector field v, i.e., $v_{i,j}$, $v_{i,jk}$, etc., are involved in the mapping $v \to \langle v, f \rangle$, and thus we are led to first-, second-, etc. order gradient theories respectively. Thus we may consider as a space $\mathcal{U}$ the cartesian product space $\mathcal{U}^{(0)} \times \mathcal{U}^{(1)} \times \ldots \times \mathcal{U}^{(m)}$ where $\mathcal{U}^m = \{v_{i,jk} \ldots$ (m-spatial derivations)$\}$.

In continuum mechanics, the first-order gradient theory is the most common, in which case the power of the system depends on the velocities v_i and the first gradients $v_{i,j}$. On the assumption that the power Π_i of the internal forces acting in Ω can be expressed in the integral form,

$$\Pi_i(v) = -\int_\Omega p_i(v)\, d\Omega \tag{3.1.7}$$

[2] On $\bar{\mathcal{U}}$ the quotient topology is considered. In this topology the filter of neighborhoods of the origin in $\bar{\mathcal{U}}$ is the image under the canonical mapping (cf., Sec. 1.1.2) of the filter of neighborhoods of the origin in $\mathcal{U}$.

we may prove that at any point of Ω

$$p_i(v) = t_{ij} D_{ij}, \qquad D_{ij} = \tfrac{1}{2}(v_{i,j} + v_{j,i}). \tag{3.1.8}$$

Indeed, $p_i(v)$ may be written at any point of Ω in the general form

$$p_i(v) = t_{ij} D_{ij} + r_{ij}\Omega_{ij} + q_i v_i, \qquad \Omega_{ij} = \tfrac{1}{2}(v_{i,j} - v_{j,i}). \tag{3.1.9}$$

Obviously, t_{ij} is a component of a symmetric tensor. If $q_i \neq 0$ in a neighborhood of a point $M \in \Omega_1 \subset \Omega$, then we can easily determine a subsystem containing M and a translational virtual velocity such that $p_i(v)$ is not zero on this neighborhood. But this contradicts (H_2). Analogously, by means of a rotational virtual velocity we find that $r_{ij} = 0$. Here $t = \{t_{ij}\}$ is called the intrinsic stress tensor [110]. The symmetric tensor $D = \{D_{ij}\}$ is the rate of deformation (or stretching) tensor. For the theory of the second gradient and the respective theories of materials with microstructure, the reader is referred to [110] [111] [197].

As a result of the preceding discussion we may consider a space $\bar{\mathcal{U}}$ (resp. $\bar{\mathcal{F}}$) as consisting of the tensors D (resp. t). Higher-order gradient theories give rise to higher-order tensors, as is obvious. Accordingly, we may enlarge the classical notion of force and think of the elements of $\mathcal{F}$ as being generalized forces. Hence the term generalized velocities may also be used for the elements of $\mathcal{U}$.

Further, we shall define the virtual power of the external and inertial forces in Ω. The external forces are classed as volume forces (e.g., gravity, electromagnetic forces, etc.) or boundary forces (e.g., contact forces) acting on the boundary Γ of Ω. In the first case they are defined as continuous linear functionals on the space $\mathcal{U}$, and in the second as continuous linear functionals on a V.S $\mathcal{U}_\gamma$. The latter space is assumed to exist and is such that it includes the traces on Γ of the elements of $\mathcal{U}$. The elements of both $\mathcal{U}$ and $\mathcal{U}_\gamma$ are not necessarily objective. Finally, the virtual power of the inertial forces is a continuous linear functional on $\mathcal{U}^{(0)}$ whose elements are not objective. If we denote by Π_v, Π_c and Π_a the three virtual powers of volume, contact and inertial forces, then (H_1) acquires at any time t and for any virtual velocity field v the form

$$\Pi_i + \Pi_v + \Pi_c = \Pi_a. \tag{3.1.10}$$

Eq. (3.1.10) may also be formulated for any subsystem of Ω occupying the domain $\Omega_1 \subset \Omega$ at time t.

In the context of a first-order gradient theory and with regard to Ω_1, we assume that $\Pi_i(v)$, $\Pi_v(v)$ and $\Pi_a(v)$ (resp. $\Pi_c(\gamma v)$) can be expressed as integrals over Ω_1 (resp. over Γ_1) of $p_i(v)$, $p_v(v)$ and $p_a(v)$ (resp. $p_c(\gamma v)$). Then, as is readily verified, the most general forms of $p_v(v)$ and $p_c(\gamma v)$ are respectively

$$p_v(v) = f_i v_i + b_{ij} D_{ij} + c_{ij}\Omega_{ij}, \qquad b_{ij} = b_{ji}, \qquad c_{ij} = -c_{ji} \tag{3.1.11}$$

and

$$p_c(\gamma v) = S_i(\gamma v)_i, \tag{3.1.12}$$

where $\{f_i\}$, $\{b_{ij}\}$ and $\{c_{ij}\}$ are respectively the volume force, the double symmetric force, and the couple tensor in Ω_1, and $\{S_i\}$ is the stress vector on Γ_1. In a purely mechanical theory[3] the density of the inertial force power is

$$p_a(v) = \rho \frac{d\hat{v}_i}{dt} v_i, \tag{3.1.13}$$

where ρ is the density of the body and $d\hat{v}/dt$ denotes the material derivative of the real velocity $\hat{v}$. If all the quantities are sufficiently smooth, we may apply the Green–Gauss theorem, and by the well-known method (cf., e.g., [112]) used for the transition from global to local forms of basic mechanical laws, we obtain the equations

$$\sigma_{ij,j} + f_i = \rho \frac{d\hat{v}_i}{dt}, \qquad \sigma_{ij} - \sigma_{ji} + c_{ij} = 0 \quad \text{in} \quad \Omega_1 \tag{3.1.14}$$

$$S_i = \sigma_{ij} n_j \quad \text{on} \quad \Gamma_1, \tag{3.1.15}$$

where $n = \{n_i\}$ is the outward unit normal vector to Γ_1,

$$\sigma_{ij} = t_{ij} - b_{ij} - c_{ij} \quad \text{in} \quad \Omega_1 \tag{3.1.16}$$

and $\sigma = \{\sigma_{ij}\}$ is the Cauchy stress tensor. Throughout this book we shall often use (3.1.11, 14) with

$$b_{ij} = c_{ij} = 0. \tag{3.1.17}$$

For static problems, $\mathcal{U}$ is the space of virtual generalized displacements, and the principle of virtual power acquires a form known as the principle of virtual work. For additional information on the principles of virtual work and virtual power, the reader is referred to the classical treatises on mechanics [6], [112], [130], [171], and [321]. The principle of virtual power does not include the first or second principles of thermodynamics, which must be postulated independently. It is worth noting, however, that the energy-balance equation can be expressed by means of the principle of virtual power if the virtual velocities also include the variation of time (see [133] and [253]).

3.1.2 Multivalued Laws and Constraints in Mechanics

With respect to a mechanical system Σ for which the triplet $\Sigma = \{\mathcal{U}, \langle v, f \rangle, \mathcal{F}\}$ is defined, we consider a generally multivalued mapping A: $\mathcal{U} \to \mathcal{F}$. This mapping introduces a mechanical law on the system Σ which may be written in the form

$$f \in A(v) \subset \mathcal{F}, \qquad \forall v \in \mathcal{U}. \tag{3.1.18}$$

The domain and range of the multivalued operator A are called respectively

[3] For the application of the virtual power principle to electromagnetic continua, we refer to [197] and to the literature given there. Note that in the presence of Maxwellian fields, b_{ij} and c_{ij} in (3.1.11) are nonzero.

the domain and range of the mechanical law. For instance, if for every $v \in \mathcal{U}$ $A(v) = \{\mathscr{C}\}$, then this mechanical law means that $f = \mathscr{C}$. Suppose that

$$f = -\operatorname{grad} W(v), \tag{3.1.19}$$

where $W: \mathcal{U} \to \mathbb{R}$ is a functional on $\mathcal{U}$ whose gradient exists for every $v \in \mathcal{U}$. This law is then called a potential law and W is the potential.

Mechanical systems are usually subjected to constraints, e.g., boundary conditions. Then only certain subsets of the spaces $\mathcal{U}$ and $\mathcal{F}$ are admissible for the mechanical system. We are thus led to the definition of a mechanical law of the form $B: \mathcal{K} \to \mathcal{F}$, $\mathcal{K} \subset \mathcal{U}$, and we say that B defines a constraint on Σ. We then write

$$f \in B(v) \subset \mathcal{F}, \qquad \forall v \in \mathcal{K}. \tag{3.1.20}$$

f is called the reaction of the constraint. Suppose, for example, that a material point in $\mathbb{R}^3$ is constrained to have permanently zero velocity. Then the reaction may have any value in $\mathbb{R}^3$ and

$$B(v) = \begin{cases} \mathbb{R}^3 & \text{if } v \in \mathcal{K} = \{0\} \\ \varnothing & \text{otherwise} \end{cases}. \tag{3.1.21}$$

The sets $\{v \mid v \in \mathcal{K}, B(v) \neq \varnothing\}$ and $\{\bigcup_v B(v) \text{ for } v \in \mathcal{K}\}$ are called respectively the domain and range of the constraint. Analogous laws and constraints are obtained if v is replaced in (3.1.18–20) by u, e.g., for static problems. Obviously, in the abstract framework with which we are working, it is not always possible to make a clear distinction between a mechanical law and a constraint.

Let us consider further the equilibrium of a system Σ subjected to certain laws and constraints which are defined on $\mathcal{U}$ and on $\mathcal{K}_j$ by the operators A_i and B_j respectively. Here $i = 1, 2, \ldots, n$ and $j = 1, 2, \ldots, m$. From the principle of virtual work we find that at the position of equilibrium

$$\begin{gathered} \sum_{i=1}^{n} f_i + \sum_{j=1}^{m} f_j = 0, \\ f_i \in A_i(u), \quad \forall u \in \mathcal{U}, \quad f_j \in B_j(u) \quad \forall u \in \mathcal{K}_j \;\; j = 1, 2, \ldots, m. \end{gathered} \tag{3.1.22}$$

Accordingly, at the position of equilibrium $u \in \bigcap_j \mathcal{K}_j$ and

$$0 \in \sum_{i=1}^{n} A_i(u) + \sum_{j=1}^{m} B_j(u). \tag{3.1.23}$$

Of special interest is the case of subdifferential laws and constraints. Then $A = -\partial\Phi$ or $B = -\partial\Phi$, where Φ is a convex, l.s.c. and proper functional on $\mathcal{U}$. Φ is called superpotential, after Moreau [211]. Then (3.1.18) and (3.1.20) take the form

$$-f \in \partial\Phi(u) \quad \text{or} \quad -f \in \partial\Phi(v). \tag{3.1.24}$$

Obviously (cf., Sec. 2.1.1) Φ needs to be defined only on a convex closed subset of $\mathcal{U}$. In Chapter 4 the superpotential law is generalized for the case of nonconvex functionals by means of a new notion, the generalized gradient of Clarke. Another generalization of (3.1.24), consisting of the consideration of laws of the form (3.1.18), where A is a maximal monotone operator, is given in [245].

By means of the conjugate functional Φ^c, which is also convex, l.s.c. and proper on $\mathcal{F}$ (cf., Sec. 2.5.2), we can write (3.1.24) (let us take the first only) in the equivalent forms

$$u \in \partial\Phi^c(-f), \tag{3.1.25}$$

$$\Phi(u) + \Phi^c(-f) + \langle u, f \rangle = 0, \tag{3.1.26}$$

$$\Phi(u) + \Phi^c(-f) + \langle u, f \rangle \leq 0, \quad \forall f \in \mathcal{F}, \quad \forall u \in \mathcal{U}. \tag{3.1.27}$$

Functionals Φ and Φ^c can be considered respectively as the potential and the complementary energy corresponding to the mechanical law or the constraint. By definition, for $u \in \mathcal{U}$ and for $f \in \mathcal{F}$, (3.1.24) and (3.1.25) are equivalent to the variational inequalities

$$\Phi(u^*) - \Phi(u) \geq -\langle f, u^* - u \rangle, \qquad \forall u^* \in \mathcal{U} \tag{3.1.28}$$

and

$$\Phi^c(-f^0) - \Phi^c(-f) \geq -\langle u, f^0 - f \rangle, \qquad \forall f^0 \in \mathcal{F}. \tag{3.1.29}$$

Let us assume further that Φ is the indicator $I_{\mathcal{K}}$ of a convex closed subset of $\mathcal{U}$. Then

$$-f \in \partial I_{\mathcal{K}}(u) \quad \text{or} \quad -f \in \partial I_{\mathcal{K}}(v) \tag{3.1.30}$$

and the constraint corresponding to (3.1.30) is called an ideal unilateral constraint. According to Sec. 2.3.1, $-f$ is an element of the outward normal cone to $\mathcal{K}$ at u or v. The term unilateral is explained by considering for $u \in \mathcal{K}$ the variational inequality

$$\langle f, u^* - u \rangle \geq 0, \qquad \forall u^* \in \mathcal{K}, \tag{3.1.31}$$

which results by definition from the first relation of (3.1.30). Indeed, if $u^* - u$ is an admissible variation of u in the sense that it satisfies (3.1.31), then the same does not occur for the variation $u - u^*$ unless $\mathcal{K}$ is a linear subspace of $\mathcal{U}$. In this case, (3.1.31) holds as an equality and thus $f \in \tilde{\mathcal{K}}^{\perp}$. From (3.1.30), the inverse relation

$$u \in \partial I^c_{\mathcal{K}}(-f) \quad \text{or} \quad v \in \partial I^c_{\mathcal{K}}(-f) \tag{3.1.32}$$

and the corresponding variational inequalities are obtained.

The ideal unilateral constraint will be illustrated by a simple example. Suppose that a material point $\mathfrak{m}$ is subjected to a force $f \in \mathbb{R}^3$ and is constrained to lie within a convex closed subset K of $\mathbb{R}^3$, the space $\mathbb{R}^3$ being referred to a Cartesian coordinate system $0x_1x_2x_3$. If $K = \{x \mid x \in \mathbb{R}^3, F(x) \leq 0\}$, where F is a continuously differentiable function on $\mathbb{R}^3$ and the contact of the material point with the boundary of K is frictionless, then the reaction force R is given by the relation

$$R = -\lambda \text{ grad } F(x), \qquad \lambda > 0, \tag{3.1.33}$$

where λ is an unknown proportionality factor. Obviously, if the material point is in relint K, then the reaction force is zero, i.e., $\lambda = 0$; otherwise $\lambda \geq 0$. This type of constraint is described by the relation

$$-R \in \partial I_K(x). \tag{3.1.34}$$

Since $\mathfrak{m}$ is in equilibrium if and only if

$$f + R = 0, \tag{3.1.35}$$

it results that

$$f \in \partial I_K(x), \tag{3.1.36}$$

and conversely. For $x \in K$, (3.1.36) is equivalent to the variational inequality

$$f_i(x_i^* - x_i) \leq 0, \qquad \forall x^* \in K, \tag{3.1.37}$$

which is the expression of the principle of virtual work.

3.1.3 Minimization Problems and Variational Inequalities Characterizing the Equilibrium Configurations

We consider first a system Σ acted upon by forces $\mathscr{f}_i$, where $i = 1, \ldots, n$, and reactions $\bar{\mathscr{f}}_j$, where $j = 1, 2, \ldots, m$. Let $\mathcal{U}$ be the space of (generalized) displacements. These are derived from the superpotentials Φ_i and $\bar{\Phi}_j$, respectively, which are defined on the space $\mathcal{U}$. Then the condition of equilibrium (3.1.23) reads

$$0 \in \sum_{i=1}^{n} \partial\Phi_i(u) + \sum_{j=1}^{m} \partial\bar{\Phi}_j(u). \tag{3.1.38}$$

Note that if $\mathcal{U}$ is the space of generalized velocities and if u is replaced in (3.1.38) by v then (3.1.38) describes the motion of Σ if the inertia forces are neglected. A solution $u \in D(\partial\Phi_1) \cap \ldots \cap D(\partial\bar{\Phi}_m)$ of (3.1.38) satisfies

$$0 \in \partial\Phi_0(u), \qquad \Phi_0(u) = \sum_{i=1}^{n} \Phi_i(u) + \sum_{j=1}^{m} \bar{\Phi}_j(u), \tag{3.1.39}$$

the converse being generally not true. A combination of Props. 2.4.3 and 2.4.4 supplies the following sufficient condition for the equivalence of (3.1.38) and (3.1.39): if (i) the gradients of l $(0 \leq l \leq m + n)$ of the superpotentials Φ_i and $\bar{\Phi}_j$ exist for every $u \in \mathcal{U}$, and if (ii) a $u_0 \in \mathcal{U}$ exists such that from the remaining $m+n-l$ functionals $m+n-l-1$ are at u_0 finite and continuous with respect to a topology compatible with the duality between $\mathcal{U}$ and $\mathcal{F}$ and (iii) the $(m+n-l)$–th functional is finite at u_0, then every solution of (3.1.38) is a solution of (3.1.39), and conversely. (3.1.39) is equivalent to the condition (cf., Sec. 2.3.1)

$$\Phi_0(u) = \min \{\Phi_0(u^*) \mid u^* \in \mathcal{U}\}. \tag{3.1.40}$$

Accordingly, u minimizes Φ_0 on $\mathcal{U}$ for every equilibrium configuration, and if (3.1.38) and (3.1.39) are equivalent, then every solution of (3.1.40) corresponds to an equilibrium configuration.

Suppose further that some of the constraints are ideal unilateral constraints, i.e., $\bar{\Phi}_j = I_{\mathcal{K}_j}$ for $1 \le j \le p < m$, and that the subgradients $\partial\Phi_i$, $1 \le i \le n$ and $\partial\bar{\Phi}_j$, $p+1 \le j \le m$ are additive. Let

$$\Phi_0(u) = \sum_{i=1}^{n} \Phi_i(u) + \sum_{j=p+1}^{m} \bar{\Phi}_j(u), \qquad \mathcal{K}_0 = \bigcap_{j=1}^{p} \mathcal{K}_j \neq \varnothing. \tag{3.1.41}$$

It follows that every equilibrium configuration is a solution of the minimization problem

$$\Phi_0(u) = \min \{\Phi_0(u^*) \mid u^* \in \mathcal{K}_0\}. \tag{3.1.42}$$

The converse is true if $\partial\Phi_0$ and $\partial I_{\mathcal{K}_0}$ are additive. For this to be so, any one of the following three conditions is sufficient:

(i) the gradient of Φ_0 exists for every $u \in \mathcal{K}_0$;
(ii) there exists $u_0 \in \mathcal{K}_0$ such that Φ_0 is finite and continuous at u_0;
(iii) int $\mathcal{K}_0 \cap D(\Phi_0) \neq \varnothing$.

The first is a consequence of Prop. 2.4.3, the second and third of Prop. 2.4.4. We leave the proof as an exercise.

Returning to (3.1.23), we assume that the first q operators $A_i (1 \le i \le q < n)$ are constant, i.e., $A_i(u) = \{\mathscr{C}_i\}$, $\forall u \in \mathcal{U}$. Thus the prescribed forces $\mathscr{C}_i$, $1 \le i \le q$ act on the system. For the remaining operators A_i and B_j we assume that $A_i = -\partial\Phi_i$, $q+1 \le i \le n$ and $B_j = -\partial\bar{\Phi}_j$, $1 \le j \le m$. Then (3.1.23) becomes

$$\mathscr{C} = \sum_{i=1}^{q} \mathscr{C}_i \in \sum_{i=q+1}^{n} \partial\Phi_i(u) + \sum_{j=1}^{m} \partial\bar{\Phi}_j(u). \tag{3.1.43}$$

Every solution of (3.1.43) is a solution of the algebraic inclusion

$$\mathscr{C} \in \partial\Phi_0(u), \qquad \Phi_0(u) = \sum_{i=q+1}^{n} \Phi_i(u) + \sum_{j=1}^{m} \bar{\Phi}_j(u). \tag{3.1.44}$$

The converse is true, if the additivity of the subdifferentials is ensured. Sufficient conditions for this can easily be formulated as previously. Relation

(3.1.44) is for $u \in \mathfrak{U}$ equivalent to the variational inequality

$$\Phi_0(u^*) - \Phi_0(u) \geq \langle \mathscr{C}, u^* - u \rangle, \qquad \forall u^* \in \mathfrak{U}. \tag{3.1.45}$$

Obviously every solution of (3.1.45) is a solution of the minimization problem

$$\Pi(u) = \min \{\Pi(u^*) \mid u^* \in \mathfrak{U}\}, \tag{3.1.46}$$

where

$$\Pi(u^*) = \Phi_0(u^*) - \langle \mathscr{C}, u^* \rangle \tag{3.1.47}$$

and conversely.

In (3.1.46) Π expresses the potential energy of the system as will be made apparent by concrete problems, in coming sections. Until now we have considered the (generalized) displacement u as the unknown in the equilibrium problem of a system. In Sec. 3.2 the equilibrium problem will be formulated with respect to the forces by means of the theory of duality.

3.1.4 Dissipative Laws. A Note on the Eigenvalue Problem for Superpotential Laws

Let us consider a mechanical element $\Sigma = \{\mathscr{U}, \langle v, f \rangle, \mathscr{F}\}$ and a mechanical law $A: \mathfrak{X} \to \mathfrak{F}$, $\mathfrak{X} \subset \mathfrak{U}$ on Σ. This law is called dissipative if for $f \in A(v)$

$$\langle v, f \rangle \leq 0. \tag{3.1.48}$$

A general class of dissipative laws is characterized by the following criterion: if $-A$ is monotone and $0 \in A(0)$, the mechanical law is dissipative. This follows immediately by setting $v_2 = 0$ and $f_2 = 0$ in the monotonicity inequality $\langle v_1 - v_2, f_1 - f_2 \rangle \leq 0$. As a consequence (3.1.24) is dissipative if $0 \in \partial\Phi(0)$.

Suppose that for a mechanical problem the equilibrium configuration is characterized by the relation

$$f \in A(u), \tag{3.1.49}$$

where A is a multivalued operator and f a given force. As in the classical theory of eigenvalue problems, we call a scalar λ an eigenvalue of A with respect to the operator $G: \mathfrak{U} \to \mathfrak{F}$, $G \in \mathscr{L}(\mathscr{U}, \mathscr{F})$, if

$$\lambda G u \in A(u). \tag{3.1.50}$$

The respective u, if any exists, is called an eigenvector of A. A subset $\mathfrak{Y} \subset \mathfrak{U}$ is termed[4] an eigenset of A if for some λ

$$A(\mathfrak{Y}) \supset \lambda G(\mathfrak{Y}). \tag{3.1.51}$$

If $A = -\partial\Phi$, the eigenvalue problem takes the form: find $\lambda \in \mathbb{R}$ and $u \in \mathfrak{U}$ such

[4] Such problems have been studied for monotone processes and finite dimensional spaces by Rockafellar [286].

that

$$\Phi(u^*) - \Phi(u) \geq -\langle \lambda G u, u^* - u \rangle, \qquad \forall u^* \in \mathcal{U}. \tag{3.1.52}$$

Eigenvalue problems for variational inequalities will be studied more extensively in Chapter 7 in the framework of the theory of von Kármán plates, and therefore we shall not concern ourselves with this problem in detail here.

For the sake of completeness, however, we state here without proof some results of [1] and [68]. Let us consider the eigenvalue problem which corresponds to the mechanical law (3.1.24), and assume that $\mathcal{U}$ and $\mathcal{F}$ coincide with a Hilbert space $\mathcal{H}$. Then the following proposition holds (for the proof, see [1]).

Proposition 3.1.1. Suppose that

(i) $\Phi(u) \geq 0$, $\forall u \in \mathcal{H}$, $\Phi(0) = 0$;
(ii) the set $A = \{u \mid u \in \mathcal{H}, \|u\|_{\mathcal{H}} \leq c, \Phi(u) \leq c\}$ is compact for every c const > 0,
(iii) there exists a closed convex cone $\mathcal{K} \subset \mathcal{H}$ with 0 as origin such that $\mathcal{K} \cap \{u \mid \Phi(u) = 0\}$ is bounded;
(iv) there exists a nonzero $u \in \mathcal{K}$ such that $\Phi(u) < \infty$, and
(v) if $P_{\mathcal{K}}: \mathcal{H} \to \mathcal{K}$ denotes the projection[5] operator from $\mathcal{H}$ onto $\mathcal{K}$, then $\Phi(P_{\mathcal{K}} u) \leq \Phi(u)$, $\forall u \in \mathcal{H}$.

If (i–v) hold, then there exist $\lambda < 0$ and $u \in \mathcal{K}$, $u \neq 0$, which satisfy the eigenvalue problem

$$-\lambda u \in \partial\Phi(u). \tag{3.1.53}$$

3.2 Superpotentials and Duality

3.2.1 The Hypothesis of Normal Dissipation

In the previous section, we introduced subdifferential laws of the form (3.1.24) between generalized velocities $v \in \mathcal{U}$ and forces $f \in \mathcal{F}$. More generally, we may introduce similar multivalued relations between "fluxes" and "forces" according to the terminology of Onsager's theory of irreversible thermodynamics [124]. We shall here denote by v the fluxes and by f the forces. For instance, v may be the heat flux vector, the chemical reaction rate, etc., and the respective force f the temperature gradient, the chemical affinity, etc. The simplest law between fluxes and forces has the linear form

$$f_i = A_{ij} v_j, \tag{3.2.1}$$

where $\mathcal{U} = \mathcal{F} = \mathbb{R}^n$ and the symmetry relations of Casimir–Onsager

$$A_{ij} = A_{ji} \tag{3.2.2}$$

[5] For any closed convex set $\mathcal{K} \subset \mathcal{X}$, where $\mathcal{X}$ is a Hilbert space, we can define a mapping of $\mathcal{X}$ into $\mathcal{X}$ by assigning to each $x \in \mathcal{X}$ an element $P_{\mathcal{K}} x \in \mathcal{K}$ closest to x, i.e., $\inf_{y \in \mathcal{K}} \|x - y\| = \|x - P_{\mathcal{K}}(x)\|$. Due to Prop. 2.2.1, $P_{\mathcal{K}} x$ exists and is unique. The nonlinear operator $P_{\mathcal{K}}$ is called the projection operator onto $\mathcal{K}$.

hold. Although (3.2.1) and (3.2.2) are verified by many physical problems, they have been much criticized, actually because of the difficulty in distinguishing between fluxes and forces in a physical system (see, e.g., [323]). (3.2.1) can also be written in the form

$$f = \tfrac{1}{2}\,\mathrm{grad}\,\mathscr{A}(v), \qquad \mathscr{A}(v) = A_{ij} v_i v_j. \tag{3.2.3}$$

Here $\mathscr{A}$ may also depend on the other thermodynamic parameters of the continuum [113]. We can express the intrinsic specific dissipation $\mathscr{A}$ in the form (see [112] and Appendix III)

$$\mathscr{A} = f_i v_i. \tag{3.2.4}$$

The second principle of thermodynamics implies that for an irreversible process

$$\mathscr{A} \geq 0. \tag{3.2.5}$$

Accordingly,

$$\mathscr{A}(v) = A_{ij} v_i v_j = f_i v_i \geq 0. \tag{3.2.6}$$

The following hypothesis of normal dissipation is a generalization of (3.2.3).

(H_1) For every real thermodynamic process there exists a superpotential, i.e., a convex, l.s.c. and proper functional $\mathscr{G}$ on $\mathfrak{U}$, such that

$$f \in \partial \mathscr{G}(v). \tag{3.2.7}$$

This generalization[6] is based on an idea of Ziegler [342] [343] [344], who proposed an orthogonality principle in order to generalize Onsager's relations. This idea has been exploited through convex analysis by Moreau [212] and Germain [112], [113], who proposed a law of the form (3.2.7). The functional $\mathscr{G}$ is called the superpotential of dissipation. From (3.2.7), the inverse law

$$v \in \partial \mathscr{G}^c(f) \tag{3.2.8}$$

results, where $\mathscr{G}^c$ is, like $\mathscr{G}$, convex, l.s.c. and proper as well. The following proposition holds.

Proposition 3.2.1. Hypothesis (H_1) is equivalent to each of the following two assertions:

(i) for a given flux v, the force f corresponding to v minimizes the quantity $\mathscr{G}^c(f^*) - v_i f_i^*$ on $\mathfrak{F}$; and
(ii) for a given force f, the flux v corresponding to f minimizes the quantity $\mathscr{G}(v^*) - v_i^* f_i$ on $\mathfrak{U}$.

[6] The laws between fluxes and forces are considered here only as phenomenological, perhaps rough descriptions, of some physical processes. In any case, we do not intend to participate here in the dispute between the Onsagerists and the non-Onsagerists.

Proof. Obvious from the definition of $\mathcal{G}^c$ (cf., Sec. 2.5.1), q.e.d.

The second principle of thermodynamics implies that the inequality

$$f_i v_i \geq 0 \tag{3.2.9}$$

must always be satisfied during a thermodynamic process. Note also that for every v (resp. f) the vector $\{f, -1\}$ (resp. $\{v, -1\}$) is normal to epi $\mathcal{G}$ (resp. epi $\mathcal{G}^c$) at the point $(v, \mathcal{G}(v))$ (resp. $(f, \mathcal{G}^c(f))$). This fact explains the name given to hypothesis (H_1). It is also noted that the normal dissipation hypothesis combined with the assumption of thermodynamic potential (cf., [112]), permits the definition of a large class of constitutive relations by means of two convex functionals, the superpotential of dissipation and the thermodynamic potential [113]. These laws must satisfy, besides the principle of objectivity, the inequality (3.2.9). In Sec. 4.3 a generalization of (H_1) for the case of nonconvex dissipation functionals is given.

3.2.2 Duality of Variational Principles

The equilibrium problem (3.1.43) gives rise to the convex minimization problem (3.1.46) with respect to the (generalized) displacements u. The theory of duality in convex optimization allows the problem to be formulated with respect to the (generalized) forces f. There are many approaches to duality in convex optimization. That proposed by Fenchel [95] and Rockafellar [285], [288] and elaborated by Ekeland and Temam [91] is used here, since it is directly applicable to the B.V.P.s studied in later chapters.

Problem (3.1.46) will be called the primal problem (or, in this section, problem I). Recall that Π is convex, l.s.c. and proper on $\mathcal{U}$. Let us further introduce a L.C.H.T.V.S. $\mathcal{Y}$, and let $\mathcal{Y}'$ be its dual space. For $p \in \mathcal{Y}$ and $p' \in \mathcal{Y}'$ we denote by $\langle\langle p, p' \rangle\rangle$ the duality pairing between $\mathcal{Y}$ and $\mathcal{Y}'$. Then a functional $P\colon \mathcal{U} \times \mathcal{Y} \to \bar{\mathbb{R}}$ is defined such that

$$P(u^*, 0) = \Pi(u^*). \tag{3.2.10}$$

Then the problem

$$\inf_{u^* \in \mathcal{U}} P(u^*, p), \tag{3.2.11}$$

also called problem I_p, is considered. Obviously, for $p = 0$ problem I_p coincides with problem I. Further, let P^c be the conjugate functional of P defined on the space $\mathcal{F} \times \mathcal{Y}'$. We consider the problem

$$\sup_{p'^* \in \mathcal{Y}'} - P^c(0, p'^*), \tag{3.2.12}$$

also called problem I^c. Problem I_p is called the perturbed form of problem

I, and I^c is called the dual problem of I with respect to the perturbation $\mathscr{p}$. Let us further denote by inf I and sup I^c the infimum and the supremum of problems I and I^c respectively. Inf I and sup I^c are the real numbers $\Pi(\mathscr{u})$ and $\sup\{-P^c(0, \mathscr{p}'^*) \mid \mathscr{p}'^* \in \mathscr{Y}'\}$. It may be easily verified by means of (2.5.3) that

$$-P^c(0, \mathscr{p}'^*) \leq P(\mathscr{u}^*, 0), \qquad \forall \mathscr{p}'^* \in \mathscr{Y}', \qquad \forall \mathscr{u}^* \in \mathscr{U}, \tag{3.2.13}$$

and, therefore, that

$$-\infty \leq \sup I^c \leq \inf I < \infty. \tag{3.2.14}$$

For $\mathscr{p} \in \mathscr{Y}$, we denote by g the function

$$g(\mathscr{p}) = \inf I_p = \inf_{\mathscr{u}^* \in \mathscr{U}} P(\mathscr{u}^*, \mathscr{p}), \tag{3.2.15}$$

and we assume that

$$P(\mathscr{u}^*, \mathscr{p}) \in \Gamma_0(\mathscr{U} \times \mathscr{Y}). \tag{3.2.16}$$

Proposition 3.2.2. If (3.2.16) holds, $g\colon \mathscr{Y} \to \bar{\mathbb{R}}$ is convex.

Proof. If $g(\mathscr{p})$ or $g(\mathscr{q})$ is $+\infty$, (2.1.2) is verified and thus g is convex. Suppose further that $g(\mathscr{p}) < \infty$ and $g(\mathscr{q}) < \infty$. We shall show that epi g is convex. Let $\mathscr{u}_1$ and $\mathscr{u}_2$ be two elements of $\mathscr{U}$ such that for $a, b \in \mathbb{R}$

$$g(\mathscr{p}) \leq P(\mathscr{u}_1, \mathscr{p}) \leq a \tag{3.2.17}$$

and

$$g(\mathscr{q}) \leq P(\mathscr{u}_2, \mathscr{q}) \leq b. \tag{3.2.18}$$

Thus $(\mathscr{p}, a)$ and $(\mathscr{q}, b)$, $a, b \in \mathbb{R}$ are two points of epi g. To show that $\lambda(\mathscr{p}, a) + (1-\lambda)(\mathscr{q}, b) \in \text{epi } g$, $0 < \lambda < 1$, we need merely to show that

$$g(\lambda\mathscr{p} + (1-\lambda)\mathscr{q}) \leq \lambda a + (1-\lambda)b. \tag{3.2.19}$$

Indeed,

$$\begin{aligned} g(\lambda\mathscr{p} + (1-\lambda)\mathscr{q}) &= \inf_{\mathscr{u}^* \in \mathscr{U}} P(\mathscr{u}^*, \lambda\mathscr{p} + (1-\lambda)\mathscr{q}) \\ &\leq P(\lambda\mathscr{u}_1 + (1-\lambda)\mathscr{u}_2, \lambda\mathscr{p} + (1-\lambda)\mathscr{q}) \\ &\leq \lambda P(\mathscr{u}_1, \mathscr{p}) + (1-\lambda)P(\mathscr{u}_2, \mathscr{q}) \\ &\leq \lambda a + (1-\lambda)b, \end{aligned}$$

q.e.d.

Proposition 3.2.3. Let us define the conjugate functional g^c of g. Then

$$g^c(\mathscr{p}') = P^c(0, \mathscr{p}'), \qquad \forall \mathscr{p}' \in \mathscr{Y}' \tag{3.2.20}$$

and

$$\sup I^c = \sup_{\mathscr{p}'^* \in \mathscr{Y}'} (-g^c(\mathscr{p}'^*)) = g^{cc}(0). \tag{3.2.21}$$

Proof. Both relations result from (3.2.12) by noting that for every p'

$$g^c(p') = \sup_{p \in \mathcal{Y}} [\langle\langle p', p\rangle\rangle - g(p)]$$

$$= \sup_{p \in \mathcal{Y}} \sup_{u^* \in \mathcal{U}} [\langle\langle p', p\rangle\rangle - P(u^*, p)] = P^c(0, p'),$$

and by recalling the definition of g^{cc}, q.e.d.

The perturbed problem $\sup\{-P^c(f, p'^*) | p'^* \in \mathcal{Y}'\}$ is associated with problem I^c. Hence the dual problem of I^c with respect to the perturbation f reads

$$\inf_{u^* \in \mathcal{U}} \{P^{cc}(u^*, 0)\} \tag{3.2.22}$$

and is denoted by I^{cc}. On the assumption (3.2.16), $P^{cc}(u^*, 0) = P(u^*, 0)$ for every $u^* \in \mathcal{U}$ (cf. (2.5.11)), and thus problem I^{cc} coincides with the primal problem I. Moreover, since $P^{ccc} = P^c$ (cf., (2.5.13)), problem I^{ccc}, which would result by continuing the dualization procedure, is identical to I^c. The following proposition gives a condition under which $\inf I = \sup I^c$.

Proposition 3.2.4. On the assumption (3.2.16), the following three conditions are equivalent to each other:

(i) $-\infty < \inf I = \sup I^c < \infty$;
(ii) $g(0)$ is finite and g is l.s.c. at $p = 0$ (normality of I [91]);
(iii) Problem I^c is normal.

Proof. Assume that (i) is true. By means of (3.2.21) and (3.2.15), (i) is written in the form

$$g(0) = g^{cc}(0) \in \mathbb{R}. \tag{3.2.23}$$

If $\bar{g}$ is the l.s.c. regularization of g (Sec. 2.5.1), then the inequality (cf., (2.5.9))

$$g^{cc} \le \bar{g} \le g \tag{3.2.24}$$

holds. Thus $g^{cc}(0) = \bar{g}(0) = g(0)$, which implies (ii). Assume that (ii) is true. From (ii) and from (3.2.24), it results that $\bar{g}(0) = g(0) \in \mathbb{R}$. $\bar{g}$ is convex and l.s.c., because of the convexity of g, and since $\bar{g}(0) \in \mathbb{R}$ it results (cf., Props. 2.5.1, 2) that $\bar{g} \in \Gamma^0(\mathcal{Y})$. Thus $\bar{g}^{cc} = \bar{g}$. The duality reverses the order of inequalities and thus

$$g^{ccc} \ge \bar{g}^c \ge g^c. \tag{3.2.25}$$

But $g^{ccc} = g^c$ and thus $g^c = \bar{g}^c$, from which it follows that $g(0) = \bar{g}(0) = \bar{g}^{cc}(0) = g^{cc}(0)$. This last equality implies (i).

It remains to be shown that (ii) is equivalent to (iii). This results directly from problem I being identical to problem I^{cc}, q.e.d.

The set of solutions of problem I^c is identical to $\partial g^{cc}(0)$. This is readily verified; indeed, if $p' \in \mathcal{Y}'$ satisfies problem I^c, then (3.2.12, 20) imply that

$$-g^c(p') \geq -g^c(q') \qquad \forall q' \in \mathcal{Y}'. \tag{3.2.26}$$

But

$$-g^c(p') = \sup_{q' \in \mathcal{Y}'} (\langle 0, q' \rangle - g^c(q')) = g^{cc}(0), \tag{3.2.27}$$

and, therefore,

$$p' \in \partial g^{cc}(0). \tag{3.2.28}$$

Proposition 3.2.5. Assume that $g(0)$ is finite. Then $\partial g(0) \neq \emptyset$ if and only if g is l.s.c. at $p = 0$ and problem I^c has at least one solution.

Proof. Since $\partial g(0) \neq \emptyset$ then $g^{cc}(0) = g(0) \in \mathbb{R}$, g is l.s.c. at 0 and $\partial g^{cc}(0) = \partial g(0) \neq \emptyset$. Indeed, if $x \in \partial g(0)$, then the affine continuous function l, such that $l(u^*) = \langle x, u^* \rangle + g(0)$, is bounded from above by g. But since g^{cc} is the Γ-regularization of g (cf., Sec. 2.5.1),

$$l(u^*) \leq g^{cc}(u^*) \leq g(u^*), \qquad \forall u^* \in \mathcal{U}. \tag{3.2.29}$$

From (3.2.29) and $l(0) = g(0)$, we find that

$$g^{cc}(0) = g(0), \tag{3.2.30}$$

and thus g is l.s.c. at 0 (cf., (3.2.24)). Moreover,

$$\partial g^{cc}(0) = \partial g(0) \neq \emptyset. \tag{3.2.31}$$

Since $\partial g^{cc}(0)$ is the set of solutions of I^c, (3.2.31) implies the existence of at least one solution of I^c.

In order to prove the converse, we show by the method followed in Prop. 3.2.4 that $g(0) = g^{cc}(0) \in \mathbb{R}$. Then, if I^c has at least one solution, $\partial g^{cc}(0) = \partial g(0) \neq \emptyset$, q.e.d.

Proposition 3.2.6. Suppose that solutions to I and I^c exist, and that

$$-\infty < \inf I = \sup I^c < \infty. \tag{3.2.32}$$

Then any solution u of I and any solution p' of I^c satisfy the relation

$$(0, p') \in \partial P(u, 0). \tag{3.2.33}$$

Conversely, if u and p' satisfy (3.2.33), then u is a solution of I, p' is a solution of I^c and (3.2.32) holds.

Proof. The equalities

$$\inf I = P(u, 0) = \sup I^c = -P^c(0, p') \tag{3.2.34}$$

hold. Thus

$$P(u, 0) + P^c(0, p') = \langle (u, 0), (0, p') \rangle = 0, \tag{3.2.35}$$

which is equivalent to (3.2.33). Conversely, if u and p' satisfy (3.2.33), then (3.2.35) and (3.2.13) imply (3.2.32) and also that u (resp. p') is a solution of problem I (resp. I^c), q.e.d.

Relation (3.2.35) is also called the extremality condition of the problem. As is concluded from Prop. 3.2.5, conditions $-\infty < g(0) < \infty$ and $\partial g(0) \neq \emptyset$ referred to as the stability conditions, are of importance, since they ensure the existence of the solution of the dual problem.

In mechanical problems we often encounter the following situation: let the (generalized) displacement space have the form $\mathcal{U}^{(0)} \times \bar{\mathcal{U}}^{(1)}$ (cf., Sec. 3.1.1) where $u \in \mathcal{U}^{(0)}$, $\varepsilon \in \bar{\mathcal{U}}^{(1)}$ and u, ε are related by a compatibility condition of the form

$$\varepsilon = \Lambda u. \tag{3.2.36}$$

ε is here termed the internal and u the external generalized displacement. With the notation of this section, let $u \in \mathfrak{U}$ and $\varepsilon \in \mathfrak{Y}$, where $\mathfrak{U}$ and $\mathfrak{Y}$ are L.C.H.T.V.S.s with duals $\mathcal{F}$ and $\mathfrak{Y}'$, and let $\Lambda \in L(\mathfrak{U}, \mathfrak{Y})$ with transpose $\Lambda' \in L(\mathfrak{Y}', \mathcal{F})$. The transpose operator maps the space $\mathfrak{Y}'$ of the internal generalized forces, denoted by σ, into the space $\mathcal{F}$ of the external generalized forces f. For instance, in the case of a deformable body and in the framework of a small deformation theory (cf., Ch. 5), u (resp. f) is the displacement (resp. the volume force) field, ε (resp. σ) is the strain (resp. stress) field and Λ is the deformator, i.e., $\Lambda u = \frac{1}{2}(u_{i,j} + u_{j,i})$. We then assume that the energy functional Π, which must be minimized (cf., (3.1.46)), can be written in the form

$$\Pi(u) = \bar{\Pi}(u, \Lambda u), \tag{3.2.37}$$

where $\bar{\Pi}: \mathfrak{U} \times \mathfrak{Y} \to \bar{\mathbb{R}}$. In this case we define the functional P by means of

$$P(u, p) = \bar{\Pi}(u, \Lambda u - p), \; p = \varepsilon. \tag{3.2.38}$$

If Π is convex, P is convex as well, and if $\bar{\Pi} \in \Gamma_0(\mathfrak{U} \times \mathfrak{Y})$, then $P \in \Gamma_0(\mathfrak{U} \times \mathfrak{Y})$. The dual problem (3.2.12) acquires the form ($p' = \sigma$) [91]

$$\sup_{p'^* \in \mathfrak{Y}'} -\bar{\Pi}^c(\Lambda' p'^*, -p'^*), \tag{3.2.39}$$

which constitutes a variational formulation with respect to the generalized forces. In order to show (3.2.39), we note that

$$\begin{aligned} P^c(0, p') &= \sup_{\substack{p^* \in \mathfrak{Y} \\ u^* \in \mathfrak{U}}} [\langle\langle p', p^* \rangle\rangle - \bar{\Pi}(u^*, \Lambda u^* - p^*)] \\ &= \sup_{u^* \in \mathfrak{U}} \sup_{\substack{q^* \in \mathfrak{Y} \\ q^* + p^* = \Lambda u^*}} [\langle\langle p', \Lambda u^* \rangle\rangle - \langle\langle p', q^* \rangle\rangle - \bar{\Pi}(u^*, q^*)] \\ &= \bar{\Pi}^c(\Lambda' p', -p'). \end{aligned} \tag{3.2.40}$$

Let us proceed to give sufficient conditions for the stability of the primal problem I and for the existence of a solution of I and I^c.

Proposition 3.2.7. Let $\bar{\Pi}$ be convex and inf I finite. Suppose that for at least one $u_0 \in \mathcal{U}$,

$$\bar{\Pi}(u_0, \Lambda u_0) < \infty \quad \text{and} \quad p \to \bar{\Pi}(u_0, p) \text{ is continuous at } \Lambda u_0. \tag{3.2.41}$$

Then the primal problem

$$\inf_{u^* \in \mathcal{U}} \bar{\Pi}(u^*, \Lambda u^*) \tag{3.2.42}$$

is stable.

Proof. The assumptions of the proposition imply that there exists a $u_0 \in \mathcal{U}$ such that $p \to P(u_0, p)$ is finite and continuous at $p = 0$. Accordingly, there exists a neighborhood $\mathcal{A}$ of 0 in $\mathcal{Y}$ such that for any $p \in \mathcal{A}$, $P(u_0, p)$ is bounded from above. Thus (cf., (3.2.15)) $g(p)$ is also bounded from above for every $p \in \mathcal{A}$. Hence g is continuous at 0 (Prop. 2.1.4) and thus $\partial g(0) \neq \varnothing$ (Prop. 2.3.2), q.e.d.

The stability of I implies that

$$\inf I = \sup I^c, \tag{3.2.43}$$

and that I^c has at least one solution p' (Props. 3.2.4, 5). The extremality condition (3.2.35) becomes

$$\bar{\Pi}(u, \Lambda u) + \bar{\Pi}^c(\Lambda' p', -p') = 0, \tag{3.2.44}$$

and (3.2.33) takes the form

$$(\Lambda' p', -p') \in \partial\bar{\Pi}(u, \Lambda u). \tag{3.2.45}$$

Proposition 3.2.8. Let $\mathcal{U}$ be a reflexive B-space, and $\bar{\Pi} \in \Gamma_0(\mathcal{U} \times \mathcal{Y})$; let condition (3.2.41) hold and

$$\bar{\Pi}(u, \Lambda u) \to \infty \quad \text{as} \quad \| u \| \to \infty, \ u \in \mathcal{U}. \tag{3.2.46}$$

Then both the primal and the dual problems have solutions which satisfy (3.2.43) and (3.2.44) (or (3.2.45)).

Proof. We apply Prop. 2.2.1 to $\bar{\Pi}(u, \Lambda u)$. This proves the existence of a solution of the primal problem. The rest is a consequence of Props. (3.2.6, 7), q.e.d.

If the contribution to Π of the internal and of the external generalized displacements can be separated–a case which we shall often encounter–then

$$\bar{\Pi}(u, \Lambda u) = A(u) + B(\Lambda u). \tag{3.2.47}$$

It follows that if A and B are convex, Π and P are convex as well, and if $A \in \Gamma_0(\mathcal{U})$ and $B \in \Gamma_0(\mathcal{Y})$, then $P \in \Gamma_0(\mathcal{U} \times \mathcal{Y})$. As can be easily verified, the dual problem (3.2.39) reads

$$\sup_{\not{p}'^* \in \mathcal{Y}'} [-A^c(-\Lambda' \not{p}'^*) - B^c(\not{p}'^*)], \tag{3.2.48}$$

and (3.2.45) is decomposed into the two relations

$$-\Lambda' \not{p}' \in \partial A(\mathscr{u}) \tag{3.2.49}$$

and

$$\not{p}' \in \partial B(\Lambda \mathscr{u}). \tag{3.2.50}$$

Similarly, the extremality relation (3.2.44) is replaced by the two relations

$$A(\mathscr{u}) + A^c(-\Lambda' \not{p}') = -\langle \Lambda' \not{p}', \mathscr{u} \rangle \tag{3.2.51}$$

and

$$B(\Lambda \mathscr{u}) + B^c(\not{p}') = \langle\!\langle \not{p}', \Lambda \mathscr{u} \rangle\!\rangle. \tag{3.2.52}$$

3.3 Subdifferential Boundary Conditions and Constitutive Laws

3.3.1 Subdifferential Boundary Conditions

Along the lines of the previous sections, we may define subdifferential boundary conditions between "energy (or power) corresponding"[7] quantities. These boundary conditions include as special cases the classical boundary conditions of mechanics. We denote by Ω an open bounded subset of $\mathbb{R}^3$ which is occupied by a deformable body. The boundary of Ω is denoted by Γ. The points $x \in \Omega$, $x = \{x_i\}$, $i = 1, 2, 3$, are referred to a Cartesian coordinate system. We denote by $S = \{S_i\}$ the stress vector on Γ. $S_i = \sigma_{ij} n_j$, where $\sigma = \{\sigma_{ij}\}$ is an appropriately defined stress tensor and $n = \{n_i\}$ is the outward unit normal vector on Γ. Often the vector S will be decomposed into a normal component S_N and a tangential component S_T with respect to Γ

$$S_N = \sigma_{ij} n_j n_i \quad \text{and} \quad S_{T_i} = \sigma_{ij} n_j - (\sigma_{ij} n_i n_j) n_i. \tag{3.3.1}$$

Analogously to S_N and S_T, u_N and u_T denote the normal and the tangential components of the displacement vector u with respect to Γ. If the velocity v is considered, v_N and v_T denote its normal and tangential components. S_N, u_N and v_N are considered as positive if they are parallel to n.

[7] In the sense that their product expresses energy or power.

A maximal monotone operator β_i: $\mathbb{R} \to \mathscr{P}(\mathbb{R})$ is introduced and a boundary condition of the form

$$-S_i \in \beta_i(u_i) \tag{3.3.2}$$

is considered in the i-th direction. By Prop. 2.6.4, a convex, l.s.c. and proper functional j_i on $\mathbb{R}$ may be determined up to an additive constant such that

$$\beta_i = \partial j_i. \tag{3.3.3}$$

Then (3.3.2) is written in the form

$$-S_i \in \partial j_i(u_i). \tag{3.3.4}$$

This relation constitutes a subdifferential boundary condition and is understood pointwise, i.e., as a relation between $-S_i(x) \in \mathbb{R}$ and $u_i(x) \in \mathbb{R}$ at every point $x \in \Gamma$. Therefore (3.3.2) constitutes a local constraint within the terminology of Sec. 3.1.2, with $\mathcal{U} = \mathbb{R}$, $\mathcal{F} = \mathbb{R}$ and $\langle S_i, u_i \rangle = S_i u_i$ (no summation). For the extension of this law to more general spaces, the reader is referred to Sec. 3.3.4. Obviously, (3.3.4) may also be written in the inverse forms

$$u_i \in \partial j_i^c(-S_i) \tag{3.3.5}$$

and

$$u_i \in \beta_i^c(-S_i), \tag{3.3.6}$$

where $\beta_i^c = \partial j_i^c$ is again a maximal monotone operator on $\mathbb{R}$ and is the inverse operator of β_i. As was pointed out in Sec. 2.6.2, the graph of β_i, referred to a Cartesian system $0xy$, is a complete nondecreasing curve in $\mathbb{R}^2$ which is generally multivalued. Thus the graph may include segments parallel to both coordinate axes (Fig. 3.3.1) which correspond to ideal plasticity, jump and locking effects. Superpotential j_i (resp. j_i^c) is called a local superpotential (resp. conjugate superpotential) and expresses the potential (resp.

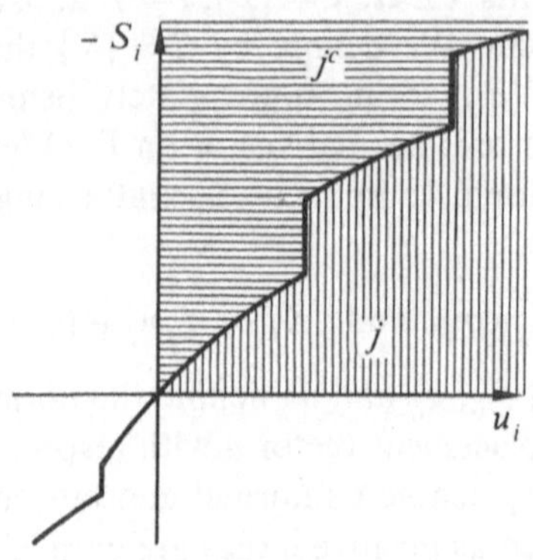

Fig. 3.3.1. Illustration of the definition of j and j^c.

complementary energy) of the local constraint. The constraint may be visualized as induced by a fictive spring of zero length at x in the ith-direction. The spring is assumed to obey the material law (3.3.2).

Analogously to (3.3.2), a boundary condition of the form

$$-S_N \in \beta_N(u_N) = \partial j_N(u_N) \tag{3.3.7}$$

may be defined.

Assume that j is a convex, l.s.c., proper functional on $\mathbb{R}^3$. Then a relation of the form

$$-S \in \partial j(u) \tag{3.3.8}$$

is defined pointwise on Γ, i.e., as a monotone relation between $S(x)$ and $u(x)$. Equivalently to (3.3.8), we may write

$$u \in \partial j^c(-S) \tag{3.3.9a}$$

and

$$j(u) + j^c(-S) = -u_i S_i. \tag{3.3.9b}$$

Obviously, for (3.3.8) $\mathcal{U} = \mathbb{R}^3$, $\mathcal{F} = \mathbb{R}^3$ and $\langle u, f\rangle = u_i f_i$. Similarly to (3.3.8), a subdifferential law

$$-S_T \in \partial j_T(u_T) \tag{3.3.10}$$

may be considered.

In some classes of mechanical problems, similar boundary conditions may be defined between S and the partial time derivative of the displacement $\partial u/\partial t$ or the velocity v, i.e.,

$$-S \in \partial j\left(\frac{\partial u}{\partial t}\right) \tag{3.3.11}$$

and

$$-S \in \partial j(v). \tag{3.3.12}$$

We shall provide examples to illustrate these boundary conditions.

(i) The classical boundary conditions $u_i = u_{i0}$ can be put in the form (3.3.2) through the operator (Fig. 3.3.2a)

$$\beta_i(u_i) = \begin{cases} \mathbb{R} & \text{if} \quad u_i = u_{i0} \\ \varnothing & \text{otherwise} \end{cases}, \tag{3.3.13}$$

or through the functional $j_i(u_i) = \{0 \text{ if } u_i = u_{i0} \text{ and } \infty \text{ otherwise}\}$. The boundary condition $S_i = C_i$ is written in the form (3.3.2) or (3.3.4) with $\beta_i(u_i) = -C_i$ (C_i given) or $j_i(u_i) = -C_i u_i$ (no summation) for every $u_i \in \mathbb{R}$ (Fig. 3.3.2b).

(ii) The boundary condition

$$-S_N = k u_N, \; k \text{ const} > 0 \tag{3.3.14}$$

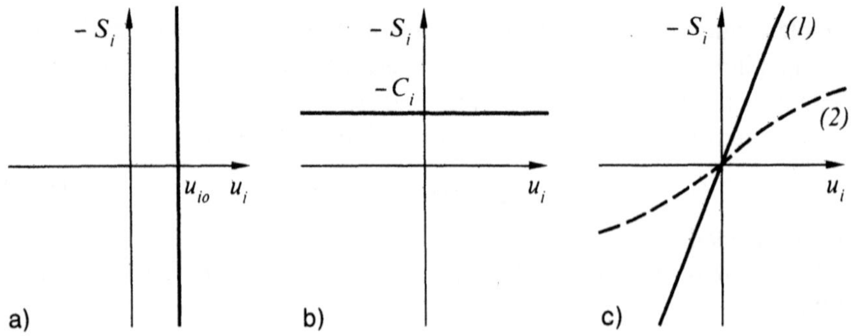

Fig. 3.3.2. Some simple boundary conditions.

may be written in the form (3.3.7) by taking

$$\beta_N(u_N) = ku_N, \qquad j_N(u_N) = \tfrac{1}{2}ku_N^2. \tag{3.3.15}$$

This is the Winkler law, which describes in a simplified manner the interaction between a deformable body and the soil. The corresponding fictive spring is termed a linear Winkler spring. Although this model is oversimplified, it finds extensive application in civil engineering. In order to improve this model, we can assume a nonlinear, monotone stress-strain relation for the Winkler spring (the dotted line in Fig. 3.3.2c) which may be written in the form (3.3.7).

(iii) The foregoing boundary condition does not describe the case in which the body loses contact with the soil (cf., [103], [241]). To do so we should consider the following law:

$$\text{if} \quad u_N < 0, \quad \text{then} \quad S_N = 0; \tag{3.3.16}$$

$$\text{if} \quad u_N \geq 0, \quad \text{then} \quad S_N + ku_N = 0, \qquad k \text{ const} > 0. \tag{3.3.17}$$

(3.3.16) corresponds to the case of noncontact and (3.3.17) to the case of contact. The regions of contact and noncontact are not known a priori, and therefore (3.3.16, 17) lead to a free B.V.P. The respective operator β_N (resp. j_N) is given by

$$\beta_N(u_N) = \begin{cases} ku_N & \text{if} \quad u_N \geq 0 \\ 0 & \text{if} \quad u_N < 0 \end{cases} \tag{3.3.18}$$

and

$$j_N(u_N) = \begin{cases} \tfrac{1}{2}ku_N^2 & \text{if} \quad u_N \geq 0 \\ 0 & \text{if} \quad u_N < 0 \end{cases} \tag{3.3.19}$$

$j_N(u_N)$ can be written compactly as $\frac{1}{2}ku_{N+}^2$, where u_{N+} denotes the positive part of u_N, i.e., $u_{N+} = \sup\{0, u_N\}$. (3.3.16, 17) are termed conditions of unilateral contact [241] for a linear Winkler law, whereas (3.3.14) is the condition of bilateral contact.

We can consider generally the operators

$$\beta_N(u_N) = \begin{cases} \beta_1(u_N) & \text{if} \quad u_N \geq 0 \\ 0 & \text{if} \quad u_N < 0 \end{cases} \tag{3.3.20}$$

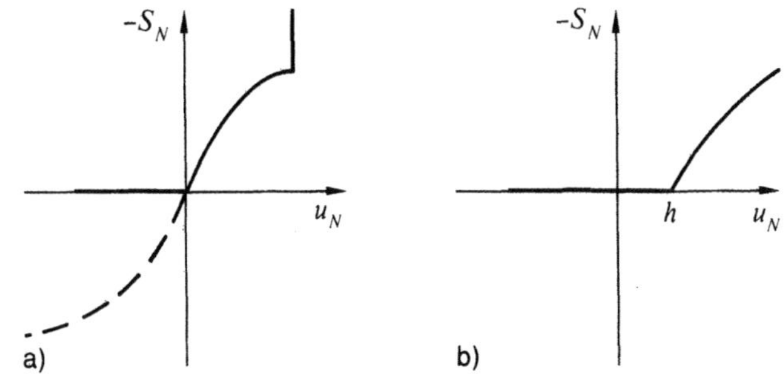

Fig. 3.3.3. Unilateral contact boundary conditions.

and

$$\beta_N(u_N) = \begin{cases} \beta_1(u_N - h) & \text{if} \quad u_N \geq h \\ 0 & \text{if} \quad u_N < h \end{cases} \tag{3.3.21}$$

Here β_1 is assumed to be a maximal monotone operator on $\mathbb{R}$ such that $0 \in \beta_1(0)$. (3.3.20) and (3.3.21) also lead to unilateral contact boundary conditions, but with a nonlinear Winkler law. In the second case, the support is at a given distance $h = h(x)$ from the body under consideration (Fig. 3.3.3). The foregoing relations are not sufficient to formulate B.V.P.s, but have to be combined with a boundary condition concerning S_T or u_T or both, e.g., $S_T = C_T$, where $C_T = C_T(x)$ is given, or $u_T = 0$, (which states that no sliding occurs), or, more generally, (3.3.10). It is also possible for β_N to change from point to point, in which case $\beta_N = \beta_N(u_N(x), x)$. Note that the uncoupling of the contact conditions in the tangential and in the normal directions is a considerable simplification of the mechanical problem.

(iv) If only the body is deformable while the support is not, then the boundary conditions of Signorini-Fichera result [99], [301]. They read (Fig. 3.3.4):

$$\begin{aligned} &\text{if} \quad u_N < 0, \quad \text{then} \quad S_N = 0; \\ &\text{if} \quad u_N = 0, \quad \text{then} \quad S_N \leq 0, \end{aligned} \tag{3.3.22}$$

or equivalently

$$S_N \leq 0, \qquad u_N \leq 0, \quad \text{and} \quad S_N u_N = 0. \tag{3.3.23}$$

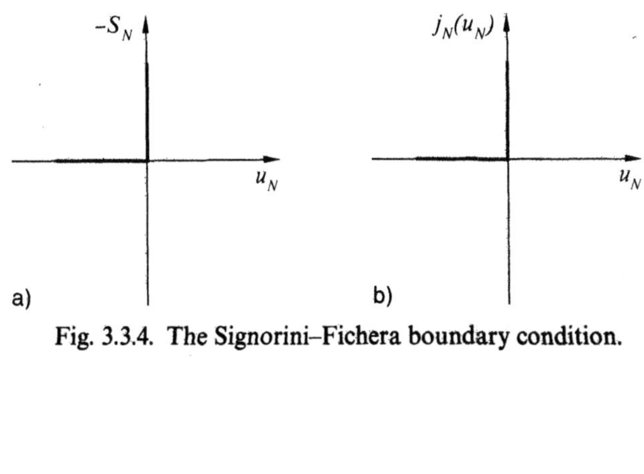

Fig. 3.3.4. The Signorini–Fichera boundary condition.

The respective operator β_N is

$$\beta_N(u_N) = \begin{cases} 0 & \text{if} \quad u_N < 0 \\ [0, +\infty) & \text{if} \quad u_N = 0, \\ \varnothing & \text{if} \quad u_N > 0 \end{cases} \tag{3.3.24}$$

and the superpotential

$$j_N(u_N) = \begin{cases} 0 & \text{if} \quad u_N \leq 0 \\ \infty & \text{if} \quad u_N > 0 \end{cases} \tag{3.3.25}$$

The contact between two deformable bodies can be described by a boundary condition similar to (3.3.22) on the assumption that the boundary displacements are sufficiently small (Fig. 3.3.5). As the two bodies cannot penetrate one another, we assume that the sum of the displacements $u_N^{(1)}$ and $u_N^{(2)}$ of the two bodies and of the existing distance between them $h = h(x)$ must be greater than, or equal to the approach u^0 of the two bodies due to a rigid body displacement. Let us denote by $\bar{u}_N$ the quantity

$$u_N^{(1)} + u_N^{(2)} + h - u^0,$$

and let R_N be the respective contact force. The contact conditions are then:

$$\begin{aligned} &\text{if} \quad \bar{u}_N > 0, \quad \text{then} \quad R_N = 0; \\ &\text{if} \quad \bar{u}_N = 0, \quad \text{then} \quad R_N \geq 0. \end{aligned} \tag{3.3.26}$$

Analogous conditions describe the static behavior of a crack in a continuum (cf., [233]).

(v) The next example concerns the friction boundary conditions. We consider the following boundary conditions:

$$\text{if} \quad |S_T| < \mu |S_N|, \quad \text{then} \quad u_{T_i} = 0, \qquad i = 1, 2, 3; \tag{3.3.27}$$

$$\begin{aligned} &\text{if} \quad |S_T| = \mu |S_N|, \quad \text{then there exists} \quad \lambda \geq 0 \\ &\text{such that} \quad u_{T_i} = -\lambda S_{T_i}, \qquad i = 1, 2, 3. \end{aligned} \tag{3.3.28}$$

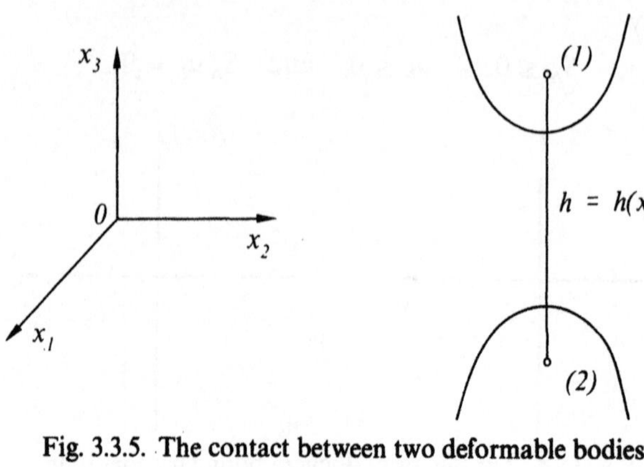

Fig. 3.3.5. The contact between two deformable bodies.

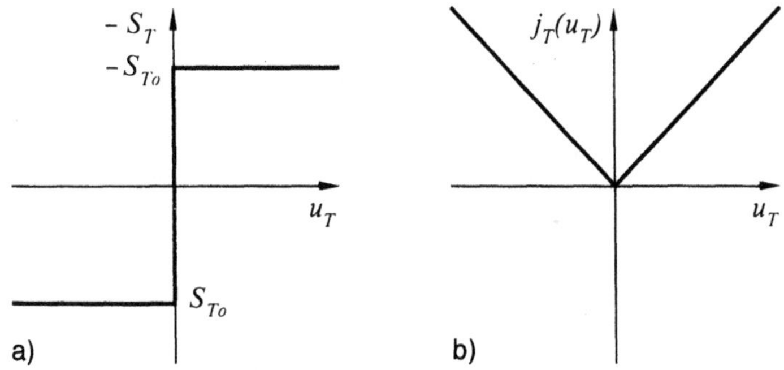

Fig. 3.3.6. The friction boundary condition.

Here $\mu = \mu(x) > 0$ denotes the coefficient of friction and $|\cdot|$ the usual $\mathbb{R}^3$-norm. If Ω is a two-dimensional body, Γ is a curve, and thus S_T, u_T may be referred to a local right-handed coordinate system (n, τ) on Γ where τ denotes the unit vector tangential to Γ. (3.3.27, 28) are written in the form

$$-S_T \in \beta_T(u_T), \tag{3.3.29}$$

where (Fig. 3.3.6)

$$\beta_T(u_T) = \begin{cases} [-\mu|S_N|, +\mu|S_N|] & \text{if } u_T = 0 \\ \mu|S_N| & \text{if } u_T > 0. \\ -\mu|S_N| & \text{if } u_T < 0 \end{cases} \tag{3.3.30}$$

Assume further that $S_N = C_N$, where C_N is given, and denote $\mu|C_N|$ by S_{T_0}. Then

$$\beta_T(u_T) = \partial(S_{T_0}|u_T|). \tag{3.3.31}$$

If Ω is a three-dimensional body, then (3.3.27, 28) can be written only in the form (3.3.10), with

$$j_T(u_T) = S_{T_0}|u_T| \tag{3.3.32}$$

We can verify that

$$j_T^c(-S_T) = \begin{cases} 0 & \text{if } |S_T| \le S_{T_0} \\ \infty & \text{otherwise} \end{cases}, \tag{3.3.33}$$

and thus (3.3.27, 28) are equivalently written in the form

$$u_T \in \partial j_T^c(-S_T). \tag{3.3.34}$$

The two dual subdifferential formulations give rise, for u_T, $S_T \in \mathbb{R}^3$ to the variational inequality

$$j_T(u_T^*) - j_T(u_T) \ge -S_{T_i}(u_{T_i}^* - u_{T_i}), \qquad \forall u_T^* \in \mathbb{R}^3 \tag{3.3.35}$$

and to

$$-u_{T_i}(S_{T_i}^0 - S_{T_i}) \le 0, \qquad \forall S_T^0 \in \mathbb{R}^3 \text{ such that } |S_T^0| \le S_{T_0} \tag{3.3.36}$$

for $|S_T| \le S_{T_0}$.

In dynamic problems, a friction law of the form

$$-S_T \in \partial j_T(v_T) = \partial(S_{T_0} | v_T |) \tag{3.3.37}$$

can be considered (Coulomb's law of friction). The tangential velocity v_T is replaced by $\partial u_T/\partial t$ if the displacements are sufficiently small.

A generalization of (3.3.37) is obtained if the superpotential (dissipation function) j_T of (3.3.37) is replaced by the function

$$j_T(v_T) = | C_N | (\mu_a^2 v_{Ta}^2 + \mu_b^2 v_{Tb}^2)^{1/2}. \tag{3.3.38}$$

Here a and b are two orthogonal directions termed orthotropy directions, which are defined at every point on the surface of the body, v_{Ta} and v_{Tb} are the components of the velocity v_T with respect to the local coordinate system (a, b), and μ_a and μ_b are the two corresponding friction coefficients. (3.3.37, 38) lead to the law of orthotropic friction [218], [202], [344], which reads:

$$\text{if} \quad \left[\left(\frac{S_{Ta}}{\mu_a}\right)^2 + \left(\frac{S_{Tb}}{\mu_b}\right)^2\right]^{1/2} < | C_N |, \quad \text{then} \quad v_{Ta} = v_{Tb} = 0; \tag{3.3.39}$$

$$\begin{gathered}\text{if} \quad \left[\left(\frac{S_{Ta}}{\mu_a}\right)^2 + \left(\frac{S_{Tb}}{\mu_b}\right)^2\right]^{1/2} = | C_N |, \quad \text{then there exists } \lambda \geq 0 \\ \text{such that} \quad v_{Ta} = -\lambda \frac{S_{Ta}}{\mu_a^2}, \qquad v_{Tb} = -\lambda \frac{S_{Tb}}{\mu_b^2}.\end{gathered} \tag{3.3.40}$$

It is possible to combine the friction boundary condition with the unilateral contact boundary condition. We then obtain the following relations:

$$\begin{gathered}\text{if} \quad u_N < 0, \quad \text{then} \quad S_N = 0 \qquad S_{T_i} = 0, \qquad i = 1, 2, 3, \\ \text{if} \quad u_N \geq 0, \quad \text{then} \quad S_N + k u_N = 0,\end{gathered} \tag{3.3.41}$$

where k is a constant > 0 and (3.3.27, 28) hold. In this case, however, it is not possible to write the boundary condition in the form (3.3.8).

Further, we give some subdifferential boundary conditions arising in the theory of plates. Ω is here an open bounded subset $\mathbb{R}^2$ defined by the middle surface of the plate. Γ denotes the boundary of Ω. The points of Ω are referred to a fixed Cartesian coordinate system $0x_1 x_2 x_3$. The x_1- and x_2-axes coincide with the middle surface of the plate, and the x_3-axis with the direction of the normal to the middle surface. The positive direction of the x_3-axis is upwards. The displacements of the plate in its plane are denoted by u_1, u_2, and vertical to its plane by ζ. By M and $\bar{Q}$, we denote respectively the bending moment and the total shearing force [116] on the boundary of the plate (Fig. 3.3.7), and we introduce boundary conditions of the form

$$M \in \beta_1\left(\frac{\partial \zeta}{\partial n}\right) = \partial j_1\left(\frac{\partial \zeta}{\partial n}\right) \tag{3.3.42}$$

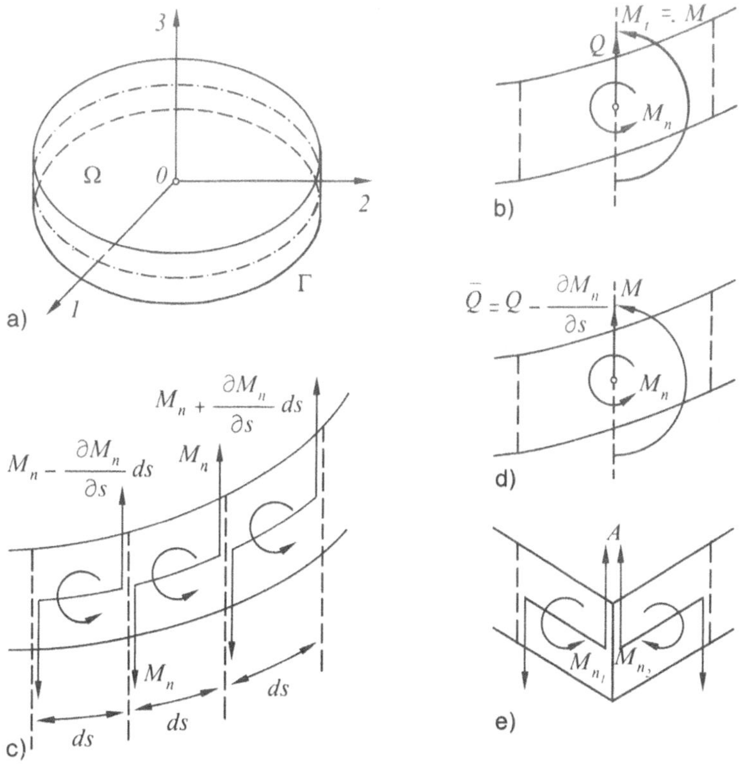

Fig. 3.3.7. Bending moments and total shearing force of the plate.

$$-\bar{Q} \in \beta_2(\zeta) = \partial j_2(\zeta), \tag{3.3.43}$$

where β_i (resp. j_i) $i = 1, 2$ is a maximal monotone operator (resp. a convex, l.s.c., proper functional) on $\mathbb{R}$. The foregoing relations must be understood in the pointwise sense.

Next, some examples on (3.3.42) are given. (3.3.43) is of the same form as (3.3.8), and, therefore, the examples for (3.3.8) may serve as examples for (3.3.43) as well.

(vi) With respect to the rotation of the plate, the friction boundary conditions are defined by the relations

$$\begin{aligned}
&\text{if } |M| < M_0, \text{ then } \frac{\partial\zeta}{\partial n} = 0,\\
&\text{if } |M| = M_0, \text{ then there exists } \lambda \geq 0\\
&\qquad \text{such that } M = \lambda \frac{\partial\zeta}{\partial n},
\end{aligned} \tag{3.3.44}$$

and can be written in the form (3.3.42), where β_1 has the form

$$\beta_1\left(\frac{\partial\zeta}{\partial n}\right) = \begin{cases} [-M_0, M_0] & \text{if } \dfrac{\partial\zeta}{\partial n} = 0 \\ M_0 & \text{if } \dfrac{\partial\zeta}{\partial n} > 0 \\ -M_0 & \text{if } \dfrac{\partial\zeta}{\partial n} < 0 \end{cases} \tag{3.3.45}$$

and M_0 is a prescribed bending moment (see Fig. 3.3.8a, and for a slight generalization Fig. 3.3.8b). The foregoing boundary condition has a useful physical meaning: it can be assumed that a plastic hinge is formed on the boundary of the plate. Then M_0 or M_1, M_2 are the plastic moments of the plate cross-section.

(vii) The boundary conditions of unilateral rotation of a plate have a form similar to the Signorini–Fichera boundary condition (3.3.22), i.e.,

$$\frac{\partial\zeta}{\partial n} \geq 0, \qquad M \leq 0, \qquad M\frac{\partial\zeta}{\partial n} = 0. \tag{3.3.46}$$

The form of β and j is evident (cf. (3.3.24, 25)).

In the theory of plates, another type of subdifferential relation can be formulated. Assume that the load vector f at every point $x \in \Omega_0 \subset \Omega$ consists of a part $\bar{f}$ which is given and of another part $\bar{\bar{f}}$ related to the displacement of that point by a relation of the form

$$-\bar{\bar{f}} \in \beta_3(\zeta) = \partial j_3(\zeta). \tag{3.3.47}$$

Here β_3 and j_3 have the same properties as β_i and j_i in (3.3.42, 43).

(viii) As an example, let us consider a plate which at points $x_0 \in \Omega_0 \subset \Omega$, $\bar{\Omega}_0 \cap \Gamma = \emptyset$, is at a distance $h = h(x)$ from a deformable support (Fig. 3.3.9). It is assumed that the support generates a reaction force which is

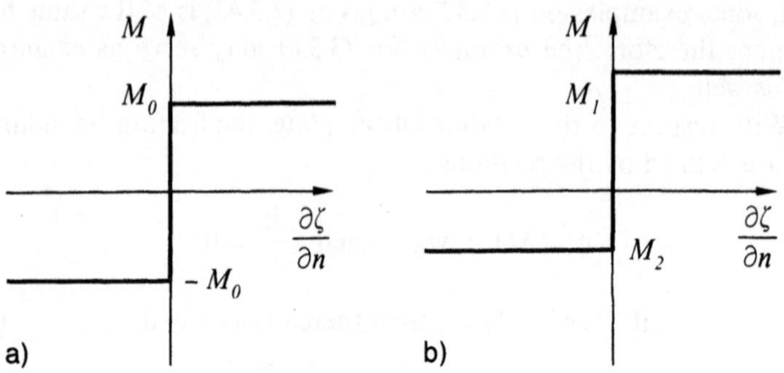

Fig. 3.3.8. The plastic-hinge boundary condition for a plate.

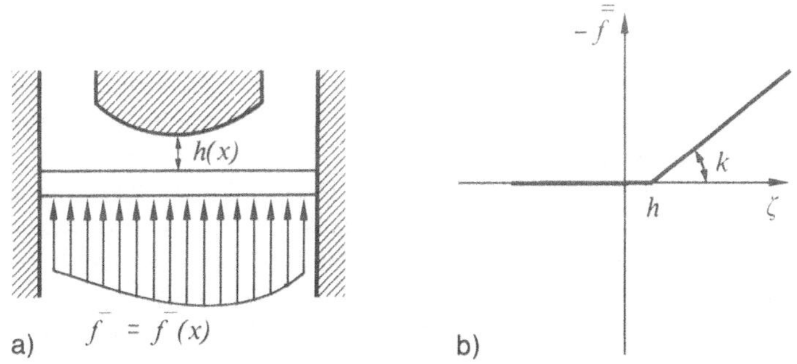

Fig. 3.3.9. Example (viii).

proportional to its deformation (Winkler support). We may then write

$$-\bar{f} \in \beta(\zeta) \quad \text{in} \quad \Omega_0 \subset \Omega, \tag{3.3.48}$$

and

$$\bar{f} = 0 \quad \text{in} \quad \Omega - \Omega_0. \tag{3.3.49}$$

β is a maximal monotone operator defined by

$$\beta(\zeta) = \begin{cases} k(\zeta - h)\ k \text{ const} > 0 & \text{if} \quad \zeta \geq h \\ 0 & \text{if} \quad \zeta < h \end{cases}. \tag{3.3.50}$$

In dynamic problems, boundary conditions of the form

$$M \in \partial j_1\left(\frac{\partial}{\partial t}\left(\frac{\partial \zeta}{\partial n}\right)\right), -\bar{Q} \in \partial j_2\left(\frac{\partial \zeta}{\partial t}\right) \tag{3.3.51}$$

may be considered, as well as reaction-displacement laws of the form

$$-\bar{f}_3 \in \partial j_3\left(\frac{\partial \zeta}{\partial t}\right). \tag{3.3.52}$$

In the framework of heat-conduction problems, subdifferential boundary conditions between the heat flux vector $q = \{q_i\}$ and the absolute temperature θ on the boundary of a body can be formulated. They read

$$q_i n_i \in \beta(\theta) \tag{3.3.53}$$

where $\beta\colon \mathbb{R} \to \mathscr{P}(\mathbb{R})$ is a maximal monotone operator. More generally, we may assume that $\beta = \beta(\theta(x, t), x, t)$. From Fourier's law,

$$q_i = -\mathscr{k}_{ij}\theta_{,j}, \tag{3.3.54}$$

where $\mathscr{k} = \{\mathscr{k}_{ij}\}$, with $\mathscr{k}_{ij} = \mathscr{k}_{ji}$, is the thermal conductivity tensor. For an isotropic material, $\mathscr{k}_{ij} = \mathscr{k}\,\delta_{ij}$. Thus (3.3.53) can also be written in the form

$$-\frac{\partial \theta}{\partial n_{\mathscr{k}}} \in \beta(\theta) = \partial j(\theta), \tag{3.3.55}$$

where j is the functional corresponding to β, and $\partial\theta/\partial n_{\mathscr{k}} = \mathscr{k}_{ij} n_i \theta_{,j}$.

The following examples illustrate (3.3.53)

(ix) Let us consider a body with a boundary which is semipermeable regarding heat conduction, i.e., it allows a heat flux into, but not out of the body. Moreover, it is assumed that the temperature on Γ changes instantaneously. Let $\theta_0 = \theta_0(x)$, $x \in \Gamma$ be the temperature just outside the boundary and $\theta = \theta(x, t)$ the body temperature. When $\theta > \theta_0$, no heat flux takes place because of the semipermeability of the boundary and thus $q_i n_i = 0$. When $\theta \leq \theta_0$, heat flux takes place and thus $q_i n_i \leq 0$. The assumption that the temperature changes instantaneously implies that $\theta = \theta_0$. This boundary condition is written, in the form (3.3.53), with β defined by (Fig. 3.3.10a),

$$\beta(\theta) = \begin{cases} 0 & \text{if } \theta > \theta_0 \\ [0, -\infty) & \text{if } \theta = \theta_0 \, . \\ \varnothing & \text{if } \theta < \theta_0 \end{cases} \tag{3.3.56}$$

(x) The boundary condition corresponding to the graph of Fig. 3.3.10b appears in temperature control problems. The temperature θ on Γ must always remain smaller than θ_1. For $\theta > \theta_1$, heat flux $q_i n_i$ takes place outwards from the body in order to maintain $\theta \leq \theta_1$. This heat flux is a monotone function of θ for $\theta_1 \leq \theta \leq \theta_2$, whereas for $\theta \geq \theta_2$ the heat flux remains constant and equal to g.

3.3.2 Subdifferential Constitutive Laws I

In this section we study constitutive laws expressed by means of subdifferential relations. Some of these laws are purely phenomenological but have proved very useful in the study of certain classes of problems in applied mechanics and engineering. This also applies for certain boundary conditions of the previous section.

First, a relation of the form

$$\sigma \in \partial w(\varepsilon), \tag{3.3.57}$$

is considered in reference to a deformable body Ω and in the framework of a small deformation theory. $\sigma = \{\sigma_{ij}\}$ is the stress tensor and $\varepsilon = \{\varepsilon_{ij}\}$ the strain tensor; w is a convex, l.s.c., proper functional on $\mathbb{R}^6$. (3.3.57) is understood here pointwise, i.e., as a relation between $\sigma(x)$ and $\varepsilon(x)$ at every point

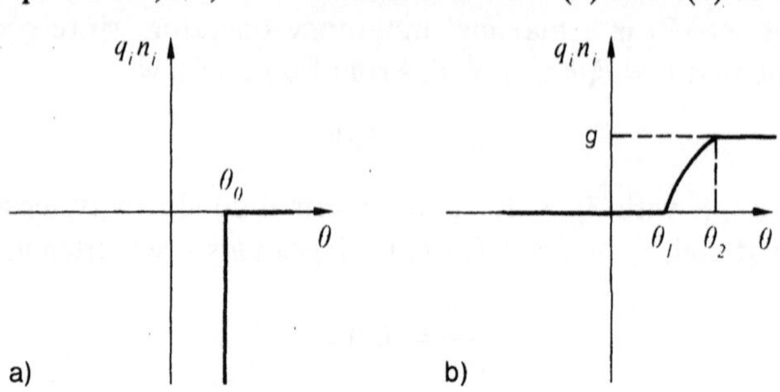

Fig. 3.3.10. Examples (ix) and (x).

$x \in \Omega$. (3.3.57) can also be written in the inverse form

$$\varepsilon \in \partial w^c(\sigma). \tag{3.3.58}$$

Functional w (resp. w^c) is the potential (resp. the complementary) energy per unit volume, and satisfies for $\varepsilon \in \mathbb{R}^6$ (resp. for $\sigma \in \mathbb{R}^6$) the variational inequality

$$w(\varepsilon^*) - w(\varepsilon) \geq \sigma_{ij}(\varepsilon_{ij}^* - \varepsilon_{ij}), \qquad \forall \varepsilon^* \in \mathbb{R}^6 \tag{3.3.59a}$$

resp.

$$w^c(\sigma^0) - w^c(\sigma) \geq \varepsilon_{ij}(\sigma_{ij}^0 - \sigma_{ij}), \qquad \forall \sigma^0 \in \mathbb{R}^6. \tag{3.3.59b}$$

Along the lines of Sec. 3.1.2, w and w^c can be seen as local superpotentials which define at every point of the body the law (3.3.57) (or (3.3.58)). Further, some examples are given.

(i) Assume that w is defined by

$$w(\varepsilon) = \tfrac{1}{2} C_{ijhk} \varepsilon_{ij} \varepsilon_{hk}, \tag{3.3.60}$$

where the components of the tensor $C = \{C_{ijhk}\}$, $i, j, h, k = 1, 2, 3$ have the symmetry property

$$C_{ijhk} = C_{jihk} = C_{khij} \tag{3.3.61a}$$

and the ellipticity property

$$C_{ijhk} \varepsilon_{ij} \varepsilon_{hk} \geq c \varepsilon_{ij} \varepsilon_{hk},$$
$$\forall \varepsilon = \{\varepsilon_{ij}\} \in \mathbb{R}^6, \qquad c \text{ const} > 0. \tag{3.3.61b}$$

In this case (3.3.57) becomes

$$\sigma_{ij} = C_{ijhk} \varepsilon_{hk}, \tag{3.3.62}$$

and this is Hooke's law of the theory of elasticity.

(ii) If w in (3.3.57) is continuously differentiable, the relation

$$\sigma_{ij} = \frac{\partial w(\varepsilon)}{\partial \varepsilon_{ij}} \tag{3.3.63}$$

is obtained, which corresponds to a nonlinear elastic material.

(iii) Consider on $\mathbb{R}$ a superpotential w defined by

$$w(\varepsilon) = \begin{cases} \tfrac{1}{2} a \varepsilon^2 + \tfrac{1}{2} b(\varepsilon - \varepsilon_0)^2, \quad a, b > 0 & \text{if } \varepsilon \geq \varepsilon_0 \\ \tfrac{1}{2} a \varepsilon^2 & \text{if } 0 \leq \varepsilon \leq \varepsilon_0 . \\ 0 & \text{if } \varepsilon < 0 \end{cases} \tag{3.3.64}$$

Then

$$\sigma = \begin{cases} a\varepsilon + b(\varepsilon - \varepsilon_0) & \text{if } \varepsilon \geq \varepsilon_0 \\ a\varepsilon & \text{if } 0 \leq \varepsilon \leq \varepsilon_0 . \\ 0 & \text{if } \varepsilon < 0 \end{cases} \tag{3.3.65}$$

This stress-strain relation (Fig. 3.3.11) describes the behavior of a nonlinear elastic rod element which cannot sustain negative stress (compression), e.g., a cable element [243].

(iv) Let

$$w(\varepsilon) = \begin{cases} \frac{1}{2}a\varepsilon^2 & \text{if } \varepsilon \le \varepsilon_0 \\ \frac{1}{2}b(\varepsilon - \varepsilon_0)^2 + (\varepsilon - \varepsilon_0)(a\varepsilon_0 + c) + \frac{1}{2}a\varepsilon_0^2 & \text{if } \varepsilon_0 \le \varepsilon \le \varepsilon_0', \\ \infty & \text{if } \varepsilon > \varepsilon_0' \end{cases} \tag{3.3.67}$$

with a, b, c, constants > 0. Then

$$\sigma \in \partial w(\varepsilon) = \begin{cases} a\varepsilon & \text{if } \varepsilon < \varepsilon_0 \\ [a\varepsilon_0, a\varepsilon_0 + c] & \text{if } \varepsilon = \varepsilon_0 \\ b(\varepsilon - \varepsilon_0) + a\varepsilon_0 + c & \text{if } \varepsilon_0 < \varepsilon < \varepsilon_0'. \\ [b(\varepsilon_0' - \varepsilon_0) + a\varepsilon_0 + c, \infty) & \text{if } \varepsilon = \varepsilon_0' \\ \varnothing & \text{if } \varepsilon > \varepsilon_0' \end{cases} \tag{3.3.68}$$

The material law (3.3.68) describes the behavior of a rod element which for $\varepsilon < \varepsilon_0$ is linear elastic, for $\varepsilon = \varepsilon_0$ presents a finite jump in the stresses, between ε_0 and ε_0' is again linear elastic, and for $\varepsilon = \varepsilon_0'$ presents an infinite jump called a locking effect [272]. Of particular interest here is the similarity of the jump effect in Fig. 3.3.12b to the Masson–Savart effect [21] indicated in Fig. 3.3.12c.

(v) Elastic ideally locking materials: let us consider a functional w defined on $\mathbb{R}^6$ by

$$w(\varepsilon) = w_0(\varepsilon) + I_K(\varepsilon), \qquad \varepsilon = \{\varepsilon_{ij}\}, \tag{3.3.69}$$

where w_0 is a continuously differentiable convex functional and I_K is the indicator of the convex closed subset of $\mathbb{R}^6$

$$K = \{\varepsilon \mid Q(\varepsilon) \le 0\}. \tag{3.3.70}$$

Q is a convex, continuously differentiable functional on $\mathbb{R}^6$ such that $0 \in K$. From (3.3.57) and (3.3.69) we obtain

$$\sigma_{ij} \in [\partial w(\varepsilon)]_{ij} = \frac{\partial w_0(\varepsilon)}{\partial \varepsilon_{ij}} + \partial I_K(\varepsilon), \tag{3.3.71}$$

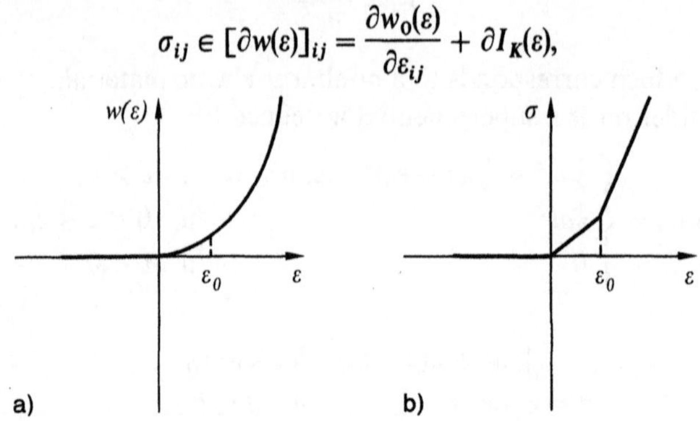

Fig. 3.3.11. A nonlinear elastic cable element.

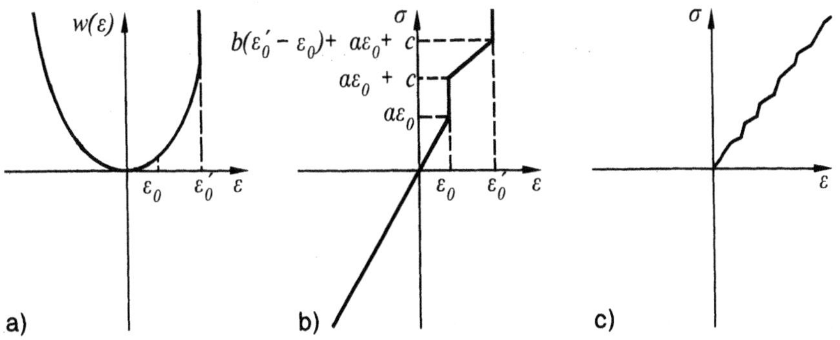

Fig. 3.3.12. Example (iv).

or

$$\sigma_{ij} = \frac{\partial w_0(\varepsilon)}{\partial \varepsilon_{ij}} + \bar{\sigma}_{ij},$$

$$\bar{\sigma}_{ij}(\varepsilon^*_{ij} - \varepsilon_{ij}) \leq 0 \quad \text{for} \quad \varepsilon \in K \quad \text{and} \quad \forall \varepsilon^* \in K. \tag{3.3.72}$$

But $\bar{\sigma} = \{\bar{\sigma}_{ij}\}$ is an element of the outward normal cone to K at the point ε, and thus we may write (3.3.72) in the equivalent form

$$\sigma_{ij} = \frac{\partial w_0(\varepsilon)}{\partial \varepsilon_{ij}} + \mu \frac{\partial Q(\varepsilon)}{\partial \varepsilon_{ij}}, \quad \mu \geq 0, \quad \mu Q(\varepsilon) = 0, \quad Q(\varepsilon) \leq 0. \tag{3.3.73}$$

Accordingly, if $Q(\varepsilon) < 0$, then $\mu = 0$, and the material behaves like a nonlinear elastic material. If $Q(\varepsilon) = 0$ then no finite or infinite increment of the stresses can cause an increase in the value of the function $Q(\varepsilon)$. This is called an ideal-locking effect [184] [272]. Two possible forms for $Q(\varepsilon)$ are [273]

$$Q(\varepsilon) = \tfrac{1}{2}\varepsilon^D_{ij}\varepsilon^D_{ij} - \kappa^2, \tag{3.3.74}$$

where $\varepsilon^D = \{\varepsilon^D_{ij}\}$ is the strain deviator, and

$$Q(\varepsilon) = (\varepsilon_{ii} - \kappa^2). \tag{3.3.75}$$

In both cases, κ is a positive material constant. Materials obeying the locking criterion defined by (3.3.75) are called materials of limited compressibility. The behavior of rubber as well as of some other types of plastic materials can be described fairly well by the stress-strain law of locking materials (cf., Fig. 3.3.13) and [127]). If w_0 is given by (3.3.60) where $C_{ijhk} \in L^\infty(\Omega)$, $i, j, h, k = 1, 2, 3$ and satisfies (3.3.61a,b), then it may easily be verified that w is a strictly convex, continuous function of ε satisfying the relation

$$w(\varepsilon) \geq c\varepsilon_{ij}\varepsilon_{ij}, \qquad \forall \varepsilon \in \mathbb{R}^6, \qquad c \text{ const} > 0. \tag{3.3.76}$$

This inequality will be used in Chapter 6 to prove the existence of a solution of a B.V.P. for elastic ideally locking materials.

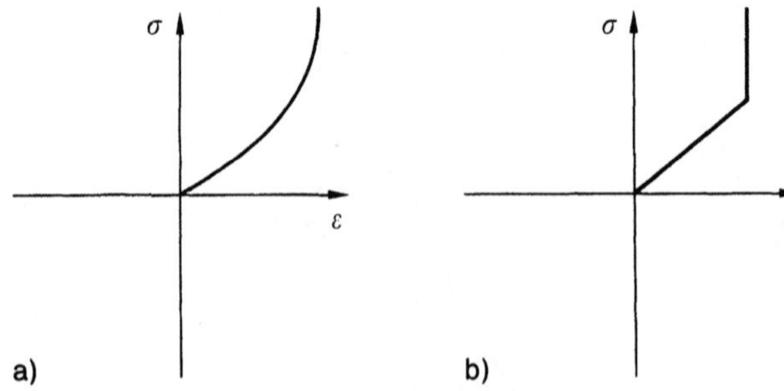

Fig. 3.3.13. A stress-strain law for a one-dimensional rubber element: (a) real (schematic) and (b) idealized law.

(vi) Polygonal stress-strain laws: in the one-dimensional examples of Figs. 3.3.14 and 3.3.15 the material is characterized by more than one modulus of elasticity. Such laws are usually simplifications of more complex stress-strain laws (dotted lines in Fig. 3.3.14b, c). The stress-strain laws of Fig. 3.3.14a, b may be considered as elastic analoga of an elastic hardening and a rigid hardening material respectively. The law of Fig. 3.3.14c is a simplification of the behavior of soil in uniaxial compression. The bimodulus elasticity law of Fig. 3.3.14d arises in fiber-reinforced composites which present different behavior in compression and in tension because of the buckling of the fibers [219]. The diagrams of Fig. 3.3.15a, b represent the behavior of a nondilatant-joint element in rock mechanics [114]. This element simulates the behavior of the interfaces in a rock mass and is termed nondilatant if the simplifying assumption of uncoupling between shearing $(\varepsilon_T, \sigma_T)$ and compression $(\varepsilon_N, \sigma_N)$ can be made. We shall present here a generalization of the law of Fig. 3.3.14a for three-dimensional continua. If $K = \{\varepsilon \mid \varepsilon \leq \varepsilon_0\}$ and $P_K(\varepsilon) = \{\varepsilon \text{ if } \varepsilon \in K, \varepsilon_0 \text{ if } \varepsilon \notin K\}$ denotes the projection operator of $\mathbb{R}^6$ onto the convex set K, then the diagram of Fig. 3.3.14a can be written in the form

$$\sigma = \varepsilon tg\phi_1 + (I - P_K)(\varepsilon)(tg\phi_2 - tg\phi_1), \tag{3.3.77}$$

where I denotes the identity operator. In the three-dimensional case, K is a

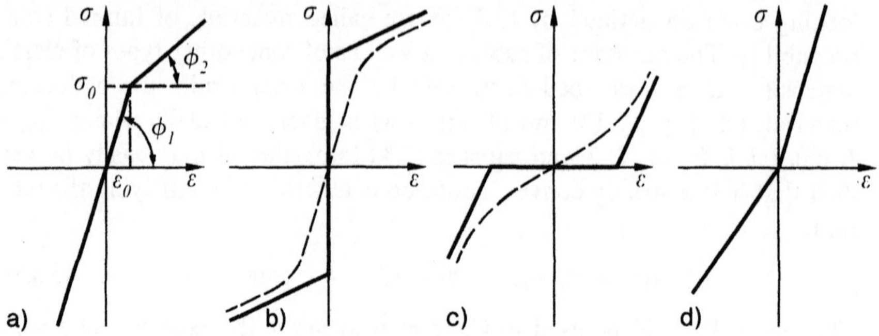

Fig. 3.3.14. One-dimensional polygonal laws.

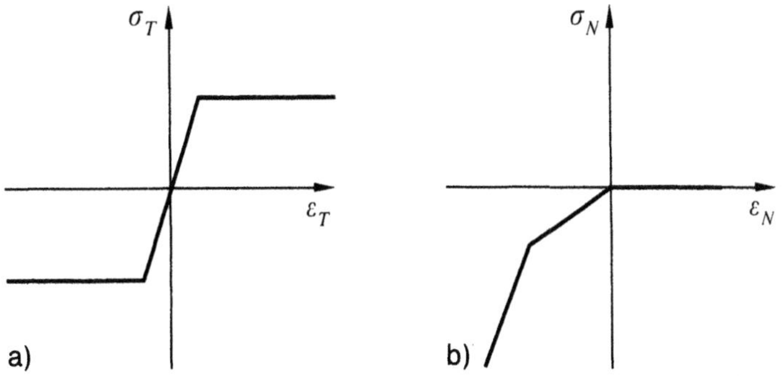

Fig. 3.3.15. Stress-strain law of a nondilatant joint element in rock mechanics.

convex, closed subset of the space $\mathbb{R}^6$ of the strain tensors ε such that $0 \in K$ and P_K is the projection operator onto K. Then the superpotential

$$w(\varepsilon) = \tfrac{1}{2}C_{ijhk}\,\varepsilon_{ij}\,\varepsilon_{hk} + \frac{1}{2\mu}(\varepsilon_{ij} - [P_K(\varepsilon)]_{ij})(\varepsilon_{ij} - [P_K(\varepsilon)]_{ij}), \qquad \mu > 0 \quad (3.3.78)$$

is considered and from this the law

$$\sigma_{ij} = C_{ijhk}\,\varepsilon_{hk} + \frac{1}{\mu}[(I - P_K)\varepsilon]_{ij} \qquad (3.3.79)$$

is obtained. Similarly, we may write the bimodulus elasticity law for a three-dimensional body. If $\mu \to 0$, (3.3.71) is obtained formally with w_0 expressed by (3.3.60). Note that (3.3.77) may also be written in the form

$$\varepsilon = \frac{1}{tg\phi_1}\sigma + (I - P_{\bar{K}})\sigma\left(\frac{1}{tg\phi_2} - \frac{1}{tg\phi_1}\right), \qquad (3.3.80)$$

where $\bar{K} = \{\sigma \mid \sigma \leq \sigma_0\}$. This leads to a complementary energy density of the general form

$$w^c(\sigma) = \tfrac{1}{2}c_{ijhk}\,\sigma_{ij}\,\sigma_{hk} + \frac{1}{2\bar{\mu}}(\sigma_{ij} - [P_{\bar{K}}(\sigma)]_{ij})(\sigma_{ij} - [P_{\bar{K}}(\sigma)]_{ij}), \qquad \bar{\mu} > 0, \quad (3.3.81)$$

where $c = \{c_{ijhk}\}$ is the inverse Hooke's tensor, which also satisfies symmetry and ellipticity conditions similar to those in (3.3.61). $\bar{K}$ is a closed convex set of the space $\mathbb{R}^6$ of the stress tensor σ, and $\bar{\mu} > 0$ is a constant. $\bar{K}$ and $\bar{\mu}$ must be compatible with K and μ. If $\bar{\mu} \to 0$, we find formally the material law

$$\varepsilon \in \partial w^c(\sigma) = \partial(\tfrac{1}{2}c_{ijhk}\,\sigma_{ij}\,\sigma_{hk} + I_{\bar{K}}(\sigma)), \qquad (3.3.82)$$

which generalizes for a three-dimensional continuum the material law depicted in Fig. 3.3.15a. Indeed, if $\bar{K} = \{\sigma \mid F(\sigma) \leq 0\}$, where $F\colon \mathbb{R}^6 \to \mathbb{R}$ is continuously differentiable, then (3.3.82) becomes

$$\varepsilon_{ij} = c_{ijhk}\sigma_{hk} + \lambda\frac{\partial F}{\partial\sigma_{ij}}, \quad \lambda \geq 0, \quad F(\sigma) \leq 0, \quad \lambda F(\sigma) = 0. \tag{3.3.83}$$

This material is the elastic analogon of an elastic, perfectly plastic material. It is also called a "holonomic" elastic perfectly plastic material, since (3.3.83) does not include the stress and strain rates (cf., Sec. 3.3.3). If in (3.3.83) $c = \{c_{ijhk}\} = 0$, we obtain the stress-strain law of a Hencky-material [275]. The three-dimensional expressions of the polygonal laws of Figs. 3.3.14b, c are obtained from (3.3.81) and (3.3.78) respectively by disregarding the elastic parts, i.e.,

$$\varepsilon_{ij} = \frac{1}{\mu}(\sigma_{ij} - [P_K(\sigma)]_{ij}) \quad \text{and} \quad \sigma_{ij} = \frac{1}{\mu}(\varepsilon_{ij} - [P_K(\varepsilon)]_{ij}). \tag{3.3.84}$$

Finally, three-dimensional generalizations of polygonal laws with more than one vertex are obtained along the lines of (3.3.78) by considering more than one convex sets and the corresponding projections. Other methods of describing polygonal stress-strain laws can be found in [186] [219]. With respect to (3.3.78) we may also verify that the application $\varepsilon \to |\varepsilon - P_K(\varepsilon)|^2$ (here $|\cdot|$ denotes the $\mathbb{R}^6$-norm) satisfies the inequality

$$0 \leq |\varepsilon - P_K(\varepsilon)|^2 - |\varepsilon^0 - P_K(\varepsilon^0)|^2 - 2(\varepsilon^0 - P_K(\varepsilon^0), \varepsilon - \varepsilon^0) \leq |\varepsilon - \varepsilon^0|^2. \tag{3.3.85}$$

Thus w is continuously differentiable, strictly convex and satisfies (3.3.76).

(vii) Deformation theory of plasticity [333] [156]: assume a material law of the form

$$\sigma_{ii} = 3\mathcal{K}\varepsilon_{ii}(u), \qquad \mathcal{K} = \frac{E}{3(1-2\nu)} \tag{3.3.86}$$

$$\sigma_{ij}^D(u) = \alpha(\Gamma)\varepsilon_{ij}^D(u), \qquad g^2 = \sigma_{ij}^D\sigma_{ij}^D, \qquad \Gamma^2 = \varepsilon_{ij}^D\varepsilon_{ij}^D. \tag{3.3.87}$$

Here $\mathcal{K}$ denotes the bulk modulus of the material, g is a single-valued function of Γ such that

$$\frac{g(\Gamma)}{\Gamma} = \alpha(\Gamma) \geq c_1, \quad \frac{dg(\Gamma)}{d\Gamma} = \beta(\Gamma) \geq c_2, \quad c_1, c_2 \text{ const} > 0 \tag{3.3.88}$$

and α is a continuously differentiable function of Γ. This law can be written in the form (3.3.57) with

$$w(\varepsilon) = \frac{E}{6(1-2\nu)}\varepsilon_{ii}^2 + \int_0^\Gamma g(\Gamma)\, d\Gamma. \tag{3.3.89}$$

It results that w is strictly convex and continuously differentiable and that it satisfies (3.3.76).

3.3.3 Subdifferential Constitutive Laws II

In this section another class of material laws having the general form

$$\sigma \in \partial w(D) \tag{3.3.90}$$

is to be studied. It is assumed to hold pointwise with respect to a deformation process. Here w is a superpotential ($w \in \Gamma_0(\mathbb{R}^6)$), $\sigma = \sigma(x, t)$ is the Cauchy stress tensor and $D = D(x, t)$ is the rate of deformation tensor at the point x.

In the Lagrangian description of motion, we may define subdifferential relations between the time rate of the Green strain tensor E and the Piola–Kirchhoff stress tensor $\boldsymbol{\Sigma}$ (cf., App. I). The time rate of the Green strain tensor in a Cartesian coordinate system has the form

$$\dot{\mathsf{E}}_{ij} = \frac{d\mathsf{E}_{ij}}{dt} = \tfrac{1}{2}(v_{i,j} + v_{j,i} + v_{k,i}u_{k,j} + v_{k,j}u_{k,i}), \tag{3.3.91}$$

where

$$v_{i,j} = \partial v_i/\partial X_j \quad \text{and} \quad v_i = \partial x_i(X, t)/\partial t = v_i(X, t).$$

In the case of small displacements and small displacement gradients, the Eulerian description coincides with the Lagrangian, and (3.3.91) implies that (cf., App. I)

$$\dot{\mathsf{E}}_{ij} = \dot{\varepsilon}_{ij} = \frac{\partial \varepsilon_{ij}}{\partial t} = D_{ij}. \tag{3.3.92}$$

In this case, (3.3.90) acquires the (nonholonomic) form

$$\sigma \in \partial w(\dot{\varepsilon}). \tag{3.3.93}$$

The inverse laws to (3.3.90, 93) are expressed by means of the conjugate superpotential w^c as

$$D \in \partial w^c(\sigma) \quad \text{and} \quad \dot{\varepsilon} \in \partial w^c(\sigma). \tag{3.3.94}$$

If w also depends on x and t, (3.3.90, 93) are written as

$$\sigma \in \partial w(D, x, t) \quad \text{and} \quad \sigma \in \partial w(\dot{\varepsilon}, x, t). \tag{3.3.95}$$

Next, some examples of the material laws (3.3.90, 93) are given.

(i) Perfectly plastic materials: we define a convex, closed subset of $\mathbb{R}^6$, $K = \{\sigma \mid F(\sigma) \le 0\}$ such that $0 \in K$, where F is a continuously differentiable function called the yield function of the material. Let us consider the law

$$D \in \partial I_K(\sigma), \tag{3.3.96}$$

which can also be written in the form (cf., (3.3.83))

$$D_{ij} = \lambda \frac{\partial F}{\partial \sigma_{ij}}, \quad \lambda \ge 0, \quad F(\sigma) \le 0, \quad \lambda F(\sigma) = 0. \tag{3.3.97}$$

We thus obtain the constitutive relations for a rigid perfectly plastic material. In the case of small deformations, D is replaced in (3.3.97) by the plastic strain-rate tensor $\dot{\varepsilon}^P$ (see Fig. 3.3.16).

(ii) In order to obtain the Prandtl–Reuss relations for an elastic perfectly plastic material it is sufficient to consider a material law of the form

$$\dot{\varepsilon} = \dot{\varepsilon}^E + \dot{\varepsilon}^P \in c\dot{\sigma} + \partial I_K(\sigma) \tag{3.3.98}$$

where $c\dot{\sigma} = \{c_{ijhk}\dot{\sigma}_{hk}\}$. Thus

$$\begin{gathered}\dot{\varepsilon}_{ij} = \dot{\varepsilon}_{ij}^E + \dot{\varepsilon}_{ij}^P = c_{ijhk}\dot{\sigma}_{hk} + \lambda\frac{\partial F}{\partial \sigma_{ij}},\\ \lambda \geq 0, \quad F(\sigma) \leq 0, \quad \lambda F(\sigma) = 0.\end{gathered} \tag{3.3.99}$$

Accordingly, $\dot{\varepsilon}^P$ considered as a vector of $\mathbb{R}^6$ is an element of the outward normal cone to K at the point σ (Fig. 3.3.17). If the boundary of K has a corner, e.g., if $K = \{\sigma \mid F_k(\sigma) \leq 0,\ k = 1, 2, \ldots, m\}$, then at the corner

$$\dot{\varepsilon}_{ij} = \dot{\varepsilon}_{ij}^E + \sum_{k=1}^{m} \lambda_k \frac{\partial F_k}{\partial \sigma_{ij}}, \quad \lambda_k \geq 0, \quad F_k(\sigma) \leq 0, \quad \lambda_k F_k(\sigma) = 0, \quad k = 1, \ldots, m. \tag{3.3.100}$$

Obviously the plastic strain-rate tensor fulfills for $\sigma \in K$ the inequality

$$\dot{\varepsilon}_{ij}^P(\sigma_{ij}^0 - \sigma_{ij}) \leq 0, \qquad \forall \sigma^0 \in K, \tag{3.3.101}$$

which is well-known as Drucker's stability postulate[8] [163] or Hill's principle of maximum plastic work [141] [142] [278] (cf., also Sec. 4.3). Concerning the possible expressions of the yield criterion $F(\sigma) \leq 0$, we refer the reader to the various treatises on plasticity. Generally, the yield criterion

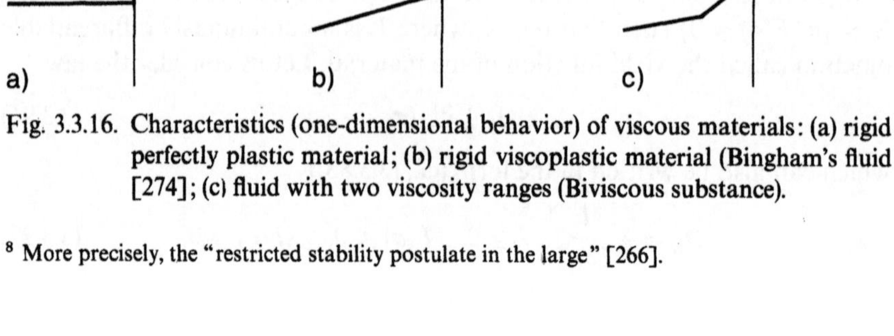

Fig. 3.3.16. Characteristics (one-dimensional behavior) of viscous materials: (a) rigid perfectly plastic material; (b) rigid viscoplastic material (Bingham's fluid [274]; (c) fluid with two viscosity ranges (Biviscous substance).

[8] More precisely, the "restricted stability postulate in the large" [266].

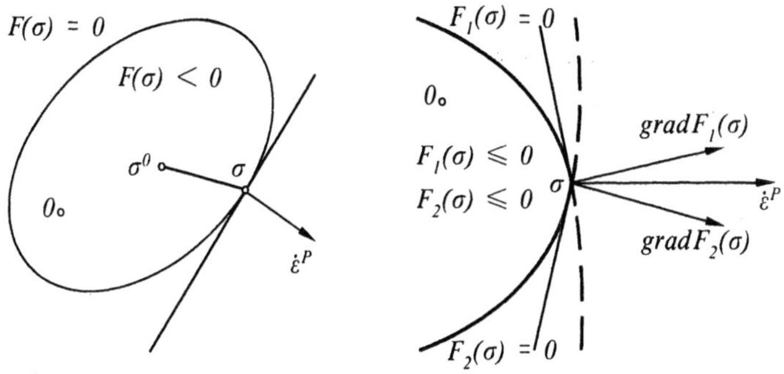

Fig. 3.3.17. Geometric interpretation of $\dot{\varepsilon}^P$.

may depend on the time and space coordinates and thus we write $K = K(x, t)$ in (3.3.97). Until now, only the case of loading has been considered; in order to include the case of unloading, we complete (3.3.99) by writing

$$\lambda = 0 \text{ if } F(\sigma) < 0, \quad \text{or if} \quad F(\sigma) = 0 \quad \text{and} \quad \dot{F}(\sigma) < 0, \tag{3.3.102}$$

and

$$\lambda \geq 0 \text{ if } F(\sigma) = 0 \quad \text{and} \quad \dot{F}(\sigma) = 0. \tag{3.3.103}$$

Accordingly, we may write the material law in the form (3.3.98), where

$$\text{for } \sigma \text{ such that} \quad F(\sigma) \leq 0, \quad \dot{\varepsilon}_{ij}^P(\sigma_{ij}^0 - \sigma_{ij}) \leq 0 \qquad \forall \sigma^0 \in K \tag{3.3.104}$$

and

$$\text{for } \sigma \text{ such that} \quad F(\sigma) = 0, \qquad \dot{\varepsilon}_{ij}^P \dot{\sigma}_{ij} = 0. \tag{3.3.105}$$

The yield surface may change shape during the deformation process (hardening effects). As a result, we may assume generally that the yield criterion is $F(\sigma, \hbar(\dot{\varepsilon}^P), A) \leq 0$, where $\hbar$ is a tensor and A a scalar.

(iii) Rigid viscoplastic materials: keeping in mind the method followed for the study of the polygonal stress-strain laws in the previous section, and with the notation used there [cf., (3.3.81-84)], we may consider the superpotential

$$J_\mu(\sigma) = \frac{1}{4\mu} |\sigma - P_K(\sigma)|^2, \qquad \mu > 0, \tag{3.3.106}$$

where $K = \{\sigma \,|\, F(\sigma) \leq 0\}$ is a convex closed subset of $\mathbb{R}^6$ such that $0 \in K$ and such that F is continuously differentiable. The constant μ is called the viscosity coefficient. The material law

$$D \in \partial J_\mu(\sigma) \tag{3.3.107}$$

is the three-dimensional generalization of the law of Fig. 3.3.16b; it describes a rigid-viscoplastic material, called also a Bingham fluid, and it reads

$$D = \frac{1}{2\mu}(\sigma - P_K(\sigma)). \tag{3.3.108}$$

Accordingly, the material behaves rigidly if $F(\sigma) < 0$ and viscously if $F(\sigma) \geq 0$. Assuming that $F(\sigma)$ has the form proposed by von Mises, i.e.,

$$F(\sigma) = \tfrac{1}{2}\sigma_{ij}^D\sigma_{ij}^D - g^2 = \sigma_{\mathrm{II}} - g^2 \leq 0, \tag{3.3.109}$$

we obtain[9]

$$\begin{aligned} D_{ij} &= 0 \quad \text{if} \quad \sigma_{\mathrm{II}}^{1/2} < g \\ D_{ij} &= \frac{1}{2\mu}\left(1 - \frac{g}{\sigma_{\mathrm{II}}^{1/2}}\right)\sigma_{ij}^D \quad \text{if} \quad \sigma_{\mathrm{II}}^{1/2} \geq g. \end{aligned} \tag{3.3.110}$$

Relation (3.3.107) can be written in the inverse form

$$\sigma \in \partial J_\mu^c(D), \tag{3.3.111}$$

from which we obtain the σ-D relation up to the factor $-p\delta_{ij}$, i.e.,

$$\begin{aligned} \sigma_{ij} &= -p\delta_{ij} + g\frac{D_{ij}}{D_{\mathrm{II}}^{1/2}} + 2\mu D_{ij} \quad \text{if} \quad D_{\mathrm{II}} \neq 0 \\ \sigma_{\mathrm{II}}^{1/2} &\leq g \quad \text{if} \quad D_{\mathrm{II}} = 0. \end{aligned} \tag{3.3.112}$$

Here $D_{\mathrm{II}} = \tfrac{1}{2}D_{ij}D_{ij}$ and p is a scalar which can be interpreted as the pressure of the fluid.

There are several materials whose behavior complies well with the above law, as for instance liquid concrete, certain greases, etc. If thermomechanical coupling is assumed, the coefficient μ is a function of temperature. For further information on Bingham's fluids, refer to [224] [274].

(iv) Elastic viscoplastic materials: we consider the material law

$$\dot{\varepsilon} = \dot{\varepsilon}^E + \dot{\varepsilon}^{VP} \in c\dot{\sigma} + \partial J_\mu(\sigma), \tag{3.3.113}$$

where J_μ is given by (3.3.106). This law describes an elastic viscoplastic body.

(v) Elastic viscous plastic materials [59]: this material results if the total strain-rate tensor is the sum of an elastic part $\dot{\varepsilon}^E$, a viscous part $\dot{\varepsilon}^V$ and an ideally plastic part $\dot{\varepsilon}^P$. We can write the constitutive equations in the form

$$\dot{\varepsilon} = \dot{\varepsilon}^E + \dot{\varepsilon}^V + \dot{\varepsilon}^P \in c\dot{\sigma} + \frac{\sigma^D}{2\mu} + \partial I_K(\sigma). \tag{3.3.114}$$

(vi) Elastic viscous viscoplastic materials [106]: analogously to (3.3.113), we have

$$\dot{\varepsilon} = \dot{\varepsilon}^E + \dot{\varepsilon}^V + \dot{\varepsilon}^{VP} \in c\dot{\sigma} + \frac{\sigma^D}{2\mu_1} + \partial J\mu_2(\sigma), \tag{3.3.115}$$

generally with $\mu_1 \neq \mu_2$.

(vii) Elastic plastic viscoplastic material[10] [59]: for this material the total strain

[9] This material is clearly incompressible, since div $v = D_{kk} = 0$.

[10] By means of subdifferentials the constitutive relations for viscoelastic plastic materials [59] can be generalized, as can the Norton-Hoff law for steady-state creep problems [105], [107].

rate is written in the form

$$\dot{\varepsilon} = \dot{\varepsilon}^E + \dot{\varepsilon}^P + \dot{\varepsilon}^{VP} \in c\dot{\sigma} + \partial I_K(\sigma) + \partial J_\mu(\sigma) \qquad (3.3.116)$$

For the loading and unloading processes of these last three material classes, see [59]. In Sec. 4.3 another important class of materials which obey a subdifferential material law, the standard generalized materials, will be studied.

In the framework of an incremental problem we may consider constitutive laws of the form

$$\dot{\sigma} \in \partial w(\dot{\varepsilon}) \quad \text{or} \quad \dot{\varepsilon} \in \partial w^c(\dot{\sigma}). \qquad (3.3.117)$$

We shall give an example of such a material law in the incremental theory of plasticity.

(viii) Assume that the continuum is discretized by means of appropriately defined finite elements. The stress and strain rates of the i–th element are denoted as $\dot{\mathbf{s}}_i$ and $\dot{\mathbf{e}}_i$. Further, an incremental elastoplastic stress-strain relation is considered [190] [191]. It is of the linear type proposed and discussed by Maier in [188] (no summation with respect to i):

$$\dot{\mathbf{e}}_i = \dot{\mathbf{e}}_i^E + \dot{\mathbf{e}}_i^P = \mathbf{F}_{0i}\dot{\mathbf{s}}_i + \dot{\mathbf{e}}_i^P, \qquad \dot{\mathbf{e}}_i^P = \mathbf{V}_i\dot{\boldsymbol{\lambda}}_i \qquad (3.3.118)$$

$$\dot{\mathbf{F}}_i = \mathbf{N}_i^T\dot{\mathbf{s}}_i - \mathbf{H}_i\dot{\boldsymbol{\lambda}}_i, \qquad \dot{\boldsymbol{\lambda}}_i \geq \mathbf{0}, \qquad \dot{\mathbf{F}}_i \leq \mathbf{0}, \qquad \dot{\boldsymbol{\lambda}}_i^T\dot{\mathbf{F}}_i = \mathbf{0}. \quad (3.3.119)$$

Here $\mathbf{F}_{0i}$ is the flexibility matrix of the i-th element, and $\mathbf{N}_i$ is the matrix formed by the gradients $\mathbf{N}_i^j = (\partial F_i^j/\partial \mathbf{s}_i^{(\alpha)})|_{\mathbf{s}_i^{(\alpha)}}$, $j = 1, \dots, m$ of the m yield functions F_i^j, which are zero at the stress state $\mathbf{s}_i^{(\alpha)}$. Matrix $\mathbf{V}_i$ is formed, similarly to $\mathbf{N}_i$, by the gradients $\mathbf{V}_i^j = (\partial R_i^j/\partial \mathbf{s}_i^{(\alpha)})|_{\mathbf{s}_i^{(\alpha)}}$ of the plastic potentials. Vector $\dot{\mathbf{F}}_i$ is formed by the yield function rates $\dot{F}_i^1, \dot{F}_i^2, \dots, \dot{F}_i^m$ and vector $\dot{\boldsymbol{\lambda}}_i$ by the m plastic multiplier rates $\dot{\lambda}_i^1, \dot{\lambda}_i^2, \dots., \dot{\lambda}_i^m$. $\mathbf{H}_i$ is the $m \times m$ interaction work-hardening matrix, and index T denotes the transpose of a vector or a matrix. The present theory covers the case of nonassociated flow laws (lack of normality, i.e., $\mathbf{V}_i \neq \mathbf{N}_i$), the case of work-hardening according to Koiter's hypothesis ($\mathbf{H}_i = \text{diag } (H_i^1, \dots, H_i^m) H_i^j > 0, \ j = 1, \dots m$) or Mandel's hypothesis ($\mathbf{H}_i$ symmetric, positive semidefinite), and the case of work-softening. The relations (3.3.118, 119) can be written by means of subdifferentials in the form

$$\dot{\mathbf{e}}_i^P \in -\mathbf{V}_i\{\partial I_K(\mathbf{H}_i\dot{\boldsymbol{\lambda}}_i - \mathbf{N}_i^T\dot{\mathbf{s}}_i)\}, \qquad K = \{\mathbf{y}_i | \mathbf{y}_i \geq \mathbf{0}\}, \qquad (3.3.120)$$

i.e., $-\dot{\mathbf{e}}_i^P$ is an element, transformed by $\mathbf{V}_i$, of the outward normal cone to K at the point $\mathbf{H}_i\dot{\boldsymbol{\lambda}}_i - \mathbf{N}_i^T\dot{\mathbf{s}}_i$. Thus the material law becomes

$$\dot{\mathbf{e}}_i \in \mathbf{F}_{0i}\dot{\mathbf{s}}_i - \mathbf{V}_i\{\partial I_K(\mathbf{H}_i\dot{\boldsymbol{\lambda}}_i - \mathbf{N}_i^T\dot{\mathbf{s}}_i)\}. \qquad (3.3.121)$$

This law could be generalized by considering additional superpotentials describing viscoplastic effects, as for example

$$\dot{\mathbf{e}}_i \in \mathbf{F}_{0i}\dot{\mathbf{s}}_i - \mathbf{V}_i\{\partial I_K(\mathbf{H}_i\dot{\boldsymbol{\lambda}}_i - \mathbf{N}_i^T\dot{\mathbf{s}}_i)\} + \sum_{\alpha=1}^{\rho} \mathbf{Q}_i^\alpha\{\partial w^\alpha(\dot{\mathbf{s}}_i)\}. \qquad (3.3.122)$$

In continuum mechanics we often encounter more complicated laws. By means of the notion of superpotential they can be modified to include discontinuities of a phenomenological nature arising in technical problems. Thus, for instance, it is well known that the constitutive relation

$$\sigma = \phi_0 I + \phi_1 D + \phi_2 D^2 + \tau, \qquad \tau \in \partial I_K(D), \tag{3.3.123}$$

where K is a convex closed subset of $\mathbb{R}^6$ and the ϕ_i's, $i = 0, 1, 2$, are functions of both the density and the principal invariants of $\dot{D}$[20], describes a Reiner–Rivlin fluid, when τ is omitted. With τ, (3.3.123) describes the same fluid as before but now subjected to a locking effect. This heuristic extension of the well-founded constitutive law of a Rivlin–Ericksen fluid poses the problem of studying multivalued laws by using the methods of rational mechanics.

3.3.4 Extension of Subdifferential Relations to Function Spaces

In the previous three sections, subdifferential boundary conditions and constitutive laws of a pointwise nature have been introduced. The superpotentials define a relation, at every point of the body Ω or of its boundary Γ, between the value of σ and ε, or of S and u, etc. Here we shall give extensions of these subdifferential relations to function spaces. This is necessary for the formulation and the study of any B.V.P. connected with subdifferential boundary conditions and material laws. From the standpoint of mechanics this is necessary for the derivation and the study of the corresponding variational principles. Roughly speaking, we are concerned in the present section with the following important question: if $w(\varepsilon(x)) \in \Gamma_0(\mathbb{R}^6)$, what will the properties of $\int_\Omega w(\varepsilon(x))d\Omega$ be? As we shall see, this question is closely connected to the theory of convex intergrands [34] [287] [290].

Assume generally that $(E, \mathcal{B}, \mu)$ is a positive measure space with $\mu(E) < \infty$. If $A: H \to \mathcal{P}(H)$ is a multivalued operator and H is a Hilbert space identified with its dual, then we can define the operator $\bar{A}$ on $L^2(E, H)$ (extension of A to $L^2(E, H)$) by setting

$$f \in \bar{A}(u) \Leftrightarrow f(x) \in A(u(x)) \qquad \mu\text{-a.e. on } E. \tag{3.3.124}$$

Furthermore, it is proved in [29] that if A is maximal monotone, $\bar{A}$ is too.

Proposition 3.3.1. Let $A = \partial\phi$, where ϕ is a convex, l.s.c., proper functional on H. For $u \in L^2(E, H)$, we define the functional

$$\Phi(u) = \begin{cases} \displaystyle\int_E \phi(u(x))\, d\mu & \text{if } \phi(u) \in L^1(E) \\ \infty & \text{otherwise} \end{cases}. \tag{3.3.125}$$

Then Φ is convex, l.s.c. and proper on $L^2(E, H)$, and $\bar{A} = \partial\Phi$, i.e., $\partial\Phi$ is the extension of $\partial\phi$ to $L^2(E, H)$.

Proof. Since ϕ is l.s.c., $\phi(u(\cdot))$ is measurable on E for every u integrable on E. Φ is obviously convex and proper. In order to prove that Φ is l.s.c. on $L^2(E, H)$, it suffices to show that epi Φ is closed, i.e., that the set $\{(u, \lambda) \mid u \in L^2(E, H), \lambda \in \mathbb{R}, \Phi(u) \leq \lambda\}$ is closed. Let $\{u_n\}$ be a sequence on $L^2(E, H) \cap D(\Phi)$ such that $\Phi(u_n) \leq \lambda$ and $u_n \to u \in L^2(E, H)$. Then there exists a subsequence, again denoted by $\{u_n\}$, such that $u_n(x) \to u(x)$ μ-a.e. on E. We shall now show that $\Phi(u) \leq \lambda$. Suppose that $a \in D(\partial\phi)$ and $b \in \partial\phi(a)$, a, $b \in H$. Then

$$g(u_n) = \phi(u_n) - \phi(a) - (b, u_n - a)_H \geq 0 \qquad \mu\text{-a.e. on } E, \tag{3.3.126}$$

and since g is l.s.c.,

$$\liminf g(u_n) \geq g(u) = \phi(u) - \phi(a) - (b, u - a)_H \qquad \mu\text{-a.e. on } E \tag{3.3.127}$$

and

$$g(u) \geq 0, \qquad \forall u \in H. \tag{3.3.128}$$

In order to apply Fatou's Lemma, it must be proved that $g(\tilde{u})$ is measurable, or, equivalently, that the set $\{x \mid x \in E, g(\tilde{u}(x)) > \alpha\}$ is measurable for every $\alpha \in \mathbb{R}$. This is true, since the set $\{\tilde{u}(x) \mid \tilde{u}(x) \in H, g(\tilde{u}(x)) > \alpha\}$ is open. Fatou's Lemma implies that

$$\begin{aligned}\int_E g(u(x))\, d\mu &\leq \int_E \liminf g(u_n(x))\, d\mu \\ &\leq \liminf \int_E g(u_n(x))\, d\mu \end{aligned} \tag{3.3.129}$$

and thus

$$\begin{aligned}\int_E \phi(u(x))\, d\mu &\leq \liminf \int_E \phi(u_n(x))\, d\mu \\ &= \liminf \Phi(u_n) \leq \lambda. \end{aligned} \tag{3.3.130}$$

Accordingly, $\Phi(u) \leq \lambda$. In order to prove that $\bar{A} = \partial\Phi$, we shall show the equivalence of the following two conditions: for u, $f \in L^2(E, H)$,

$$\text{(i)} \quad \Phi(v) - \Phi(u) \geq \int_E (f, v - u)_H\, d\mu \qquad \forall v \in L^2(E, H); \tag{3.3.131}$$

$$\begin{aligned}\text{(ii)} \quad &\phi(v(x)) - \phi(u(x)) \geq (f(x), v(x) - u(x))_H \\ &\mu\text{-a.e.} \quad \text{on} \quad E, \qquad \forall v \in L^2(E, H). \end{aligned} \tag{3.3.132}$$

From (ii), (i) results immediately. Let us now show the converse. Let S be a μ-measurable subset of E. We set in (3.3.131)

$$v = \begin{cases} u(x) & \text{on} \quad E - S \\ w(x) & \text{on} \quad S \end{cases}$$

and there results

$$\int_S (\phi(w(x)) - \phi(u(x)) - (f(x), w(x) - u(x))_H \, d\mu \geq 0, \tag{3.3.133}$$

which implies (ii) since (3.3.133) holds for every S, q.e.d.

For the sake of completeness we also prove the next proposition.

Proposition 3.3.2. Each of the variational inequalities (3.3.131, 132) is equivalent to the following condition:

(iii) there exists $A \subset E$ with $\mu(A) = 0$ such that

$$\phi(v) - \phi(u(x)) \geq (f(x), v - u(x))_H, \qquad \forall x \in E - A, \qquad \forall v \in H. \tag{3.3.134}$$

Proof. To show the equivalence of (ii) to (iii), let us introduce in $[L^2(E, H)]^2$the monotone graphs $\mathcal{G}(A_1)$ and $\mathcal{G}(A_2)$ of the operators A_1 and A_2, defined by the condition that $f \in A_1(u)$ (resp. $f \in A_2(u)$) if and only if f and u satisfy (3.3.132) (resp. 3.3.134). Obviously, $\mathcal{G}(A_2) \subset \mathcal{G}(A_1)$. We shall show that A_2 is maximal monotone, and thus that $\mathcal{G}(A_2) = \mathcal{G}(A_1)$. By Prop. 2.6.2, it suffices to show that the operator $I + A_2$ is surjective,i.e., that for every $f \in L^2(E, H)$ there exists $u \in L^2(E, H)$ such that $f \in (A_2 + I)u$. The variational inequality

$$(u, v - u)_H + \phi(v) - \phi(u) \geq (f, v - u)_H, \qquad \forall v \in H \tag{3.3.135}$$

admits for every $f \in H$ a solution $u \in H$. Indeed, $\Pi(u) = \frac{1}{2}\|u\|_H^2 + \phi(u) - (f, u)_H$ has a minimum on H (cf., Prop. 2.2.1). Moreover, this minimum is unique and the application $f \to u$ maps strongly convergent sequences $\{f_n\}$ into weakly convergent sequences $\{u_n\}$ in H [28]. Thus for every $f \in L^2(E, H)$ there exists u such that

$$(u(x), v - u(x))_H + \phi(v) - \phi(u(x)) \geq (f(x), v - u(x))_H \qquad \mu\text{-a.e. on } E, \qquad \forall v \in H. \tag{3.3.136}$$

It results [28] that u is measurable and $u \in L^2(E, H)$, q.e.d.

If $E = \Omega$ (connected, with the Lebesgue measure) and $H = \mathbb{R}^6$, then Prop. 3.3.1 implies that the functional W defined on $[L^2(\Omega)]^6$ by

$$W(\varepsilon) = \begin{cases} \int_\Omega w(\varepsilon(x)) \, d\Omega & \text{if} \quad w(\varepsilon) \in L^1(\Omega) \\ \infty & \text{otherwise} \end{cases} \tag{3.3.137}$$

is convex, l.s.c. and proper if w has the same properties on $\mathbb{R}^6$. Moreover, the equivalence of (i) to (ii) previously proved, implies that

$$\sigma \in \partial W(\varepsilon), \qquad \varepsilon, \sigma \in [L^2(\Omega)]^6 \tag{3.3.138}$$

if and only if for $\sigma(x), \varepsilon(x) \in \mathbb{R}^6$

$$\sigma(x) \in \partial w(\varepsilon(x)) \qquad \text{a.e. in } \Omega. \tag{3.3.139}$$

Analogously, the relations

$$\varepsilon(x) \in \partial w^c(\sigma(x)) \qquad \text{a.e. in } \Omega \tag{3.3.140}$$

and

$$w^c(\sigma(x)) + w(\varepsilon(x)) = \varepsilon_{ij}(x)\sigma_{ij}(x) \qquad \text{a.e. in } \Omega \tag{3.3.141}$$

hold if and only if

$$\varepsilon \in \partial W^c(\sigma) \qquad \varepsilon, \sigma \in [L^2(\Omega)]^6 \tag{3.3.142}$$

and (cf., Prop. 2.5.5.)

$$W^c(\sigma) + W(\varepsilon) = (\varepsilon, \sigma) \tag{3.3.143}$$

hold respectively, where

$$W^c(\sigma) = \begin{cases} \displaystyle\int_\Omega w^c(\sigma(x))\, d\Omega & \text{if } \; w^c(\sigma) \in L^1(\Omega) \\ \infty & \text{otherwise} \end{cases}. \tag{3.3.144}$$

Functionals W and W^c are the potential and the complementary energies of the internal forces of the body under consideration. The pointwise material laws (3.3.90) and (3.3.93) can likewise be extended to $[L^2(\Omega)]^6$.

In order to extend the subdifferential boundary conditions of Sec. 3.3.1, we apply Prop. 3.3.1 for $E = \Gamma$ (with the usual Lebesgue measure) and $H = \mathbb{R}^3$, if for instance (3.3.8) is considered, or $H = \mathbb{R}$, in the case of (3.3.7) or (3.3.55), etc. Here we shall construct the extension of $-S(x) \in \partial j(u(x))$ to $L^2(\Gamma)$. A method similar to what follows could be used for the other types of subdifferential boundary conditions. The functional[11]

$$\Phi(u) = \begin{cases} \displaystyle\int_\Gamma j(u(x))\, d\Gamma & \text{if } \; j(u) \in L^1(\Gamma) \\ \infty & \text{otherwise} \end{cases} \tag{3.3.145}$$

is convex, l.s.c. and proper on $[L^2(\Gamma)]^3$. For $u, S \in [L^2(\Gamma)]^3$ the relation

$$-S \in \partial\Phi(u) \tag{3.3.146}$$

holds if and only if

$$-S(x) \in \partial j(u(x)) \qquad \text{a.e. on } \Gamma. \tag{3.3.147}$$

(3.3.146, 147) are obviously equivalent to the inverse relations

$$u \in \partial\Phi^c(-S) \tag{3.3.148}$$

and

$$u(x) \in \partial j^c(-S(x)) \qquad \text{a.e. on } \Gamma, \tag{3.3.149}$$

[11] For simplicity's sake we shall often omit x in those expressions which are considered to hold pointwise.

as they are to the relations

$$\Phi(u) + \Phi^c(-S) = -\int_\Gamma u_i(x)S_i(x)\, d\Gamma \tag{3.3.150}$$

and

$$j(u(x)) + j^c(-S(x)) = -u_i(x)S_i(x) \qquad \text{a.e. on } \Gamma, \tag{3.3.151}$$

where

$$\Phi^c(-S) = \begin{cases} \int_\Gamma j^c(-S(x))\, d\Gamma & \text{if } j^c(-S) \in L^1(\Gamma) \\ \infty & \text{otherwise} \end{cases}. \tag{3.3.152}$$

We shall often assume that $u \in [H^1(\Omega)]^3$. Then $u \to \Phi(u|_\Gamma)$ is a convex, l.s.c. and proper functional on $[H^1(\Omega)]^3$ due to the continuity of the trace application. Let us denote by $\bar{\Phi}$ the restriction $\Phi|_{[H^{1/2}(\Gamma)]^3}$ of Φ to $[H^{1/2}(\Gamma)]^3$, assuming that $\bar{\Phi}$ on $H^{1/2}(\Gamma)$ is not identically equal to $+\infty$. Then $\partial\Phi$ and $\partial\bar{\Phi}$ define maximal monotone graphs in

$$[L^2(\Gamma)]^3 \times [L^2(\Gamma)]^3 \quad \text{and} \quad [H^{1/2}(\Gamma)]^3 \times [H^{-1/2}(\Gamma)]^3$$

respectively. In many B.V.P.s we shall encounter the boundary condition written for $u \in [H^1(\Omega)]^3$ and $S \in [H^{-1/2}(\Gamma)]^3$ in the form

$$\Phi(\gamma v) - \Phi(\gamma u) \geq \langle -S, \gamma v - \gamma u \rangle, \quad \forall v \in [H^1(\Omega)]^3. \tag{3.3.153a}$$

Since the trace application $v \to \gamma v$ is surjective from $[H^1(\Omega)]^3$ onto $[H^{1/2}(\Gamma)]^3$, (3.3.153a) implies that

$$\bar{\Phi}(v) - \bar{\Phi}(\gamma u) \geq \langle -S, v - \gamma u \rangle, \qquad \forall v \in [H^{1/2}(\Gamma)]^3, \tag{3.3.153b}$$

which is equivalent to

$$-S \in \partial\bar{\Phi}(\gamma u) \quad \text{or} \quad \gamma u \in \partial\bar{\Phi}^c(-S). \tag{3.3.154}$$

Recall that $\langle \cdot, \cdot \rangle$ denotes the duality pairing between $[H^{1/2}(\Gamma)]^3$ and $[H^{-1/2}(\Gamma)]^3$ which coincides with the integral $\int_\Gamma S_i(v_i - u_i)\, d\Gamma$ if $S \in [L^2(\Gamma)]^3$. Henceforth, the bar of Φ and the γ will be omitted if no ambiguity occurs. The expression (3.3.154) is a weak formulation of the boundary condition (3.3.8) since only if $S \in [L^2(\Gamma)]^3$ is the complete equivalence ensured. Props 3.3.1 and 3.3.2 can be generalized when H is a Banach space X, with the dual X', and when $u \in L^p(E, X)$ and $f \in L^{p'}(E, X')$, where $1 \leq p \leq \infty$ and p' is the conjugate exponent to p [28]. Additional propositions on this subject are proved in [126].

Suppose further that Φ is a proper functional on $[H^{1/2}(\Gamma)]^3$. We say that Φ is decomposable if two proper functionals Φ_N on $H^{1/2}(\Gamma)$ and Φ_T on $H_T(\Gamma)$ (cf., Sec. 1.4.3) can be determined such that

$$\Phi(v) = \Phi_N(v_N) + \Phi_T(v_T), \qquad \forall v \in [H^{1/2}(\Gamma)]^3, \tag{3.3.155}$$

where

$$v = \{v_i\}, \qquad v_i = v_N n_i + v_{T_i}.$$

Proposition 3.3.3. Suppose that Γ is sufficiently regular[12] and that the functional Φ is proper and decomposable on $[H^{1/2}(\Gamma)]^3$. If Φ is convex (resp. l.s.c.), then Φ_N and Φ_T are convex (resp. l.s.c.) as well, and conversely. Moreover,

$$\Phi^c(T) = \Phi_N^c(T_N) + \Phi_T^c(T_T), \qquad \forall T \in [H^{-1/2}(\Gamma)]^3, \tag{3.3.156}$$

where

$$\langle T, v\rangle = \langle T_N, v_N\rangle_{1/2} + \langle T_T, v_T\rangle_{H_T}, \qquad \forall v \in [H^{1/2}(\Gamma)]^3. \tag{3.3.157}$$

Proof. Since Φ is l.s.c., then for any sequence $\{v_n\}$ converging strongly to v in $[H^{1/2}(\Gamma)]^3$ as $n \to \infty$

$$\begin{aligned}\Phi_N(v_N) + \Phi_T(v_T) = \Phi(v) &\le \liminf \Phi(v_n)\\ &= \liminf\, [\Phi_N(v_{nN}) + \Phi_T(v_{nT})].\end{aligned} \tag{3.3.158}$$

By considering the sequence $\{v_n\} = \{v_{nN}\, n + v_T^0\}$, where v_T^0 is fixed such that $\Phi_T(v_T^0) < \infty$, we obtain the lower semicontinuity of Φ_N. We may argue similarly for Φ_T. The converse is true since $[H^{1/2}(\Gamma)]^3$ is isomorphic to $H_T(\Gamma) \times H^{1/2}(\Gamma)$. The proof of convexity is immediate. (3.3.156) results by noting that

$$\begin{aligned}\Phi^c(T) &= \sup_{v \in [H^{1/2}(\Gamma)]^3} [\langle T, v\rangle - \Phi(v)]\\ &= \sup_{v_N \in H^{1/2}(\Gamma)} [\langle T_N, v_N\rangle_{1/2} - \Phi_N(v_N)] + \sup_{v_T \in H_T} [\langle T_T, v_T\rangle_{H_T} - \Phi_T(v_T)]\\ &= \Phi_N^c(T_N) + \Phi_T^c(T_T), \qquad \text{q.e.d.}\end{aligned}$$

Further, we shall study the weak formulations of some boundary conditions.

(i) Suppose that $c_T \in H_T$ and $C_N \in H^{-1/2}(\Gamma)$ are given and let us consider the boundary conditions

$$u_T = c_T \quad \text{and} \quad S_N = C_N. \tag{3.3.159}$$

We define a functional Φ on $[H^{1/2}(\Gamma)]^3$ by (cf., 3.3.155)

$$\Phi_N(v_N) = -\langle C_N, v_N\rangle_{1/2} \quad \text{for} \quad v_N \in H^{1/2}(\Gamma) \tag{3.3.160a}$$

and

$$\Phi_T(v_T) = \begin{cases} 0 & \text{if} \quad v_T = c_T \\ \infty & \text{if} \quad v_T \neq c_T \end{cases} \tag{3.3.160b}$$

[12] It suffices that Γ be $C^{1,1}$-regular [150].

for $v_T \in H_T$. Φ belongs to the class $\Gamma_0([H^{1/2}(\Gamma)]^3)$ and is decomposable. By Prop. (3.3.3), (3.3.156) holds, and since

$$\Phi_N^c(T_N) = \begin{cases} 0 & \text{if} \quad T_N = -C_N \\ \infty & \text{otherwise} \end{cases} \tag{3.3.160c}$$

for $T_N \in H^{-1/2}(\Gamma)$ and

$$\Phi_T^c(T_T) = \langle T_T, c_T \rangle_{H_T} \quad \text{for} \quad T_T \in H'_T, \tag{3.3.160d}$$

it results that

$$\Phi^c(T) = \begin{cases} \langle T_T, c_T \rangle_{H_T} & \text{if} \quad T_N = -C_N \\ \infty & \qquad \text{otherwise} \end{cases} \tag{3.3.160e}$$

By means of Φ and Φ^c, (3.3.159) can be put in the form (3.3.154). If $C_N \in L^2(\Gamma)$, then the conditions (3.3.159) are equivalent to the pointwise boundary conditions

$$u_T(x) = c_T(x) \quad \text{and} \quad S_N(x) = C_N(x) \quad \text{a.e. on } \Gamma. \tag{3.3.161}$$

(ii) Let $C_T \in H'_T$ be given. We consider the Signorini–Fichera boundary condition

$$u_N \leq 0 \quad \text{a.e. on } \Gamma, \quad \langle S_N, u_N \rangle_{1/2} = 0 \quad S_N \leq 0 \quad \text{in} \quad H^{-1/2}(\Gamma) \tag{3.3.162}$$

and

$$S_T = C_T. \tag{3.3.163}$$

It should be noted that the last of (3.3.162) is by definition (cf., (1.3.17)) equivalent to

$$\langle S_N, v_N \rangle_{1/2} \geq 0, \qquad \forall v_N \in H^{1/2}(\Gamma) \qquad v_N \leq 0 \qquad \text{a.e. on } \Gamma. \tag{3.3.164}$$

Further, the functional Φ is defined on $[H^{1/2}(\Gamma)]^3$ by (3.3.155), where for

$v_N \in H^{1/2}(\Gamma)$

$$\Phi_N(v_N) = \begin{cases} 0 & \text{if} \quad v_N \leq 0 \quad \text{a.e. on } \Gamma \\ \infty & \text{if} \quad v_N > 0 \quad \text{on } \Gamma_1 \subset \Gamma \text{ with mes } \Gamma_1 > 0 \end{cases} \tag{3.3.165}$$

and for $v_T \in H_T$

$$\Phi_T(v_T) = -\langle C_T, v_T \rangle_{H_T}. \tag{3.3.166}$$

Then we have

$$\Phi_N^c(T_N) = \begin{cases} 0 & \text{if} \quad T_N \leq 0 \\ \infty & \text{otherwise} \end{cases} \tag{3.3.167}$$

for $T_N \in H^{-1/2}(\Gamma)$ and

$$\Phi_T^c(T_T) = \begin{cases} 0 & \text{if } \; T_T = -C_T \\ \infty & \text{otherwise} \end{cases} \tag{3.3.168}$$

for T_T in H'_T. Then (3.3.156) yields $\Phi^c(T)$. If $S_N \in [L^2(\Gamma)]^3$, (3.3.162) is equivalent to the pointwise condition (3.3.22). The case of unilateral contact with a deformable support can be treated in the same manner.

(iii) Let $S_{T_0} \in L^\infty(\Gamma)$ with essinf $S_{T_0} > 0$. We denote by $\mathfrak{J}$, $\mathfrak{J}'$, $\mathfrak{J}_T$ and $\mathfrak{J}'_T$ the injections

$$\mathfrak{J}: H^{1/2}(\Gamma) \to L^2(\Gamma),\ \mathfrak{J}': L^2(\Gamma) \to H^{-1/2}(\Gamma),$$

$$\mathfrak{J}_T: H_T \to L_T = \{v \in [L^2(\Gamma)]^3, \quad v_i n_i = 0 \quad \text{a.e. on } \Gamma\}$$

$$\text{and} \quad \mathfrak{J}'_T: L_T \to H'_T.$$

Obviously, $v \to \{v_N, v_T\}$ is an isometry from $[L^2(\Gamma)]^3$ onto $L^2(\Gamma) \times L_T$. Let us now consider the following boundary conditions:

$$\text{if} \quad S_T \in H'_T, \quad S_T = \mathfrak{J}'_T \tilde{S}_T, \quad \tilde{S}_T \in L_T \quad \text{and} \quad |\tilde{S}_T| < S_{T_0} \tag{3.3.169}$$

a.e. on Γ, then $u_T = 0$, with $u_T \in H_T$;

$$\text{if} \quad S_T \in H'_T, \quad S_T = \mathfrak{J}'_T \tilde{S}_T, \quad \tilde{S}_T \in L_T \quad \text{and} \quad |\tilde{S}_T| = S_{T_0}$$

a.e. on Γ, then there exists $\lambda \geq 0$ such that

$$u_T = -\lambda \tilde{S}_T, \qquad u_T \in H_T; \tag{3.3.170}$$

$$S_N = C_N \quad \text{in} \quad H^{-1/2}(\Gamma). \tag{3.3.171}$$

It is easily verified that (3.3.169, 170) are equivalent to

$$-\tilde{S}_T \in \partial F_T(\mathfrak{J}_T u_T), \qquad S_T = \mathfrak{J}'_T \tilde{S}_T, \tag{3.3.172}$$

where

$$F_T(v) = \int_\Gamma S_{T_0} |v| \, d\Gamma \quad \text{for} \quad v \in L_T. \tag{3.3.173}$$

Another equivalent formulation of (3.3.172) is the following:

$$\begin{aligned} &S_T = \mathfrak{J}'_T \tilde{S}_T, \ |\tilde{S}_T| \leq S_{T_0} \quad \text{a.e. on } \Gamma, \\ &S_{T_0} |u_T| + \tilde{S}_{T_i} u_{T_i} = 0 \qquad \text{a.e. on } \Gamma. \end{aligned} \tag{3.3.174}$$

The functional F_T^c is defined on L_T by

$$F_T^c(T) = \begin{cases} 0 & \text{if } \; \tilde{T} \in L_T, \ |\tilde{T}| \leq S_{T_0} \quad \text{a.e. on } \Gamma \\ \infty & \text{otherwise} \end{cases} \tag{3.3.175}$$

In order to write (3.3.169, 170) in the form (3.3.154), we set $\Phi_T = F_T \circ \mathfrak{J}_T$;

then

$$\Phi_T^c(T) = \begin{cases} 0 & \text{if } T \in H_T' \text{ with } T = \mathfrak{Z}_T' \tilde{T}, \quad \tilde{T} \in L_T \\ & \text{and } |\tilde{T}_T| \leq S_{T_0} \quad \text{a.e. on } \Gamma, \\ \infty & \text{otherwise} \end{cases} \tag{3.3.176}$$

and hence (3.3.169, 170) acquires the equivalent form (cf., (2.4.13))

$$-S_T \in \partial\Phi_T(u_T) \quad \text{or} \quad u_T \in \partial\Phi_T^c(-S_T) \tag{3.3.177}$$

on $H_T \times H_T'$ for $S_T = \mathfrak{Z}_T' \tilde{S}_T$. In certain cases by means of the special features of a superpotential we can derive useful results. Here by making use of the special structure of Φ_T we find from (3.3.177) that

$$\frac{S_T}{S_{T_0}} \in [L^\infty(\Gamma)]^3 \quad \text{and} \quad \left|\frac{S_T}{S_{T_0}}\right|_\infty \leq 1. \tag{3.3.178}$$

even if $S_T \notin$ range of $\mathfrak{Z}_T'$ [84]. Indeed, from

$$\langle S_T, v_T - u_T \rangle_{H_T} + \int_\Gamma S_{T_0}(|v_T| - |u_T|)\, d\Gamma \geq 0, \quad \forall v_T \in H_T \tag{3.3.179}$$

we obtain for $u_T \in H_T$

$$\langle S_T, v \rangle + \int_\Gamma S_{T_0} |v|\, d\Gamma - \left(\langle S_T, u_T \rangle_{H_T} + \int_\Gamma S_{T_0} |u_T|\, d\Gamma \right) \geq 0$$

$$\forall v \in [H^{1/2}(\Gamma)]^3, \tag{3.3.180}$$

since $|v| \geq |v_T|$ and $\langle S_T, v_T \rangle_{H_T} = \langle S_T, v \rangle$. Replacing v in (3.3.180) by $\pm\lambda v_1$, $\lambda \geq 0$ and setting $\lambda = 0$ and $\lambda \to \infty$, we find that

$$\langle S_T, u_T \rangle_{H_T} + \int_\Gamma S_{T_0} |u_T|\, d\Gamma \leq 0 \tag{3.3.181}$$

and

$$|\langle S_T, v \rangle| \leq \int_\Gamma S_{T_0} |v|\, d\Gamma \quad \forall v \in [H^{1/2}(\Gamma)]^3 \tag{3.3.182}$$

respectively. (3.3.182) implies that the linear form $v \to \langle S_T S_{T_0}^{-1}, S_{T_0} v \rangle$ is continuous on $[H^{1/2}(\Gamma)]^3$ for the topology induced by $[L^1(\Gamma)]^3$ (when the latter space is equipped with the norm $\int_\Gamma S_{T_0} |v|\, d\Gamma$). But $[H^{1/2}(\Gamma)]^3$ is dense in $[L^1(\Gamma)]^3$ and thus (3.3.178) is obtained. Accordingly

$$|S_T| \leq S_{T_0} \quad \text{a.e. on } \Gamma. \tag{3.3.183}$$

(3.3.181) and (3.3.183) imply that

$$S_{T_i} u_{T_i} + S_{T_0} |u_T| = 0 \quad \text{a.e. on } \Gamma, \tag{3.3.184}$$

which, together with (3.3.183), yields the pointwise relations (3.3.27, 28).

If the superpotential law is dependent on x and perhaps on t, as happens for instance in elastoplasticity if $K = K(x, t)$, then Prop. 3.3.1 does not apply. For the global formulation of the pointwise subdifferential material law, elements of the theory of convex integrands [152], [287], [290] are needed. Let us state without proof some results of the application of this theory to the material law

$$\sigma(x) \in \partial w_D(x, D(x)). \tag{3.3.185}$$

We define the functional

$$W(D) = \begin{cases} \displaystyle\int_\Omega w(x, D(x))\, d\Omega & \text{if} \quad w(., D(.)) \in L^1(\Omega) \\ \infty & \text{otherwise} \end{cases}. \tag{3.3.186}$$

If w is a normal convex integrand on $\Omega \times \mathbb{R}^6$ in the sense of Rockafellar and a growth condition holds, then $W \in \Gamma_0([L^2(\Omega)]^6)$ and $\sigma \in \partial W(D)$, where $\sigma \in [L^2(\Omega)]^6$, if and only if $\sigma(x) \in \partial w(x, D(x))$ a.e. on Ω. If W depends on t as well, we may replace Ω by $\Omega \times (0, T)$ in (3.3.186). In this case, analogous results are obtained.

Chapter 4

Variational Inequalities and Multivalued Convex and Nonconvex Problems in Mechanics

In the first section of this chapter two general types of variational inequalities are studied and a general scheme is proposed for the derivation of variational inequality "principles". The second section deals with the study of problems of coexistent phases, using the notion of subdifferentiability. Also minimum problems are derived for Gibbsian states. In the third section we attempt to generalize the notion of superpotential for nonconvex energy functionals through the concepts of the generalized gradient of Clarke and of the derivate container of Warga. Hemivariational inequalities are derived, substationarity properties are proved and certain classes of material laws and boundary conditions leading to such problems are discussed. This chapter closes with a section concerning the study of inequality problems in terms of multivalued integral equations.

4.1 Two General Variational Inequalities and the Derivation of Variational Inequality "Principles" in Mechanics

4.1.1 Variational Inequalities of the Fichera Type

Let us consider for $u, v \in V$ where V is a real Hilbert space, a symmetric, continuous bilinear form $a(u, v)$ and a linear form (f, v). Assume that $f \in V'$, and let H be a pivot Hilbert space such that

$$V \subset H \subset V'$$

holds and the injections are continuous and dense. We denote the duality pairing by $(.,.)$, the norm on H by $|\cdot|$ and on V by $\|\cdot\|$. The following problem is now considered: find $u \in K$ such that

$$a(u, v-u) - (f, v-u) \geq 0, \qquad \forall v \in K \tag{4.1.1}$$

where K is a convex closed subset of V. This variational inequality will be called hereafter a variational inequality of the Fichera type, after the mathematician who has studied it extensively [96], [97], [99].

It is known that there exists a symmetric, bounded linear operator $A: V \to V'$ such that

$$a(u, v) = (Au, v), \qquad \forall u, v \in V. \tag{4.1.2}$$

By Prop. 2.2.2, the solution $u \in K$ of (4.1.1), if any exists, is a solution of the minimization problem

$$\Pi(u) = \min \{\Pi(v) \mid v \in K\}, \tag{4.1.3}$$

where $\Pi(v) = \frac{1}{2}(Av, v) - (f, v)$, and conversely. It is worth noting here that the equivalence of (4.1.3) to (4.1.1) is proved only on condition that $a(u, u) \geq 0$, which ensures the convexity of $a(v, v)$. First, we assume that $a(u, v)$ is coercive on V, i.e.,

$$a(u, u) \geq c \| u \|^2, \qquad \forall u \in V, \qquad c \text{ const} > 0. \tag{4.1.4}$$

Proposition 4.1.1. Suppose that (4.1.4) holds. Then the variational inequality (4.1.1) admits a unique solution.

Proof. The coercivity of $a(u, v)$ guarantees that (2.2.3) holds, q.e.d.

Next we study the semicoercive case. Let

$$\ker A = \{v \mid v \in V, Av = 0\}, \tag{4.1.5}$$

and let Q (resp. $\tilde{Q}$) be the orthogonal projector of V onto $\ker A$ in the topology of H (resp. of V). We set $P = I - Q$ (resp. $\tilde{P} = I - \tilde{Q}$), where I denotes the identity mapping, and we assume that

$$a(v, v) \geq c \,| Pv |^2, \qquad \forall v \in V \qquad c \text{ const} > 0. \tag{4.1.6}$$

Obviously, $\ker A = \{v \mid v \in V \; a(v, v) = 0\}$. We assume that

$$\ker A \text{ is finite–dimensional} \tag{4.1.7}$$

and that

$$c_1 (a(u, u)^{1/2} + | u |) \leq \| u \| \leq c_2 (a(u, u)^{1/2} + | u |) \qquad c_1, c_2 \text{ const} > 0. \tag{4.1.8}$$

Thus, $a(u, u)^{1/2}$ is a seminorm on V and $|||u||| = a(u, u)^{1/2} + |u|$ is a norm on V equivalent to $\|u\|$. Let U be a subset of V which contains nonzero elements and let for $u \in U \; u \neq 0$

$$p(u, U) = \sup_t \left\{ t \middle| t \geq 0 \, \frac{tu}{\| u \|} \in U \right\}. \tag{4.1.9}$$

From (4.1.6) and (4.1.8) we obtain

$$a(v, v) \geq c\|Pv\|^2 \qquad \forall v \in V \qquad c \text{ const} > 0 \tag{4.1.10}$$

and that $\|Pv\|$ is equivalent to $\|\tilde{P}v\|$ for $v \in V$. The kernel of f

$$\ker f = \{v \mid v \in V, (f, v) = 0\},$$

is now considered. Denote by L the intersection of ker A and ker f and let L_1 be a subspace of ker A such that

$$\ker A = L \oplus L_1. \tag{4.1.11}$$

Next, let us denote by $\bar{Q}$ and $\bar{\bar{Q}}$ the orthogonal projectors of V onto L and L_1 respectively in the topology of V, and by $\bar{P}$ the operator $I - \bar{Q}$. Obviously, $\bar{\bar{Q}} = \bar{P} - \tilde{P}$ and $\bar{\bar{Q}}v$ is orthogonal to $\tilde{P}v$. With respect to $u_0 \in K$, we denote by K_{u_0} the set $\{v \mid v \in V, v + u_0 \in K\}$. We prove the following proposition (cf., [99]).

Proposition 4.1.2. Assume that (4.1.6–8) hold and that a $u_0 \in K$ exists such that

(i) $(f, \rho) < 0$ for $\rho \in \ker A \cap K_{u_0}$, (4.1.12)
$p(\bar{Q}\rho, \bar{Q}(\ker A \cap K_{u_0})) = \infty$, and
(ii) the set $\bar{P}(K_{u_0})$ is closed in V. (4.1.13)

Then (4.1.1) has a solution.

Proof. Since (4.1.1) is equivalent to (4.1.3), it suffices to consider the minimization problem. Denote by I the infimum of $\Pi(v)$ on K and let $\{u_n\}$ be a minimizing sequence for $\Pi(v)$ in K, i.e., for $n \to \infty$

$$\Pi(u_n) \to \inf_K \Pi(v) = I. \tag{4.1.14}$$

Obviously, $\{w_n\} = \{u_n - u_0\}$ is a minimizing sequence for the functional $\Pi(v + u_0)$ and for $v \in K_{u_0}$. As in the proof of Prop. 2.2.1, we shall show that the sequence $\{\bar{P}w_n\}$ is bounded in V. Suppose that $\|\bar{P}w_n\| \to \infty$ as $n \to \infty$. Moreover,

$$c\|Pw_n\|^2 \leq a(w_n, w_n) = 2\Pi(u_n) - 2\Pi(u_0) + 2(f, w_n) - 2a(u_0, w_n), \tag{4.1.15}$$

from which, by setting $z_n = w_n/\|\bar{P}w_n\|$, we obtain

$$c\|\bar{P}w_n\|^2\|Pz_n\|^2 \leq c_1 + 2(\|f\|_{V'} + \|Au_0\|_{V'})\|\bar{P}w_n\|\,\|\bar{P}z_n\|. \tag{4.1.16}$$

From (4.1.16), we find by means of (4.1.8) that

$$\lim_{n\to\infty}\|Pz_n\| = \lim_{n\to\infty}\|\tilde{P}z_n\| = 0. \tag{4.1.17}$$

But

$$\|\bar{P}z_n\|^2 = \|\tilde{P}z_n\|^2 + \|\bar{\bar{Q}}z_n\|^2 = 1. \tag{4.1.18}$$

Hence, $\lim_{n\to\infty} \|\bar{\bar{Q}}z_n\|^2 = 1$, and $\|\bar{\bar{Q}}z_n\| \leq c$. Since $\bar{\bar{Q}}z_n \in \ker A$, which is finite-dimensional by assumption, we may extract a subsequence, again denoted by $\{z_n\}$, such that $\{\bar{\bar{Q}}z_n\}$ converges strongly to a vector $\rho \in \bar{\bar{Q}}(\ker A)$ with $\|\rho\| = 1$. We shall show that for $t > 0$

$$t\rho \in \bar{P}(K_{u0}). \tag{4.1.19}$$

Indeed, $w_n \in K_{u0}$ and $\{0\} \in K_{u0}$ imply that for any $t > 0$ and for n sufficiently large, $tz_n \in K_{u0}$. Thus $t\bar{P}z_n \in \bar{P}(K_{u0})$, and so (4.1.13) implies (4.1.19). If $a \in K_{u0}$ such that $\bar{P}a = \rho$, we have $a = \rho + (a - \rho), \rho \in \bar{\bar{Q}}(\ker A)$ and $a - \rho \in \bar{Q}(\ker A)$. Accordingly, $a \in \ker A \cap K_{u0}$ and $(f, \rho) = (f, \rho) + (f, a - \rho) = (f, a) < 0$ because of (4.1.12). From (4.1.8, 15) it results that

$$0 \leq \liminf (\|\bar{P}w_n\| \, \|Pz_n\|^2) \leq \limsup \left[\frac{2}{\|\bar{P}w_n\|}[\Pi(u_n) - \Pi(u_0)] + 2(f, \bar{P}z_n) - 2a(u_0, \tilde{P}z_n)\right] = 2(f, \rho) < 0 \tag{4.1.20}$$

which leads to a contradiction. Thus $\{\bar{P}w_n\}$ is bounded in V. Accordingly, we may extract a subsequence, denoted by $\{w_n\}$, such that $\{\bar{P}w_n\}$ weakly converges to a limit $\bar{P}w$. Since $\bar{P}(K_{u0})$ is closed, it is weakly closed as well, and thus $\bar{P}w \in \bar{P}(K_{u0})$. To complete the proof, we show that for $u = w + u_0 \in K$

$$\Pi(u) = I. \tag{4.1.21}$$

Indeed, $(f, \bar{Q}w) = 0$ and thus

$$\Pi(w_n + u_0) = \Pi(\bar{P}w_n + u_0). \tag{4.1.22}$$

It is well known that for a bounded, nonnegative, symmetric bilinear form $a(u, v)$ on V and for $v_n \to v$ weakly in V, the inequality[1]

$$\liminf_{n\to\infty} a(v_n, v_n) \geq a(v, v) \tag{4.1.23}$$

holds. Thus

$$I = \lim_{n\to\infty} \Pi(\bar{P}w_n + u_0) \geq \Pi(\bar{P}w + u_0) = \Pi(u), \tag{4.1.24}$$

which with (4.1.14) implies (4.1.21), q.e.d.

[1] This relation can be easily verified. Let $v_n \to v$ weakly in V. Then $\liminf_{n\to\infty} a(v_n, v_n) = \liminf_{n\to\infty} \{a(v, v) + 2a(v, v_n - v) + a(v_n - v, v_n - v)\} \geq a(v, v)$ since $a(v, v_n - v) \to 0$ and $a(v_n - v, v_n - v) \geq 0$.

(4.1.1) has a solution if K is bounded, in which case (4.1.6) may be replaced by the condition

$$a(v, v) \geq 0 \qquad \forall v \in V. \tag{4.1.25}$$

The following proposition supplies a necessary condition for the existence of a solution of (4.1.1).

Proposition 4.1.3. Assume that (4.1.1) has a solution. Then for any $u_0 \in K$ and any $\rho \in \ker A \cap K_{u0}$ such that $p(\bar{Q}\rho, \bar{Q}\,(\ker A \cap K_{u_0})) = \infty$, the condition

$$(f, \rho) < 0 \tag{4.1.26}$$

must be satisfied.

Proof. Suppose that u is a solution of (4.1.1). Then we set in (4.1.1) $v = \lambda\rho + u_0$, $\lambda > 0$, where $\lambda\rho$ is such that $\lambda\rho \in \ker A \cap K_{u0}$ for $\lambda > 0$. We thus obtain

$$a(u, u_0 - u) + (f, u - u_0) \geq \lambda(f, \rho) \neq 0, \tag{4.1.27}$$

which implies (4.1.26), q.e.d.

Further, let us consider the quotient space $V/\ker A$.

Proposition 4.1.4. If u is a solution of (4.1.1), then $[u] \in V/\ker A$ is uniquely determined. Every other solution u' of (4.1.1) can be put in the form $u' = u + \rho$ such that

$$(f, \rho) = 0 \quad \text{and} \quad \rho \in \ker A \cap K_u. \tag{4.1.28}$$

Proof. Suppose that $u_1 \in K$ and $u_2 \in K$ are two solutions of (4.1.1). By setting $u = u_1$, $v = u_2$ and $u = u_2$, $v = u_1$ in (4.1.1), we deduce by means of (4.1.10) that

$$c\,\|P(u_1 - u_2)\|^2 \leq 0, \tag{4.1.29}$$

which implies $Pu_1 = Pu_2$, i.e., the uniqueness of Pu or of $[u]$. If u' minimizes Π on K, then $[u] = [u']$, and thus $u' = u + \rho$. But

$$\Pi(u') = \Pi(u) - (f, \rho), \tag{4.1.30}$$

and accordingly (4.1.28) holds. Conversely, if $u' = u + \rho$ and (4.1.28) holds, then u' is another solution of (4.1.1), q.e.d.

A generalization of the present theory for A nonsymmetric can be found in [99]. The foregoing propositions will now be applied to the study of the Signorini–Fichera B.V.P. in the theory of elasticity.

Let us consider a linear elastic body which in its undeformed state occupies an open, bounded, connected subset Ω of $\mathbb{R}^3$. Ω is referred to a fixed Cartesian coordinate system $0x_1x_2x_3$ and its boundary Γ is assumed to be regular (cf., Sec. 1.3.1). Let $\Gamma = \Gamma_U \cup \Gamma_S \cup \Gamma_\varnothing$ where Γ_U and Γ_S are disjoint open sets and mes $\Gamma_\varnothing = 0$. We assume for Γ_S that mes $\Gamma_S > 0$, and for Γ_U we consider the cases mes $\Gamma_U > 0$ and $\Gamma_U = \varnothing$. On Γ_S the Signorini–Fichera boundary condition (3.3.22), combined with the tangential boundary condition $S_T = C_T$, where $C_T \in [L^2(\Gamma_S)]^3$, is assumed to hold. On Γ_U the displacement u is taken to be zero, for the sake of simplicity. For mes $\Gamma_U > 0$ the B.V.P. gives rise to a variational inequality (cf. Sec. 4.1.3 and 6.2.1) of the form (4.1.1) with

$$\begin{aligned} V = V_0 = \{v \mid v \in [H^1(\Omega)]^3, \quad v = 0 \quad \text{a.e. on } \Gamma_U\}, \\ H = [L^2(\Omega)]^3, \qquad K = \{v \mid v \in V_0, \quad v_N \leq 0 \quad \text{a.e. on } \Gamma_S\} \end{aligned} \tag{4.1.31}$$

and

$$a(u, v) = \int_\Omega C_{ijhk}\, \varepsilon_{ij}(u) \varepsilon_{hk}(v)\, d\Omega, \qquad i, j = 1, 2, 3, \tag{4.1.32}$$

where $C = \{C_{ijhk}\}$ is the linear-elasticity tensor with $C_{ijhk} \in L^\infty(\Omega)$, $\varepsilon = \{\varepsilon_{ij}\}$ is the strain tensor related to the displacement vector $u = \{u_i\}$ by

$$\varepsilon_{ij}(u) = \tfrac{1}{2}(u_{i,j} + u_{j,i}), \tag{4.1.33}$$

and

$$(f, v) = \int_\Omega f_i v_i\, d\Omega, \tag{4.1.34}$$

where $f = \{f_i\}$ represents the volume force vector.

One of the consequences of Korn's first inequality (1.4.28) is that

$$a(u, u) \geq c \| u \|_1^2, \qquad \forall u \in V_0, \qquad c \text{ const} > 0, \tag{4.1.35}$$

as is evident from (4.1.32) and from the properties (3.3.61) of the elasticity tensor. Then $(a(u, u))^{1/2}$ is an equivalent norm on V_0. By Prop. 4.1.1, the variational inequality of the Signorini–Fichera problem has for mes $\Gamma_U > 0$ a unique solution $u \in V_0$. Further, we assume that $\Gamma_U = \varnothing$. Then $V = [H^1(\Omega)]^3$ and ker $a(v, v) = \mathcal{R}$, where $\mathcal{R}$ is the space of rigid-body displacements and $a(v, v)$ is semicoercive[2] on V. Let us now consider the quotient

[2] $(a(v, v))^{1/2}$ is a seminorm on V.

space $\dot{V} = V/\mathscr{R}$. Korn's second inequality (1.4.29) implies[3] that

$$a([u], [u]) \geq c \| [u] \|_{\dot{V}}^2, \qquad \forall [u] \in \dot{V}, \qquad c \text{ const} > 0 \tag{4.1.36}$$

where

$$a([u], [v]) = a(u, v), \qquad \forall u \in [u], \qquad \forall v \in [v]. \tag{4.1.37}$$

Korn's second inequality also implies that for $\mu > 0$ a $c > 0$ exists such that

$$a(u, u) + \mu |u|_2^2 \geq c \|u\|_1^2, \qquad \forall u \in [H^1(\Omega)]^3. \tag{4.1.38}$$

Note that (4.1.36) is equivalent to (4.1.10) where $P = I - Q$ and Q is the orthogonal projector of $[H^1(\Omega)]^3$ onto $\mathscr{R}$. Let

$$\mathscr{R}_1 = \mathscr{R} \cap K$$

and let (4.1.39)

$$\mathscr{R}_2 = \{\rho \mid \pm \rho \in \mathscr{R}_1\}.$$

The following proposition is a direct consequence of Props. 4.1.2–4.

Proposition 4.1.5. Suppose that

$$(f, \rho) \leq 0, \qquad \forall \rho \in \mathscr{R}_1 \tag{4.1.40}$$

and

$$(f, \rho) < 0, \qquad \forall \rho \in \mathscr{R}_1 - \mathscr{R}_2. \tag{4.1.41}$$

[3] In order to verify (4.1.36), it suffices to show that (cf., (4.1.6))

$$\varepsilon(u) = \int_\Omega \varepsilon_{ij}(u)\varepsilon_{ij}(u)\, d\Omega \geq c_1 |u - Q'u|_2^2 \qquad c_1 \text{ const} > 0, \qquad \forall u \in [H^1(\Omega)]^3, \tag{4.1.35a}$$

where Q' is the orthogonal projector of $[L^2(\Omega)]^3$ onto $\mathscr{R}$ in the L^2-norm. Indeed, because of (1.4.29) we can take $(\int_\Omega \varepsilon_{ij}(u)\varepsilon_{ij}(u)\, d\Omega + |u|_2^2)^{1/2}$ as an equivalent norm on $[H^1(\Omega)]^3$. Thus (4.1.36) is equivalent to $\varepsilon(u) \geq c[\inf_{r \in \mathscr{R}} |u - r|_2^2 + \varepsilon(u)]$, c a constant > 0. Because (cf., (1.1.6)) $\inf |v - r|_2^2 = |P'v|_2^2$, $P' = I - Q'$, it results from (4.1.35a) that (4.1.36) holds.

If (4.1.35a) were not true, a sequence $\{\bar{u}_n\} = \{u_n/|u_n - Q'u_n|_2\}$ could be determined such that $\varepsilon(\bar{u}_n) \to 0$ and $|\bar{u}_n - Q'\bar{u}_n|_2 = 1$. We set $\bar{u}_n - Q'\bar{u}_n = P'\bar{u}_n = \bar{v}_n \in [H^1(\Omega)]^3$, and we find from (1.4.29) that $\|\bar{v}_n\|_1$ is bounded. Therefore, a subsequence, again denoted by $\{\bar{v}_n\}$, can be extracted such that $\bar{v}_n \to \bar{v}$ weakly in $[H^1(\Omega)]^3$ (cf. Sec. 1.2.3), and strongly in $[L^2(\Omega)]^3$ due to (1.3.53). It can be easily verified (cf., 4.1.23) that for $\bar{v}_n \to \bar{v}$ weakly in $[H^1(\Omega)]^3$

$$\liminf \int_\Omega \varepsilon_{ij}(\bar{v}_n)\varepsilon_{ij}(\bar{v}_n)\, d\Omega \geq \int_\Omega \varepsilon_{ij}(\bar{v})\varepsilon_{ij}(\bar{v})\, d\Omega, \tag{4.1.35b}$$

and therefore $\varepsilon(\bar{v}) = 0$, implying that $\bar{v} \in \mathscr{R}$. But $\bar{v}_n = \bar{u}_n - Q'\bar{u}_n = P'\bar{u}_n$, which is an element of the orthogonal complement $\mathscr{R}^\perp$ of $\mathscr{R}$ (with respect to the L^2-norm). Therefore, $\bar{v} \in \mathscr{R} \cap \mathscr{R}^\perp$ and thus $\bar{v} = 0$. On the other hand, the strong convergence of $\bar{v}_n$ to $\bar{v}$ in $[L^2(\Omega)]^3$ implies that $|\bar{v}| = 1$, which leads to a contradiction. Thus (4.1.35a) is true, q.e.d.

Then the variational inequality of the Signorini–Fichera problem has a solution u. Any other solution u_1 is given by $u_1 = u + \rho$, where

$$\rho \in \mathscr{R}, \qquad (f, \rho) = 0, \qquad u + \rho \in K. \tag{4.1.42}$$

Proof. We apply Prop. 4.1.2 for $u_0 = 0$. Then it suffices to show that $\bar{P}(K)$ is closed ($\bar{Q}$ is the projector of V onto $\mathscr{R}_2$), as is indeed the case, q.e.d.

Other applications of the foregoing theory as well as some regularity results concerning (4.1.1) are given by Fichera in [99].

4.1.2 Variational Inequalities of Other Types

In this section we shall study variational inequalities of the form

$$a(u, v - u) + \Phi(v) - \Phi(u) \geq (f, v - u), \qquad \forall v \in V, \qquad \text{for } u \in V, \tag{4.1.43}$$

where Φ is a convex, l.s.c., proper functional on $V, f \in V'$ and $a(u, v)$ is again a bilinear, symmetric, continuous form on V (for this type of inequalities see [27] [28]). If $\Phi = I_K$, where I_K is the indicator of a convex, closed subset of V, then (4.1.43) coincides with (4.1.1). If $\Phi = \tilde{\Phi} + I_K$, then for $u \in K$ (4.1.43) is equivalent to the inequality

$$a(u, v - u) + \tilde{\Phi}(v) - \tilde{\Phi}(u) \geq (f, v - u), \qquad \forall v \in K. \tag{4.1.44}$$

If (4.1.43) has a solution u, then

$$\Pi(u) = \min\{\Pi(v) \mid v \in V\}, \tag{4.1.45}$$

where

$$\Pi(v) = \tfrac{1}{2}a(v, v) + \Phi(v) - (f, v), \tag{4.1.46}$$

and conversely. This easily results from Prop. 2.2.2, the only condition being that

$$a(u, u) \geq 0, \qquad \forall u \in V. \tag{4.1.47}$$

Here we shall study a more general type of variational inequality involving nonlinear pseudomonotone operators. This type includes (4.1.43) as a special case. Let V be a real Hilbert space and V' its dual space. The functional framework of (4.1.1) holds. We seek a $u \in V$ such that for $f \in V'$

$$(Tu, v - u) + \Phi(v) - \Phi(u) \geq (f, v - u), \qquad \forall v \in V. \tag{4.1.48}$$

Here $(\cdot, \cdot)$ denotes the duality pairing between V and V', and T is generally a nonlinear operator. Now the following proposition is given (cf., Schatzman [296] and Potier–Ferry [268]).

Proposition 4.1.6. Assume that

(i) the norm $\|v\|$ on V is equivalent to $p(v) + |v|$, where $p(v)$ and $|v|$ are a seminorm and a norm on V and on H respectively;

(ii) $\mathcal{Q} = \{q \mid q \in V,\ p(q) = 0\}$ is a finite-dimensional subspace of V, Q is the orthogonal projector of V onto $\mathcal{Q}$ with respect to $|\cdot|$, $P = I - Q$, and

$$|Pv| \leq cp(v), \qquad \forall v \in V, \qquad c \text{ const} > 0; \tag{4.1.49}$$

(iii) Φ is a convex, l.s.c. functional on V such that $D(\Phi) \neq \emptyset \Phi \not\equiv -\infty$;

(iv) the operator $T: V \to V'$ is pseudomonotone (cf., Sec. 1.4.3);

(v) $(Tv, q) = 0, \qquad \forall v \in V \quad \text{and} \quad \forall q \in \mathcal{Q};$ (4.1.50)

and

(vi) there exist $b > 1$ such that

$$T(v, v) \geq c(p(v))^b \qquad c \text{ const} > 0 \tag{4.1.51}$$

holds for every v such that $p(v)$ is sufficiently large.

(a) If assumptions (ii) and (v) hold, the relation

$$f|_{\mathcal{Q}} \in \overline{R(\partial\Phi_{u_0})}, \qquad \forall u_0 \in D(\Phi) \tag{4.1.52}$$

is a necessary condition for the existence of a solution of (4.1.48).

(b) If assumptions (i) through (vi) are valid, then a sufficient condition for the existence of a solution of (4.1.48) is that there exists at least one $u_0 \in D(\Phi)$ such that

$$f|_{\mathcal{Q}} \in \operatorname{relint} R(\partial\Phi_{u_0}). \tag{4.1.53}$$

Proof. (a) If we set $v - u = q \in \mathcal{Q}$ in (4.1.48), it results, because of (4.1.50), that

$$\Phi(u + q) - \Phi(u) \geq (f, q), \tag{4.1.54}$$

and thus

$$f \in \partial\Phi_u(0). \tag{4.1.55}$$

Taking into account that $\overline{R(\partial\Phi_u)}$ is independent of u, we derive from (4.1.55) the necessary condition (4.1.52), q.e.d.

(b) Let γ be a positive real and assume that $\Phi(0) = 0$. We seek a solution u of (4.1.48) such that $\|u\| < \gamma$. (4.1.48) has such a solution if [27][140] for $\|v\| = \gamma$

$$(T(v), v) + \Phi(v) > (f, v), \quad \forall v \in V \quad \text{such that} \quad \|v\| = \gamma. \tag{4.1.56}$$

Let us now consider the space $\tilde{\mathcal{Q}} \oplus \mathcal{Q}_2$ where $\tilde{\mathcal{Q}}$ and $\mathcal{Q}_2$ are defined in Prop. 2.4.6. Due to (i) and (ii) $p(v)$ is a norm on $\tilde{\mathcal{Q}}$ equivalent to the norm $\|v\|$. From Prop. 2.4.6 and from (4.1.51), we find that for every $q_2 \in \mathcal{Q}_2$ and every

$\bar{u} \in \tilde{\mathscr{Q}}$ with $\| \bar{u} \|$ sufficiently large, the inequality

$$(T(\bar{u} + q_2), \bar{u} + q_2) + \Phi(\bar{u} + q_2) - (f, \bar{u} + q_2)$$
$$\geq c \| \bar{u} \|^b - \| f \|_{V'} \| \bar{u} \| + c_1 - c_2 \| \bar{u} \| + c_3 \| q_2 \| \quad (4.1.57)$$

holds. Thus we obtain (4.1.56) for $\| \bar{u} \|$ and $\| q_2 \|$ sufficiently large, proving the existence of the solution of (4.1.48) for $v \in \tilde{\mathscr{Q}} \oplus \mathscr{Q}_2$. By means of Prop. 2.4.6 and of (4.1.50), we conclude the same for $v \in V$. If $\Phi(0) \neq 0$, then an appropriate translation leads to the case just studied, q.e.d.

4.1.3 The Derivation of Variational Inequality "Principles" in Mechanics

A variational "principle" is a statement[4] that a solution of an operator equation with given boundary and/or initial conditions fulfills a variational equality or a variational inequality. In the former case, we shall call the corresponding problem bilateral, and in the latter unilateral. For B.V.P.s leading to variational inequalities, the term inequality problem is used throughout this book. The variational equality or inequality formulations of a mechanical problem correspond, from the standpoint of mechanics, to the principle of virtual power or of virtual work in its inequality form. As was pointed out by Fourier [171] the inequality form of the principle is due to the fact that the variations of some quantities of the problem are unilateral. For instance, if (4.1.1) held for $u, v \in V$, then the substitution $v - u = \pm w \in V$ would lead to a variational equality. But the fact that $u, v \in K$, where K is a closed and convex subset of V, does not permit the foregoing substitution. This is also the case in the variational inequality (4.1.43), because of the term $\Phi(v) - \Phi(u)$.

In applied mechanics much attention is paid to the derivation of variational principles. Tonti [318] [319] came to the conclusion that the search for a variational principle in equality form can be considered as the search for a potential for a given vector field. Of course, the function spaces chosen, especially the duality pairing between the space of the (generalized) velocities or displacements and the space of the (generalized) forces, strongly influences the form of the variational equalities. An analogous situation arises in a unilateral problem. Here the search for a variational principle can be viewed as the search for a superpotential for a given vector field, as has already been stated in [245]. We shall give a generalization for nonconvex potentials in Sec. 4.3.

To briefly elaborate, let us consider a deformable body. With the notation of Sec. 1.4.3, and on the assumption of small deformations, we can

[4] The term "principle" is used in the literature of mechanics for historical reasons (see, e.g., [130]).

write the relation

$$\int_\Omega \sigma_{ij}(u)\varepsilon_{ij}(v-u)\,d\Omega + \int_\Omega \sigma_{ij,j}(u)(v_i-u_i)\,d\Omega$$

$$= \langle \sigma_{ij}n_j, (v_i-u_i)\rangle, \qquad \forall v \in [H^1(\Omega)]^3 \quad (4.1.58)$$

for

$$u \in [H^1(\Omega)]^3, \quad \sigma_{ij} \in L^2(\Omega), \quad \sigma_{ij,j} \in L^2(\Omega), \quad i, j = 1, 2, 3.$$

This relation, which is obtained from the operator equations of the problem by applying the Green–Gauss theorem, is the expression of the principle of virtual work for the body when it is considered to be free, i.e., with no constraint on its boundary Γ. Note that for the derivation of (4.1.58) only the strain-displacement relation (4.1.33) is used. Suppose as well that the boundary condition (3.3.8) holds. Moreover, let the equilibrium condition

$$\sigma_{ij,j} + f_i = 0 \qquad (4.1.59)$$

hold, where $f \in [L^2(\Omega)]^3$ is the volume force vector. Thus (4.1.58, 59) and (3.3.145) lead to the variational inequality: Find $u \in [H^1(\Omega)]^3$ such that

$$\Phi(v) - \Phi(u) + \int_\Omega \sigma_{ij}(u)\varepsilon_{ij}(v-u)\,d\Omega$$

$$\geq \int_\Omega f_i(v_i-u_i)\,d\Omega, \qquad \forall v \in [H^1(\Omega)]^3. \quad (4.1.60)$$

We assume that a superpotential w yields $\sigma = \{\sigma_{ij}\}$ through (3.3.57). From (4.1.60) and (3.3.59a) the variational inequality

$$\Phi(v) - \Phi(u) + W(v) - W(u) \geq \int_\Omega f_i(v_i-u_i)\,d\Omega \qquad \forall v \in [H^1(\Omega)]^3 \quad (4.1.61)$$

results, where W is given by (3.3.137). The solution u of (4.1.61) will be sought in $[H^1(\Omega)]^3$. From the above, we conclude that in order to formulate a variational inequality principle for a mechanical problem it is necessary, first, to define a functional framework in which the Green–Gauss theorem holds, and, second, to determine the superpotential which describes the physical phenomenon. The dependence of the form of the variational inequality on the chosen functional framework, especially on the duality pairing, will be made apparent from the following example.

Let H be a Hilbert space and V a Banach space with the dual V'. It is assumed that $V \subset H \subset V'$, where the injections are continuous and dense. The same also holds for $V_1 = L^p(0T, V)$, $H_1 = L^2(0T, H)$ and $V_1' = L^{p'}(0T, V')$, where $2 \leq p \leq \infty$, when the spaces have the usual norms. Consider a multivalued operator problem of the form: find u: $[0, T] \to V$ with

$u(t) \in K$ a.e. on $[0, T]$ such that

$$f - \frac{du}{dt} + Au + \lambda u \in \partial I_K(u) + \partial\Phi(u) \text{ a.e. on } [0, T], \quad u(0) = u_0 \in V, \tag{4.1.62}$$

where K is a closed convex subset of V, $\Phi \in \Gamma_0(V)$, $\lambda \in \mathbb{R}$, $A: V \to V'$, $f \in V'$ and $du/dt \in V'$ a.e. on $[0, T]$. We now introduce on $V_1' \times V_1$ the duality pairing

$$\langle f, u \rangle = \int_0^T (f, u) e^{-2\lambda t}\, dt \tag{4.1.63}$$

($(\cdot, \cdot)$ denotes the duality pairing on $V \times V'$) and we define the functional $\tilde{\Phi}$ on V_1 by setting

$$\tilde{\Phi}(u) = \begin{cases} \int_0^T e^{-2\lambda t}\Phi(u)dt & \text{if } \Phi(u) \in L^1(0, T) \\ \infty & \text{otherwise} \end{cases}. \tag{4.1.64}$$

Due to (4.1.63), (4.1.62) gives rise to the following problem by integrating formally from 0 to T: find

$$u \in V_1 \quad \text{such that } u(t) \in K \text{ a.e. on } [0, T]$$

and

$$\left\langle \frac{dv}{dt}, v - u \right\rangle + \tilde{\Phi}(v) - \tilde{\Phi}(u) - \lambda\langle v, v - u\rangle - \langle Au + f, v - u\rangle$$

$$\geq -\tfrac{1}{2}|v(0) - u_0|_H^2, \ \forall v \in V_1 \text{ with } \frac{dv}{dt} \in V_1', \ v(t) \in K \text{ a.e. on } [0, T]. \tag{4.1.65}$$

Because of the general features of the superpotentials, these variational inequality "principles" describe much more realistic, and, therefore, complicated physical phenomena, than do the "principles" in variational equality form.

4.2 Coexistent Phases. The Morphology of Material Phases

4.2.1 Neoclassical Processes and Gibbsian States. Rules for Coexistent Phases

By means of the notion of the subdifferential, the morphology and stability of material phases[5] can be studied. Here we follow principally the ideas developed in [78] by Dunn and Fosdick, who, along the lines of the work of Gibbs [115] and Coleman and Greenberg [54–55], have formu-

[5] By "phase" we mean, taking water as an example, the solid, liquid and gaseous states or phases. Note that in technical problems, e.g., for an elastic ideally plastic structure, the term coexistent phases often means the simultaneous appearance of elastic and ideally plastic regions.

lated a general theory on the coexistent phases of a one-component material system.

By Ω we denote a subset of $\mathbb{R}^3$ occupied by the body in question in its reference configuration (at $t = 0$), which is assumed as fixed. The boundary Γ of Ω is appropriately smooth. With respect to Ω a process Π is defined. This is by definition an ordered set of functions $\{\eta, \chi, \varepsilon, \theta, \sigma, q, f, \mathfrak{Q}\}$ of the position $X \in \Omega$ (see App. I) and time $t \in \mathbb{R}$. Moreover, Π is such as to satisfy the equations expressing the balance of energy and the balance of linear momentum as well as the Clausius–Duhem inequality, which are written here as

$$\frac{d}{dt}\int_{\Omega'} (\varepsilon + \tfrac{1}{2}\, v_i v_i)\, dm = \int_{\Gamma_t'} (v_i \sigma_{ij} n_j - q_i n_i)\, d\Gamma + \int_{\Omega'} (f_i v_i + \mathfrak{Q})\, dm, \qquad (4.2.1)$$

$$\frac{d}{dt}\int_{\Omega'} v_i\, dm = \int_{\Gamma_t'} \sigma_{ij} n_j\, d\Gamma + \int_{\Omega'} f_i\, dm, \qquad i = 1, 2, 3 \qquad (4.2.2)$$

$$\frac{d}{dt}\int_{\Omega'} \eta\, dm \geq \int_{\Omega'} \frac{\mathfrak{Q}}{\theta}\, dm - \int_{\Gamma_t'} \frac{q_i n_i}{\theta}\, d\Gamma \qquad (4.2.3)$$

respectively. Here $\Omega' \subset \Omega$, with the boundary Γ'; $\eta = \eta(X, t)$, $\varepsilon = \varepsilon(X, t)$ and $f = f(X, t)$ denote the entropy, internal energy and body force vectors per unit mass respectively; m is the mass (measure) of the body; $\theta = \theta(X, t)$ is the temperature, assumed to be positive; $\sigma = \{\sigma_{ij}\}$, $\sigma_{ij} = \sigma_{ij}(X, t)$ is the symmetric stress tensor; $q = \{q_i\}$, $q_i = q_i(X, t)$ is the heat flux vector; $\mathfrak{Q} = \mathfrak{Q}(X, t)$ is the radiant heating per unit mass; and $n = \{n_i\}$ is the outward unit normal vector to Γ_t'. If $\chi = \chi(X, t) \in \mathbb{R}^3$ is the motion, then we define $\Omega_t' = \chi(\Omega', t)$, $\Gamma_t' = \chi(\Gamma', t)$ and $\Gamma_t = \chi(\Gamma, t)$. The volume occupied by Ω' at time t is denoted by $V_t(\Omega')$. With respect to the motion χ, a positive-valued function $v : \Omega \times \mathbb{R} \to \mathbb{R}_+$ such that

$$V_t(\Omega') = \int_{\Omega'} v(X, t)\, dm \qquad (4.2.4)$$

is introduced. It is the specific volume per unit mass. The set of all processes Π which can take place with respect to the material of Ω is denoted by $\Pi(\Omega)$.

Further following [78] a general constitutive hypothesis is proposed. Being of an abstract nature, it permits us to consider large classes of material laws. The material need be neither fluid nor solid, and arbitrary types of fading memory [322] are admissible. The constitutive hypothesis is defined with respect to a subset $\Pi'(\Omega)$ of $\Pi(\Omega)$ having as elements the neoclassical processes. These processes obey certain weak conditions, making possible the study of the changes in entropy, energy and specific volume. The constitutive hypothesis is based on the new notions of static site manifold and rest map. A process $\Pi \in \Pi'(\Omega)$ is called neoclassical if

(i) $f_i(X, t) = \mathfrak{Q}(X, t) = 0, \quad i = 1, 2, 3, \quad \forall X \in \Omega, \quad \forall t \geq 0$ and

(4.2.5)

(ii) there exist two real numbers ϑ' and p', respectively called ambient temperature and pressure, such that

$$\frac{d}{dt}\int_{\Omega} \eta \, dm \geq -\int_{\Gamma_t} \frac{q_i n_i}{\theta} \, d\Gamma \geq -\frac{1}{\vartheta'}\int_{\Gamma_t} q_i n_i \, d\Gamma \tag{4.2.6}$$

and

$$-p' \frac{d}{dt} V_t(\Omega) = -p' \frac{d}{dt}\int_{\Omega} v \, dm = \int_{\Gamma_t} v_i \sigma_{ij} n_j \, d\Gamma. \tag{4.2.7}$$

Here $-\int_{\Gamma_t} q_i n_i \, d\Gamma$ is the total heating of the body and is denoted by $\mathfrak{R}(t)$, and $\int_{\Gamma_t} v_i \sigma_{ij} n_j \, d\Gamma$ is the total work on the boundary and is denoted by $\mathfrak{W}(t)$. From (4.2.1), (4.2.3) and (4.2.6) it results that

$$\frac{d}{dt}\int_{\Omega} \eta \, dm \geq \frac{\mathfrak{R}}{\vartheta'} \tag{4.2.8}$$

and

$$\frac{d}{dt}\int_{\Omega} (\varepsilon + p'v + \tfrac{1}{2} v_i v_i) \, dm = \mathfrak{R}(t) \tag{4.2.9}$$

for every $\Pi \in \Pi'(\Omega)$ and every $t > 0$. With respect to any process $\Pi \in \Pi'(\Omega)$, we define for $X \in \Omega$ the function $\mathcal{T}(X, \cdot) = \{\eta(X, \cdot), v(X, \cdot), \varepsilon(X, \cdot)\} : [0, \infty) \to \mathbb{R}^3$ and the set of all these triads

$$\mathcal{D} = \bigcup_{\Pi \in \Pi'(\Omega)} \mathcal{T}(X, t) \tag{4.2.10}$$

is called the set of all dynamically realizable triads. Let $\mathcal{G}$ be a continuous surface in $\mathbb{R}^3$, and let $p \in \mathcal{G}$. We then define the function

$$\mathrm{I}_{\mathcal{G}}(p) = \begin{cases} 0 & \text{if } p \in \mathcal{G} \\ \infty & \text{otherwise} \end{cases} \tag{4.2.11}$$

and the set

$$\partial(p) = \{g \mid g \in \mathbb{R}^3, \langle g, a\rangle = -1, -g \in \partial I_{\mathcal{G}}(p)\}, \tag{4.2.12}$$

where a is the vector (1, 0, 0) and $\langle \cdot, \cdot \rangle$ denotes (in this section) the inner product in $\mathbb{R}^3$. Equivalently, we can write

$$\partial(p) = \{g \mid g \in \mathbb{R}^3, \langle g, a\rangle = -1, \langle p^* - p, g\rangle \geq 0, \quad \forall p^* \in \mathcal{G}\}. \tag{4.2.13}$$

If $\partial(p) \neq \emptyset$, then $\partial(p)$ is closed and convex, as may be easily shown. Furthermore, if $\mathcal{G}$ is smooth at p, then $\partial(p)$ consists of the normal $\tilde{g}$ to $\mathcal{G}$ at p

with $\langle \tilde{g}, a \rangle = -1$. Henceforth, we shall write

$$g \in \partial(\mathscr{G}) \tag{4.2.14}$$

if at least one point $p \in \mathscr{G}$ exists such that $g \in \partial(p)$.

We can now formulate the constitutive hypothesis (see Figs. 4.2.1, 2).

(H_1) Assume that there exists a continuous surface $\mathscr{G}$ in $\mathbb{R}^3$ and a continuous mapping $r : \mathscr{D} \to \mathscr{G}$ such that

(i) $$\langle r(s) - s, g \rangle \leq 0, \quad \forall s \in \mathscr{D}, \quad \forall g \in \partial(\mathscr{G}), \quad s = (\eta, v, \varepsilon), \tag{4.2.15}$$

and

(ii) $$r(s) = s, \qquad \forall s \in \mathscr{D} \cap \mathscr{G}. \tag{4.2.16}$$

$\mathscr{G}$ and r are respectively called the static site manifold and the rest mapping for Ω. A point $p \in \mathscr{G}$ is called a thermostatic phase point for Ω. A point $p \in \mathscr{G}$ such that $g \in \partial(p)$ is called a thermostatic support point for $\mathscr{G}$. These notions apply to many classes of materials as, for instance, to linearly-viscous compressible fluids, where $\mathscr{D} \equiv \mathscr{G}$ and r is the identity mapping; they also apply to a class of materials with fading memory [54] [55].

Further, let $\mathfrak{C}_p$ be a set of piecewise continuous, integrable functions $\tilde{p} : \Omega \to \mathscr{G}$ with $\tilde{p}(X) = (\tilde{\eta}(X), \tilde{v}(X), \tilde{\varepsilon}(X)) \in \mathscr{G}$, which we call states. Here the tilde ($\sim$) is used to denote that $\tilde{p}, \tilde{\eta}, \tilde{v}, \tilde{\varepsilon}$ are functions of X only. Ω can be divided into subregions Ω_i such that in the interior of each Ω_i $\tilde{p}$ is continuous. The collection of all continuity points of p is denoted by Ω_c. The set $\tilde{p}(\Omega_c) \subset \mathscr{G}$ is called the effective range of $\tilde{p}$ and any element of $\tilde{p}(\Omega_c)$ is called the effective value of $\tilde{p}$. The set of all $X \in \Omega_c$ such that for $\tilde{p} \in \mathfrak{C}_p$, $\tilde{p}(X) = \hat{p} \in \mathscr{G}$ is called a $\hat{p}$-phase for Ω. For $\beta \in \mathbb{R}$ and $\gamma \in \mathbb{R}$, let

$$\mathfrak{C}_p(\beta, \gamma) = \left\{ \tilde{p} = (\tilde{\eta}, \tilde{v}, \tilde{\varepsilon}) \in \mathfrak{C}_p \,\middle|\, \int_\Omega \tilde{v}\, dm = \beta, \int_\Omega \tilde{\varepsilon}\, dm = \gamma \right\}. \tag{4.2.17}$$

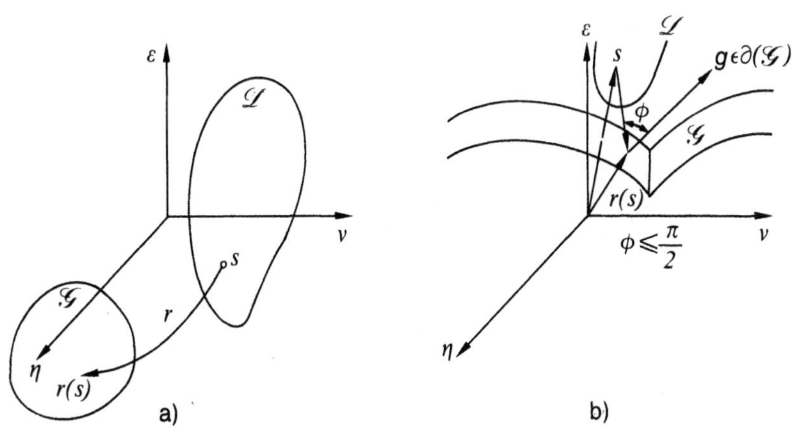

Fig. 4.2.1. Interpretation of the constitutive hypothesis.

A state $\tilde{p}_0 = (\tilde{\eta}_0, \tilde{\nu}_0, \tilde{\varepsilon}_0) \in \mathfrak{C}_p(\beta, \gamma)$ is called by definition a Gibbsian state if

$$\int_\Omega \tilde{\eta}_0 \, dm = \max\left\{\int_\Omega \tilde{\eta} \, dm \,|\, \tilde{p} \in \mathfrak{C}_p(\beta, \gamma)\right\}. \tag{4.2.18}$$

Further, the projector $\mathscr{A}$ defined by $\mathscr{A}p = \mathscr{A}(\tilde{p}_1, \tilde{p}_2, \tilde{p}_3) = (\tilde{p}_2, \tilde{p}_3)$ for every $\tilde{p} \in \mathbb{R}^3$ is introduced. Thus (4.2.17) is written as

$$\mathfrak{C}_p(\beta, \gamma) = \left\{\tilde{p} \in \mathfrak{C}_p \,|\, \mathscr{A}\left(\int_\Omega \tilde{p} \, dm\right) = (\beta, \gamma)\right\} \tag{4.2.19}$$

and (4.2.18) as

$$\left\langle a, \int_\Omega \tilde{p}_0 \, dm\right\rangle \geq \left\langle a, \int_\Omega \tilde{p} \, dm\right\rangle, \qquad \forall \tilde{p} \in \mathfrak{C}_p(\beta, \gamma). \tag{4.2.20}$$

This last inequality characterizes the Gibbsian state $\tilde{p}_0$. If for $\mathscr{G}$ continuous (resp. continuously differentiable) the mapping (resp. the differential of the mapping) $(\tilde{\nu}(\cdot), \tilde{\varepsilon}(\cdot))$ is a homeomorphism (resp. an isomorphism) for every $\tilde{p}'$ in a neighborhood of $\tilde{p}$ on $\mathscr{G}$, then we say[6] that $\mathscr{G}$ is properly (resp. perfectly) $(\tilde{\nu}, \tilde{\varepsilon})$-projectable at $\tilde{p}$. More simply, $\mathscr{G}$ is properly $(\tilde{\nu}, \tilde{\varepsilon})$-projectable at $\tilde{p}$ if a neighborhood of $\tilde{p}$ on $\mathscr{G}$ exists which projected on the $(\tilde{\nu}, \tilde{\varepsilon})$-plane does not "fold back on itself" (see [78], Fig. 1); $\mathscr{G}$ is perfectly $(\tilde{\nu}, \tilde{\varepsilon})$-projectable at $\tilde{p}$ if the η-th component of the normal vector to $\mathscr{G}$ at $\tilde{p}$ is nonzero. The definitions and notions introduced above permit the formulation of necessary and sufficient conditions under which $\partial(\mathscr{G}) \neq \emptyset$ and a Gibbsian state exists. We shall not concern ourselves in detail with this subject here, since our main task is to study the results of the constitutive hypothesis. However, for completeness we state here the principal results and for the proofs we refer the reader to [78].

(a) If β and γ are given, $\tilde{p} \in \mathfrak{C}_p(\beta, \gamma)$ is Gibbsian, $\tilde{p}$ is continuous at $X \in \Omega$, and $\mathscr{G}$ is perfectly $(\tilde{\nu}, \tilde{\varepsilon})$-projectable at $\tilde{p}(X)$, then there exists a unique $g \in \mathbb{R}^3$ such that

$$g \in \partial(\tilde{p}(X)). \tag{4.2.21}$$

(b) For the existence of a Gibbsian state in $\mathfrak{C}_p(\beta, \gamma)$, it is necessary that there exist (i) $\tilde{p}(X_\mu)$, $X_\mu \in \Omega_c$, $\mu = 1, 2, \ldots, m$, (due to (a), $\partial(\tilde{p}(X_\mu)) \neq \emptyset$, if $\mathscr{G}$ is perfectly $(\tilde{\nu}, \tilde{\varepsilon})$-projectable at $\tilde{p}(X_\mu)$) and (ii) $\alpha_\mu \in (0, 1)$ with $\sum_{\mu=1}^m \alpha_\mu = 1$ such that $(\beta, \gamma)/m(\Omega)$ is the $\mathscr{A}$-image of $\sum_{\mu=1}^m \alpha_\mu \tilde{p}(X_\mu)$.

(c) Every Gibbsian state $\tilde{p} \in \mathfrak{C}_p(\beta, \gamma)$ with $\tilde{p}(X)$ non-m-a.e.-constant for $X \in \Omega$ can be associated with a Gibbsian state $\tilde{p}_s \in \mathfrak{C}_p(\beta, \gamma)$ which is a step function of X. The effective range of $\tilde{p}_s$ is a subset of $\tilde{p}(\Omega_c)$.

[6] All the foregoing definitions can be extended in an abstract manner to general T.V.S. s, as is obvious from the mathematical notions involved. Then $\langle \cdot, \cdot \rangle$ denotes a duality pairing.

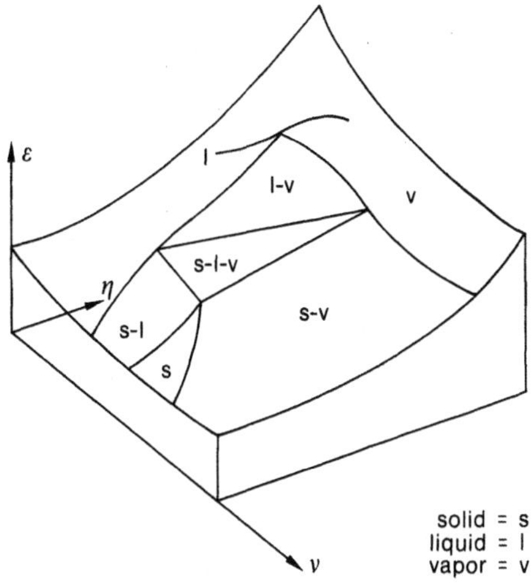

Fig. 4.2.2. The form of the $\mathcal{G}$-surface for water.

(d) If $\tilde{p}_0$ is a thermostatic support point for $\mathcal{G}$ and $\mathcal{A}\tilde{p}_0 = (\beta, \gamma)/m(\Omega)$, then the constant function with value $\tilde{p}_0$ is a Gibbsian state in $\mathcal{C}_p(\beta, \gamma)$. If, furthermore, for at least one $g \in \partial(\tilde{p}_0)$

$$\langle \tilde{p}_0 - \tilde{p}, g \rangle < 0, \qquad \forall \tilde{p} \in \mathcal{G} - \{\tilde{p}_0\},^7 \tag{4.2.22}$$

then $\tilde{p}_0$ is m-a.e. the unique Gibbsian state in $\mathcal{C}_p(\beta, \gamma)$. If $\tilde{p}_0$ does not satisfy (4.2.22), there exists another Gibbsian state in $\mathcal{C}_p(\beta, \gamma)$, namely, a step function which is not m-a.e. equal to $\tilde{p}_0$ in any part of it.

(e) Suppose that $\mathcal{C}_p(\beta, \gamma)$ contains a Gibbsian state $\tilde{p}$ with the property that $\mathcal{G}$ is perfectly (v, ε)-projectable at the images $\tilde{p}(X) \in \mathcal{G}$ of the continuity points X of $\tilde{p}$. Such a state is termed normal. Then the images $\tilde{p}(X)$ are thermostatic support points for $\mathcal{G}$ having a single, common supporting hyperplane. The $\mathcal{A}$-image of their convex hull is $(\beta, \gamma)/m(\Omega)$. The converse is also true: if (β, γ) is such that $(\beta, \gamma)/m(\Omega)$ is the $\mathcal{A}$-image of the convex hull of a set $\{\tilde{p}_j\}$ of thermostatic support points for $\mathcal{G}$ with the property that $\bigcap_j \partial(\tilde{p}_i) \neq \varnothing$, then Gibbsian states exist in $\mathcal{C}_p(\beta, \gamma)$. The number of distinct Gibbsian states is at least equal to the number of distinct sets of thermostatic support points with the above properties.

(f) For any thermostatic support point $\tilde{p} \in \mathcal{G}$ and for every $g \in \partial(\tilde{p})$, we define the set

$$\mathcal{S}(\tilde{p}, g) = \{p \mid p \in \mathcal{G}, \quad \langle \tilde{p} - p, g \rangle = 0\}, \tag{4.2.23}$$

called the g-th planar slice of $\mathcal{G}$ at $\tilde{p}$, which consists of the intersection of $\mathcal{G}$ with the hyperplane through $\tilde{p}$ which is normal to g. We say that $\mathcal{S}(\tilde{p}, g)$ is affinely independent if we cannot determine a finite set of points $\tilde{p}_\mu \in \mathcal{S}(\tilde{p}, g)$

[7] I.e., the hyperplane through $\tilde{p}_0$ and normal to g, has no other point in common with $\mathcal{G}$.

and $\alpha_\mu \in \mathbb{R}$ not all zero, $\mu = 1, 2, \ldots, m$, such that

$$\sum_{\mu=1}^{m} \alpha_\mu \tilde{p}_\mu = 0, \qquad \sum_{i=1}^{m} \alpha_\mu = 1.$$

The following results concern the number of coexistent phases of Ω. Phase points belonging to the effective range of at least one normal Gibbsian state are said to be coexistent. If $\tilde{p} \in \mathfrak{C}_p(\beta, \gamma)$ is a normal Gibbsian state, then there exists another Gibbsian state $\tilde{p}_s \in \mathfrak{C}_p(\beta, \gamma)$ which is a step function with no more than three effective values which belong to $\tilde{p}(\Omega_c)$. A step function which is a normal Gibbsian state and which is uniquely determined up to a rearrangement[8] cannot have more than three effective values (i.e., there exist at most three coexistent phases). Suppose that $\mathscr{G}$ is everywhere perfectly (v, ε)-projectable, every planar slice of $\mathscr{G}$ is affinely independent, and $\mathfrak{C}_p(\beta, \gamma)$ contains a Gibbsian state $\tilde{p}$. Then $\tilde{p}$ is a step function unique up to a rearrangement with at most three effective values.

4.2.2 Minimum Problems for Gibbsian States

Besides the maximization problem (4.2.18) any Gibbsian state of $\mathfrak{C}_p$ satisfies various other extremum problems. Some of them will be formulated in this section.

Let $\tilde{p} \in \mathfrak{C}_p$ be a normal Gibbsian state. Then there exists a unique $g \in \mathbb{R}^3$ such that (4.2.21) holds for every $X \in \Omega_c$. Thus $\langle g, \tilde{p}(X)\rangle$ is constant for $X \in \Omega_c$, i.e.,

$$\langle g, \tilde{p}(X')\rangle = \langle g, \tilde{p}(X'')\rangle, \qquad \forall X', X'' \in \Omega_c. \tag{4.2.24}$$

All the points of $\tilde{p}(\Omega_c)$ have a common tangent plane normal to g. Let $p\colon U \to \mathbb{R}^3$, with $U \subset \mathbb{R}^2$, be a C^1-local parametrization of $\mathscr{G}$ at $\tilde{p}(X)$; if $\mathrm{a}(\tau)$, is a C^1-curve in U with $p(\mathrm{a}(0)) = \tilde{p}(X)$, $\mathrm{a}' = d\mathrm{a}(\tau)/d\tau|_{\tau=0}$, then the relation

$$\left\langle g, \frac{\partial p(\mathrm{a})}{\partial \mathrm{a}}\mathrm{a}' \right\rangle = 0 \quad \text{at} \quad \mathrm{a} = p^{-1}(\tilde{p}(X)), \qquad \forall X \in \Omega_c \tag{4.2.25}$$

holds, as may easily be verified. If we take $g = (-1, \bar{\hbar}, \bar{k})$ (recall that $\langle g, a\rangle = -1$, $a = (1, 0, 0)$), (4.2.21) becomes

$$\begin{gathered} \bar{\hbar}(\tilde{v}(X) - \tilde{v}) + \bar{k}(\tilde{\varepsilon}(X) - \tilde{\varepsilon}) - (\tilde{\eta}(X) - \tilde{\eta}) \le 0, \\ \forall (\tilde{\eta}, \tilde{v}, \tilde{\varepsilon}) \in \mathscr{G}, \qquad \forall X \in \Omega_c. \end{gathered} \tag{4.2.26}$$

Similarly, (4.2.24) and (4.2.25) respectively take the forms

$$\begin{aligned} &\bar{\hbar}(\tilde{v}(X') - \tilde{v}(X'')) + \bar{k}(\tilde{\varepsilon}(X') - \tilde{\varepsilon}(X'')) \\ &\qquad - (\tilde{\eta}(X') - \tilde{\eta}(X'')) = 0, \qquad \forall X', X'' \in \Omega_c \end{aligned} \tag{4.2.27}$$

[8] A step function $p_1 \in \mathfrak{C}_p$ is called a rearrangement of another step function $p_0 \in \mathfrak{C}_p$ if $p_1(\Omega_{c1}) = p_0(\Omega_{c0})$ and $m(p_1^{-1}(\tilde{s})) = m(p_0^{-1}(\tilde{s}))$, $\forall \tilde{s} \in p_0(\Omega_{c0})$.

and

$$\bar{h}\frac{\partial}{\partial \mathrm{a}}\tilde{v}(X)+\bar{k}\frac{\partial}{\partial \mathrm{a}}\tilde{\varepsilon}(X)=\frac{\partial}{\partial \mathrm{a}}\tilde{\eta}(X), \qquad \forall X\in\Omega_c. \tag{4.2.28}$$

In the latter formula $(\partial/\partial \mathrm{a})\tilde{v}(X)=(\partial/\partial \mathrm{a})v(p^{-1}(\tilde{p}(X)))$, and similarly for the other derivatives.

We define the thermostatic temperature

$$\bar{\vartheta}=\begin{cases}(1/\bar{k}) & \text{if } \bar{k}\neq 0\\ \pm\infty & \text{if } \bar{k}=0\end{cases} \tag{4.2.29}$$

and the thermostatic pressure

$$\bar{p}=\begin{cases}(\bar{h}/\bar{k}) & \text{if } \bar{k}\neq 0\\ \pm\infty & \text{if } \bar{k}=0 \text{ and } \bar{h}\neq 0\end{cases}, \tag{4.2.30}$$

and we assume henceforth that $\bar{\vartheta}>0$, or, equivalently, $0<\bar{k}<\infty$. Then (4.2.26–28) are written respectively in the forms

$$-\bar{\vartheta}(\tilde{\eta}-\tilde{\eta}(X))+\bar{p}(\tilde{v}-\tilde{v}(X))+(\tilde{\varepsilon}-\tilde{\varepsilon}(X))\geq 0, \qquad \forall\tilde{p}=(\tilde{\eta},\tilde{v},\tilde{\varepsilon})\in\mathscr{G}, \tag{4.2.31}$$

$$-\bar{\vartheta}(\tilde{\eta}(X')-\tilde{\eta}(X''))+\bar{p}(\tilde{v}(X')-\tilde{v}(X''))+(\tilde{\varepsilon}(X')-\tilde{\varepsilon}(X''))=0 \tag{4.2.32}$$

and

$$\bar{p}\frac{\partial}{\partial \mathrm{a}}\tilde{v}(X)+\frac{\partial}{\partial \mathrm{a}}\tilde{\varepsilon}(X)=\bar{\vartheta}\frac{\partial}{\partial \mathrm{a}}\tilde{\eta}(X) \tag{4.2.33}$$

for every $X, X', X''\in\Omega_c$. Further, similarly to (4.2.17), we use the notations

$$\mathfrak{C}_p(\alpha)=\left\{(\tilde{\eta},\tilde{v},\tilde{\varepsilon})\in\mathfrak{C}_p\,\middle|\,\int_\Omega\tilde{\eta}\,dm=\alpha\in\mathbb{R}\right\},$$

$$\mathfrak{C}_p(\beta)=\left\{(\tilde{\eta},\tilde{v},\tilde{\varepsilon})\in\mathfrak{C}_p\,\middle|\,\int_\Omega\tilde{v}\,dm=\beta\in\mathbb{R}\right\},$$

$$\mathfrak{C}_p(\alpha,\beta)=\left\{(\tilde{\eta},\tilde{v},\tilde{\varepsilon})\in\mathfrak{C}_p\,\middle|\,\int_\Omega\tilde{\eta}\,dm=\alpha,\ \int_\Omega\tilde{v}\,dm=\beta,\ \alpha,\beta\in\mathbb{R}\right\},$$

$$\mathfrak{C}_p(\alpha,\gamma)=\left\{(\tilde{\eta},\tilde{v},\tilde{\varepsilon})\in\mathfrak{C}_p\,\middle|\,\int_\Omega\tilde{\eta}\,dm=\alpha,\ \int_\Omega\tilde{\varepsilon}\,dm=\gamma,\ \alpha,\gamma\in\mathbb{R}\right\}.$$

Until now we have assumed that $\tilde{p}$ is a normal Gibbsian state. Next it is assumed, more generally, that for $\tilde{p} \in \mathbb{C}_p$ at least one $g \in \mathbb{R}^3$ exists such that

$$g \in \partial(\tilde{p}(X)), \qquad \forall X \in \Omega_c. \tag{4.2.34}$$

This state is termed properly supported. A consequence of this assumption is that $\tilde{p}(\cdot)$ is a Gibbsian state in $\mathbb{C}_p(\mathscr{A}(\int_\Omega \tilde{p}\, dm))$. For this to be true it merely suffices that the static site manifold be continuous. As in the previous case of normal Gibbsian states, we call $\bar{\vartheta}$ the thermostatic temperature and $\bar{p}$ the thermostatic pressure of the properly supported state $\tilde{p}$ if $g = (-1, \bar{h}, \bar{k})$ satisfies (4.2.34). To every element of the set $\partial(\tilde{p}(\cdot))$ there corresponds a pair $\{\bar{\vartheta}, \bar{p}\}$, i.e., we have multiplicity of $\bar{p}$ and $\bar{\vartheta}$ for a properly supported state.

Proposition 4.2.1. Every normal Gibbsian state $\tilde{p} \in \mathbb{C}_p$, with $\bar{\vartheta} > 0$, satisfies the following minimum problems:

$$\text{(i) } \min\left\{\int_\Omega \tilde{\varepsilon}\, dm \,|\, \tilde{p} \in \mathbb{C}_p(\alpha, \beta)\right\}, \tag{4.2.35}$$

$$\text{(ii) } \min\left\{\int_\Omega (\tilde{\varepsilon} - \bar{\vartheta}\tilde{\eta} + \bar{p}\tilde{v})\, dm \,|\, \tilde{p} \in \mathbb{C}_p\right\}, \tag{4.2.36}$$

$$\text{(iii) } \min\left\{\int_\Omega (\tilde{\varepsilon} - \bar{\vartheta}\tilde{\eta})\, dm \,|\, \tilde{p} \in \mathbb{C}_p(\beta)\right\}, \tag{4.2.37}$$

$$\text{(iv) } \min\left\{\int_\Omega (\tilde{\varepsilon} + \bar{p}\tilde{v})\, dm \,|\, \tilde{p} \in \mathbb{C}_p(\alpha)\right\}, \tag{4.2.38}$$

and if $\bar{p} > 0$,

$$\text{(v) } \min\left\{\int_\Omega \tilde{v}\, dm \,|\, \tilde{p} \in \mathbb{C}_p(\alpha, \gamma)\right\}. \tag{4.2.39}$$

Every properly supported state $\tilde{p} \in \mathbb{C}_p$, such that $\bar{\vartheta} > 0$ for some $g = (-1, \bar{h}, \bar{k})$ satisfying (4.2.34), satisfies (4.2.35–39) similarly. If $\hat{\tilde{p}} \in \mathbb{C}_p$ is another solution to one of the minimum problems (4.2.35–39), then

$$g \in \partial\hat{\tilde{p}}(X)), \qquad \forall X \in \hat{\Omega}_c. \tag{4.2.40}$$

Proof. If $\tilde{p}$ is a normal Gibbsian state, then (4.2.35–39) result directly from (4.2.31) by integrating over Ω. If $\tilde{p}$ is properly supported, (4.2.35–39) are obtained from (4.2.34). Relation (4.2.40) is equivalent to the inequality

$$\langle g, \hat{\tilde{p}}(X) - p\rangle \le 0, \qquad \forall X \in \hat{\Omega}_c, \qquad \forall p \in \mathscr{G}. \tag{4.2.41}$$

Suppose that (4.2.41) does not hold for some $\bar{X} \in \hat{\Omega}_c$. Then by means of (4.2.34) we obtain the inequality

$$\langle g, \tilde{p}(X_1) - \hat{\tilde{p}}(X) \rangle \leq 0, \qquad \forall X_1 \in \Omega_c, \forall X \in \hat{\Omega}_c, \tag{4.2.42}$$

where

$$\langle g, \tilde{p}(X_1) - \hat{\tilde{p}}(X) \rangle < 0 \qquad \forall X_1 \in \Omega_c \tag{4.2.43}$$

on a neighborhood of $\bar{X}$. (4.2.43) implies that

$$\int_\Omega (\tilde{\varepsilon} - \hat{\tilde{\varepsilon}})\, dm - \bar{\vartheta} \int_\Omega (\tilde{\eta} - \hat{\tilde{\eta}})\, dm + \bar{p} \int_\Omega (\tilde{v} - \hat{\tilde{v}})\, dm < 0 \tag{4.2.44}$$

where $\tilde{p} = (\tilde{\eta}, \tilde{v}, \tilde{\varepsilon})$ and $\hat{\tilde{p}} = (\hat{\tilde{\eta}}, \hat{\tilde{v}}, \hat{\tilde{\varepsilon}})$. (4.2.44) contradicts the assumption that $\hat{p}$ is a solution of one of the minimization problems (4.2.35–39), q.e.d.

Obviously, $\hat{\tilde{p}}$ is also Gibbsian in $\mathrm{C}_p(\mathscr{A}(\int_\Omega \hat{p}\, dm))$. It is worth noting that relations (4.2.26, 27) also imply that, since $\bar{k}$ and $\bar{\ell}$ in (4.2.29, 30) are reals, no normal Gibbsian state may correspond to $\bar{\vartheta} = 0$, or, if $\bar{\vartheta} \neq \pm\infty$, to $\bar{p} = \pm\infty$. Therefore, we have assumed that $\bar{\vartheta} > 0$. The integrals in (4.2.36) and (4.2.37) are called, after Dunn and Fosdick, the Gibbs "semifree" energy and the Helmholtz "semifree" energy, respectively, and the integral in (4.2.38) the "semi-enthalpy" so drawing our attention to their correspondence to the notions of classical thermostatics.

4.2.3 Comparison of Gibbsian States. Some Results of the Dynamic Problem

In this section we assume that $\tilde{p}_1$ and $\tilde{p}_2$ are two properly supported states. Furthermore, let $\partial(\tilde{p}_1) \neq \partial(\tilde{p}_2)$ and suppose that $g_1 \in \partial(\tilde{p}_1)$, $g_1 \notin \partial(\tilde{p}_2)$, $g_2 \in \partial(\tilde{p}_2)$ or that $g_1 \in \partial(\tilde{p}_1)$, $g_2 \in \partial(\tilde{p}_2)$ and $g_2 \notin \partial(\tilde{p}_1)$. Thus we may write the relations

$$\langle g_1, \tilde{p}_1(X) - \tilde{p}_2(Y) \rangle \leq 0, \qquad \forall X \in \Omega_{1c}, \qquad \forall Y \in \Omega_{2c} \tag{4.2.45}$$

and

$$\langle g_2, \tilde{p}_2(Y) - \tilde{p}_1(X) \rangle \leq 0, \qquad \forall X \in \Omega_{1c}, \qquad \forall Y \in \Omega_{2c}, \tag{4.2.46}$$

where Ω_{1c} (resp. Ω_{2c}) denotes the collection of the continuity points of p_1 (resp. p_2). Either (4.2.45) holds strictly for some $Y_0 \subset \Omega_{2c}$, or (4.2.26) for some $X_0 \in \Omega_{1c}$. For $\tilde{p}_i = (\tilde{\eta}_i, \tilde{v}_i, \tilde{\varepsilon}_i)$, $\tilde{g}_i = (-1, \bar{k}_i, \bar{\ell}_i)$, $\bar{p}_i = \bar{k}_i/\bar{\ell}_i$ $(i = 1, 2)$ and on the additional assumption that $\bar{\vartheta}_i = 1/\bar{\ell}_i$ is finite and positive, (4.2.45, 46) become

$$\int_\Omega (\tilde{\varepsilon}_2 - \tilde{\varepsilon}_1)\, dm + \bar{p}_1 \int_\Omega (\tilde{v}_2 - \tilde{v}_1)\, dm - \bar{\vartheta}_1 \int_\Omega (\tilde{\eta}_2 - \tilde{\eta}_1)\, dm \geq 0 \tag{4.2.47}$$

and

$$\int_{\Omega} (\tilde{\varepsilon}_1 - \tilde{\varepsilon}_2)\, dm + \bar{p}_2 \int_{\Omega} (\tilde{v}_1 - \tilde{v}_2)\, dm - \bar{\vartheta}_2 \int_{\Omega} (\tilde{\eta}_1 - \tilde{\eta}_2)\, dm \geq 0, \tag{4.2.48}$$

where either inequality holds strictly. From (4.2.47, 48), we readily obtain the inequalities

$$(\bar{p}_1 - \bar{p}_2)\left(\int_{\Omega} (\tilde{v}_1 - \tilde{v}_2)\, dm\right) < (\bar{\vartheta}_1 - \bar{\vartheta}_2)\int_{\Omega} (\tilde{\eta}_1 - \tilde{\eta}_2)\, dm \tag{4.2.49}$$

and

$$(\bar{p}_1\bar{\vartheta}_2 - \bar{p}_2\bar{\vartheta}_1)\left(\int_{\Omega} (\tilde{v}_1 - \tilde{v}_2)\, dm\right) < (\bar{\vartheta}_1 - \bar{\vartheta}_2)\int_{\Omega} (\tilde{\varepsilon}_1 - \tilde{\varepsilon}_2)\, dm. \tag{4.2.50}$$

If in addition $\bar{p}_1 > 0$ and $\bar{p}_2 > 0$, the inequality

$$(\bar{p}_1\bar{\vartheta}_2 - \bar{p}_2\bar{\vartheta}_1)\left(\int_{\Omega} (\tilde{\eta}_1 - \tilde{\eta}_2)\, dm\right) < (\bar{p}_1 - \bar{p}_2)\int_{\Omega} (\tilde{\varepsilon}_1 - \varepsilon_2)\, dm \tag{4.2.51}$$

results. The cases

$$\{\bar{\vartheta}_i > 0, \bar{p}_i < 0\}, \quad \{\bar{\vartheta}_i < 0, \bar{p}_i < 0\},$$
$$\{\bar{\vartheta}_i < 0, \bar{p}_i > 0\}, \quad i = 1, 2, \quad \{\bar{\vartheta}_1 > 0 \cdot \text{or} \quad < 0, \quad \bar{\vartheta}_2 = +\infty \quad \tilde{p} \in \mathfrak{C}_p(\beta)\}$$

may be treated in the same way.

Suppose first that $\bar{\vartheta}_i \neq \pm\infty$, $i = 1, 2$. Then if $\bar{\vartheta}_i > 0$ (resp. $\bar{\vartheta}_i < 0$), $i = 1, 2$,

$$(\bar{\vartheta}_1 - \bar{\vartheta}_2)\int_{\Omega} (\tilde{\eta}_1 - \tilde{\eta}_2)\, dm \geq (\text{resp.} \leq)(\bar{p}_1 - \bar{p}_2)\int_{\Omega} (\tilde{v}_1 - \tilde{v}_2)\, dm \tag{4.2.52}$$

and

$$(\bar{\vartheta}_1 - \bar{\vartheta}_2)\int_{\Omega} (\tilde{\varepsilon}_1 - \tilde{\varepsilon}_2)\, dm \geq (\text{resp.} \geq)(\bar{p}_1\bar{\vartheta}_2 - \bar{p}_2\bar{\vartheta}_1)\int_{\Omega} (\tilde{v}_1 - \tilde{v}_2)\, dm. \tag{4.2.53}$$

If, furthermore, $\bar{\vartheta}_1\bar{\vartheta}_2 > 0$ and $\bar{\vartheta}_i\, \bar{p}_i > 0$, (resp. $\bar{\vartheta}_i\, \bar{p}_i < 0$) $i = 1, 2$ (no addition), then

$$(\bar{p}_1\bar{\vartheta}_2 - \bar{p}_2\bar{\vartheta}_1)\int_{\Omega} (\tilde{\eta}_1 - \tilde{\eta}_2)\, dm \leq (\text{resp.} \geq)(\bar{p}_1 - \bar{p}_2)\int_{\Omega} (\tilde{\varepsilon}_1 - \tilde{\varepsilon}_2)\, dm. \tag{4.2.54}$$

If g_1 corresponding to $\{\bar{\vartheta}_1, \bar{p}_1\}$ is not an element of $\partial(\tilde{p}_2)$, or if g_2 corresponding to $\{\bar{\vartheta}_2, \bar{p}_2\}$ does not belong to $\partial(\tilde{p}_1)$, then the foregoing inequalities are strict.

Further, let $\tilde{p}^{\infty} = (\tilde{\eta}^{\infty}, \tilde{v}^{\infty}, \tilde{\varepsilon}^{\infty}) \in \mathfrak{C}_p(\beta)$ be a properly supported state, which may have a nonfinite thermostatic temperature, i.e., $\bar{\vartheta} = \pm\infty$. Then, if $\tilde{p}_1$ and $\tilde{p}_2$ are two properly supported states with the properties that the

$\bar{\vartheta}$'s corresponding to $\tilde{p}_2$ (resp. to $\tilde{p}_1$) are negative finite (resp. positive finite) and that

$$\tilde{p}_i \in \mathfrak{C}_p(\beta), \qquad \beta = \int_\Omega \tilde{v}_i \, dm, \qquad i = 1, 2,$$

then

$$\int_\Omega (\tilde{\eta}_i - \tilde{\eta}^\infty) \, dm < 0 \quad \text{and} \quad \int_\Omega (\tilde{\varepsilon}_1 - \tilde{\varepsilon}^\infty) \, dm < 0, \quad \int_\Omega (\tilde{\varepsilon}_2 - \tilde{\varepsilon}^\infty) \, dm > 0. \tag{4.2.55}$$

The foregoing comparison relations (4.2.47–55) have some important consequences. If, for instance $\bar{\vartheta}_1 > 0$ $\bar{\vartheta}_2 > 0$ and $\bar{p}_1 = \bar{p}_2$ or $\int_\Omega (\tilde{v}_1 - \tilde{v}_2) \, dm = 0$, then (4.2.52) implies that

$$(\bar{\vartheta}_1 - \bar{\vartheta}_2)\left(\int_\Omega (\tilde{\eta}_1 - \tilde{\eta}_2) \, dm\right) \geq 0. \tag{4.2.56}$$

If $\bar{\vartheta}_1 = \bar{\vartheta}_2 > 0$, or $\int_\Omega (\tilde{\eta}_1 - \tilde{\eta}_2) \, dm = 0$, then we obtain from (4.2.52) the inequality

$$(\bar{p}_1 - \bar{p}_2)\left(\int_\Omega (\tilde{v}_1 - \tilde{v}_2) \, dm\right) \leq 0. \tag{4.2.57}$$

(4.2.56, 57) are strict whenever (4.2.52) is too.

Inequality (4.2.56) is a global form of the well-known principle of Van't Hoff and (4.2.57) of the principle of Le Chatelier[9] (see e.g. [321]). By means of the notation $\bar{V}_i = \int_\Omega \tilde{v}_i \, dm$, $i = 1, 2$, (4.2.57) implies that

$$(\bar{p}_1 - \bar{p}_2)(\bar{V}_1 - \bar{V}_2) \leq 0 \qquad (\text{and} < 0 \text{ if } \bar{p}_1 \neq \bar{p}_2) \tag{4.2.58}$$

for any two properly-supported states $\tilde{p}_1, \tilde{p}_2 \in \mathfrak{C}_p$ with $\bar{\vartheta}_1 = \bar{\vartheta}_2 > 0$. Another result of (4.2.52, 53) is that if $\bar{V}_1 = \bar{V}_2$ and $\bar{\vartheta}_i > 0$ (resp. < 0), $i = 1, 2$, then

$$(\bar{\vartheta}_1 - \bar{\vartheta}_2)(\bar{H}_1 - \bar{H}_2) \geq 0 \qquad (\text{resp.} \leq 0) \tag{4.2.59}$$

and

$$(\bar{\vartheta}_1 - \bar{\vartheta}_2)(\bar{E}_1 - \bar{E}_2) \geq 0 \qquad (\text{resp.} \geq 0) \tag{4.2.60}$$

with

$$\bar{H}_i = \int_\Omega \tilde{\eta}_i \, dm \quad \text{and} \quad \bar{E}_i = \int_\Omega \tilde{\varepsilon}_i \, dm, \; i = 1, 2.$$

[9] Usually both principles are found in differential forms. These forms are only locally applicable and result, as is evident, from the global forms derived here, under stronger assumptions.

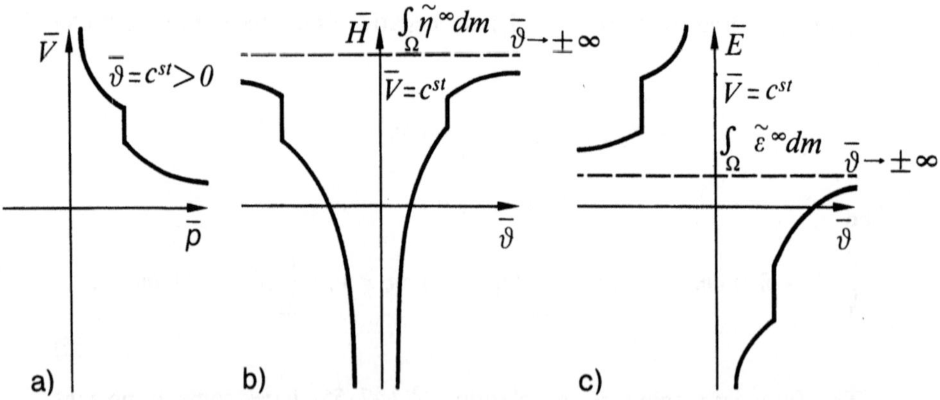

Fig. 4.2.3. [78] (a) $\bar{p}-\bar{V}$ diagram for $\bar{\vartheta}>0$ and constant; (b) $\bar{\vartheta}-\bar{H}$ diagram for $\bar{V}$ constant; (c) $\bar{\vartheta}-\bar{E}$ diagram for $\bar{V}$ constant. (Properly supported states)

Fig. 4.2.3a gives a diagram of the variation of pressure $\bar{p}$ (assumed to be positive) with the total volume $\bar{V}$ for a set of Gibbsian states having fixed temperature $\bar{\vartheta} > 0$. By taking into account (4.2.55), we give in Fig. 4.2.3b, c a diagram of the variation of the total entropy $\bar{H}$ and total energy $\bar{E}$ with respect to $\bar{\vartheta}$ for a set of Gibbsian states having a constant total volume $\bar{V}$.

We shall now consider a neoclassical process $\Pi \in \Pi(\Omega)$. For Π the triplet (η, v, ε) is a function of X and t and generally does not belong to $\mathscr{G}$. We assume that for $t = t_0$ $\{\eta(\cdot, t_0), v(\cdot, t_0), \varepsilon(\cdot, t_0)\} \in \mathfrak{C}_p$ and is Gibbsian. In this case, we say that Π passes through a Gibbsian state. We denote by $\mathfrak{P}(\vartheta', p')$ the subset of $\Pi(\Omega)$, consisting of those neoclassical processes with ambient temperature ϑ' and ambient pressure p'.

Proposition 4.2.2. Let $\Pi \in \mathfrak{P}(\vartheta', p')$, where $\bar{\vartheta} \neq 0$ and $\bar{p} \neq 0$ are given numbers, and suppose that for a time $\tau \in [0, \infty)$, $s_\tau = (\eta(\cdot, \tau), v(\cdot, \tau), \varepsilon(\cdot, \tau))$ is a properly supported state with $\bar{\vartheta} = \vartheta'$ and $\bar{p} = p'$. Then

(a)
$$\frac{1}{\vartheta'}(\mathfrak{K}(t) - \mathfrak{K}(\tau)) \leq \int_\Omega \langle s_\tau - s_t, g' \rangle \, dm \leq 0, \quad \forall t \geq \tau, \tag{4.2.61}$$

where $s_t = (\eta(\cdot, t), v(\cdot, t), \varepsilon(\cdot, t))$, $\mathfrak{K}(t) = \frac{1}{2}\int_\Omega v_i v_i \, dm$ is kinetic energy of Ω in process Π at time t and

$$g' = (-1, \mathscr{h}', \mathscr{k}'), \tag{4.2.62}$$

with $\mathscr{h}'$, $\mathscr{k}'$ corresponding to $\bar{\vartheta} = \vartheta'$ and to $\bar{p} = p'$.

(b) If $\vartheta' > 0$,

$$\mathfrak{K}(t) \leq \mathfrak{K}(\tau), \qquad \forall t > \tau, \tag{4.2.63}$$

and if $\vartheta' < 0$,

$$\mathfrak{K}(t) \geq \mathfrak{K}(\tau), \qquad \forall t > \tau. \tag{4.2.64}$$

(c) If $\vartheta' > 0$, and $\mathfrak{K}(\tau) = 0$, then

$$\mathfrak{K}(t) = 0, \qquad \forall t \geq \tau \tag{4.2.65}$$

and

$$\langle s_t(Y), g' \rangle = \langle r(s_t(Y)), g' \rangle = \langle s_\tau(X), g' \rangle, \qquad \forall X \in \Omega_{\tau c}, \qquad \forall Y \in \Omega_{tc}, \tag{4.2.66}$$

where $\Omega_{\tau c}$ (resp. Ω_{tc}) denotes the collection of all the continuity points of s_τ (resp. s_t). Moreover, for all $t \geq \tau$

$$r(s_t(Y)) \in \mathbb{S}(s_\tau(\cdot), g') \quad \forall Y \in \Omega_{tc}. \tag{4.2.67}$$

(d) If, in addition, $p' > 0$ and $r(s) = (\tilde{\eta}, \tilde{v}, \tilde{\varepsilon}) \in \mathscr{G}$ is such that

$$\eta \leq \tilde{\eta}, \quad v \geq \tilde{v}, \quad \varepsilon \geq \tilde{\varepsilon}, \quad \forall s = (\eta, v, \varepsilon) \in \mathscr{D}, \tag{4.2.68}$$

then (4.2.66, 67) hold, and, additionally

$$s_t(Y) = r(s_t(Y)), \qquad \forall Y \in \Omega_{tc} \tag{4.2.69}$$

for all $t \geq \tau$.

Proof. (a) From (4.2.8, 9) we obtain

$$\frac{d}{dt}\left(\int_\Omega \left\{-\eta + \frac{p'}{\vartheta'} v + \frac{1}{\vartheta'} \varepsilon\right\} dm + \frac{1}{\vartheta'} \int \mathfrak{K}(t)\right) \leq 0, \qquad \forall t \geq 0, \tag{4.2.70}$$

which becomes

$$\frac{d}{dt}\left(\int_\Omega \langle s_t, g' \rangle \, dm + \frac{1}{\vartheta'} \mathfrak{K}(t)\right) \leq 0, \qquad \forall t \geq 0. \tag{4.2.71}$$

Integrating from τ to t, we obtain the first inequality of (4.2.61). In order to show that

$$\int_\Omega \langle s_\tau - s_t, g' \rangle \, dm \leq 0, \qquad \forall t \geq \tau,$$

we assume that s_τ is properly supported and that the pair $(\bar{\vartheta}, \bar{p})$ of s_τ is identical to (ϑ', p'). Thus

$$g' \in \partial(s_\tau(X)), \qquad \forall X \in \Omega_{\tau c} \tag{4.2.72}$$

and since $s_t(Y) \in \mathscr{D}$ and $r(s_t(Y)) \in \mathscr{G}$,

$$\langle r(s_t(Y)) - s_t(Y), g' \rangle \leq 0, \qquad \forall Y \in \Omega. \tag{4.2.73}$$

Hence

$$\langle s_\tau(X), g' \rangle \leq \langle r(s_t(Y)), g' \rangle \leq \langle s_t(Y), g' \rangle, \quad \forall X \in \Omega_{\tau c}, \quad \forall Y \in \Omega, \tag{4.2.74}$$

which completes the proof of (4.2.61).

(b) An obvious result of (4.2.61).

(c) For $\mathfrak{K}(\tau) = 0$ and since $\vartheta' > 0$ (4.2.61) implies the relation (4.2.65) and the equality

$$\int_\Omega \langle s_\tau - s_t, g' \rangle \, dm = 0, \qquad \forall t \geq \tau. \tag{4.2.75}$$

(4.2.66) results from (4.2.74, 75). Since $r(s_t(Y)) \in \mathscr{G}$, $\forall Y \in \Omega$, (4.2.67) holds as well.

(d) The left-hand side of (4.2.66) is written in the form

$$(\tilde{\eta} - \eta) + \mathscr{k}'(v - \tilde{v}) + \mathscr{k}'(\varepsilon - \tilde{\varepsilon}) = 0. \tag{4.2.76}$$

Due to (4.2.68) and to $p' > 0$, (4.2.76) implies that $\eta = \tilde{\eta}$, $v = \tilde{v}$ and $\varepsilon = \tilde{\varepsilon}$, i.e., $s = r(s)$, which implies (4.2.69), q.e.d.

It is noteworthy that if $\vartheta' < 0$ and $\mathfrak{K}(\tau) \neq 0$ (4.2.64) implies that the body Ω will never be at rest during the process Π for any $t > \tau$, regardless of the eventual development of dissipation mechanisms. This justifies, in some way, why in continuum mechanics a negative temperature is not considered.

Finally, we state without proof (see [78]) a conclusion concerning the number of coexistent phases in the dynamic case.

If $\mathbb{S}(s_\tau, g')$ is affinely independent, it contains at most three points p_i, $i = 1, 2, \ldots, n$ with $n \leq 3$. The mass growth of the p_i-th phase $m(s_t^{-1}(p_i))$, where $p_i \in \mathbb{S}(s_\tau, g')$ is for $t \geq \tau$ uniquely determined by the total heating $\mathfrak{R}(t)$.

4.3 Nonconvex Superpotentials

4.3.1 Introduction and Brief Survey of the Basic Mathematical Properties

The notion of superpotential was introduced by J. J. Moreau [211] for convex but generally nondifferentiable and nonfinite energy functionals. As we have seen in Chapter 3, the superpotential leads to variational inequalities and to corresponding convex minimization problems. The variational inequalities can be interpreted, from the standpoint of mechanics, as expressions of the principle of virtual power (or work). However, as is well known from the classical treatises of mechanics (cf., e.g., [6] [130]) this principle is of general validity and has no connection with the notion of convexity. The subdifferential boundary conditions introduced in Sec. 3.3.1, define nondecreasing relations, as, e.g., (3.3.7) does between $-S_N$ and u_N at each point of the boundary of the body. There is, however, a large class of contact problems, e.g., contact between an elastic structure and a granular medium, where the interaction between the two bodies is idealized by a nonmonotone $S_N - u_N$ curve giving rise to a nonconvex energy function. There are also cases in the theory of plasticity where the yield surface is not convex. This happens if there is an interaction between elastic and plastic deformations, i.e. if the elastic properties depend on the plastic deformation (see, e.g., [239] [151] [341] [240]). It has been experimentally verified [341] that in such a case the yield surface can be concave, as Ilyushin has already remarked [239] [151]. According to Green and Naghdi [123] a star-shaped yield surface is admissible.[10] Similar star-shaped domains for the admissible stresses have recently been considered in soil mechanics [298].

These categories of problems substantially confirm the need for a generalization of the superpotential in the case of lack of convexity [260]. This generalization requires an extension of the subgradient $\partial f(x)$ of convex analysis for the case of nonconvex functionals. A contribution to this subject has been made by Clarke [45] [46], who extended $\partial f(x)$ to arbitrary l.s.c. functions $f\colon \mathbb{R}^n \to (-\infty, +\infty]$. This new concept, called the "generalized gradient" and denoted here by $\bar{\partial} f(x)$, has been successfully applied to many problems in optimal control, in the calculus of variations and in nonconvex optimization[11] [45–51] [144] [145] [292] [293] [12] [14] [37] [175]. Clarke, by defining $\bar{\partial} f$ for Lipschitz functions, introduced the notion of the normal cone to a closed set. He then applied this notion to the epigraphs of l.s.c. functions, thus arriving at the definition of $\bar{\partial} f$ for l.s.c. functions. A straightforward definition of the generalized gradient in terms of appropriately defined directional derivatives for arbitrary functions has quite recently been achieved by Rockafellar [292] [293], who introduced the notions of upper subderivative[12] and of substationarity. In this section, we shall give a summary of the basic mathematical notions to be used. For proofs, we refer the reader to [47] [293].

Let X be a L.C.H.T.V.S., X' its dual, $\langle \cdot, \cdot \rangle$ the duality pairing between X and X' and S a T.S. First we define certain new limit notions to be used in defining the upper subdifferential. Suppose generally that $A\colon S \to X$ is a multivalued operator. Then as $\tilde{s} \to s$, the set liminf $A(\tilde{s})$ consists of all $x \in X$ such that for every $Y \in F_x$ there exists $U \in F_s$, and for every $\tilde{s} \in U$, $Y \cap A(\tilde{s}) \neq \varnothing$. We denote here by F_x (resp. F_s a filter of neighborhoods of x (resp. of s). If liminf $A(\tilde{s}) = A(s)$ for all $s \in S$, then the multivalued mapping A is called l.s.c. The definition of limsup $A(\tilde{s})$ is obtained in the same way. For a function $g\colon S \times X \to [-\infty, +\infty]$, we define the expression "limsupinf." as

$$h(s, x) = \operatorname*{limsupinf}_{\tilde{s} \to s\ \tilde{x} \to x} g(\tilde{s}, \tilde{x}) = \sup_{Y \in F_x} \inf_{U \in F_s} \sup_{\tilde{s} \in S} \inf_{\tilde{x} \in Y} g(\tilde{s}, \tilde{x}). \tag{4.3.1}$$

For a set $C \in X$ and for $x \in X$, the tangent cone $T_C(x)$ to C at x is by definition

$$T_C(x) = \operatorname*{liminf}_{\substack{\tilde{x} \to_C x \\ \mu \to 0+}} \frac{1}{\mu} (C - \{\tilde{x}\}); \tag{4.3.2}$$

[10] As is proposed in [123], the convexity of the yield surface can be considered as an additional constitutive assumption.

[11] A survey on this subject may be found in [293].

[12] To be consistent with our earlier terminology (cf., Sec. 1.4.2) we shall hereafter use the term upper subdifferential.

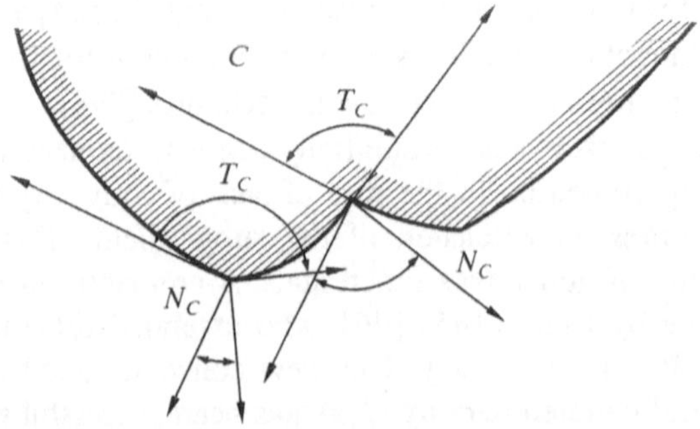

Fig. 4.3.1. Tangent and normal cones.

here $\tilde{x} \to_C x$ means that $\tilde{x} \to x$ with $\tilde{x} \in C$. Equivalently,

$$T_C(x) = \{y \mid y \in X, \quad \text{for} \quad \mu_n \to 0_+, \text{and} \quad x_n \to_C x, \quad \text{there exists } y_n \to y \text{ with } y_n + \mu_n x_n \in C\}. \tag{4.3.3}$$

Obviously, $T_{\bar{C}}(x) = T_C(x)$ for every $x \in \bar{C}$. It may be proved that $T_C(x)$ is a convex closed cone which contains 0. The normal cone $N_C(x)$ to C at x is defined as (Fig. 4.3.1)

$$N_C(x) = \{x' \mid x' \in X', \langle z, x' \rangle \leq 0, \forall z \in T_C(x)\}. \tag{4.3.4}$$

Obviously,

$$T_C(x) = \{y \mid y \in X, \langle y, z' \rangle \leq 0, \forall z' \in N_C(x)\}. \tag{4.3.5}$$

If C is convex, the definition (4.3.4) coincides with the definition (2.3.11) of the normal cone to a convex set. Let f be a function on X with values in $[-\infty, +\infty]$. The upper subdifferential $f^{\uparrow}(x, y)$ of f at $x(f(x)$ finite) with respect to y is defined by

$$f^{\uparrow}(x, y) = \operatorname{limsupinf} \frac{f(\tilde{x} + \mu\tilde{y}) - \tilde{a}}{\mu}, \tag{4.3.6}$$

where the "limsupinf" is formed[13] as $(\tilde{x}, \tilde{a}) \to (x, f(x))$ with $f(\tilde{x}) \leq \tilde{a}$, $\mu \to 0_+$ and $\tilde{y} \to y$. If f is l.s.c. at x, then (4.3.6) is simplified and the limsupinf is formed as $\tilde{x} \to x, f(\tilde{x}) \to f(x)$, $\tilde{y} \to y$ and $\mu \to 0_+$. Function $g\colon y \to f^{\uparrow}(x, y)$ is convex, l.s.c. and positively homogeneous (i.e., $g(ay) = ag(y)$, $\forall y \in X$ and $a > 0$). Moreover,

$$\operatorname{epi} g = T_{\operatorname{epi} f}(x, f(x)). \tag{4.3.7}$$

[13] It is the limsupinf defined in (4.3.1), where S is $\operatorname{epi} f \times [0, \infty)$ and

$$g(\tilde{x}, \tilde{a}, \mu, \tilde{y}) = \begin{cases} (f(\tilde{x} + \mu\tilde{y}) - \tilde{a})/\mu & \text{if } \mu > 0 \\ -\infty & \text{if } \mu = 0 \end{cases}$$

If f is convex, then

$$f^{\uparrow}(x, y) = \liminf_{\tilde{y}' \to y} \tilde{f}'(x, \tilde{y}), \qquad \forall y \in X, \tag{4.3.8}$$

where $\tilde{f}'(x, y)$ is the one-sided directional Gâteaux-differential of f at x in the direction y (cf., Sec. 2.3.2). If, additionally, there exist $a > 0$ and $\mu > 0$ such that $f(\tilde{x}) \leq a$ for every $\tilde{x}$ belonging to a neighborhood of $x + \mu y$, then

$$f^{\uparrow}(x, y) = \tilde{f}'(x, y). \tag{4.3.9}$$

Analogously to $f^{\uparrow}(x, y)$, the lower subdifferential $y \to f^{\downarrow}(x, y)$ is again defined by (4.3.6) but now with the "limsupinf" replaced by "liminfsup" which is formed as $(\tilde{x}, \tilde{a}) \to (x, f(x))$ with $\tilde{a} \leq f(\tilde{x})$, $\mu \to 0_+$ and $\tilde{y} \to y$. If f is Lipschitzian on a neighborhood of x, then

$$f^{\uparrow}(x, y) = -f^{\downarrow}(x, -y) = f^0(x, y), \qquad \forall y \in X \tag{4.3.10}$$

where $y \to f^0(x, y)$ is the directional differential of Clarke at x in the direction y and is

$$f^0(x, y) = \limsup \frac{f(\tilde{x} + \mu y) - \tilde{a}}{\mu}. \tag{4.3.11}$$

Here the "limsup" is formed as $(\tilde{x}, \tilde{a}) \to (x, f(x))$ with $\tilde{a} \geq f(\tilde{x})$ and $\mu \to 0_+$. We recall that f is said to be Lipschitzian at x if a neighborhood U of x exists such that f is finite on U, and for some continuous seminorm p on X

$$|f(x_1) - f(x_2)| \leq cp(x_1 - x_2), \qquad c \text{ const} > 0, \qquad \forall x_1, x_2 \in U. \tag{4.3.12}$$

f is Lipschitzian at x if it is continuously differentiable at x, or if it is convex (or concave) and finite at x, or if f is the linear combination of Lipschitzian functions at x.

We shall give two equivalent definitions of the generalized gradient $\bar{\partial} f(x)$ of $f: X \to [-\infty, +\infty]$ at $x \in X$ for $f(x)$ finite:

(i) $$\bar{\partial} f(x) = \{x' \mid x' \in X', f^{\uparrow}(x, x_1 - x) \geq \langle x', x_1 - x \rangle, \quad \forall x_1 \in X\} \tag{4.3.13}$$

and

(ii) $$\bar{\partial} f(x) = \{x' \mid x' \in X', (x', -1) \in N_{\text{epi} f}(x, f(x))\}. \tag{4.3.14}$$

$\bar{\partial} f(x) = \varnothing$ if $f^{\uparrow}(x, 0) = -\infty$, otherwise $\bar{\partial} f(x) \neq \varnothing$ and for every $y \in X$

$$f^{\uparrow}(x, y) = \sup \{\langle y, x' \rangle \mid x' \in \bar{\partial} f(x)\}. \tag{4.3.15}$$

Generally, $\bar{\partial} f(x)$ is a convex closed subset of X' for the $\Sigma(\mathscr{X}', \mathscr{X})$-topology. If, moreover, $f^{\uparrow}(x, y)$ is finite for every y, then $\bar{\partial} f(x)$ is a nonempty $\Sigma(\mathscr{X}', \mathscr{X})$-compact subset of X', and conversely; in this case, (4.3.15) is valid with "sup" replaced by "max". If I_C denotes the indicator of the set C, then it may easily be verified that

$$\bar{\partial} I_C(x) = N_C(x) \tag{4.3.16}$$

and

$$I_C^{\uparrow}(x, y) = I_{Tc(x)}(y). \tag{4.3.17}$$

If f is convex (resp. concave and bounded below on a neighborhood of x), then

$$\bar{\partial} f(x) = \partial f(x) \tag{4.3.18}$$

resp.

$$\bar{\partial} f(x) = -\partial(-f)(x) \tag{4.3.19}$$

at every x where f is finite. For grad $f(\cdot)$ continuous at x,

$$\bar{\partial} f(x) = \{\operatorname{grad} f(x)\}. \tag{4.3.20}$$

It is easily verified through (4.3.6) and (4.3.13) that if f has at x a local minimum, then

$$0 \in \bar{\partial} f(x). \tag{4.3.21}$$

Further, we restrict ourselves to the case of f Lipschitz and X a Banach space. Let us suppose that $f, g \colon X \to \mathbb{R}$ are Lipschitz functions; then

$$\bar{\partial}(f + g)(x) \subset \bar{\partial} f(x) + \bar{\partial} g(x) \tag{4.3.22}$$

and

$$\bar{\partial}(\lambda f)(x) = \lambda \bar{\partial} f(x) \quad \text{for} \quad \lambda \in \mathbb{R}. \tag{4.3.23}$$

Results analogous to (2.4.8) for $\bar{\partial} f$ can be found in [292]. Let X and Y be two Banach spaces such that $X \subset Y$, where the injection is continuous and dense and X is reflexive. If f is defined on Y and its restriction to X is f_1, then

$$\bar{\partial} f_1(x) \subset \bar{\partial} f(x), \qquad \forall x \in X. \tag{4.3.24}$$

If f is convex, (4.3.24) holds as an equality [37]. Also of importance is the case of $X = \mathbb{R}^n$ and f Lipschitz on X, which has been studied primarily by Clarke. Then $\bar{\partial} f(x)$ may be defined as the convex hull of all points y of the form

$$y = \lim_{i \to \infty} \operatorname{grad} f(x_i), \tag{4.3.25}$$

where $\{x_i\}$ is a sequence which converges to x and is such that grad $f(x_i)$ exists at each x_i [46].

The notion of substationarity [292] is equally important. A direction $y \in X$ (X may be a L.C.H.T.V.S.) is a direction of approximately uniform descent (d.a.u.d.) at $x \in X$ if for a $\rho > 0$ and every neighborhood U of y there exist a neighborhood V of x and positive numbers δ and λ such that for every $0 < \mu < \lambda$ and every $\tilde{x} \in V$ with $f(\tilde{x}) - f(x) \leq \delta$ the relation

$$\inf_{\tilde{y} \in U} f(\tilde{x} + \mu\tilde{y}) \leq f(\tilde{x}) - \mu\rho \tag{4.3.26}$$

holds. f is assumed to be l.s.c. and finite at x. f is called substationary at x if no d.a.u.d. exists at x. Local minima and a large class of local maxima are substationarity points, but the converse is not always true. We denote by G the function

$$G(x_1) = f(x_1) - \langle x_1, z' \rangle, \qquad z' \in X', \qquad x_1 \in X. \tag{4.3.27}$$

This function provides another equivalent definition of $\bar{\partial} f(x)$ for f l.s.c. and $f(x)$ finite:

$$\bar{\partial} f(x) = \{z' \mid G(x_1) \text{ is substationary at } x_1 = x\}. \tag{4.3.28}$$

We say that y is a substationarity point of f with respect to a closed set K if $f + I_K$ is substationary at x.

Let us close this section with the following applications. Suppose that f is a maximum-type function, i.e., $f = \max\,\{\phi_1, \ldots, \phi_m\}$, where $\phi_i = \phi_i(x)$ $i = 1, \ldots, m$, $x \in \mathbb{R}^n$ are smooth functions (Fig. 4.3.2). We denote the sets $\{x \mid \phi_i = f\}$ by A_i. It is easy to verify that

$$\bar{\partial} f(x) = \{\text{grad}\ \phi_i(x)\} \qquad \text{if } x \in A_i,$$

$$\bar{\partial} f(x) = \text{co}\ \{\text{grad}\ \phi_i(x), \text{grad}\ \phi_j(x)\} \qquad \text{if } x \in A_i \cap A_j \tag{4.3.29a}$$

and

$$\bar{\partial} f(x) = \text{co}\ \{\text{grad}\ \phi_i(x), \text{grad}\ \phi_j(x), \text{grad}\ \phi_k(x)\} \qquad \text{if } x \in (A_i \cap A_j) \cap A_k, \qquad \text{etc.}$$

If (4.3.9) holds for every $y \in X$, f is called $\bar{\partial}$-regular. This is the case if f is convex or a maximum-type function. (4.3.22) holds as an equality if f and g are $\bar{\partial}$-regular. Another result is also of importance to the applications: if

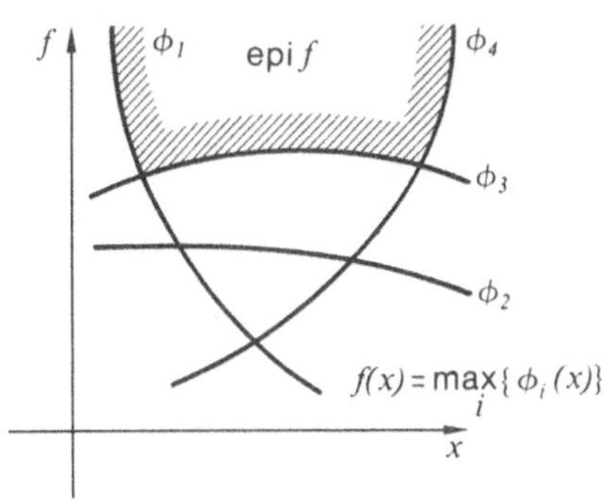

Fig. 4.3.2. A maximum type function.

$C = \{x \in \mathbb{R}^n \mid f(x) \leq 0\}$, then at a point x_0 with $f(x_0) = 0$

$$N_C(x_0) \subset \{\lambda x' \mid x' \in \mathbb{R}^n, \lambda \geq 0, x' \in \bar{\partial} f(x_0)\}, \tag{4.3.29b}$$

whenever f is Lipschitzian on a neighborhood of x_0 and $0 \notin \bar{\partial} f(x_0)$. If, moreover, f is $\bar{\partial}$-regular at x_0, then (4.3.29b) holds as an equality.

4.3.2 Nonconvex Superpotentials. Hemivariational Inequalities and Substationarity Principles

With respect to a mechanical system Σ characterized by the triplet $\Sigma = \{\mathcal{U}, \langle u, f\rangle, \mathcal{F}\}$, we can define a mechanical law between the generalized forces f and the generalized displacements u of the form (cf., Sec. 3.1.2)

$$-f \in \bar{\partial}\Phi(u), \tag{4.3.30}$$

where Φ is an extended real-valued functional defined on $\mathcal{U}$. We shall call Φ a nonconvex superpotential or a superpotential in the sense of Clarke. This mechanical law is by definition equivalent to the inequality

$$\Phi^{\uparrow}(u, u^* - u) \geq \langle -f, u^* - u\rangle, \qquad \forall u^* \in \mathcal{U} \tag{4.3.31}$$

for $u \in \mathcal{U}$, and to the inclusion

$$(-f, -1) \in N_{\operatorname{epi}\Phi}(u, \Phi(u)). \tag{4.3.32}$$

(4.3.31) will henceforth be called a hemivariational inequality. Obviously, if Φ is convex, (4.3.31) coincides with (3.1.28). For Φ Lipschitzian, $\Phi^{\uparrow}$ is replaced in (4.3.31) by Φ^0. An extension of the conjugacy theory of convex analysis to nonconvex energy functions does not yet exist. However, we also consider mechanical laws of the form

$$u \in \bar{\partial}\Phi_1(-f) \tag{4.3.33}$$

where, it must be pointed out, (4.3.33) and (4.3.30) are not generally related. (4.3.33) leads to the hemivariational inequality

$$\Phi_1^{\uparrow}(-f, -f^0 + f) \geq \langle -u, f^0 - f\rangle, \qquad \forall f^0 \in \mathcal{F}, \qquad f \in \mathcal{F}. \tag{4.3.34}$$

If Φ is the indicator $I_{\mathcal{C}}$ of a set $\mathcal{C}$, then (4.3.30) implies

$$-f \in \bar{\partial} I_{\mathcal{C}}(u) = N_{\mathcal{C}}(u), \tag{4.3.35}$$

i.e., $-f$ is an element of the normal cone to $\mathcal{C}$ at u. Analogous expressions are obtained if generalized velocities are taken into account.

In order to illustrate (4.3.35), we place ourselves in the framework of (3.1.33). We assume now that $K = \{x \mid f_i(x) \leq 0 \; i = 1, \ldots, m\}$ is a closed but not convex subset of $\mathbb{R}^3$; the f_i's are assumed to be continuously differentiable. K can be written as $\{x \mid F(x) \leq 0\}$, where $F = \max\{f_i, \ldots, f_m\}$. F is Lipschitzian and $\bar{\partial}$-regular. As is obvious, the reaction force R satisfies

(3.1.33) on any smooth part of the boundary of K. On any corner of $K, -R$ is a nonnegative linear combination of the gradients of those f_i's which are zero on the corner. Accordingly, (4.3.29a, b) imply that, if $0 \notin \bar{\partial} F(x)$, the relation

$$-R \in N_K(x) \tag{4.3.36}$$

holds. As a result, for equilibrium it is necessary and sufficient that the inclusion

$$f \in \bar{\partial} I_K(x) \tag{4.3.37}$$

holds. (4.3.37) is equivalent to the inequality (cf. (4.3.17))

$$I_{T_K(x)}(x^* - x) \geq f_i(x_i^* - x_i), \qquad \forall x^* \in \mathbb{R}^3, \qquad \text{and for } x \in \mathbb{R}^3. \tag{4.3.38}$$

(4.3.38) reduces to (3.1.37) if $(x^* - x) \in T_K(x)$. At a re-entrant corner of K $I_{T_K(x)}(x^* - x)$ may become ∞ if $(x^* - x) \notin T_K(x)$.

Suppose now that an equilibrium condition of a form similar to (3.1.38) holds, i.e.,

$$0 \in \sum_{i=1}^{n} \bar{\partial}\Phi_i(u) + \sum_{j=1}^{m} \bar{\partial}\bar{\Phi}_j(u). \tag{4.3.39}$$

Let the Φ_i's and $\bar{\Phi}_j$'s be Lipschitzian. Obviously, if

$$0 \in \bar{\partial}\Phi_0(u), \qquad \Phi_0(u) = \sum_{i=1}^{n} \Phi_i(u) + \sum_{j=1}^{m} \bar{\Phi}_j(u), \tag{4.3.40}$$

then (4.3.39) holds, but the converse is not always true; u is called a critical point of Φ_0. Accordingly, for the present mechanical system, u is an equilibrium configuration if u is a critical point of Φ_0. Recall that if u is a local minimum or maximum of Φ_0 then u is a critical point of Φ_0.

We now place ourselves in the framework of Sec. 3.3. Let j, j_1 and w, w_1 be extended real-valued functionals. The first two are defined on $\mathbb{R}^3$ and the second two on $\mathbb{R}^6$. On the boundary Γ of Ω and in the pointwise sense, boundary conditions of the form

$$-S \in \bar{\partial} j(u) \tag{4.3.41}$$

are considered. Of course, if j changes from point to point on Γ, (4.3.41) must be replaced by the relation $-S \in \bar{\partial}_u j(x, u)$ (generalized gradient with respect to u). In Ω, a material law of the form

$$\sigma \in \bar{\partial} w(\varepsilon) \tag{4.3.42}$$

is assumed to hold pointwise. In some problems, instead of (4.3.41, 42) the relations

$$u \in \bar{\partial} j_1(-S) \tag{4.3.43}$$

and

$$\varepsilon \in \bar{\partial} w_1(\sigma) \tag{4.3.44}$$

may arise. All the subdifferential boundary conditions and material laws of Sections 3.3.1 and 3.3.2 can be written similarly for nonconvex energy functionals, merely by replacing ∂ by $\bar{\partial}$. As in the convex case, these relations may be extended to a separable Hilbert space. The following proposition holds (cf., the notions of Sec. 3.3.4).

Proposition 4.3.1. Let H be a separable Hilbert space and $j: E \times H \to \mathbb{R}$ be a functional such that (i) $x \to j(x, u)$ is measurable for each u (ii) $u \to j(x, u)$ is a Lipschitz function for each x and (iii) $x \to j(x, 0)$ is finitely integrable. Let $p = 2$ and $c \geq 0$ exist such that for every (x, u) and for every $f \in \bar{\partial}_u j(x, u)$ the estimate

$$|f| \leq c(1 + |u|^{p-1}) \tag{4.3.45}$$

holds. Then for $u \in L^2(E, H)$ the functional $\Phi: L^2(E, H) \to \mathbb{R}$ defined by

$$\Phi(u) = \int_E j(x, u(x))\, d\mu \tag{4.3.46}$$

is Lipschitzian on every bounded subset of $L^2(E, H)$. Every element $f \in L^2(E, H)$ of $\bar{\partial}\Phi(u)$ satisfies the relation

$$f(x) \in \bar{\partial}_u j(x, u(x)) \qquad \mu-\text{a.e. on } E, \tag{4.3.47}$$

and if $j_u^0(x, u, v) = \tilde{j}_u'(x, u, v)$ $\mu-$a.e. on E for every $v \in H$ ($\bar{\partial}$-regularity), then the converse is also true.

For the proof of this proposition, the reader is referred to [14], where the more general case $u \in L^p(E, H)$, $p \geq 1$ with H a separable Banach space is studied. In this context, see also [37]. The above proposition can be written as

$$\bar{\partial}\Phi(u) \subset \int_E \bar{\partial}_u j(x, u(x))\, d\mu(x), \tag{4.3.48}$$

where the equality holds when j is $\bar{\partial}$-regular. This is the case e.g., if j is continuously differentiable or convex. If $E = \Omega$ and $H = \mathbb{R}$, then (4.3.48) combined with (4.3.24) permits the calculation of $\bar{\partial}\Phi$ on $\mathring{H}^1(\Omega)$.

Further we "formally" derive some hemivariational inequalities. As in Sec. 4.1.3, a deformable body will be considered such that the boundary condition (4.3.41) holds and (4.3.42) describes the material behavior. Assuming that all the quantities are sufficiently smooth, (4.1.58) is written for $u \in U_{ad}$ in the form

$$\int_\Omega \sigma_{ij}(u)\varepsilon_{ij}(v - u)\, d\Omega + \int_\Omega f_i(v_i - u_i)\, d\Omega = \int_\Gamma S_i(v_i - u_i)\, d\Gamma, \qquad \forall v \in U_{ad}. \tag{4.3.49}$$

Here U_{ad} denotes the set of sufficiently smooth, kinematically admissible displacements and $v - u$ denotes a kinematically admissible variation of the displacements. Then the inequalities

$$j^{\uparrow}(u, v - u) \geq -S_i(v_i - u_i) \qquad \forall v \in \mathbb{R}^3 \tag{4.3.50}$$

and

$$w^{\uparrow}(\varepsilon(u), \varepsilon(v) - \varepsilon(u)) \geq \sigma_{ij}(\varepsilon_{ij}(v) - \varepsilon_{ij}(u)) \qquad \forall \varepsilon(v) \in \mathbb{R}^6 \tag{4.3.51}$$

are considered. (4.3.49–51) imply for $u \in U_{ad}$ the hemivariational inequality

$$\int_\Omega w^{\uparrow}(\varepsilon(u), \varepsilon(v - u))\, d\Omega + \int_\Gamma j^{\uparrow}(u, v - u)\, d\Gamma \geq \int_\Omega f_i(v_i - u_i)\, d\Omega, \qquad \forall v \in U_{ad}, \tag{4.3.52}$$

on the assumption that the integrals exist and are finite. If the functional framework of (4.1.58) is adopted, we are led (cf., (4.1.61)) to the following problem.

Problem 1. Find $u \in [H^1(\Omega)]^3$ such as to satisfy (4.3.52) for every $v \in [H^1(\Omega)]^3$.

Obviously, every solution of the initial problem is a solution to Problem 1, but not conversely. If, instead of (4.3.41, 42), relations (4.3.43, 44) hold, then we may proceed analogously by considering variations of the stresses. Instead of (4.3.49), the relation

$$\int_\Omega \varepsilon_{ij}(u)(\tau_{ij} - \sigma_{ij})\, d\Omega + \int_\Omega u_i(\tau_{ij,j} - \sigma_{ij,j})\, d\Omega = \int_\Gamma u_i(\tau_{ij}n_j - \sigma_{ij}n_j)\, d\Gamma \tag{4.3.53}$$

is considered. Let us denote by Σ_{ad} the set of all sufficiently smooth stress fields $\tau = \{\tau_{ij}\}$ that satisfy the condition of equilibrium (τ is called a statically admissible stress field). From (4.3.53) and for $\sigma \in \Sigma_{ad}$, the hemivariational inequality

$$\int_\Omega w_1^{\uparrow}(\sigma, \tau - \sigma)\, d\Omega + \int_\Gamma j_1^{\uparrow}(-S, -T + S)\, d\Gamma \geq 0, \qquad \forall \tau \in \Sigma_{ad} \tag{4.3.54}$$

is obtained. Within the functional framework of (4.1.58), we are led to the following problem.

Problem 1′. Find $\sigma \in \Sigma_{ad} = \{\sigma \mid \sigma \in [L^2(\Omega)]^6, \sigma_{ij,j} + f_i = 0\}$ such as to satisfy (4.3.54) for every $\tau \in \Sigma_{ad}$.

For dynamic problems, on the assumption of small displacements, the inertia forces $\rho(\partial^2 u_i/\partial t^2)$ must be added, and initial conditions for u and $\partial u/\partial t$ must be considered. Thus the motion of the body is governed by the inequality (cf., (4.3.52))

$$\int_\Omega \rho \frac{\partial^2 u_i}{\partial t^2}(v_i - u_i)\, d\Omega + \int_\Omega w^\uparrow(\varepsilon(u), \varepsilon(v-u))\, d\Omega$$

$$+ \int_\Omega j^\uparrow(u, v-u)\, d\Omega \geq \int_\Omega f_i(v_i - u_i)\, d\Omega, \qquad \forall v \in U_{ad}, \quad (4.3.55)$$

where $\quad u:[0T] \to U_{ad}$ and $\dfrac{\partial^2 u_i}{\partial t^2}:[0T] \to L^2(\Omega)$.

Suppose now that the material is linearly elastic. Then

$$\int_\Omega w^\uparrow(\varepsilon(u), \varepsilon(v-u))\, d\Omega = \int_\Omega C_{ijhk}\, \varepsilon_{ij}(u)\varepsilon_{hk}(v-u)\, d\Omega = a(u, v-u) \quad (4.3.56a)$$

and

$$\int_\Omega w_1^\uparrow(\sigma, \tau - \sigma)\, d\Omega = \int_\Omega c_{ijhk}\, \sigma_{ij}(\tau_{ij} - \sigma_{ij})\, d\Omega = A(\sigma, \tau - \sigma), \quad (4.3.56b)$$

where $a(u, v)$ and $A(\sigma, \tau)$ denote the bilinear forms of linear elasticity.

Further, let us consider the potential energy functional

$$\Pi(v) = \tfrac{1}{2}a(v, v) + \int_\Gamma j(u)\, d\Gamma - \int_\Omega f_i v_i\, d\Omega \quad (4.3.57a)$$

and the complementary energy functional

$$\Pi^c(\tau) = \tfrac{1}{2}A(\tau, \tau) + \int_\Gamma j_1(-T)\, d\Gamma, \quad (4.3.57b)$$

also assuming that j and j_1 are l.s.c. The two following problems are considered.

Problem 2. Find $u \in [H^1(\Omega)]^3$ such that

$$\Pi(u) \text{ is substationary at } u \text{ with respect to } [H^1(\Omega)]^3. \quad (4.3.58a)$$

Problem 2′. Find $\sigma \in \Sigma_{ad}$ such that

$$\Pi^c(\tau) \text{ is substationary at } \sigma \text{ with respect to } \Sigma_{ad}. \quad (4.3.58b)$$

We shall investigate the relation between the solutions of Problem 1 and Problem 2 for a linearly elastic material.

Problem 2 is equivalent (cf., 4.3.28) to the assertion that

$$f \in \partial\Pi_1(u), \qquad u \in [H^1(\Omega)]^3, \qquad f \in [L^2(\Omega)]^3, \quad (4.3.59)$$

where

$$\Pi_1(u) = \tfrac{1}{2}a(u, u) + \int_\Gamma j(u)\, d\Gamma.$$

Suppose that $\int_\Gamma j(u)\, d\Gamma$ is Lipschitzian. Then by (4.3.22) $f - \frac{1}{2}$ grad $a(u, u) \in \bar{\partial}\int_\Gamma j(u)\, d\Gamma$, from which we obtain Problem 1, if for $\Phi(u) = \int_\Gamma j(u)\, d\Gamma$ the relation

$$\Phi^\uparrow(u, v) = \int_\Gamma j^\uparrow(u, v)\, d\Gamma, \tag{4.3.60}$$

holds. This is verified if j satisfies the assumptions of Prop. 4.3.1 and is $\bar{\partial}$-regular. Conversely, suppose that $u \in [H^1(\Omega)]^3$ is a solution to Problem 1 and that j is $\bar{\partial}$-regular. Then (4.3.60) is valid and $(\frac{1}{2}$ grad $a(u, u)) + \bar{\partial}\int_\Gamma j(u)\, d\Gamma = \bar{\partial}\Pi_1(u)$, which implies (4.3.59), i.e., that u is a solution of Problem 2. Analogous substationarity problems may be formulated if on Γ classical boundary conditions are given and in Ω the material law (4.3.42) (resp. (4.3.44)) holds. Then

$$\Pi(v) = \int_\Omega w(v)\, d\Omega - \int_\Omega f_i v_i\, d\Omega \tag{4.3.61a}$$

resp.

$$\Pi^c(\tau) = \int_\Omega w_1(\tau)\, d\Omega \tag{4.3.61b}$$

is substationary at u (resp. σ) relative to U_{ad} (resp. to Σ_{ad}).

The remarks in Sec. 4.1.3 may be modified to cover the nonconvex case as well. In fact, a hemivariational inequality may be considered as an expression of the principle of virtual work in inequality form, and the search for a hemivariational inequality can be viewed as the search for a nonconvex superpotential for a given vector field.

The research on the generalized gradient and its application is indeed still in full development. Only a few B.V.P.s connected with the operator $\bar{\partial}f$ (therefore giving rise to hemivariational inequalities) have been studied until now (see, e.g., [37]). In this context most of the problems are still open.

We shall now provide examples to illustrate certain applications of the generalized gradient.

(i) We first consider a one-dimensional element having the strain energy function of Fig. 4.3.3a. f is Lipschitzian and the graph of $\bar{\partial}f$ is given in Fig. 4.3.3b; it includes vertical jumps and is not monotone-increasing as would occur in the convex case (dotted line). The stress-strain diagram of Fig. 4.3.3c may be observed in performing an experiment of uniaxial monotonic loading of a granular medium (e.g., rock, soil). The form of the diagram is the result of localized phenomena of brittle fracture and crack formation. By means of (4.3.42) such a law is generalized for a three-dimensional body. The same can be said for Fig. 4.3.3d, which gives the stress-strain diagram

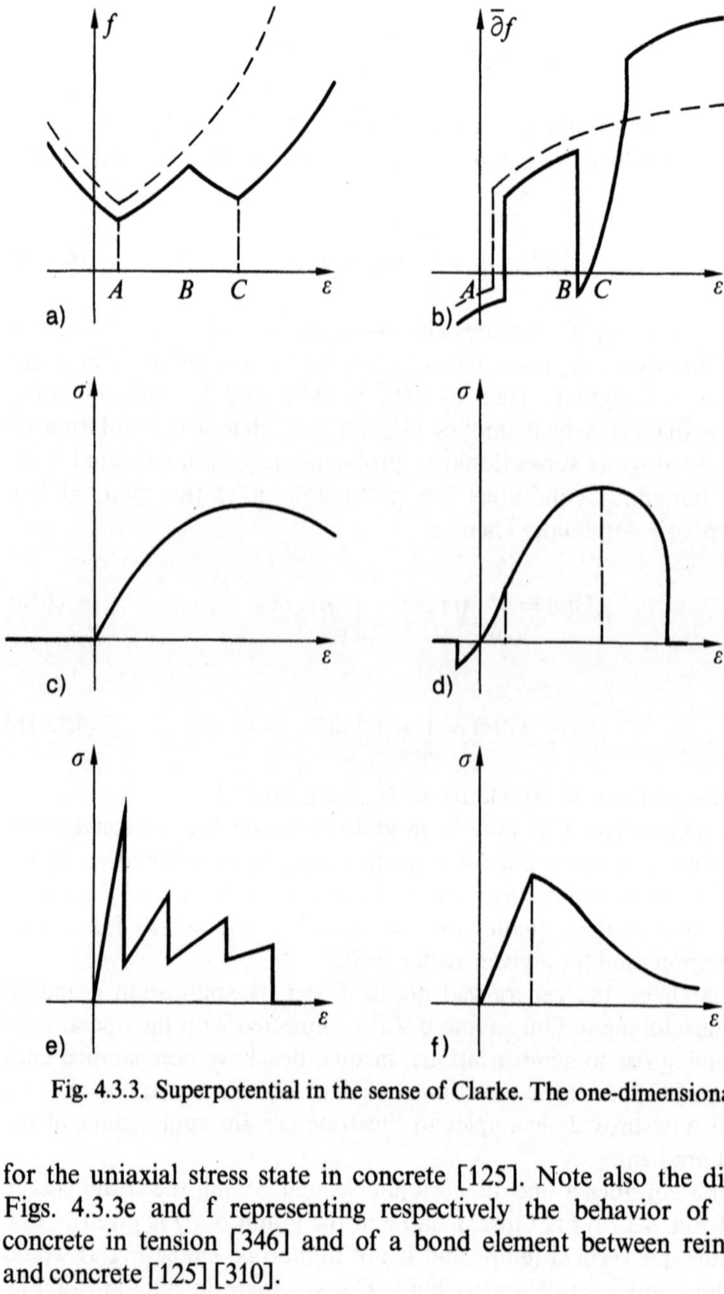

Fig. 4.3.3. Superpotential in the sense of Clarke. The one-dimensional case.

for the uniaxial stress state in concrete [125]. Note also the diagrams of Figs. 4.3.3e and f representing respectively the behavior of reinforced concrete in tension [346] and of a bond element between reinforcement and concrete [125] [310].

(ii) If L is a closed subset of the strain space and $I_L(\varepsilon)$ is the indicator of L, i.e.,

$$I_L(\varepsilon) = \begin{cases} 0 & \text{if} \quad \varepsilon \in L \\ \infty & \text{if} \quad \varepsilon \notin L, \end{cases} \tag{4.3.62}$$

then the relation

$$\sigma - C\varepsilon \in \bar{\partial} I_L(\varepsilon) = N_L(\varepsilon) \tag{4.3.63}$$

generalizes (3.3.71) for a locking criterion defined by a closed but generally nonconvex surface in the strain space. Here C denotes the elasticity tensor, and $C\varepsilon = \{C_{ijhk}\varepsilon_{hk}\}$. Indeed, if the boundary of L is smooth, a relation of the form (3.3.73) results, which justifies our conjecture. A material law of the form

$$\varepsilon - c\sigma \in \bar{\partial} I_K(\sigma), \tag{4.3.64}$$

where K is a closed subset of the stress space, describes within the framework of a holonomic theory of plasticity, an elastoplastic material but with a nonconvex yield function. If $\sigma \in \operatorname{relint} K$ the material is elastic; otherwise it is plastic. (4.3.64) generalizes the holonomic elastic perfectly plastic law (3.3.83) when the yield criterion is nonconvex. If $K = \{\sigma \mid F(\sigma) \leq k^2\}$ and F is continuously differentiable, then (4.3.64) implies (cf., (4.3.29b)) that

$$\begin{gathered} \varepsilon_{ij} - c_{ijhk}\sigma_{hk} = \lambda\left(\frac{\partial F}{\partial \sigma_{ij}}\right), \\ \lambda \geq 0,\ F(\sigma) \leq 0, \qquad \lambda F(\sigma) = 0. \end{gathered} \tag{4.3.65}$$

This last example seems to be of academic interest only, since in the theory of plasticity the yield surface is almost always assumed to be convex. However, if the elastic properties change with plastic deformation then a nonconvex yield surface cannot be excluded. This happens, actually, in the context of the flow theory of plasticity.

(iii) Let us introduce a material law of the form

$$\dot{\varepsilon} \in \bar{\partial} w(\sigma). \tag{4.3.66}$$

To describe a situation in which (4.3.66) may result, let us consider a subset K of the stress space defined by the yield criterion $F(\sigma) \leq k^2$ such that $0 \in K$. We shall now slightly modify Drucker's "pictorial proof" [77] with which he justified the convexity of the yield surface. Let us consider a stress state $\sigma^* = \{\sigma^*_{ij}\}$ on or inside the yield surface, and an external agency which causes a change of the stresses until a stress state $\sigma = \{\sigma_{ij}\}$ is reached such that $F(\sigma) = k^2$. Suppose further that the agency produces a plastic-strain increment $d\varepsilon^p$ and a corresponding stress increment $d\sigma$ ($B\Gamma$ in Fig. 4.3.4) directed outwards from K. Further, the loading releases $d\sigma$ and returns the stresses to the initial stress state σ^* along the elastic loading path $B\Delta O$. If the loading path $OAB\Delta O$ lies entirely in K or on the boundary of K, then, due to the reversibility of elastic deformation, the work done over the cycle $OAB\Delta O$ is zero, if the elastic properties do not change. However, we assume here that due to the plastic deformation the elastic properties undergo change, and we denote by $\tilde{W}$ the nonrecoverable elastic work. It is assumed

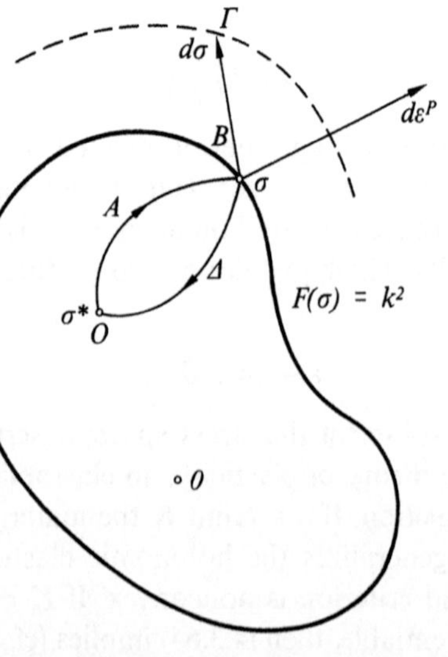

Fig. 4.3.4. Justification of the material law (4.3.66).

that $\tilde{W}$ depends on σ, on σ^* and on the path followed and that for $\sigma = \sigma^*$, $\tilde{W} = 0$. The work produced over the cycle $OAB\Gamma\Delta O$ is

$$(\sigma_{ij} - \sigma_{ij}^*)\, d\varepsilon_{ij}^p + d\varepsilon_{ij}^p\, d\sigma_{ij} + \tilde{W}(\sigma, \sigma^*, \text{path}) > 0. \tag{4.3.67}$$

This inequality holds for every $\sigma^* \in K$ and for every path $OAB\Delta O$ which does not pass out of K. (4.3.67) is also valid for $\sigma^* = \sigma$. Accordingly, $d\varepsilon_{ij}^p\, d\sigma_{ij} > 0$. But this expression can be made as small as we wish, and thus (4.3.67) implies the inequality

$$(\sigma_{ij} - \sigma_{ij}^*)\, d\varepsilon_{ij}^p + \tilde{W}(\sigma, \sigma^*, \text{path}) \geq 0. \tag{4.3.68}$$

Now, let us assume that (i) $\tilde{W}$ depends only on σ and $\sigma^* - \sigma$, and (ii) a function $W_1\colon \mathbb{R}^6 \to [-\infty, +\infty]$ exists such that

$$W_1^{\uparrow}(\sigma, \sigma^* - \sigma) = \begin{cases} \tilde{W}(\sigma, \sigma^* - \sigma) & \text{if} \quad \sigma^*, \sigma \in K \\ \infty & \text{otherwise} \end{cases} \tag{4.3.69}$$

(4.3.68) thereby yields for $\sigma \in K$ the relation

$$d\varepsilon^p \in \bar{\partial} W_1(\sigma). \tag{4.3.70}$$

In place of the increment $d\varepsilon^p$, we may introduce the rate $\dot{\varepsilon}^p$. Thus a law of the form (4.3.66) results.

(iv) Assume that the contact between a linear elastic body and a deformable halfspace is idealized by the unilateral contact condition holding on $\Gamma_S \subset \Gamma$:

$$\begin{aligned} &\text{if} \qquad u_N < 0, \qquad S_N = 0; \\ &\text{if} \qquad u_N \geq 0, \qquad S_N + k(u_N) = 0. \end{aligned} \tag{4.3.71}$$

Here the graph $(k(u_N), u_N)$ is similar to that depicted in Fig. 4.3.3c. This boundary condition may be written in the form

$$-S_N \in \bar{\partial} j(u_N), \qquad j(u_N) = \int_0^{u_{N+}} k(\alpha)\, d\alpha \tag{4.3.72}$$

with $u_{N+} = \sup\{0, u_N\}$.

(v) Let M be a measurable subset of the real line $\mathbb{R}$ such that for every open and nonempty subset I of $\mathbb{R}$, mes $(I \cap M)$ and mes $(I \cap (\mathbb{R} - M))$ are positive. Let (cf., [292])

$$g(\varepsilon) = \begin{cases} +b_1 & \text{if } \varepsilon \in M \\ -b_2 & \text{if } \varepsilon \notin M \end{cases} \tag{4.3.73}$$

and

$$f(\varepsilon) = \int_0^{\varepsilon} g(e)\, de. \tag{4.3.74}$$

Then f is Lipschitzian and it can be verified by means of (4.3.25) that

$$\bar{\partial} f(\varepsilon) = [-b_2, b_1] \tag{4.3.75}$$

for every $\varepsilon \in \mathbb{R}$, i.e., we obtain an infinite number of jumps (Fig. 4.3.5). The combination of (4.3.75) with a classical material law yields an "ideally sawtooth-formed" stress-strain diagram.

As has already been mentioned, there is no generalization of the conjugacy transformation within the framework of the theory of the generalized gradient, and so no duality propositions can be formulated. At this point, however, we wish to mention that the conjugacy of quasi-convex functions has already been studied (see, e.g., [60]), as has the duality of nonconvex

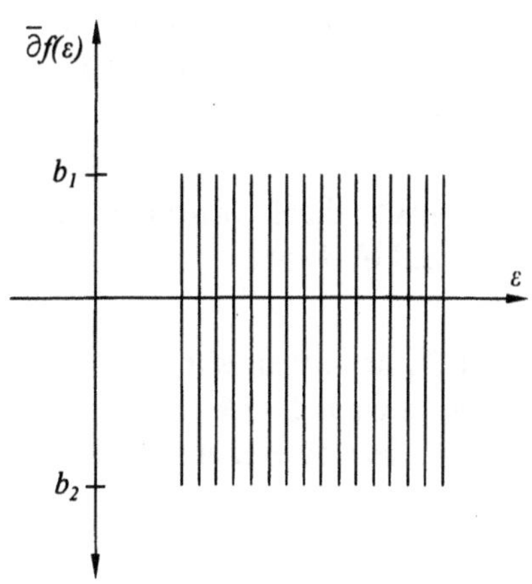

Fig. 4.3.5. The graph of (4.3.75).

functions when they may be expressed as the difference of convex, l.s.c. and proper functionals [317].

4.3.3 Generalizations of the Hypothesis of Normal Dissipation

In Sec. 3.2.1 we considered nonlinear relationships between fluxes $\mathscr{v}$ and forces $\mathscr{f}$ which can be expressed by a convex superpotential in the form (3.2.7). By means of a nonconvex superpotential, the following more general hypothesis can be formulated [258] [261].

(H) For every real thermodynamic process, an extended, real-valued, l.s.c. functional $\mathscr{G}_1$ on $\mathscr{U} = \mathbb{R}^n$ exists such that the flux $\mathscr{v} \in \mathscr{U} = \mathbb{R}^n$ associated with the force $f \in \mathscr{F} = \mathbb{R}^n$ satisfies the relation

$$\mathscr{f} \in \bar{\partial}\mathscr{G}_1(v). \tag{4.3.76}$$

The functional $\mathscr{G}_1$ will be called the generalized superpotential of dissipations. By the second principle of thermodynamics, the inequality

$$\mathscr{f}_i \mathscr{v}_i \geq 0 \tag{4.3.77}$$

must always be fulfilled (cf., App. III). (4.3.76) is by definition equivalent to both the hemivariational inequality

$$\mathscr{G}_1^{\uparrow}(\mathscr{v}, \mathscr{v}^* - \mathscr{v}) \geq \langle \mathscr{f}, \mathscr{v}^* - \mathscr{v} \rangle, \qquad \forall \mathscr{v}^* \in \mathbb{R}^n \tag{4.3.78}$$

and the relation

$$(\mathscr{f}, -1) \in N_{\operatorname{epi} \mathscr{G}_1}(\mathscr{v}, \mathscr{G}_1(\mathscr{v})) \tag{4.3.79}$$

i.e., $(\mathscr{f}, -1)$ is normal in the sense of Clarke[14] to epi $\mathscr{G}_1$ at the point $(\mathscr{v}, \mathscr{G}_1(\mathscr{v}))$. If, for instance, $\mathscr{G}_1^{\uparrow}(\mathscr{v}, -\mathscr{v}) \leq 0$, then (4.3.77) is also satisfied. The following propositions hold.

Proposition 4.3.2. To any flux $\mathscr{v}$ there corresponds at least a force $\mathscr{f}$ if and only if vector 0 is not a d.a.u.d of $\mathscr{G}_1$ at $\mathscr{v}$.

Proof. If $\tilde{\mathscr{v}}$ is a d.a.u.d. of $\mathscr{G}_1$ at $\mathscr{v}$, then $\mathscr{G}_1^{\uparrow}(\mathscr{v}, \tilde{\mathscr{v}}) < 0$ and conversely, as we may easily verify. If 0 is not a d.a.u.d. of $\mathscr{G}_1$ at $\mathscr{v}$, then $\mathscr{G}_1^{\uparrow}(\mathscr{v}, 0) \geq 0$ and thus $\bar{\partial}\mathscr{G}_1(\mathscr{v}) \neq \varnothing$. If $\bar{\partial}\mathscr{G}_1(\mathscr{v}) \neq \varnothing$, then by means of (4.3.13) $\mathscr{G}_1^{\uparrow}(\mathscr{v}, 0) \geq 0$, i.e., 0 is not a d.a.u.d. of $\mathscr{G}_1$ at $\mathscr{v}$, q.e.d.

[14] If $\mathscr{C}$ is a given set, it is possible that $N_{\mathscr{C}}(x) \cap \mathscr{C} \neq \varnothing$.

Proposition 4.3.3. The generalized hypothesis of normal dissipation is equivalent to the following three statements:

(i) if the force vector f is given, the actual flux vector v renders the functional

$$\mathscr{F}(v^*) = \mathscr{G}_1(v^*) - \langle v^*, f \rangle \text{ substationary;} \tag{4.3.80}$$

(ii) $0 \in \bar{\partial}\mathscr{F}(v)$; and (4.3.81)

(iii) $\mathscr{F}^{\uparrow}(v, v^*) \geq 0, \qquad \forall v^* \in \mathbb{R}^n.$ (4.3.82)

Proof. (i) Results from (4.3.28).

(ii) Is obtained from (4.3.13) and the definition of the upper subdifferential.

(iii) Results from (4.3.13, 81), q.e.d.

If the dissipation mechanism is given by a relation of the form

$$v \in \bar{\partial}\mathscr{G}_2(f), \tag{4.3.83}$$

then for a given flux v the actual force f is a substationarity point of the functional

$$\tilde{\mathscr{F}}(f^0) = \mathscr{G}_2(f^0) - \langle v, f^0 \rangle.$$

The normal dissipation hypothesis, as formulated for the convex case in Sec. 3.2.1, when combined with some results of thermodynamics for elastoplastic and elastic viscoplastic materials [196] leads to the definition of the generalized standard materials [128], [129]. To this class belong plastic and viscoplastic materials resulting from convex superpotentials. Nonconvex superpotentials extend this class of materials for nonconvex energy functionals.[15] We thus obtain a class of materials which we call generalized standard materials in the sense of Clarke.

With the notations introduced in App. IV, we define a generalized standard material in the sense of Clarke as an elastic viscoplastic material for which (4.3.83) is fulfilled with $v = ((\tilde{d}F_p/dt)F_p^{-1}, -\tilde{d}\{a_j\}/dt)$ and $f = \{R, \{A_j\}\}$, i.e.,

$$\left(\frac{\tilde{d}F_p}{dt}F_p^{-1}, \frac{-\tilde{d}\{a_j\}}{dt}\right) \in \bar{\partial}\mathscr{G}_2(R, \{A_j\}). \tag{4.3.84}$$

If $\mathscr{G}_2$ is convex, l.s.c. and proper, then (4.3.84) describes a generalized standard material as defined in [128] [129]. Suppose that $\mathscr{G}_2 = I_K$, where I_K is

[15] It is worth noting that Clarke's generalized gradient permits generalization of the constitutive hypothesis of Sec. 4.2.1. Then certain of the minimum problems of Prop. 4.2.1 hold as substationarity problems.

the indicator of a nonempty, convex, closed subset $\mathcal{K}$ of $\mathbb{R}^n$ where $0 \in \mathcal{K}$ and n is the dimension of $(R, \{A_j\})$. Then (4.3.84) is equivalent to the variational inequality

$$\operatorname{tr}\left(\left(\frac{\check{d}F_p}{dt}F_p^{-1}\right)^T(R - R^*)\right) - \frac{\check{d}a_j}{dt}(A_j - A_j^*) \geq 0, \qquad \forall (R^*, \{A_j^*\}) \in \mathcal{K} \tag{4.3.85}$$

for $(R, \{A_j\}) \in \mathcal{K}$, which generalizes Hill's principle of maximum plastic work [141]. The case in point corresponds to a plastic material.

If $\mathcal{G}_2 = I_{\mathcal{K}}$, where $\mathcal{K}$ is a nonempty, closed subset of $\mathbb{R}^n$ with $0 \in \mathcal{K}$, then (4.3.84) can be written in the form

$$\left(\frac{\check{d}F_p}{dt}F_p^{-1}, -\frac{\check{d}\{a_j\}}{dt}\right) \in N_{\mathcal{K}}(R, \{A_j\}), \tag{4.3.86}$$

i.e., normality in the sense of Clarke holds. In this case the inequality

$$I_{T_{\mathcal{K}}(R,\{A_j\})}((R^*, \{A_j^*\}) - (R, \{A_j\})) \geq \operatorname{tr}\left\{\left(\frac{\check{d}F_p}{dt}F_p^{-1}\right)^T(R^* - R)\right\} - \frac{\check{d}a_j}{dt}(A_j^* - A_j), \qquad \forall (R^*, \{A_j^*\}) \in \mathbb{R}^n \tag{4.3.87}$$

results, because of (4.3.17). Accordingly, (4.3.87) coincides with (4.3.85) if and only if

$$(R^*, \{A_j^*\}) - (R, \{A_j\}) \in T_{\mathcal{K}}(R, \{A_j\}).$$

If we exclude re-entrant corners, we may confirm the validity of Hill's principle for a nonconvex yield surface. (4.3.28) implies that (4.3.86) is equivalent to the fact that

$$F(R^*, A^*) = -\operatorname{tr}\left(\left(\frac{\check{d}F_p}{dt}F_p^{-1}\right)^T R^*\right) + A_j^*\frac{\check{d}a_j}{dt} \tag{4.3.88}$$

is substationary with respect to $\mathcal{K}$ at the point $(R, \{A_j\})$.

We further assume that $\mathcal{G}_2(R, A) = \Phi(R, A)$, where Φ is a continuously differentiable functional on $\mathbb{R}^n$. Then (4.3.84) is written in the form

$$\frac{\check{d}F_p}{dt}F_p^{-1} = \frac{\partial\Phi}{\partial R} \tag{4.3.89}$$

and

$$-\frac{\check{d}a_j}{dt} = \frac{\partial\Phi}{\partial A_j} \qquad j = 1, \ldots, l. \tag{4.3.90}$$

(4.3.89) extends the viscoplastic-potential theory of Rice, while the second relation concerns the behavior of the internal parameters. The function

$\mathscr{G}_2 = \Phi + I_{\mathcal{K}}$ corresponds to an elastic viscoplastic material with instantaneous plastic deformation.

Let us assume for the material law (4.3.86) that $\mathcal{K} = \{(R, \{A_j\}) \mid F_i(R, \{A_j\}) \leq 0, i = 1, \ldots, m\}$, where the F_i's are continuously differentiable functionals on $\mathbb{R}^n$. Then (4.3.86) is equivalent to the expressions

$$\frac{\tilde{d}F_p}{dt} F_p^{-1} = \sum_{i=1}^{m} \lambda_i \frac{\partial F_i}{\partial R}, \tag{4.3.91}$$

$$-\frac{\tilde{d}a_j}{dt} = \sum_{i=1}^{m} \lambda_i \frac{\partial F_i}{\partial A_j} \qquad j = 1, \ldots, l \tag{4.3.92}$$

and

$$\lambda_i F_i = 0, \quad \lambda_i \geq 0, \quad F_i \leq 0, \quad i = 1, 2, \ldots, m, \tag{4.3.93}$$

if $0 \notin \bar{\partial}\tilde{F}(R, A)(\tilde{F} = \max\{F_1, \ldots, F_m\})$. Suppose that the deformations are sufficiently small and isothermal; then (4.3.84) acquires the form[16]

$$(\dot{\varepsilon}^p, -\{\dot{a}_j\}) \in \bar{\partial}\mathscr{G}_2(\sigma, \{A_j\}), \tag{4.3.94}$$

generalizing the plasticity law (3.3.94).

Until now we have extended Moreau's superpotential for nonconvex energy functionals by means of the generalized gradient of Clarke. There is another possible extension of the superpotential by means of a notion proposed by Warga (see, e.g., [327–332]), that of the derivate container. To define this new notion, let f be a Lipschitz functional on an open subset V of $\mathbb{R}^n$. As derivate container of f at $x \in V$ is defined a bounded family

$$\{\Lambda^\varepsilon f(x) \mid \varepsilon > 0, \quad x \in V\} \tag{4.3.95}$$

of nonempty closed subsets of $\mathscr{L}(\mathbb{R}^n, \mathbb{R})$ having the properties:

(i) $\Lambda^\varepsilon f(x) \subset \Lambda^{\varepsilon'} f(x)$ if $\varepsilon' > \varepsilon$; and (4.3.96)

(ii) for every compact subset $V' \subset V$, there exists a subsequence $\{f_i\}$ of continuously differentiable functionals with some neighborhood of V' as domain of definition. This subsequence has the property that as $i \to \infty$

$$f_i \to f \tag{4.3.97}$$

uniformly on V' and for every $\varepsilon > 0$ there exist $i(\varepsilon, V')$ and $\delta(\varepsilon, V') > 0$ such that for $i \geq i(\varepsilon, V')$, $x_1 \in V'$ and $|x_1 - x| \leq \delta(\varepsilon, V')$

$$\operatorname{grad} f_i(x) \in \Lambda^\varepsilon f(x_1) \tag{4.3.98}$$

[16] In [196] the convexity of both the yield surface and the viscoplastic potential are proved, first for a monocrystal and then for a polycrystalline aggregate. However, we may accept their nonconvexity by generalizing the assumptions of the proofs in [196] as we have done in Sec. 4.3.2 with Drucker's pictorial proof of the convexity of the yield surface. This is explained physically by alloy irregularities, elastic-plastic interaction and the granular character of the material [201]. In this last case, the viscoplastic potential fails to be convex.

holds.

We then define the intersection $\Lambda f(x)$ of all the derivate containers $\Lambda^\varepsilon f(x)$, i.e.,

$$\Lambda f(x) = \bigcap_{\varepsilon > 0} \Lambda^\varepsilon f(x). \tag{4.3.99}$$

We may now consider laws relating a flux v and a force f of the form [260]

$$f \in \Lambda \tilde{\mathscr{G}}_1(v), \tag{4.3.100}$$

where $\tilde{\mathscr{G}}_1$ has the set $\mathfrak{V} \subset \mathbb{R}^n$ as its domain of definition. Functional $\tilde{\mathscr{G}}_1$ is called the superpotential in the sense of Warga, and (4.3.100) constitutes another generalization of the normal dissipation hypothesis. It is shown in [331] that if $\tilde{\mathscr{G}}_1$ is convex or concave at v or if $\tilde{\mathscr{G}}_1 : \mathbb{R} \to \mathbb{R}$, then

$$\bar{\partial} \tilde{\mathscr{G}}_1(v) \subset \Lambda \tilde{\mathscr{G}}_1(v), \tag{4.3.101}$$

i.e., $\bar{\partial}\tilde{\mathscr{G}}_1$ is the "best" derivate container. This is not true in the general case, where (4.3.101) may hold in the reverse direction[17].

4.4 The Integral Inclusion Approach to Inequality Problems

It is well known that for most problems in mechanics which can be formulated in terms of differential equations, there also exists another formulation in terms of integral equations. There is extensive literature concerning this formulation, especially in the theory of elasticity, to which we refer the reader.[18] In the theory of elasticity, there are two basic methods for the formulation of integral equations, the direct and the indirect methods. The first uses Green's formula, or, equivalently the Betti–Maxwell theorem, and yields functional equations [167] [168] [160] [39] or, by means of an appropriate limit procedure, integral equations [283] [61] [62]. In the second method, one determines layers of singular solutions of the equations of elastostatics for the infinite (in most cases) elastic medium such that the boundary conditions of the problem are satisfied.[19] An integral equation approach may be formulated for the inequality problems as well; however, due to the multivalued operators which arise, we do not obtain integral equations, but rather integral inclusions [262]. To describe the method briefly, suppose, for instance, that we have an elastic body with the boundary condition

$$-S \in \bar{\partial} f(u) \tag{4.4.1}$$

[17] In [32] a useful extension of the notion of differentiation for functions including jumps is given. In this case, the derivative has a nonlocal form. It is still an open problem whether this notion could be applied to nonlocal mechanical theories.

[18] See e.g., [252], [199], [137], [279], [36], [63], [282], [221].

[19] See e.g., [135-6], [198], [238], [280], [334].

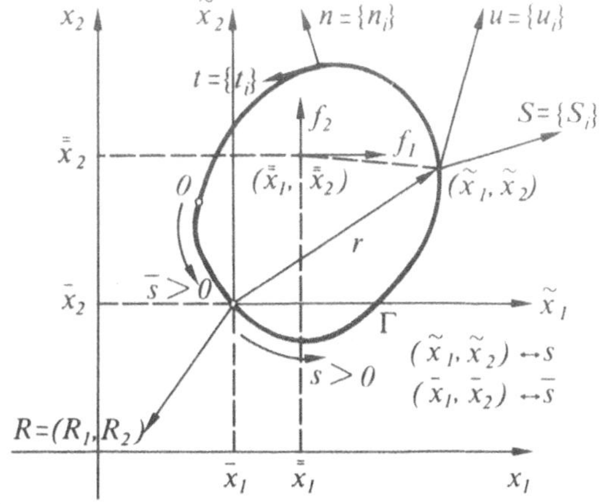

Fig. 4.4.1. The derivation of an integral inclusion for a plane elastoplastic problem.

on the boundary Γ. Should the boundary condition $S=0$ or $u=0$ on Γ be satisfied, then we would express S and u as functions of an unknown vector function, q for instance, and we would have

$$S = \mathfrak{D}_S(q), \qquad u = \mathfrak{D}_u(q), \tag{4.4.2}$$

where $\mathfrak{D}_S$ and $\mathfrak{D}_u$ are appropriately defined operators. We would then have to solve the resulting integral or functional equation $\mathfrak{D}_S(q) = 0$ or $\mathfrak{D}_u(q) = 0$. Similarly in the case of (4.4.1) q is a solution of the inclusion

$$-\mathfrak{D}_S(q) \in \bar{\partial} f(\mathfrak{D}_u(q)). \tag{4.4.3}$$

We shall illustrate the method by means of an example. Let Ω be a subset of $\mathbb{R}^2$ occupied by a two-dimensional linear elastic body (Fig. 4.4.1) subjected to the forces f_i $(i = 1, 2)$. We shall express u and S as functions of an unknown force layer on Γ, $R = (R_1, R_2)$. It may be found (see, e.g., [136]) that u (resp. S) at $(\tilde{x}_1, \tilde{x}_2)$ due to R at $(\bar{x}_1, \bar{x}_2)$ is given by the expression $(\nu' = 1/\nu)$

$$u = \begin{bmatrix} u_1 \\ u_2 \end{bmatrix} = \frac{1}{8\pi G\nu'}\left\{-(3\nu' - 1)\ln r \begin{bmatrix} 1 & 0 \\ 0 & 1 \end{bmatrix} + \frac{\nu' + 1}{r^2}\begin{bmatrix} \tilde{x}_1^2 & \tilde{x}_1\tilde{x}_2 \\ \tilde{x}_2\tilde{x}_1 & \tilde{x}_2^2 \end{bmatrix} + \frac{1}{2}\begin{bmatrix} 1 & 0 \\ 0 & 1 \end{bmatrix}\right\}\begin{bmatrix} R_1 \\ R_2 \end{bmatrix} \tag{4.4.4}$$

(resp.)

$$S = \begin{bmatrix} S_1 \\ S_2 \end{bmatrix} = -\frac{1}{4\pi\nu'}\left\{\frac{n_i\tilde{x}_i}{r^2}\left\{(\nu' - 1)\begin{bmatrix} 1 & 0 \\ 0 & 1 \end{bmatrix} + \frac{2(\nu' + 1)}{r^2}\begin{bmatrix} \tilde{x}_1^2 & \tilde{x}_1\tilde{x}_2 \\ \tilde{x}_2\tilde{x}_1 & \tilde{x}_2^2 \end{bmatrix}\right\} + \frac{t_i\tilde{x}_i}{r^2}(\nu' - 1)\begin{bmatrix} 0 & -1 \\ 1 & 0 \end{bmatrix}\right\}\begin{bmatrix} R_1 \\ R_2 \end{bmatrix}. \tag{4.4.5}$$

If on Γ the points $(\tilde{x}_1, \tilde{x}_2)$ and $(\bar{x}_1, \bar{x}_2)$ correspond to the arc length s and $\bar{s}$ respectively, then we may write (4.4.4, 5) in the abbreviated form

$$u_i^{(1)}(s) = [u,_R(s, \bar{s})]_{ij} R_j(\bar{s}) \text{ resp. } S_i^{(1)}(s) = [S,_R(s, \bar{s})]_{ij} R_j(\bar{s}). \quad (4.4.6)$$

Because of the external loading at $(\bar{\bar{x}}_1, \bar{\bar{x}}_2)$, we have an additional part of $u_i(s)$ (resp. $S_i(s)$) of the form

$$u_i^{(0)}(s) = (u,_R(\tilde{x}_1, \tilde{x}_2, \bar{\bar{x}}_1, \bar{\bar{x}}_2))_{ij} f_j$$

resp.

$$S_i^{(0)}(s) = (S,_R(\tilde{x}_1, \tilde{x}_2, \bar{\bar{x}}_1, \bar{\bar{x}}_2))_{ij} f_j$$

Integrating[20] over Γ and over Ω, we find that

$$u_i(s) = \int_\Gamma u_i^{(1)}(s)\, ds + \int_\Omega u_i^{(0)}(s)\, d\Omega + u_i^{(r)} = (\mathfrak{D}_u(R))_i, \quad (4.4.8)$$

where $u_i^{(r)}$ is a rigid body displacement, and

$$S_i(s) = \tfrac{1}{2} R_i(s) + \oint_\Gamma^* S_i^{(1)}(s)\, ds + \int_\Omega S_i^{(0)}(s)\, d\Omega = (\mathfrak{D}_S(R))_i, \quad (4.4.9)$$

where $\oint^*$ denotes the Cauchy principal value integral. Thus, if on Γ the boundary condition (4.4.1) holds, then we must solve the inclusion (4.4.3) with q replaced by R, and with $\mathfrak{D}_u$ and $\mathfrak{D}_S$, the integral operators given in (4.4.8) and (4.4.9). We may also consider as the unknown of the problem a layer of edge dislocations, or both layers of forces and layers of edge dislocations on complementary parts of the boundary. In every case, other operators $\mathfrak{D}_u$ and $\mathfrak{D}_S$, and therefore different forms of integral inclusions are obtained. There exists a great variety of methods in the theory of elasticity which lead to the operators $\mathfrak{D}_u$ and $\mathfrak{D}_S$, and, generally speaking, each of these procedures leads to another type of integral or functional inclusion. Another attempt to formulate integral inclusions for unilateral problems in the theory of elasticity, by using Green's functions, is given in [252]. The integral inclusions lead to interesting mathematical problems. Thus, in the case of the Signorini–Fichera problem (cf., Sec. 4.1.1), (4.4.3) is clearly equivalent to a linear complementarity problem of the functional form

$$\mathfrak{D}_{S_N}(q) \leq 0,\ \mathfrak{D}_{u_N}(q) \leq 0,\ \mathfrak{D}_{S_N}(q)\mathfrak{D}_{u_N}(q) = 0,\ \mathfrak{D}_{S_T}(q) = 0 \quad (4.4.10)$$

which is still an open problem (cf. also [352]).

Here we shall not study these problems, which remain open concerning both their mathematical study and numerical treatment. Note also that the mathematical study of the multivalued integral equations, based on fixed-point theorems for multivalued operators has only very recently begun (see, e.g., [4] [35] [38] [117] [311]).

[20] The integration is combined with a limit procedure. Therefore, in (4.4.9) the Cauchy principal value arises (see, e.g., [134]).

Chapter 5

Friction Problems in the Theory of Elasticity

In this chapter we shall concentrate our attention on two typical inequality problems, one static and one dynamic, formulated within the framework of linear elasticity. They are the friction B.V.P.s, which were first studied by G. Duvaut and J. L. Lions [83] [84]. Because of the boundary conditions expressing the friction phenomenon, variational inequalities arise both in the static and dynamic cases. After deriving these variational inequalities, we shall study the existence and the uniqueness of the solution in an appropriate functional setting. For the static case, the propositions of minimum potential and complementary energy are derived and their duality is proved. Finally, some other types of friction B.V.P.s are also considered, and the corresponding variational inequalities are derived.

5.1 The Static B.V.P.

5.1.1 The Classical Formulation

We consider an elastic body which in its undeformed state occupies an open, bounded, connected subset $\Omega \subset \mathbb{R}^3$. The boundary Γ of Ω is assumed to be "regular" and the points $x \in \bar{\Omega}$ are referred to a fixed Cartesian coordinate system $0x_1x_2x_3$. The boundary Γ consists of three open, disjoint parts Γ_U, Γ_F and Γ_S, i.e. $\Gamma = \bar{\Gamma}_U \cup \bar{\Gamma}_F \cup \bar{\Gamma}_S$. On Γ_U the displacements are prescribed, that is

$$u_i = U_i, \qquad U_i = U_i(x) \quad \text{on} \quad \Gamma_U, \tag{5.1.1}$$

as are the surface forces on Γ_F, i.e.,

$$S_i = \sigma_{ij} n_j = F_i, \qquad F_i = F_i(x) \quad \text{on} \quad \Gamma_F \tag{5.1.2}$$

where $i, j = 1, 2, 3$. On Γ_S friction boundary conditions are assumed to hold. In the notation of Sec. 3.3 they read:

$$\text{if} \qquad |S_T| < \mu |S_N|, \quad \text{then} \quad u_{T_i} = 0, \tag{5.1.3}$$

if $\quad |S_T| = \mu |S_N|$, there exists $\lambda \geq 0$ such that $u_{T_i} = -\lambda S_{T_i}$ (5.1.4)

and

$$S_N = C_N, \tag{5.1.5}$$

where $\mu = \mu(x) > 0$ is the friction coefficient and $C_N = C_N(x)$ is a given boundary force distribution. It is possible to assume that $\Gamma_U = \varnothing$ or $\Gamma_F = \varnothing$, or both. Obviously, $\Gamma_S \neq \varnothing$. Assuming small deformations, the equations of equilibrium and compatibility of the B.V.P. read

$$\sigma_{ij,j} + f_i = 0 \quad \text{in} \quad \Omega \tag{5.1.6}$$

and

$$\varepsilon_{ij} = \tfrac{1}{2}(u_{i,j} + u_{j,i}) \quad \text{in} \quad \Omega, \tag{5.1.7}$$

where $f = \{f_i\}$ denotes the volume force vector. Assuming linear elasticity, we may write the constitutive law in the form

$$\sigma_{ij} = C_{ijhk}\,\varepsilon_{hk} \quad \text{in} \quad \Omega, \tag{5.1.8}$$

where $i, j, h, k = 1, 2, 3$ and $C_{ijhk} \in L^\infty(\Omega)$ are the components of the elasticity tensor C. They fulfill the symmetry property (3.3.61a) and the ellipticity property (3.3.61b). In the relations (5.1.1-8), all the functions are assumed to be appropriately smooth[1]. These relations constitute the classical formulation of the B.V.P. From Eqs. (5.1.1, 6–8), we obtain by means of the Green-Gauss theorem the relation

$$a(u, v-u) = \int_\Omega f_i(v_i - u_i)\, d\Omega + \int_{\Gamma_F} S_i(v_i - u_i)\, d\Gamma$$
$$+ \int_{\Gamma_S} [S_N(v_N - u_N) + S_{T_i}(v_{T_i} - u_{T_i})]\, d\Gamma \tag{5.1.9}$$

for every v such that $v_i = U_i$ on Γ_U. Here

$$a(u, v) = \int_\Omega C_{ijhk}\,\varepsilon_{ij}(u)\varepsilon_{hk}(v)\, d\Omega. \tag{5.1.10}$$

From (5.1.3–5), the inequality

$$S_{T_i}(v_{T_i} - u_{T_i}) + \mu |C_N| (|v_T| - |u_T|) \geq 0 \quad \text{on} \quad \Gamma_S \tag{5.1.11}$$

results, which, combined with (5.1.9), implies the variational inequality

$$a(u, v-u) + \Phi(v) - \Phi(u) \geq (f, v-u)$$
$$+ \int_{\Gamma_F} F_i(v_i - u_i)\, d\Gamma + \int_{\Gamma_S} C_N(v_N - u_N) \tag{5.1.12}$$

[1] In this context, the term "regular functions" will also be used.

for every v such that $v_i = U_i$ on Γ_U. Here

$$(f, v) = \int_\Omega f_i v_i \, d\Omega \tag{5.1.13}$$

and

$$\Phi(v) = \int_{\Gamma_S} \mu |C_N| |v_T| \, d\Gamma. \tag{5.1.14}$$

The quadratic form $\frac{1}{2}a(u, u)$ denotes the elastic energy, and $\Phi(u)$ denotes the friction energy on the boundary of the body.

5.1.2 The Variational Formulation

In order to study the problem rigorously, we consider u, v as elements of $[H^1(\Omega)]^3$ and we assume that $f_i \in L^2(\Omega)$, $F_i \in L^2(\Gamma_F)$, $C_N \in L^2(\Gamma_S)$, $\mu|C_N| \in L^\infty(\Gamma_S)$ and $U_i \in \dot{H}^{1/2}(\Gamma_U)$. Then Φ is a convex, continuous (due to the trace theorem) and nondifferentiable functional on $[H^1(\Omega)]^3$. Concerning the definition of the set of the boundary displacements, we assume that $U_i \in \dot{H}^{1/2}(\Gamma_U)$, where $\dot{H}^{1/2}(\Gamma_U)$ denotes a set such that for each $U = \{U_i\}$ a $w \in [H^1(\Omega)]^3$ exists with $w|_{\Gamma_U} = U$.

It would be natural to suppose that U belongs to the restriction to Γ_U of the elements of $H^{1/2}(\Gamma)$. Let us examine this possibility. $H^{1/2}(\Gamma_U)$ denotes the restriction of $H^{1/2}(\Gamma)$ to Γ_U. A basic criterion for the definition of the space of U_i's is the applicability of the trace theorem. To illustrate the resulting questions concerning the appropriate regularity of the U_i's we extend $H^{1/2}(\Gamma_U)$ by zero to the entire boundary. This extension should be an element of $H^{1/2}(\Gamma)$: but this is not true, since the extension may generally include jumps, e.g., at the ends of Γ_U. (Strictly speaking, the extension belongs to $H^{1/2-\varepsilon}(\Gamma)$, where $\varepsilon > 0$). Concerning the exact definition of the space of U_i, we consider specifically the case $\Omega \subset \mathbb{R}^2$. Let us define on Γ_U a Hilbert scale of spaces (see Sec. 1.2.4) with $X_0 = L^2(\Gamma_U)$ and $X_1 = \mathring{H}^1(\Gamma_U)$. We then consider the space $X_{1/2}$, which we denote by $\mathring{\tilde{H}}^{1/2}(\Gamma_U)$. Here $\mathring{H}^s(\Gamma_U)$ is the closure of $D(\Gamma_U)$ in $H^s(\Gamma_U)$ for $s \geq 0$. It holds that $\mathring{\tilde{H}}^{1/2}(\Gamma_U) \subset \mathring{H}^{1/2}(\Gamma_U) = H^{1/2}(\Gamma_U)$ [181]. Note that if $X_1 = H^1(\Gamma_U)$, then $X_{1/2} = H^{1/2}(\Gamma_U)$. If $U_i \in \mathring{\tilde{H}}^{1/2}(\Gamma_U)$, then its extension by zero $U_i^* \in H^{1/2}(\Gamma)$ and the corresponding norms are equivalent (see [335], Lemma 1.1). Accordingly, we can apply the trace theorem to U, and therefore there exists a $w \in [H^1(\Omega)]^3$ such that $w|_\Gamma = U^*$ on Γ_U and $w|_\Gamma = 0$ on $\Gamma - \Gamma_U$. For further information on these questions, see [335], [263]. Several other questions, of a rather technical nature, with respect to the boundary conditions are either not treated here or are still open. For instance (cf., also [159]), compatibility conditions must hold on the "interface" between the different boundary parts. Thus, e.g., "between" Γ_U and Γ_S the prescribed displacements on Γ_U must comply with the friction conditions.

We shall study here the cases mes $\Gamma_U > 0$ and $\Gamma_U = \emptyset$. After introducing the admissible set

$$U_{ad} = \{v \mid v \in [H^1(\Omega)]^3,\ v_i = U_i \quad \text{on} \quad \Gamma_U\} \tag{5.1.15}$$

for the displacements, we consider the following problem.

Problem 1. Find $u \in U_{ad}$ such that

$$a(u, v-u) + \Phi(v) - \Phi(u) \geq (f, v-u) + \int_{\Gamma_F} F_i(v_i - u_i)\, d\Gamma + \int_{\Gamma_S} C_N(v_N - u_N)\, d\Gamma \qquad \forall v \in U_{ad}. \tag{5.1.16}$$

If, more generally, $F_i \in H^{-1/2}(\Gamma_F)$ and $C_N \in H^{-1/2}(\Gamma_S)$, then the integrals $\int_{\Gamma_F} F_i(v_i - u_i)\, d\Gamma$ and $\int_{\Gamma_S} C_{N_i}(v_{N_i} - u_{N_i})\, d\Gamma$ must be replaced by $\langle F, v - u\rangle_{\Gamma_F}$ and $\langle C_N, v_N - u_N\rangle_{\Gamma_S}$ respectively (cf., Sec. 1.4.3).

We will now discuss in what sense the solution of the variational inequality (5.1.16), also called the variational solution of the B.V.P., satisfies the boundary conditions and the equation of equilibrium. Eqs. (5.1.7, 8) define ε_{ij} and σ_{ij}. From (5.1.16), by setting $v - u = \pm\phi \in [D(\Omega)]^3$ we get (5.1.6) in the sense of distributions on Ω. Variational equality (5.1.9) follows from (5.1.6) by applying the generalized Green–Gauss formula (cf., (1.4.37)) in the chosen functional framework. (5.1.16) and (5.1.9) imply the inequality

$$\int_{\Gamma_S} [\mu |C_N|(|v_T| - |u_T|)\, d\Gamma + \langle S_T, v_T - u_T\rangle_{\Gamma_S} + \langle S_N - C_N, v_N - u_N\rangle_{\Gamma_S} + \langle S - F, v - u\rangle_{\Gamma_F} \geq 0, \qquad \forall v \in U_{ad}. \tag{5.1.17}$$

From (5.1.17), by setting $v_i - u_i = \pm\phi_i \in H^{1/2}(\Gamma)$ with support in the int Γ_F, we obtain the boundary condition (5.1.2) in the sense of traces, i.e., as an equality in $H^{-1/2}(\Gamma)$. When $v \in H^1(\Omega)$, $v_N \in H^{1/2}(\Gamma)$ and $v_T \in H_T$ (cf., (1.4.38)). From (5.1.17), by substituting $v_{T_i} = u_{T_i}$ and $S_i = F_i$, we obtain for $v_N - u_N = \pm\phi_N \in H^{1/2}(\Gamma)$ with support in int Γ_S the boundary condition (5.1.5) in the sense of traces. Then (5.1.17) becomes

$$\int_{\Gamma_S} \mu |C_N|(|v_T| - |u_T|)\, d\Gamma + \langle S_T, v_T - u_T\rangle_{\Gamma_S} \geq 0, \qquad \forall v \in U_{ad}, \tag{5.1.18}$$

which is a weak form of (5.1.3, 4) on $H_T \times H'_T$. Relations (5.1.3, 4) result from (5.1.18) (cf., Sec. 3.3.4).

In order to overcome the difficulties connected with the nondifferentiability of Φ, especially in the case of the numerical treatment of the problem, we consider the regularized problem, which arises when the nondifferentiable term Φ in (5.1.16) is replaced by a differentiable functional Φ_ε depending on $\varepsilon > 0$. For our purposes, we introduce the regularized[2] functional Φ_ε

[2] Similarly, the functional $\Phi_\varepsilon(v) = \int_{\Gamma_S} \mu |C_N|(\sqrt{v_T^2 + \varepsilon^2})\, d\Gamma$, $\varepsilon > 0$ can be considered.

defined on $[H^1(\Omega)]^3$ by

$$\Phi_\varepsilon(v) = \int_{\Gamma_S} \frac{\mu |C_N| \, |v_T|^{1+\varepsilon}}{(1+\varepsilon)} \, d\Gamma, \qquad \varepsilon > 0. \tag{5.1.19}$$

Φ_ε is differentiable and convex. Moreover, for $\varepsilon \to 0$ the relations

$$\Phi_\varepsilon(v) \to \Phi(v), \qquad \forall v \in [H^1(\Omega)]^3 \tag{5.1.20}$$

and

$$\liminf \Phi_\varepsilon(u_\varepsilon) \geq \Phi(u) \quad \text{for} \quad u_\varepsilon \to u \quad \text{weakly in} \quad U_{ad} \tag{5.1.21}$$

are easily verified. We also set $\tilde{v} = v - w$ and $\tilde{u} = u - w$, where $w|_\Gamma = U$ (cf., (1.3.46)) on Γ_U, and

$$\tilde{u}, \tilde{v} \in V_0 = \{v \mid v \in [H^1(\Omega)]^3, v = 0 \quad \text{on} \quad \Gamma_U\}. \tag{5.1.22}$$

Then the regularized problem reads:

Problem 1ε. Find $\tilde{u}_\varepsilon \in V_0$ such that

$$a(\tilde{u}_\varepsilon, \tilde{v} - \tilde{u}_\varepsilon) + \Phi_\varepsilon(\tilde{v} + w) - \Phi_\varepsilon(\tilde{u}_\varepsilon + w) \geq (f, \tilde{v} - \tilde{u}_\varepsilon)$$

$$+ \int_{\Gamma_F} F_i(\tilde{v}_i - \tilde{u}_{\varepsilon i}) \, d\Gamma + \int_{\Gamma_S} C_N(\tilde{v}_N - \tilde{u}_{\varepsilon N}) \, d\Gamma - a(w, \tilde{v} - \tilde{u}_\varepsilon), \qquad \forall \tilde{v} \in V_0. \tag{5.1.23}$$

Because of the differentiability of Φ_ε, and by setting $\tilde{v} - \tilde{u}_\varepsilon = \pm \lambda v_1$, we find that (5.1.23) is equivalent to the variational equality (cf., the proof of Prop. 2.2.2)

$$a(\tilde{u}_\varepsilon, \tilde{v}_1) + \left\langle \frac{d\Phi_\varepsilon(\tilde{u}_\varepsilon + w)}{dv}, \tilde{v}_1 \right\rangle_{\Gamma_S} = (f, \tilde{v}_1)$$

$$+ \int_{\Gamma_F} F_i \tilde{v}_{1i} \, d\Gamma + \int_{\Gamma_S} C_N \tilde{v}_N \, d\Gamma - a(w, \tilde{v}_1) \qquad \forall \tilde{v}_1 \in V_0. \tag{5.1.24}$$

The variational inequality (5.1.16) expresses the principle of virtual work for the friction B.V.P. under consideration. Because of the unilateral character of the friction boundary condition, we would naturally expect this principle to have an inequality form (cf., [130] [171]).

5.2 Existence and Uniqueness Propositions

5.2.1 Equivalent Minimum Problem. The case mes $\Gamma_U > 0$

The solution of the B.V.P., if any exists, is a solution of a minimization problem. This is shown by the following proposition.

Proposition 5.2.1. The variational inequality (5.1.16) is equivalent to the minimization problem

$$\Pi(u) = \min \{\Pi(v) \mid v \in U_{ad}\}, \tag{5.2.1}$$

where

$$\Pi(v) = \tfrac{1}{2}a(v, v) + \Phi(v) - (f, v) - \int_{\Gamma_F} F_i v_i \, d\Gamma - \int_{\Gamma_S} C_N v_N \, d\Gamma. \quad (5.2.2)$$

This holds both for mes $\Gamma_U > 0$ and for $\Gamma_U = \varnothing$. In the latter case $U_{ad} = [H^1(\Omega)]^3$.

Proof. Results by direct application of Prop. 2.2.2, in which we take as f_1 the functional $v \to \frac{1}{2}a(v, v) - (f, v) - \langle F, v \rangle_{\Gamma_F} - \langle C_N, v_N \rangle_{\Gamma_S}$ and as f_2 the functional $v \to \Phi(v)$. By means of (1.4.21), it can be verified that

$$\langle \operatorname{grad} f_1(u), v - u \rangle = a(u, v - u) - (f, v - u)$$

$$- \int_{\Gamma_F} F_i(v_i - u_i) \, d\Gamma - \int_{\Gamma_S} C_N(v_N - u_N) \, d\Gamma, \quad (5.2.3)$$

q.e.d.

Note that in the foregoing proof only the symmetry of $a(u, v)$, and the fact that $a(u, u) \geq 0$, $\forall u \in [H^1(\Omega)]^3$ have been used. The latter ensures the convexity of Π and the former allows the derivation of (5.2.3). Functional Π expresses the potential energy of the body and is the sum of the elastic energy, the energy of the external loads and the nondifferentiable friction energy. Proposition 5.2.1 may be regarded as a generalization of the proposition of minimum potential energy of classical elasticity. Further, we shall proceed to the proof of the existence and uniqueness of the solution. First, the case mes $\Gamma_U > 0$ is studied.

Proposition 5.2.2. If mes $\Gamma_U > 0$, there exists a solution of (5.2.1) and it is uniquely determined.

Proof. Because of the continuity and convexity of Π on U_{ad}, it suffices to show (cf., Prop. 2.2.1) that

$$\Pi(v) \to \infty \quad \text{for} \quad \|v\|_1 \to \infty, \qquad v \in U_{ad}. \quad (5.2.4)$$

The assumptions about U imply that there exists a $w \in [H^1(\Omega)]^3$ such that $w = U$ on Γ_U. Further, the transformation $\tilde{v} = v - w \in V_0$ is performed, as in (5.1.23). Π is thus written as

$$\Pi(v) = \tfrac{1}{2}a(\tilde{v}, \tilde{v}) + a(\tilde{v}, w) + \tfrac{1}{2}a(w, w) + \Phi(\tilde{v} + w)$$

$$- (f, \tilde{v} + w) - \int_{\Gamma_F} F_i(\tilde{v} + w)_i \, d\Gamma - \int_{\Gamma_S} C_N(\tilde{v}_N + w) \, d\Gamma. \quad (5.2.5)$$

Since mes $\Gamma_U > 0$, the first Korn inequality holds, and implies (cf., Sec. 4.1.1) that

$$a(\tilde{v}, \tilde{v}) \geq c \|\tilde{v}\|_1, \qquad \forall \tilde{v} \in V_0, \qquad c \text{ const.} > 0. \quad (5.2.6)$$

Because of the nonnegativity of $\Phi(\tilde{v} + w)$ and the continuity of the linear forms, (5.2.5) with (5.2.6) implies (5.2.4). Accordingly, the existence of at least one solution is guaranteed. Because of the strict convexity of Π, this solution is unique, q.e.d.

5.2.2 Study of the Case $\Gamma_U = \varnothing$

If $\Gamma_U = \varnothing$, then the existence proof is more complicated since the body may undergo a rigid-body displacement. In order to derive a necessary condition for the existence of the solution, we prove the following proposition.

Proposition 5.2.3. If $\Gamma_U = \varnothing$, the variational inequality (5.1.16) is equivalent to the problem: find $u \in [H^1(\Omega)]^3$ such that

$$\left| a(u, v) - (f, v) - \int_{\Gamma_F} F_i v_i \, d\Gamma - \int_{\Gamma_S} C_N v_N \, d\Gamma \right| \leq \Phi(v), \qquad \forall v \in [H^1(\Omega)]^3 \tag{5.2.7}$$

where (5.2.7) holds for $v = u$ as an equality.

Proof. By substituting $v = 0$ and $v = 2u$ in (5.1.16), there results

$$a(u, u) + \Phi(u) = (f, u) + \int_{\Gamma_F} F_i u_i \, d\Gamma + \int_{\Gamma_S} C_N u_N \, d\Gamma, \tag{5.2.8}$$

and thus (5.1.16) is equivalent to (5.2.8) and to

$$a(u, v) + \Phi(v) \geq (f, v) + \int_{\Gamma_F} F_i v_i \, d\Gamma + \int_{\Gamma_S} C_N v_N \, d\Gamma, \quad \forall v \in [H^1(\Omega)]^3. \tag{5.2.9}$$

Replacing v by $-v$ in (5.2.9), we obtain (5.2.7), q.e.d.

This proposition allows for the derivation of a necessary condition for the existence of a solution. Indeed, if in (5.2.7) we put $v = r \in \mathcal{R}$ where $\mathcal{R}$ denotes the space of rigid-body displacements and r is nonzero (cf., 1.4.27), then it follows that

$$\Phi(r) \geq \left| (f, r) + \int_{\Gamma_F} F_i r_i \, d\Gamma + \int_{\Gamma_S} C_N r_N \, d\Gamma \right|, \quad \forall r \in \mathcal{R}, \quad r \neq 0, \tag{5.2.10}$$

which is a necessary condition for the existence of an equilibrium configuration. (5.2.10) can take the equivalent form

$$\Phi(r) \geq \int_{\Omega} f_i r_i \, d\Omega + \int_{\Gamma_F} F_i r_i \, d\Gamma + \int_{\Gamma_S} C_N r_N \, d\Gamma, \quad \forall r \in \mathcal{R}, \quad r \neq 0. \tag{5.2.11}$$

The existence of a solution in the case $\Gamma_U = \varnothing$ will be shown on the stronger assumption that

$$\Phi(r) > (f, r) + \int_{\Gamma_F} F_i r_i \, d\Gamma + \int_{\Gamma_S} C_N r_N \, d\Gamma, \quad \forall r \in \mathscr{R}, \quad r \neq 0. \tag{5.2.12}$$

Proposition 5.2.4. If $\Gamma_U = \varnothing$ and if (5.2.12) holds, there exists at least one solution of the variational inequality (5.1.16).

Proof. The variational inequality (5.1.16) is equivalent to the minimization problem (5.2.1), in which the minimum is now sought over $[H^1(\Omega)]^3$. We shall therefore apply Prop. 2.2.1 in the quotient space $[H^1(\Omega)]^3/\mathscr{R}$. Let Q' be the orthogonal projector of $[H^1(\Omega)]^3$ onto $\mathscr{R}$ in the $[L^2(\Omega)]^3$-norm and $\mathscr{R}^\perp$ the orthogonal complement of $\mathscr{R}$. Then we set $P' = I - Q'$ (I is the identity operator) and

$$v = \bar{v} + r, \qquad r \in \mathscr{R}, \qquad \bar{v} = v - Q'v = P'v \in \mathscr{R}^\perp. \tag{5.2.13}$$

From (4.1.36), the relation

$$a(\bar{v}, \bar{v}) \geq c \| \bar{v} \|_1^2, \qquad c \text{ const} > 0 \qquad \forall \bar{v} \in \mathscr{R}^\perp \tag{5.2.14}$$

can be verified (cf., also Sec. 1.1.2). The continuity of the linear forms, condition (5.2.14) and the sufficient condition (5.2.12), written in the form

$$(f, r) + \int_{\Gamma_F} F_i r_i \, d\Gamma + \int_{\Gamma_S} C_N r_N \, d\Gamma$$

$$\leq \Phi(r) - c|r|_2, \qquad c \text{ const} > 0, \qquad \forall r \in \mathscr{R}, \qquad r \neq 0, \tag{5.2.15}$$

permit us to verify that

$$\Pi(v) = \Pi(\bar{v} + r) = \tfrac{1}{2}a(\bar{v}, \bar{v}) + \Phi(\bar{v} + r) - (f, \bar{v} + r) - \int_{\Gamma_F} F_i(\bar{v}_i + r_i) \, d\Gamma$$

$$- \int_{\Gamma_S} C_N(\bar{v}_N + r_N) \, d\Gamma \geq \tfrac{1}{2}a(\bar{v}, \bar{v}) + \Phi(\bar{v} + r) - \Phi(r) - (f, \bar{v}) - \int_{\Gamma_F} F_i v_i \, d\Gamma$$

$$- \int_{\Gamma_S} C_N \bar{v}_N \, d\Gamma + c|r|_2 \geq c_1(\|\bar{v}\|_1^2 + |r|_2) - c\|\bar{v}\|_1; \qquad c_1 \text{ const} > 0. \tag{5.2.16}$$

In deriving (5.2.16) we also considered that the functional Φ is convex, finite and continuous on $[H^1(\Omega)]^3$, and therefore (Prop. 2.3.2) $\partial\Phi(u) \neq \varnothing$ for every $u \in [H^1(\Omega)]^3$. Thus

$$\Phi(\bar{v} + r) - \Phi(r) \geq \langle X, \bar{v} \rangle_1 \geq -c\|\bar{v}\|_1, \qquad c \text{ const} \geq 0. \tag{5.2.17}$$

But $(\|\bar{v}\|_1^2 + |r|_2^2)^{1/2}$ is a norm equivalent to the standard norm $\|v\|_1$ on $[H^1(\Omega)]^3$. Indeed, $\|v\|_1$ is equivalent to the norm $(\varepsilon(v) + |v|_2^2)^{1/2}$. Furthermore, $\varepsilon(v) + |v|_2^2 = \varepsilon(\bar{v}) + |\bar{v}|_2^2 + |r|_2^2$, and $(\varepsilon(\bar{v}) + |\bar{v}|_2^2)^{1/2}$ is a norm equivalent to $\|\bar{v}\|_1$. Accordingly, as $\|v\|_1 \to \infty$, either $\|\bar{v}\|_1 \to \infty$ or $|r|_2 \to \infty$ or both, and thus (5.2.16) implies that $\Pi(v) \to \infty$, q.e.d.

It is to be noted that not only the two cases considered here, mes $\Gamma_U > 0$ and $\Gamma_U = \varnothing$, exist. For instance, if we have prescribed displacements at some points of the boundary with zero measure, then $\Gamma_U \neq \varnothing$ but mes $\Gamma_U = 0$. In this case a modification of the functional framework is necessary. In the following proposition, the uniqueness of the solution is investigated.

Proposition 5.2.5. If $\Gamma_U = \varnothing$, then the stresses and strains are uniquely determined. The displacements are unique up to a rigid-body displacement.

Proof. Let u_1 and u_2 be two solutions of (5.1.16). We set in (5.1.16) first $u = u_2$, $v = u_1$ and then $u = u_1$, $v = u_2$. It follows that

$$a(u_1 - u_2, u_1 - u_2) \leq 0, \tag{5.2.18}$$

and thus

$$u_1 - u_2 \in \mathcal{R}. \tag{5.2.19}$$

Accordingly, the two solutions differ by a rigid-body displacement and thus

$$\varepsilon(u_1) = \varepsilon(u_2), \qquad \sigma(u_1) = \sigma(u_2), \tag{5.2.20}$$

q.e.d.

5.2.3 Further Properties of the Solution

The solution u of (5.1.16) depends on the coefficient of friction μ and on $|C_N|$. We denote by b the term $\mu |C_N|$ and by u_b the corresponding solution on the assumption that mes $\Gamma_U > 0$. The following proposition reveals a feature of this dependence.

Proposition 5.2.6. Suppose that $b \in L^\infty(\Gamma_S)$. Then the application $b \to u_b$ is Lipschitzian (cf., Sec. 1.3.1).

Proof. Let $b_1, b_2 \in L^\infty(\Gamma_S)$ and let u_{b_1}, u_{b_2} be the corresponding solutions of (5.1.16). Setting $v = u_{b_2}$ and $v = u_{b_1}$ respectively in (5.1.16) and adding, we obtain the relation

$$a(u_{b_2} - u_{b_1}, u_{b_2} - u_{b_1}) + \int_{\Gamma_S} (b_2 - b_1)(|u_{Tb_2}| - |u_{Tb_1}|)\, d\Gamma \leq 0, \tag{5.2.21}$$

which implies, due to the coerciveness of $a(v, v)$ on V_0,

$$c\|u_{b_2} - u_{b_1}\|_1^2 \leq \|b_2 - b_1\|_{L^2(\Gamma_S)} \|u_{b_2} - u_{b_1}\|_{[L^2(\Gamma_S)]^3}. \tag{5.2.22}$$

From (1.3.54), we obtain

$$\| u_{b_2} - u_{b_1} \|_1 \leq c \| b_2 - b_1 \|_{L^2(\Gamma_S)}, \qquad c \text{ const} > 0, \tag{5.2.23}$$

q.e.d.

It can be proved as well that if b has a constant value on Γ_S, then the application $b \to \int |u_{Tb}| \, d\Gamma$ is Lipschitz decreasing and $b \to a(u_b, u_b)$ is convex decreasing (cf., [120]). The regularized problem (5.1.24) admits a unique solution if mes $\Gamma_U > 0$, because of Prop. 5.2.2. The relation between u and u_ε is given by the next proposition.

Proposition 5.2.6. Suppose that mes $\Gamma_U > 0$. Then as $\varepsilon \to 0$,

$$u_\varepsilon \to u \quad \text{in} \quad U_{ad} \tag{5.2.24}$$

Proof. The monotonicity of $d\Phi_\varepsilon/dv$ implies, that

$$\left\langle \frac{d\Phi_\varepsilon}{dv}(\tilde{u}_\varepsilon + w), \tilde{u}_\varepsilon + w \right\rangle_{\Gamma_S} \geq 0. \tag{5.2.25}$$

From (5.1.24), because of the coerciveness of $a(u, u)$, we obtain $\| u_\varepsilon \|_1 \leq c$. Accordingly, a subsequence $\{u_\varepsilon\}$ may be considered such that $u_\varepsilon \to u$ in U_{ad} weakly. In order to show that u is the solution of (5.1.16), we take the limit in (5.1.23) for $\varepsilon \to 0$. From (5.1.20, 21, 23) the inequality (5.1.16) is obtained; $\lim a(u_\varepsilon, u_\varepsilon) = a(u, u)$, as $\varepsilon \to 0$, implies (5.2.24), q.e.d.

Up to this point, only variational principles concerning displacements have been studied. In the next section, variational principles for the stresses will be derived and their relation to the foregoing variational principles will be discussed.

5.3 Dual Formulation. Complementary Energy

5.3.1 Minimization of the Complementary Energy

First we introduce the space

$$H = \{\tau \mid \tau = \{\tau_{ij}\}, \tau_{ij} = \tau_{ji}, \tau_{ij} \in L^2(\Omega), i, j = 1, 2, 3\} \tag{5.3.1}$$

and the admissible set of the stresses

$$\Sigma_{ad} = \{\tau \mid \tau \in H, \tau_{ij,j} + f_i = 0 \text{ in } \Omega, T_N = C_N \text{ and}$$
$$|T_T| \leq \mu |C_N| \text{ on } \Gamma_S, T_i = F_i \text{ on } \Gamma_F\}, \tag{5.3.2}$$

where T, T_N and T_T denote the vectors corresponding to S, S_N and S_T, $T_N \in H^{-1/2}(\Gamma)$, $T_T \in H'_T(\Gamma)$ (cf., (1.4.39) and Sec. 1.4.3), whereas the restriction of T_T to Γ_S is an element of $[L^\infty(\Gamma_S)]^3$. The set Σ_{ad} is a convex closed subset of the space Σ equipped with the norm (1.4.35). We also need the inverse relation to (5.1.8), which reads

$$\varepsilon_{ij} = c_{ijhk}\sigma_{hk}. \tag{5.3.3}$$

Tensor c enjoys the same symmetry and ellipticity properties as tensor C. Σ_{ad} is clearly nonempty either if mes $\Gamma_U \neq 0$ or if $\Gamma_U = \varnothing$ and (5.2.12) hold. Then a solution u of (5.1.16) or (5.2.1), i.e. the displacement field, exists, and from it a stress field $\sigma(u) \in \Sigma_{ad}$ can be derived.

Proposition 5.3.1. Suppose that (5.1.16) has a solution u. Then the stress field $\sigma(u)$ satisfies the variational inequality

$$A(\sigma, \tau - \sigma) \geq \langle U, T - S\rangle_{\Gamma_U}, \qquad \forall \tau \in \Sigma_{ad} \tag{5.3.4}$$

with

$$A(\sigma, \tau) = \int_\Omega c_{ijhk}\sigma_{ij}\tau_{hk}\, d\Omega. \tag{5.3.5}$$

Proof. Let us denote by $\varepsilon(u) \in [L^2(\Omega)]^6$ the strain field corresponding to u, from which $\sigma(u)$ is obtained through (5.1.8). Multiplying (5.1.7) by $\tau_{ij} - \sigma_{ij}$, where $\tau, \sigma \in \Sigma_{ad}$, and applying the Green–Gauss theorem (cf., (1.4.45)), we get

$$A(\sigma, \tau - \sigma) = \langle U, T - S\rangle_{\Gamma_U} + \langle u_T, T_T - S_T\rangle_{\Gamma_S}, \qquad \forall \tau \in \Sigma_{ad}. \tag{5.3.6}$$

But the friction boundary condition implies on $H'_T \times H_T$ (see (3.3.177)) the relation

$$\langle u_T, T_T - S_T\rangle_{\Gamma_S} \geq 0, \tag{5.3.7}$$

which with (5.3.6) yields (5.3.4), q.e.d.

From the physical point of view, the variational inequality (5.3.4) expresses the principle of complementary virtual work. Roughly speaking, the inequality form of this principle is due to the inequality subsidiary condition $|S_T| \leq \mu |C_N|$.

Proposition 5.3.2. Suppose that $\Sigma_{ad} \neq \varnothing$. Then the variational inequality (5.3.4) has a unique solution.

Moreover, (5.3.4) is equivalent to the minimization problem

$$\Pi^c(\sigma) = \min\{\Pi^c(\tau) \mid \tau \in \Sigma_{ad}\}, \tag{5.3.8}$$

where

$$\Pi^c(\tau) = \tfrac{1}{2}A(\tau, \tau) - \langle U, T\rangle_{\Gamma_U}. \tag{5.3.9}$$

Proof. The bilinear form $A(\tau, \tau)$ is coercive on H due to the ellipticity of $\{c_{ijhk}\}$. But due to the continuity of $\bar{\gamma}$ (Sec. 1.4.3) and the fact that $\tau \in \Sigma_{ad}$,

$$\Pi^c(\tau) = \tfrac{1}{2}A(\tau, \tau) - \langle U, T\rangle_{\Gamma_U} \geq c_1 \|\tau\|_H^2 - c_2\|\tau\|_\Sigma + c_1(|f|_2^2 - |f|_2^2)$$
$$= c_1\left(\|\tau\|_H^2 + \int_\Omega \tau_{ij,j}\tau_{ik,k}\, d\Omega\right) - c_2\|\tau\|_\Sigma + c_3. \tag{5.3.10}$$

The proof results directly as an application of Propositions 2.2.1 and 2.2.2, q.e.d.

Functional Π^c is the complementary energy of the body; its minimum is sought over a convex closed set. The minimization problem (5.3.8) can be viewed as a generalization of the well-known principle of minimum complementary energy of classical elasticity [236].

5.3.2 Duality

Further, the existing duality between the problems of minimum potential and minimum complementary energy will be investigated in the spirit of the general duality theory of mechanical systems, which was developed in Sec. 3.2.2.

We write (5.2.1) in the equivalent form

$$\bar{\Pi}(u) = \min\{\bar{\Pi}(u) \mid u \in [H^1(\Omega)]^3\}, \tag{5.3.11}$$

where $\bar{\Pi}(u) = \Pi(u) + I_{U_{ad}}(u)$. We recall that $I_{U_{ad}}$ is the indicator of the set U_{ad}. In relation (3.2.47), defining the primal problem, A is taken to be

$$A(u) = \bar{A}(u) + I_{U_{ad}}(u)$$
$$= \Phi(u) - (f, u) - \int_{\Gamma_F} F_i u_i\, d\Gamma - \int_{\Gamma_S} C_N u_N\, d\Gamma + I_{U_{ad}}(u). \tag{5.3.12}$$

We shall show that for $\sigma \in H$

$$A^c(-\Lambda'\sigma) = I_{\Sigma_{ad}}(\sigma) - \langle S, U\rangle_{\Gamma_U}, \tag{5.3.13}$$

where Λ' is the transpose operator of $\Lambda: u \to \Lambda(u) = \{\varepsilon_{ij}(u)\} \in [L^2(\Omega)]^6$. Indeed,

$$\begin{aligned}
A^c(-\Lambda'\sigma) &= \sup_{[H^1(\Omega)]^3} [(-\Lambda'\sigma, v) - A(v)] \\
&= \sup_{[H^1(\Omega)]^3} [(-\sigma, \Lambda v) - A(v)] \\
&= (-\sigma, \Lambda w) + (f, w) + \langle F, w\rangle_{\Gamma_F} + \langle C_N, w_N\rangle_{\Gamma_S} \\
&\quad + \sup_{V_0} [(-\sigma, \Lambda\tilde{v}) + (f, \tilde{v}) + \langle F, \tilde{v}\rangle_{\Gamma_F} \\
&\quad + \langle C_N, \tilde{v}_N\rangle_{\Gamma_S} - \Phi(\tilde{v} + w)].
\end{aligned} \tag{5.3.14}$$

Here we have used the transformation $\tilde{v} = v - w$, where $\tilde{v} \in V_0$ and $w|_{\Gamma_U} = U$. But $V_0 \cap [D(\bar{\Omega})]^3$ is dense in V_0 for the H^1-norm (cf. the proof of Theorems 1.1 and 1.3 of [312]). Thus in (5.3.14) we may replace the supremum on V_0 by the supremum over $V_0 \cap [D(\bar{\Omega})]^3$, which clearly is larger than or equal to the supremum on $[D(\Omega)]^3$. Accordingly,

$$A^c(-\Lambda'\sigma) \geq \sup_{v \in [D(\Omega)]^3} [-(\sigma, \Lambda v) + (f, v)] + (-\sigma, \Lambda w) + (f, w) + \langle F, w \rangle_{\Gamma_F} + \langle C_N, w_N \rangle_{\Gamma_S}. \quad (5.3.15)$$

But for $v \in [D(\Omega)]^3$

$$-(\sigma, \Lambda v) + (f, v) = (\sigma_{ij,j} + f_i, v), \quad (5.3.16)$$

and thus the supremum in (5.3.15) is 0 if $\sigma_{ij,j} + f_i = 0$ in Ω, and ∞ otherwise. From this last relation, by means of the Green–Gauss theorem (cf., (1.4.37)), we obtain

$$(\sigma, \Lambda v) = \langle S, v \rangle_\Gamma + (f, v), \qquad \forall v \in [H^1(\Omega)]^3, \quad (5.3.17)$$

and thus

$$\begin{aligned} A^c(-\Lambda'\sigma) &= (-\sigma, \Lambda w) + (f, w) + \int_{\Gamma_F} F_i w_i \, d\Gamma + \int_{\Gamma_S} C_N w_N \, d\Gamma \\ &\quad + \sup_{V_0 \cap D(\bar{\Omega})} \left[-\langle S, \tilde{v} \rangle_{\Gamma - \Gamma_U} + \int_{\Gamma_F} F_i \tilde{v}_i \, d\Gamma \right. \\ &\quad \left. + \int_{\Gamma_S} C_N \tilde{v}_N \, d\Gamma - \Phi(\tilde{v} + w) \right] \qquad (5.3.18) \\ &= \sup_{V_0 \cap D(\bar{\Omega})} -[\langle S - F, \tilde{v} \rangle_{\Gamma_F} + \langle S_N - C_N, \tilde{v}_N \rangle_{\Gamma_S} \\ &\quad + (\langle S_T, \tilde{v}_T + w_T \rangle_{\Gamma_S} + \Phi(\tilde{v} + w))] + \langle S_T, w_T \rangle_{\Gamma_S} \\ &\quad + (-\sigma, \Lambda w) + (f, w) + \int_{\Gamma_F} F_i w_i \, d\Gamma + \int_{\Gamma_S} C_N w_N \, d\Gamma. \end{aligned}$$

But this supremum is not less than the supremum over the linear subspace $\{v \mid v \in V_0 \cap D(\bar{\Omega}), v = 0 \text{ on } \Gamma_S\}$, the latter supremum being zero if $S_i = F_i$ on Γ_F, and ∞ otherwise. Similarly, we obtain $S_N = C_N$ on Γ_S. Moreover, (cf., (3.3.176-8))

$$\sup_{V_0 \cap D(\bar{\Omega})} -(\langle S_T, \tilde{v}_T + w_T \rangle_{\Gamma_S} + \Phi(\tilde{v} + w)) = \begin{cases} 0 & \text{if } |S_T| \leq \mu |C_N| \\ \infty & \text{otherwise,} \end{cases} \quad (5.3.19)$$

and thus we see that

$$A^c(-\Lambda'\sigma) = \begin{cases} (-\sigma, \Lambda w) + (f, w) + \int_{\Gamma_F} F_i w_i \, d\Gamma + \int_{\Gamma_S} C_N w_N \, d\Gamma + \langle S_T, w_T \rangle_{\Gamma_S} & \text{if } \sigma \in \Sigma_{ad} \\ \infty \quad \text{otherwise} \end{cases} \tag{5.3.20}$$

which, by means of the Green–Gauss theorem, is equivalent to (5.3.13).

Because of Hooke's law, $B(\Lambda u)$ in (3.3.47) reads

$$B(\Lambda u) = \frac{1}{2} \int_\Omega C_{ijhk} \varepsilon_{ij}(u) \varepsilon_{hk}(u) \, d\Omega, \tag{5.3.21}$$

and thus

$$B^c(\sigma) = \sup_{v \in [H^1(\Omega)]^3} [(\sigma, \Lambda v) - B(\Lambda v)] = \tfrac{1}{2} A(\sigma, \sigma), \quad \text{if} \quad \sigma_{ij} = C_{ijhk} \varepsilon_{hk}. \tag{5.3.22}$$

Thus the dual problem $\sup_H [-A^c(-\Lambda'\sigma) - B^c(\sigma)]$ is identical to the minimum problem (5.3.8) of the complementary energy.

The following propositions illustrate the relation between the primal and dual problems.

Proposition 5.3.3. The solution u of the primal problem and the solution σ of the dual problem satisfy the relation

$$\Pi(u) + \Pi^c(\sigma) = 0 \tag{5.3.23}$$

and are related by the conditions

$$\sigma_{ij} = C_{ijhk} \varepsilon_{hk}(u) \tag{5.3.24}$$

$$S_{T_i} u_{T_i} + \mu |C_N| \, |u_T| = 0. \tag{5.3.25}$$

Proof. (5.3.23) results directly from Prop. 3.2.7. It can be immediately verified that the extremality conditions (3.2.52, 51) have the form (5.3.24, 25), q.e.d.

We can thus formulate the following relation for $u \in U_{ad}$ and $\sigma \in \Sigma_{ad}$:

$$\Pi(v) \geq \Pi(u) = -\Pi^c(\sigma) \geq -\Pi^c(\tau), \qquad \forall v \in U_{ad}, \qquad \forall \sigma \in \Sigma_{ad}. \tag{5.3.26}$$

If $\Gamma_U = \varnothing$ and (5.2.12) holds, then U_{ad} is replaced in (5.3.26) by $[H^1(\Omega)]^3$. Relation (5.3.26) generalizes a similar result of classical linear elasticity.

Proposition 5.3.4. The conditions

(i) u minimizes the potential energy Π over U_{ad}, σ minimizes the complementary energy Π^c over Σ_{ad} and (5.3.23) holds, and

(ii) $u \in U_{ad}$ and $\sigma \in \Sigma_{ad}$ satisfy the extremality condition (5.3.24,25)

are equivalent to each other.

Proof. Follows directly by applying Prop. 3.2.6, q.e.d.

The saddle-point theory also can be applied to the problem being considered (see, e.g., [91]). Thus we can characterize the solution (u, σ) as a saddle-point of an appropriately defined Lagrangian.

5.4 The Dynamic B.V.P.

5.4.1 Classical and Variational Formulations of the Problem

The dynamic behavior of an elastic body, with frictional conditions imposed on a part of its boundary, is described, on the assumptions of small displacements and small deformations by the relations

$$\sigma_{ij,j} + f_i = \rho u_i'' \quad \text{in} \quad \Omega \times (0, T), \tag{5.4.1}$$

$$\varepsilon_{ij}(u) = \tfrac{1}{2}(u_{i,j} + u_{j,i}) \quad \text{in} \quad \Omega \times (0, T), \tag{5.4.2}$$

$$\sigma_{ij} = C_{ijhk}\,\varepsilon_{hk} \quad \text{in} \quad \Omega \times (0, T), \tag{5.4.3}$$

$$u_i = U_i \quad \text{on} \quad \Gamma_U \times (0, T), \tag{5.4.4}$$

$$S_i = F_i \quad \text{on} \quad \Gamma_F \times (0, T), \tag{5.4.5}$$

$$\text{if} \quad |S_T| < \mu |S_N|, \quad \text{then} \quad u_T' = 0 \quad \text{on} \quad \Gamma_S \times (0, T) \tag{5.4.6}$$

$$\text{if} \quad |S_T| = \mu |S_N|, \quad \text{there exists} \quad \lambda \geq 0$$

$$\text{such that} \quad u_T' = -\lambda S_T \quad \text{on} \quad \Gamma_S \times (0, T), \tag{5.4.7}$$

and

$$S_N = C_N \quad \text{on} \quad \Gamma_S \times (0, T). \tag{5.4.8}$$

Here, the time ranges over the interval $[0, T]$, and the prime denotes the time derivative $\partial/\partial t$. The density of the body $\rho = \rho(x)$, $x \in \Omega$, is assumed to be a positive function of $L^\infty(\Omega)$ (with essinf $\rho(x) > 0$). The prescribed forces and displacements f, F, C_N and U are functions of x and t. All the functions in (5.4.1–8) are assumed initially to be appropriately smooth. To the above relations the initial conditions

$$u_i = u_{0i}, \qquad u_i' = u_{1i} \quad \text{for} \quad x \in \Omega \quad \text{and} \quad t = 0 \tag{5.4.9}$$

are added. Here u_{0i} and u_{1i} are given functions of x expressing the displacement and velocity field at $t = 0$. Obviously, u_0 and u_1 must be compatible with the kinematical boundary condition on Γ_U. The appearance of u'_T instead of u_T in (5.4.6, 7) necessitates consideration of variational expressions with respect to the velocities. Then the sets

$$U_{ad} = \{v \mid v \in [H^1(\Omega)]^3,\ v_i = U'_i \text{ on } \Gamma_U\} \tag{5.4.10}$$

are introduced and the following variational formulation of the B.V.P. is considered: find u: $[0, T] \to [H^1(\Omega)]^3$ with $u'(t) \in U_{ad}$ and $u''(t) \in [L^2(\Omega)]^3$ to satisfy the variational inequality[3]

$$(\rho u'', v - u') + a(u, v - u') + \Phi(v) - \Phi(u') \geq (f, v - u')$$

$$+ \int_{\Gamma_F} F_i(v_i - u'_i)\, d\Gamma + \int_{\Gamma_S} C_N(v_N - u'_N)\, d\Gamma, \qquad \forall v \in U_{ad} \tag{5.4.11}$$

and the initial conditions (5.4.9).

This variational inequality results from (5.4.1–8) by multiplying (5.4.1) by $v - u'$, integrating over Ω, and applying formally the Green-Gauss theorem. Then the boundary conditions are taken into account, as in the static problem. At this point, we would like to call attention to the fact that the final functional framework of the problem has not yet been stated, and therefore (5.4.11) constitutes only a "preliminary" variational form of the problem. Further, after considering a more precise dependence of the data on time, we will investigate in what sense the variational solution satisfies the relations (5.4.1–9). The variational inequality (5.4.11) expresses physically the principle of virtual power in its inequality form. For the dynamic problem, the cases mes $\Gamma_U > 0$ and $\Gamma_U = \varnothing$ can be studied together, on the additional assumption that if $\Gamma_U = \varnothing$ then relation (5.2.12) holds for all $t > 0$. (5.2.12) ensures that no rigid-body motion incompatible with the boundary conditions will appear. Then the subspace

$$V_0 = \{v \mid v \in [H^1(\Omega)]^3,\ v = 0 \text{ on } \Gamma_U\} \tag{5.4.12}$$

is introduced. By means of the substitutions $\tilde{v} = v - w$ and $\tilde{u} = u - w$, where w: $[0, T] \to [H^1(\Omega)]^3$ is a function such that $w(t)|_{\Gamma_U} = U(t) \in [\dot{H}^{1/2}(\Gamma_U)]^3$ (cf. also [181] Vol. 2, Ch. 4 Theorems 2.1 and 5.3), we get the following form of the problem.

Problem 1. Find a function $\tilde{u}$: $[0, T] \to V_0$ with $\tilde{u}'(t) \in V_0$ and $\tilde{u}''(t) \in [L^2(\Omega)]^3$ such that

$$(\rho \tilde{u}'', \tilde{v} - \tilde{u}') + a(\tilde{u}, \tilde{v} - \tilde{u}') + \Phi(\tilde{v} + w')$$

$$- \Phi(\tilde{u}' + w') \geq (\psi, \tilde{v} - \tilde{u}'), \qquad \forall \tilde{v} \in V_0, \tag{5.4.13}$$

[3] More precisely, we could write $u'(t, \cdot) \in U_{ad}$, etc.

and for $t = 0$

$$\tilde{u} = \tilde{u}_0 = u_0 - w(0), \qquad \tilde{u}' = \tilde{u}_1 = u_1 - w'(0). \tag{5.4.14}$$

Here

$$(\psi, \tilde{v} - \tilde{u}') = (f, \tilde{v} - \tilde{u}') + \int_{\Gamma_F} F_i(\tilde{v}_i - \tilde{u}'_i)\, d\Gamma$$
$$+ \int_{\Gamma_S} C_N(\tilde{v}_N - \tilde{u}'_N)\, d\Gamma - (\rho w'', \tilde{v} - \tilde{u}') - a(w, \tilde{v} - \tilde{u}'). \tag{5.4.15}$$

5.4.2 Existence of Solution

In order to prove the existence of the solution, we consider a smoothened form of the variational inequality (5.4.13), the regularized problem. This form results by replacing Φ by Φ_ε in (5.4.13), where Φ_ε is given by (5.1.19). The following problem is considered.

Problem 1ε. Find a function $\tilde{u}_\varepsilon$: $[0, T] \to V_0$ with $\tilde{u}'_\varepsilon(t) \in V_0$ and $\tilde{u}''_\varepsilon(t) \in [L^2(\Omega)]^3$ which satisfies the variational equality

$$(\rho \tilde{u}''_\varepsilon, \tilde{v}) + a(\tilde{u}_\varepsilon, \tilde{v}) + \left\langle \frac{d\Phi_\varepsilon}{dv}(\tilde{u}'_\varepsilon + w'), \tilde{v} \right\rangle_{\Gamma_S} = (\psi, \tilde{v}), \qquad \forall \tilde{v} \in V_0 \tag{5.4.16}$$

and the initial conditions (5.4.14).

First, the solution of problem (5.4.13, 14) is "constructed" as the limit of the solution $\tilde{u}_\varepsilon$ of (5.4.14, 16) as $\varepsilon \to 0$, on the assumption that $\tilde{u}_\varepsilon$ exists. It is then shown that $\tilde{u}_\varepsilon$ indeed exists.

Proposition 5.4.1. Suppose that

$$f, f', f'' \in L^2(0T, [L^2(\Omega)]^3), \tag{5.4.17}$$

$$C_N, C'_N, C''_N \in L^2(0T, [L^2(\Gamma_S)]^3), \tag{5.4.18}$$

$$F, F', F'' \in L^2(0T, [L^2(\Gamma_F)]^3), \tag{5.4.19}$$

$$w, w', w'', w''', w'''' \in L^2(0T, [H^1(\Omega)]^3), \tag{5.4.20}$$

$$w(0) \in [H^2(\Omega)]^3, \tag{5.4.21}$$

$$\mu |C_N| \text{ is independent of } t, \tag{5.4.22}$$

$$\tilde{u}_0 \in [H^2(\Omega)]^3, \qquad \tilde{u}_1 \in [H^1(\Omega)]^3, \tag{5.4.23}$$

$$\tilde{u}_{1T} + w'_T(0) = 0 \quad \text{on} \quad \Gamma_S, \tag{5.4.24}$$

and

$$\tilde{v} \to a(w(0), \tilde{v}) - \langle S_T(0), \tilde{v}_T \rangle_{\Gamma_S} + \int_\Omega [C_{ijhk}\, \varepsilon_{hk}(\tilde{u}_0)]_{,j}\, \tilde{v}_i\, d\Omega \tag{5.4.25}$$

is a continuous linear functional on V_0 for the $[L^2(\Omega)]^3$-topology. Under these conditions the solution $\tilde{u}_\varepsilon$ of the regularized problem has the following properties: for $\varepsilon \to 0$

$$\tilde{u}_\varepsilon \to \tilde{u} \quad \text{in} \quad L^\infty(0T, V_0) \quad \text{weakly}_*, \tag{5.4.26}$$

$$\tilde{u}'_\varepsilon \to \tilde{u}' \quad \text{in} \quad L^\infty(0T, V_0) \quad \text{weakly}_*, \tag{5.4.27}$$

$$\tilde{u}''_\varepsilon \to \tilde{u}'' \quad \text{in} \quad L^\infty(0T, [L^2(\Omega)]^3) \quad \text{weakly}_*; \tag{5.4.28}$$

u is a solution of problem (5.4.13, 14).

Proof. (a) Assume (cf., Prop. 5.4.2) that a unique $\tilde{u}_\varepsilon \in L^\infty(0T, V_0)$ exists with

$$\tilde{u}'_\varepsilon \in L^\infty(0T, V_0), \qquad \tilde{u}''_\varepsilon \in L^\infty(0T, [L^2(\Omega)]^3). \tag{5.4.29}$$

(b) Let us consider (5.4.16) but with $\tilde{u}'_\varepsilon + w'$ in place of $\tilde{v}$. Because of the monotonicity of $d\Phi_\varepsilon/dv$, we then obtain the relation

$$(\rho\tilde{u}''_\varepsilon, \tilde{u}'_\varepsilon + w') + a(\tilde{u}_\varepsilon, \tilde{u}'_\varepsilon + w') \leq (\psi, \tilde{u}'_\varepsilon + w'), \tag{5.4.30}$$

which reduces to

$$\frac{1}{2}\frac{d}{dt}\left(\int_\Omega \rho(\tilde{u}'_\varepsilon)^2\, d\Omega + a(\tilde{u}_\varepsilon, \tilde{u}_\varepsilon)\right) \leq (\psi, \tilde{u}'_\varepsilon + w') - (\rho\tilde{u}''_\varepsilon, w') - a(\tilde{u}_\varepsilon, w'). \tag{5.4.31}$$

From (5.4.31) combined with (4.1.38), and through integration from 0 to t, we obtain[4]

$$\begin{aligned} |\tilde{u}'_\varepsilon|_2^2 + c_1 \|\tilde{u}_\varepsilon\|_1^2 - c_2 |\tilde{u}_\varepsilon|_2^2 \leq |\tilde{u}_1|_2^2 + c\|\tilde{u}_0\|_1^2 \\ + 2\int_0^t (\psi, \tilde{u}'_\varepsilon + w')\,dt - 2(\rho\tilde{u}'_\varepsilon, w') + 2(\rho\tilde{u}_1, w'(0)) \\ + 2\int_0^t (\rho\tilde{u}'_\varepsilon, w'')\,dt - 2\int_0^t a(\tilde{u}_\varepsilon, w')\,dt. \end{aligned} \tag{5.4.32}$$

Since $\psi, \psi' \in L^2(0T, V'_0)$, there follows from (5.4.32), by means of the inequality

$$|\tilde{u}_\varepsilon|_2^2 \leq c\int_0^t |\tilde{u}'_\varepsilon|_2^2\,dt + |\tilde{u}_0|_2^2, \tag{5.4.33}$$

[4] Actually, we should write $\int_0^t (\psi(t,\cdot), (\tilde{u}'_\varepsilon + w')(t,\cdot))\,dt$, etc.

the relation

$$|\tilde{u}'_\varepsilon|_2^2 + \|\tilde{u}_\varepsilon\|_1^2 \le c\left(1 + \int_0^t (|\tilde{u}'_\varepsilon|_2^2 + \|\tilde{u}_\varepsilon\|_1^2)\,dt\right)$$

$$+ 2\int_0^t (\psi, \tilde{u}'_\varepsilon)\,dt \le c\left(1 + \int_0^t (|\tilde{u}'_\varepsilon|_2^2 + \|\tilde{u}_\varepsilon\|_1^2)\,dt\right) + 2(\psi, \tilde{u}_\varepsilon) \quad (5.4.34)$$

$$- 2(\psi(0), \tilde{u}_0) + 2\int_0^t \|\psi'\|_{V'_0}\|\tilde{u}_\varepsilon\|_1\,dt.$$

Thus, the relation

$$|\tilde{u}'_\varepsilon|_2^2 + \|\tilde{u}_\varepsilon\|_1^2 \le c\left(1 + \int_0^t (|\tilde{u}'_\varepsilon|_2^2 + \|\tilde{u}_\varepsilon\|_1^2)\,dt\right) \quad (5.4.35)$$

is obtained, from which it results, by applying Gronwall's inequality [29], [326], that

$$|\tilde{u}'_\varepsilon|_2 \le c, \qquad \|\tilde{u}_\varepsilon\|_1 \le c \quad (5.4.36)$$

(c) From (5.4.16), after differentiating with respect to t and by setting $\tilde{v} = \tilde{u}''_\varepsilon + w''$, we get the relation

$$(\rho\tilde{u}'''_\varepsilon, \tilde{u}''_\varepsilon + w'') + a(\tilde{u}'_\varepsilon, \tilde{u}''_\varepsilon + w'')$$

$$+ \left\langle\left(\frac{d\Phi_\varepsilon}{dv}(\tilde{u}'_\varepsilon + w')\right)', \tilde{u}''_\varepsilon + w''\right\rangle_{\Gamma_S} = (\psi', \tilde{u}''_\varepsilon + w''). \quad (5.4.37)$$

Note that the time differentiation is formal. The procedure which follows remains valid if the difference quotients are considered instead of the time derivatives. Because of the monotonicity of $d\Phi_\varepsilon/dv$, it results by means of (5.4.22) that the inequality

$$\left\langle\left(\frac{d\Phi_\varepsilon}{dv}(\tilde{u}'_\varepsilon + w')\right)', \tilde{u}''_\varepsilon + w''\right\rangle_{\Gamma_S} \ge 0 \quad (5.4.38)$$

holds. This can be easily verified by writing the time derivatives as the limits of the respective difference quotients. Due to (5.4.38), (5.4.37) gives rise to the inequality

$$\frac{1}{2}\frac{d}{dt}\left(\int_\Omega \rho(\tilde{u}''_\varepsilon)^2\,d\Omega + a(\tilde{u}'_\varepsilon, \tilde{u}'_\varepsilon)\right)$$

$$\le (\psi', \tilde{u}''_\varepsilon + w'') - (\rho\tilde{u}'''_\varepsilon, w'') - a(\tilde{u}'_\varepsilon, w''). \quad (5.4.39)$$

From (5.4.39), after integrating from 0 to t, by means of (4.1.38) and of

$$|\tilde{u}'_\varepsilon|_2^2 \leq c \int_0^t |\tilde{u}''_\varepsilon|_2^2 \, dt + |\tilde{u}_1|_2^2, \tag{5.4.40}$$

we obtain the inequality

$$|\tilde{u}''_\varepsilon|_2^2 + \|\tilde{u}'_\varepsilon\|_1^2 \leq c\left(\|\tilde{u}_1\|_1^2 + |\tilde{u}''_\varepsilon(0)|_2 + \int_0^t |\tilde{u}''_\varepsilon|_2^2 \, dt\right)$$
$$+ 2\int_0^t (\psi', \tilde{u}''_\varepsilon) \, dt - 2\int_0^t (\rho\tilde{u}'''_\varepsilon, w'') \, dt. \tag{5.4.41}$$

Because of (5.4.17–21), $\psi'' \in L^2(0T, V'_0)$, and thus we can write the relation

$$\int_0^t (\psi', \tilde{u}''_\varepsilon) \, dt = (\psi', \tilde{u}'_\varepsilon) - (\psi'(0), \tilde{u}_1) - \int_0^t (\psi'', \tilde{u}'_\varepsilon) \, dt$$
$$\leq c(\|\psi'\|_{V'_0} \|\tilde{u}'_\varepsilon\|_1 + 1) + \int_0^t \|\psi''\|_{V'_0} \|\tilde{u}'_\varepsilon\|_1 \, dt. \tag{5.4.42}$$

From (5.4.41, 42), by means of the relation

$$\int_0^t (\rho\tilde{u}'''_\varepsilon, w'') \, dt = (\rho\tilde{u}''_\varepsilon, w'') - (\rho\tilde{u}''_\varepsilon(0), w''(0)) - \int_0^t (\rho\tilde{u}''_\varepsilon, w''') \, dt \tag{5.4.43}$$

we get the inequality

$$|\tilde{u}''_\varepsilon|_2^2 + \|\tilde{u}'_\varepsilon\|_1^2 \leq c(\|\tilde{u}_1\|_1^2 + |\tilde{u}''_\varepsilon(0)|_2^2) + c\int_0^t (\|\tilde{u}'_\varepsilon\|_1^2 + |\tilde{u}''_\varepsilon|_2^2) \, dt. \tag{5.4.44}$$

(d) By setting $t = 0$ in (5.4.16), and using (5.4.24) we obtain the relation

$$(\rho\tilde{u}''_\varepsilon(0), \tilde{v}) + a(\tilde{u}_0, \tilde{v}) = (\psi(0), \tilde{v}) = (f(0), \tilde{v}) + \int_{\Gamma_F} F_i(0)\tilde{v}_i \, d\Gamma$$
$$+ \int_{\Gamma_S} C_N(0)\tilde{v}_N \, d\Gamma - (\rho w''(0), \tilde{v}) - a(w(0), \tilde{v}). \tag{5.4.45}$$

But

$$a(\tilde{u}_0, \tilde{v}) = \int_\Omega C_{ijhk}\,\varepsilon_{ij}(\tilde{u}_0)\varepsilon_{hk}(\tilde{v}) \, d\Omega$$
$$= -\int_\Omega (C_{ijhk}\,\varepsilon_{hk}(\tilde{u}_0))_{,j}\tilde{v}_i \, d\Omega \tag{5.4.46}$$
$$+ \int_{\Gamma_F} F_i(0)\tilde{v}_i \, d\Gamma + \int_{\Gamma_S} C_N(0)\tilde{v}_N \, d\Gamma + \langle S_T(0), \tilde{v}_T\rangle_{\Gamma_S},$$

and, therefore,

$$(\rho\tilde{u}_\varepsilon''(0), \tilde{v}) - \int_\Omega (C_{ijhk}\,\varepsilon_{hk}(\tilde{u}_0))_{,j}\, v_i \, d\Omega + \langle S_T(0), \tilde{v}_T\rangle_{\Gamma_S}$$
$$= (f(0), \tilde{v}) - (\rho w''(0), \tilde{v}) + a(w(0), \tilde{v}). \quad (5.4.47)$$

From (5.4.25, 47) and using (5.4.17, 21), we find that

$$|\tilde{u}_\varepsilon''(0)|_2 \le c. \quad (5.4.48)$$

(5.4.44, 48) imply that

$$|\tilde{u}_\varepsilon''|_2^2 + \|\tilde{u}_\varepsilon'\|_1^2 \le c\left(1 + \int_0^t (|\tilde{u}_\varepsilon''|_2^2 + \|\tilde{u}_\varepsilon'\|_1^2)\, dt\right), \quad (5.4.49)$$

from which, through Gronwall's inequality,

$$|\tilde{u}_\varepsilon''|_2 \le c, \qquad \|\tilde{u}_\varepsilon'\|_1 \le c. \quad (5.4.50)$$

(e) From (5.4.36) and (5.4.50), it is concluded that the sequences $\{\tilde{u}_\varepsilon\}$, $\{\tilde{u}_\varepsilon'\}$ and $\{\tilde{u}_\varepsilon''\}$ are bounded, the first two in the space $L^\infty(0T, V_0)$ and the last one in $L^\infty(0T, [L^2(\Omega)]^3)$. Thus we can select subsequences, again denoted by $\{u_\varepsilon\}$, $\{\tilde{u}_\varepsilon'\}$ and $\{\tilde{u}_\varepsilon''\}$, which are weakly$_*$ convergent. We denote the limits by u, u' and u'' respectively. In the next step, it will be shown that u, u' and u'' fulfill the variational inequality (5.4.13) and the initial conditions (5.4.14).

(f) The variational equality (5.4.16) holds if $\tilde{v}$ is replaced by $\tilde{v} - \tilde{u}_\varepsilon'$. From this relation, there results by means of

$$\Phi_\varepsilon(\tilde{v} + w') - \Phi_\varepsilon(\tilde{u}_\varepsilon' + w') \ge \left\langle \frac{d\Phi(\tilde{u}_\varepsilon' + w')}{dv}, \tilde{v} - \tilde{u}_\varepsilon' \right\rangle_{\Gamma_S}, \quad (5.4.51)$$

the inequality

$$(\rho\tilde{u}_\varepsilon'', \tilde{v} - \tilde{u}_\varepsilon') + a(\tilde{u}_\varepsilon, \tilde{v} - \tilde{u}_\varepsilon') + \Phi_\varepsilon(\tilde{v} + w')$$
$$- \Phi_\varepsilon(\tilde{u}_\varepsilon' + w') \ge (\psi, \tilde{v} - \tilde{u}_\varepsilon'), \qquad \forall \tilde{v} \in V_0. \quad (5.4.52)$$

We assume, henceforth, that v depends on t and, more precisely, that $\tilde{v} \in L^2(0T, V_0)$. Integrated from 0 to T, (5.4.52) reduces to

$$G_{1\varepsilon} \ge G_{2\varepsilon}, \quad (5.4.53)$$

with

$$G_{1\varepsilon} = \int_0^T [(\rho\tilde{u}_\varepsilon'', \tilde{v}) + a(\tilde{u}_\varepsilon, \tilde{v}) + \Phi_\varepsilon(\tilde{v} + w') - (\psi, \tilde{v} - \tilde{u}_\varepsilon')]\, dt \quad (5.4.54a)$$

and

$$G_{2\varepsilon} = \int_0^T ((\rho\tilde{u}_\varepsilon'', \tilde{u}_\varepsilon') + a(\tilde{u}_\varepsilon, \tilde{u}_\varepsilon') + \Phi_\varepsilon(\tilde{u}_\varepsilon' + w'))\, dt. \quad (5.4.54b)$$

But

$$G_{2\varepsilon} = \frac{1}{2}\left[\int_\Omega \rho(\tilde{u}'_\varepsilon(T))^2\, d\Omega + a(\tilde{u}_\varepsilon(T), \tilde{u}_\varepsilon(T))\right] - \frac{1}{2}\left[\int_\Omega \rho(\tilde{u}_1)^2\, d\Omega + a(\tilde{u}_0, \tilde{u}_0)\right] + \int_0^T \Phi_\varepsilon(\tilde{u}'_\varepsilon + w')\, dt, \quad (5.4.55)$$

and, thus, as $\varepsilon \to 0$

$$\liminf G_{2\varepsilon} \geq \frac{1}{2}\left[\int_\Omega \rho(\tilde{u}'(T))^2\, d\Omega + a(\tilde{u}(T), \tilde{u}(T))\right] - \frac{1}{2}\left[\int_\Omega \rho(\tilde{u}_1)^2\, d\Omega + a(\tilde{u}_0, \tilde{u}_0)\right] + \int_0^T \Phi(\tilde{u}' + w')\, dt \quad (5.4.56)$$

$$= \int_0^T [(\rho\tilde{u}'', \tilde{u}') + a(\tilde{u}, \tilde{u}') + \Phi(\tilde{u}' + w')]\, dt$$

because of (5.1.21). On the other hand, as $\varepsilon \to 0$

$$\lim G_{1\varepsilon} = \int_0^T [(\rho\tilde{u}'', \tilde{v}) + a(\tilde{u}, \tilde{v}) + \Phi(\tilde{v} + w') - (\psi, \tilde{v} - \tilde{u}')]\, dt. \quad (5.4.57)$$

If one takes into account that

$$\lim G_{1\varepsilon} = \liminf G_{1\varepsilon} \geq \liminf G_{2\varepsilon}, \quad (5.4.58)$$

then (5.4.57) combined with (5.4.56) yields the variational inequality

$$\int_0^T [(\rho\tilde{u}'', \tilde{v} - \tilde{u}') + a(\tilde{u}, \tilde{v} - \tilde{u}') + \Phi(v + w') - \Phi(\tilde{u}' + w') - (\psi, \tilde{v} - \tilde{u}')]\, dt \geq 0, \qquad \forall \tilde{v} \in L^2(0T, V_0). \quad (5.4.59)$$

(5.4.59) can be considered as the expression of Hamilton's Principle for the present problem.[5] Moreover, u and u' satisfy the initial conditions (5.4.14) because of (5.4.26–28).

(g) From (5.4.59), the pointwise inequality (5.4.13) will be derived. To this end, we consider the sequence $\{\theta_K\}$ of the intervals of the form $\theta_K = (t_0 - 1/K, t_0 + 1/K) \subset (0, T)$, and we put in (5.4.59) $\tilde{v} = \tilde{u}'$, $\forall t \notin \theta_K$ and $\tilde{v} = \tilde{\tilde{v}}$, $\forall t \in \theta_K$. Here $\tilde{\tilde{v}}$ denotes a fixed element of V_0. Then (5.4.59) reduces to

$$\frac{1}{|\theta_K|}\int_{\theta_K} [(\rho\tilde{u}'', \tilde{\tilde{v}} - \tilde{u}') + a(\tilde{u}, \tilde{\tilde{v}} - \tilde{u}') + \Phi(\tilde{\tilde{v}} + w') - \Phi(\tilde{u}' + w') - (\psi, \tilde{\tilde{v}} - \tilde{u}')]\, dt \geq 0. \quad (5.4.60)$$

[5] The procedure leading to (5.4.59) greatly resembles the procedure used in analytical mechanics or in linear elastodynamics for the derivation of the classical Hamilton's Principle.

Further, we let $K \to \infty$ and we apply (1.4.10). Then

$$\frac{1}{|\theta_K|} \int_{\theta_K} [(\rho\tilde{u}'', \tilde{v}) + a(\tilde{u}, \tilde{v}) - (\psi, \tilde{v})]\, dt$$
$$\to (\rho\tilde{u}''(t_0), \tilde{v}) + a(\tilde{u}(t_0), \tilde{v}) - (\psi(t_0), \tilde{v}) \quad (5.4.61)$$

for every $t_0 \notin \theta' \subset (0, T)$ with mes $\theta' = 0$, and

$$\frac{1}{|\theta_K|} \int_{\theta_K} [(\rho\tilde{u}'', \tilde{u}') + a(\tilde{u}, \tilde{u}') + \Phi(\tilde{u}' + w') + (\psi, \tilde{u}')]\, dt$$
$$\to (\rho\tilde{u}''(t_0), \tilde{u}'(t_0)) + a(\tilde{u}(t_0), \tilde{u}'(t_0)) + \Phi(\tilde{u}'(t_0) + w'(t_0)) + (\psi(t_0), \tilde{u}') \quad (5.4.62)$$

for every $t_0 \notin \theta'' \subset (0, T)$ with mes $\theta'' = 0$. From (5.4.60–62), it results that for $t_0 \notin \theta' \cup \theta''$ we can take in (5.4.60) the limit for $K \to \infty$. Thus the variational inequality (5.4.13) results, q.e.d.

We shall briefly discuss in what sense the variational solution of the problem satisfies the relations (5.4.1–8). We proceed as in the static case. Thus, it is easily concluded that a variational solution satisfies (5.4.1) in the sense of distributions over Ω, and (5.4.4–8) in the sense of traces. Because $\tilde{u} \in L^\infty(0T, V_0)$ and $\tilde{u}' \in L^\infty(0T, V_0)$, it follows by the continuity theorem (Sec. 1.4.1) that u: $[0, T] \to V_0$ is continuous, and thus the initial condition $\tilde{u} = \tilde{u}_0$ for $t = 0$ makes sense. The same applies to the other boundary conditions as well.

5.4.3 The Regularized Problem

In order to complete the proof of the existence of a solution of (5.4.13), it remains to be shown that u_ε exists.

Proposition 5.4.2. On the same assumptions as in Prop. 5.4.1, and with the initial conditions (5.4.14), (5.4.16) admits a solution $\tilde{u}_\varepsilon \in L^\infty(0T, V_0)$ such that

$$\tilde{u}'_\varepsilon \in L^\infty(0T, V_0) \quad \text{and} \quad \tilde{u}''_\varepsilon \in L^\infty(0T, [L^2(\Omega)]^3). \quad (5.4.63)$$

Proof. The method of Galerkin's approximation is used. Let $\{v_1 \ldots, v_m \ldots\}$ be a sequence of elements from V_0 such that

(i) for every m the elements $\{v_1, \ldots, v_m\}$ are linearly independent and span the finite-dimensional vector space V_m;
(ii) the linear combinations $\sum_{i=1}^m a_i v_i$, with $a_i \in \mathbb{R}$, $i = 1, 2, \ldots, m$, m finite, are dense in V_0.

Due to the separability of V_0, there exists such a sequence, also called a Galerkin basis. Without loss of generality we can take $\tilde{u}_0, \tilde{u}_1 \in V_{m0}$ for a fixed m_0. The following initial value problem is now considered for the functions g_{m_i} ($i = 1, \ldots, m$, $m \geq m_0$) satisfying the relation

$$\tilde{u}_{\varepsilon m}(t) = \sum_{i=1}^{m} g_{m_i}(t) v_i \in V_m \qquad \text{a.e. on } (0, T). \tag{5.4.64}$$

$$(\rho \tilde{u}''_{\varepsilon m}, v_p) + a(\tilde{u}_{\varepsilon m}, v_p) + \left\langle \frac{d\Phi_\varepsilon}{dv}(\tilde{u}'_{\varepsilon m} + \tilde{w}'), v_p \right\rangle_{\Gamma_S} = (\psi, v_p), \qquad 1 \leq p \leq m \tag{5.4.65}$$

$$\tilde{u}_{\varepsilon m} = \tilde{u}_0, \qquad \tilde{u}'_{\varepsilon m} = \tilde{u}_1 \quad \text{for} \quad t = 0. \tag{5.4.66}$$

As is known from the theory of differential equations, there exist for every $m \geq m_0$ a $t_{m0} \in [0, T]$ and continuous functions g_{m_i} on $[0, t_m]$, $i = 1, \ldots, m$, which fulfill Eqs. (5.4.65, 66). Actually, $t_m = T$, as results from the following *a priori* estimates. From (5.4.65) there results the variational equality

$$(\rho \tilde{u}''_{\varepsilon m}, \tilde{v}) + a(\tilde{u}_{\varepsilon m}, \tilde{v}) + \left\langle \frac{d\Phi_\varepsilon}{dv}(\tilde{u}'_{\varepsilon m} + w'), \tilde{v} \right\rangle_{\Gamma_S} = (\psi, \tilde{v}), \qquad \forall \tilde{v} \in V_m \tag{5.4.67}$$

The rest of the proof is similar to that of Prop. 5.4.1. Putting $\tilde{v} = \tilde{u}'_{\varepsilon m} + w'$ in (5.4.67), we obtain a relation analogous to (5.4.30), from which the relations

$$|\tilde{u}'_{\varepsilon m}|_2 \leq c, \qquad \|\tilde{u}_{\varepsilon m}\|_1 \leq c \tag{5.4.68}$$

are obtained. From (5.4.65), by differentiating with respect to t we get for $\tilde{u}_{\varepsilon m}$ the inequality (5.4.44). Further, the relation

$$|\tilde{u}''_{\varepsilon m}(0)|_2 \leq c \tag{5.4.69}$$

is verified, and thus

$$|\tilde{u}''_{\varepsilon m}|_2 \leq c, \qquad \|\tilde{u}'_{\varepsilon m}\|_1 \leq c. \tag{5.4.70}$$

Accordingly, for $m \to \infty$

$$\tilde{u}_{\varepsilon m} \to \tilde{u}_\varepsilon \quad \text{in} \quad L^\infty(0T, V_0) \quad \text{weakly}_*, \tag{5.4.71}$$

$$\tilde{u}'_{\varepsilon m} \to \tilde{u}'_\varepsilon \quad \text{in} \quad L^\infty(0T, V_0) \quad \text{weakly}_*, \tag{5.4.72}$$

$$\tilde{u}''_{\varepsilon m} \to \tilde{u}''_\varepsilon \quad \text{in} \quad L^\infty(0T, [L^2(\Omega)]^3) \quad \text{weakly}_*. \tag{5.4.73}$$

u_ε fulfills the initial conditions (5.4.14), as results from (5.4.71–73) by means of (5.4.66). It remains to be shown that u_ε is a solution of (5.4.16). From (5.4.67), it may be concluded that the sequence $\{(d\Phi/dv)(u'_{\varepsilon m} + w')\}$ is bounded in the space $L^\infty(0T, V'_0)$. Thus for $m \to \infty$

$$\frac{d\Phi_\varepsilon}{dv}(\tilde{u}'_{\varepsilon m} + w') \to y_\varepsilon \quad \text{in} \quad L^\infty(0T, V'_0) \quad \text{weakly}_*. \tag{5.4.74}$$

From (5.4.65), by keeping p fixed and by taking the limit for $m \to \infty$ we obtain the relation

$$(\rho \tilde{u}''_\varepsilon, v_p) + a(\tilde{u}_\varepsilon, v_p) + \langle y_\varepsilon, v_p \rangle_{\Gamma_S} = (\psi, v_p). \tag{5.4.75}$$

But the linear combinations $\sum_{p=1}^{m} g_{mp} v_p$ are dense in V_0 and, therefore,

$$(\rho\tilde{u}_\varepsilon'', \tilde{v}) + a(\tilde{u}_\varepsilon, \tilde{v}) + \langle y_\varepsilon, \tilde{v}\rangle_{\Gamma_S} = (\psi, \tilde{v}) \quad \forall \tilde{v} \in V_0. \tag{5.4.76}$$

To show that

$$y_\varepsilon = \frac{d\Phi_\varepsilon}{dv}(\tilde{u}_\varepsilon' + w'), \tag{5.4.77}$$

a function $z \in L^2(0T, V_0)$ is considered. Then the monotonicity of $d\Phi_\varepsilon/dv$ implies the inequality

$$A_m = \int_0^T \left\langle \frac{d\Phi_\varepsilon}{dv}(\tilde{u}_{\varepsilon m}' + w') - \frac{d\Phi_\varepsilon}{dv}(z + w'), \tilde{u}_{\varepsilon m}' - z \right\rangle_{\Gamma_S} dt \geq 0. \tag{5.4.78}$$

The expression A_m can be calculated from (5.4.65), and has the form

$$A_m = -\int_0^T [(\rho\tilde{u}_{\varepsilon m}'', \tilde{u}_{\varepsilon m}') + a(\tilde{u}_{\varepsilon m}, \tilde{u}_{\varepsilon m}') - (\psi, \tilde{u}_{\varepsilon m}')]\, dt$$
$$-\int_0^T \left\langle \frac{d\Phi_\varepsilon}{dv}(\tilde{u}_{\varepsilon m}' + w'), z \right\rangle_{\Gamma_S} dt - \int_0^T \left\langle \frac{d\Phi_\varepsilon}{dv}(z + w'), \tilde{u}_{\varepsilon m}' - z \right\rangle_{\Gamma_S} dt \tag{5.4.79}$$
$$= -\tfrac{1}{2}|\rho^{\frac{1}{2}}\tilde{u}_{\varepsilon m}'(T)|_2^2 + \tfrac{1}{2}|\rho^{\frac{1}{2}}\tilde{u}_1|_2^2 - \tfrac{1}{2}a(\tilde{u}_{\varepsilon m}(T), \tilde{u}_{\varepsilon m}(T)) + \tfrac{1}{2}a(\tilde{u}_0, \tilde{u}_0) + \int_0^T (\psi, \tilde{u}_{\varepsilon m}')\, dt$$
$$-\int_0^T \left\langle \frac{d\Phi_\varepsilon}{dv}(\tilde{u}_{\varepsilon m}' + w'), z \right\rangle_{\Gamma_S} dt - \int_0^T \left\langle \frac{d\Phi_\varepsilon}{dv}(z + w'), \tilde{u}_{\varepsilon m}' - z \right\rangle_{\Gamma_S} dt.$$

But since $\tilde{u}_{\varepsilon m}'(T) \to \tilde{u}_\varepsilon'(T)$ and $\tilde{u}_{\varepsilon m}(T) \to \tilde{u}_\varepsilon(T)$ weakly in V_0, we can easily verify that

$$\limsup\,(-|\rho^{\frac{1}{2}}\tilde{u}_{\varepsilon m}'(T)|_2^2) \leq -|\rho^{\frac{1}{2}}\tilde{u}_\varepsilon'(T)|_2^2, \tag{5.4.80}$$

$$\limsup - a(\tilde{u}_{\varepsilon m}(T), \tilde{u}_{\varepsilon m}(T)) \leq -a(\tilde{u}_\varepsilon(T), \tilde{u}_\varepsilon(T)) \tag{5.4.81}$$

and thus as $m \to \infty$

$$0 \leq \limsup A_m \leq -\int_0^T [(\rho\tilde{u}_\varepsilon'', \tilde{u}_\varepsilon') + a(\tilde{u}_\varepsilon, \tilde{u}_\varepsilon') - (\psi, \tilde{u}_\varepsilon')]\, dt$$
$$-\int_0^T \langle y_\varepsilon, z\rangle_{\Gamma_S}\, dt - \int_0^T \left\langle \frac{d\Phi_\varepsilon}{dv}(z + w'), \tilde{u}_\varepsilon' - z \right\rangle_{\Gamma_S} dt. \tag{5.4.82}$$

This last relation, combined with

$$\int_0^T [(\rho\tilde{u}_\varepsilon'', \tilde{u}_\varepsilon') + a(\tilde{u}_\varepsilon, \tilde{u}_\varepsilon') - (\psi, \tilde{u}_\varepsilon')]\, dt = -\int_0^T (y_\varepsilon, \tilde{u}_\varepsilon')\, dt, \tag{5.4.83}$$

(which results from (5.4.76)), gives

$$\int_0^T \left\langle y_\varepsilon - \frac{d\Phi_\varepsilon}{dv}(z + w'), \tilde{u}_\varepsilon' - z \right\rangle_{\Gamma_S} dt \geq 0 \tag{5.4.84}$$

for every $z \in L^2(0T, V_0)$. We put $z = \tilde{u}'_\varepsilon - \lambda \bar{z}$, $\lambda > 0$ and $\bar{z} \in L^2(0T, V_0)$, and we let $\lambda \to 0$. The relation (5.4.84) reduces to

$$\int_0^T \left\langle y_\varepsilon - \frac{d\Phi_\varepsilon}{dv}(\tilde{u}'_\varepsilon + w'), z \right\rangle_{\Gamma_S} dt \geq 0, \qquad \forall z \in L^2(0T, V_0), \tag{5.4.85}$$

which implies finally (5.4.77), q.e.d.

5.4.4 The Uniqueness of the Solution

Assume that Prop. 5.4.1 holds and that a solution u of (5.4.13, 14) exists. The following proposition asserts the uniqueness of the solution.

Proposition 5.4.3. The displacement field u which satisfies (5.4.13, 14) is uniquely determined.

Proof. Arguing by contradiction, we assume that u_1 and u_2 are two solutions of (5.4.13, 14). Then through the substitutions $\tilde{u} = u_1$, $\tilde{v} = u_2$, and $\tilde{u} = u_2$, $\tilde{v} = u_1$ and through addition it results that

$$-(\rho(u''_1 - u''_2), u'_1 - u'_2) - a(u_1 - u_2, u'_1 - u'_2) \geq 0 \tag{5.4.86}$$

or

$$\frac{d}{dt}\left[\int_\Omega \rho(u'_1 - u'_2)^2 \, d\Omega + a(u_1 - u_2, u_1 - u_2)\right] \leq 0. \tag{5.4.87}$$

The initial conditions imply that $u_1 - u_2 = 0$ and $u'_1 - u'_2 = 0$ for $t = 0$, which, combined with (5.4.87), yields the uniqueness of the solution, q.e.d.

5.5 A Note on Other Types of Friction Problems

The mathematical model which we have studied in the previous sections concerns a simplified case of Coulomb's friction. Indeed, the normal force S_N on Γ_S is not given in most problems arising in real situations and must be calculated. If we assume that the normal displacement is prescribed, for instance,

$$u_N = 0 \quad \text{on} \quad \Gamma_S, \tag{5.5.1}$$

and if $u \in U'_{ad} = \{v \mid v \in [H^1(\Omega)]^3, v = U \text{ on } \Gamma_U, u_N = 0 \text{ on } \Gamma_S\}$, then the variational inequality

$$a(u, v - u) + \int_{\Gamma_S} \mu |S_N| (|v_T| - |u_T|) \, d\Gamma \geq (f, v - u) + \int_{\Gamma_F} F_i(v_i - u_i) \, d\Gamma, \qquad \forall v \in U'_{ad} \tag{5.5.2}$$

is formally obtained. Because of the appearance of S_N in (5.5.2), where the dependence of S_N on u is not given, further study of (5.5.2) is still an open problem. An inequality analogous to (5.5.2) holds for the dynamic problem.

If instead of (5.5.1) the boundary condition $-S_N \in \partial j(u_N)$ holds (e.g., the unilateral contact boundary condition), it can easily be shown that the solution u satisfies the variational inequality

$$a(u, v-u) + \int_{\Gamma_S} \mu |S_N|(|v_T| - |u_T|)\, d\Gamma + \Phi(v_N) - \Phi(u_N)$$

$$\geq (f, v-u) + \int_{\Gamma_F} F_i(v_i - u_i)\, d\Gamma, \quad \forall v \in U''_{ad}, \quad (5.5.3)$$

where

$$U''_{\text{ad}} = \{v \mid v \in [H^1(\Omega)]^3,\ v = U \quad \text{on} \quad \Gamma_U\}$$

and

$$\Phi(v_N) = \begin{cases} \displaystyle\int_{\Gamma_S} j(u_N(x))\, d\Gamma & \text{if} \quad j(u_N) \in L^1(\Gamma_S) \\ \infty \ \text{otherwise.} \end{cases} \quad (5.5.4)$$

Further, we assume that the Signorini boundary conditions hold on Γ_S with respect to u_N. In this case, $\Phi(u_N) = I_K(v_N)$ (the indicator of the convex closed set $\{v_N \mid v_N \in H^{1/2}(\Gamma_S),\ v_N \leq 0\}$), and the problem reduces to the solution of the variational inequality (5.5.3), but without the terms $\Phi(v_N)$ and $\Phi(u_N)$ and with the condition $v_N \leq 0$ included in U''_{ad}. The existence of the solution for this problem has recently been studied [90], [232]. It is proved in [232], using a fixed-point method, that a solution exists for small values of the friction coefficient. If S_N is assumed to be known, (5.5.3) reduces to a variational inequality of the form (2.2.9) which has for $a(u, u)$ coercive (mes $\Gamma_U > 0$), a unique solution. For the semicoercive case ($\Gamma_U = \emptyset$), see Sec. 6.2.

Thus far we have considered only isotropic friction defined by a condition of the form $|S_T| \leq \mu |C_N|$. The friction law considered hitherto recalls holonomic plasticity laws, and therefore many attempts have already been made to generalize it in the spirit of modern plasticity theory (associated and nonassociated friction laws; see, e.g., [202]). Here the case of orthotropic friction is briefly touched upon.

We denote the principal orthotropy axes, on the tangential plane at a point of Γ_S, by T_1, T_2 and let S_{T_1}, S_{T_2} and u_{T_1}, u_{T_2} be the components of S_T and u_T along these axes. The corresponding coefficients of friction are denoted by μ_1, μ_2. We then introduce a static friction law of the form:

$$\text{if} \quad \left[\left(\frac{S_{T_1}}{\mu_1}\right)^2 + \left(\frac{S_{T_2}}{\mu_2}\right)^2\right]^{1/2} < |C_N|, \quad \text{then } u_T = 0; \quad (5.5.5)$$

if $[(S_{T_1}/\mu_1)^2 + (S_{T_2}/\mu_2)^2]^{1/2} = |C_N|$, then there exists $\lambda \geq 0$ such that

$$u_{T_1} = -\frac{\lambda S_{T_1}}{\mu_1^2}, \qquad u_{T_2} = -\frac{\lambda S_{T_2}}{\mu_2^2}. \quad (5.5.6)$$

(5.5.6) constitutes a holonomic plasticity law with the yield criterion

$$F(S) = \left[\left(\frac{S_{T_1}}{\mu_1}\right)^2 + \left(\frac{S_{T_2}}{\mu_2}\right)^2\right]^{1/2} - |C_N| \leq 0. \quad (5.5.7)$$

In the dynamic problem, u_T is replaced by $\partial u_T/\partial t$. Then the variational inequalities of both the static and dynamic problems retain the same form, but with the function Φ given by

$$\Phi(u) = \int_{\Gamma_S} |C_N|((\mu_1 u_{T_1})^2 + (\mu_2 u_{T_2})^2)^{1/2}\, d\Gamma. \tag{5.5.8}$$

The dissipation function Φ for the dynamic problem is obtained analogously. The proofs given in this chapter for the friction laws (5.1.3–5) and (5.4.6–8) can be adapted, with slight modifications, to the respective orthotropic friction laws. Furthermore, a friction law connected with a general limit condition of the form $f(S_{T_1}, S_{T_2}, C_N) \leq 0$ can also be studied by the method employed in Sec. 6.2. Thus any case of friction anisotropy can be considered.

Chapter 6

Subdifferential Constitutive Laws and Boundary Conditions

In this chapter, the following three problems are studied regarding a deformable body undergoing small deformations.

(i) First, it is assumed that the body obeys a material law of the form $\sigma \in \partial w(\varepsilon)$. Classical boundary conditions hold on the boundary.
(ii) The body is linear elastic and on its boundary subdifferential relations of the form $-S \in \partial j(u)$ must be satisfied.
(iii) Both the material law and the boundary conditions are of a subdifferential nature.

The corresponding dynamic problems are formulated as well, and attention is also paid to the nonmonotone laws described by a superpotential. Concerning the mechanical meaning of the B.V.P.s studied here, the reader is referred to Chap. 3. In preparing this chapter we have based our work on the following references: G. Duvaut and J. L. Lions [84], G. Fichera [99], M. Frémond [103], R. Hünlich and J. Naumann [150], F. Lené [177], M. Schatzman [296], among others. The variational inequalities formulated here often arise in the framework of a technical theory, for instance, in the static or dynamic analysis of nonlinear elastic bodies with unilateral contact or friction boundary conditions. Bridge bearings, plate and pile foundations, retaining walls, etc. are typical examples of this kind of problem.

6.1 Subdifferential Material Laws and Classical Boundary Conditions

6.1.1 Formulation of the Problem

The body under consideration occupies an open bounded, connected, subset Ω of $\mathbb{R}^3$, when it is not deformed. Ω is referred to a fixed Cartesian coordinate system $0x_1x_2x_3$. The boundary Γ of Ω is divided into two open, disjoint parts, Γ_U and Γ_F, i.e., $\Gamma = \overline{\Gamma}_U \cup \overline{\Gamma}_F$. Using the notation of Chapters 3 and 4, we consider the following B.V.P.:

$$\sigma_{ij,j} + f_i = 0 \quad \text{in} \quad \Omega, \tag{6.1.1}$$

$$\varepsilon_{ij}(u) = \tfrac{1}{2}(u_{i,j} + u_{j,i}) \quad \text{in} \quad \Omega, \tag{6.1.2}$$

$$\sigma \in \partial w(\varepsilon) \quad \text{in} \quad \Omega, \tag{6.1.3}$$

$$u_i = U_i \quad \text{on} \quad \Gamma_U, \tag{6.1.4}$$

$$S_i = F_i \quad \text{on} \quad \Gamma_F. \tag{6.1.5}$$

Here $U_i = U_i(x)$, $x \in \Gamma_U$, $F_i = F_i(x)$, $x \in \Gamma_F$, (6.1.3) holds pointwise, and w is a convex, l.s.c., proper functional on $\mathbb{R}^6$. Moreover, it is assumed that for every ε

$$c_2(\varepsilon_{ij}\varepsilon_{ij}) \leq w(\varepsilon) \leq c_1(\varepsilon_{ij}\varepsilon_{ij}), \tag{6.1.6}$$

with c_1 and c_2 positive constants.

Assuming that in (6.1.1–5) all the functions are sufficiently smooth and applying the Green–Gauss theorem, we obtain the relation

$$\int_\Omega \sigma_{ij}[\varepsilon_{ij}(v) - \varepsilon_{ij}(u)]\, d\Omega = \int_\Omega f_i(v_i - u_i)\, d\Omega + \int_{\Gamma_F} F_i(v_i - u_i)\, d\Gamma \tag{6.1.7}$$

for every v such that $v_i = U_i$ on Γ_U. (6.1.3) implies the variational inequality

$$w(\varepsilon(v)) - w(\varepsilon(u)) \geq \sigma_{ij}[\varepsilon_{ij}(v) - \varepsilon_{ij}(u)] \tag{6.1.8}$$

holding pointwise on Ω for every $\varepsilon(v) = \{\varepsilon_{ij}(v)\} \in \mathbb{R}^6$. We shall give the variational formulation of this B.V.P. in an appropriately defined subspace of $[H^1(\Omega)]^3$. To this end, we assume that $f \in [L^2(\Omega)]^3$, $U \in [\dot{H}^{1/2}(\Gamma_U)]^3$ and $F \in [L^2(\Gamma_F)]^3$ (cf. Sec. 5.1.2), and we introduce the admissible set for the displacements

$$U_{ad} = \{v \mid v \in [H^1(\Omega)]^3,\ v = U \quad \text{on} \quad \Gamma_U\}. \tag{6.1.9}$$

Further, the functional

$$W(\varepsilon) = \begin{cases} \displaystyle\int_\Omega w(\varepsilon)\, d\Omega & \text{if} \quad w(\varepsilon) \in L^1(\Omega) \\ \infty & \text{otherwise} \end{cases} \tag{6.1.10}$$

is defined for $\varepsilon \in [L^2(\Omega)]^6$. According to Prop. 3.3.1. W is a convex, l.s.c. and proper functional on $[L^2(\Omega)]^6$ extending the pointwise relation (6.1.3) for σ, $\varepsilon \in [L^2(\Omega)]^6$. From (6.1.7, 8) and (6.1.10), we are led to the following problem.

Problem 1. Find a displacement field $u \in U_{ad}$ such that

$$W(\varepsilon(v)) - W(\varepsilon(u)) \geq (f, v - u) + \int_{\Gamma_F} F_i(v_i - u_i)\, d\Gamma, \quad \forall v \in U_{ad}. \tag{6.1.11}$$

U_{ad} is a closed, convex subset of $[H^1(\Omega)]^3$, and W, as a function of u, is convex, proper and l.s.c. Note that this last statement is a consequence of the continuity of the mapping ε: $[H^1(\Omega)]^3 \to [L^2(\Omega)]^6$. The variational inequality (6.1.11) may be considered as a definition of the B.V.P.; its solution satisfies (6.1.1–5) in a generalized sense. We make this clear by considering the case of w being continuously differentiable. Then (6.1.11) yields the equality

$$\int_\Omega \frac{\partial w}{\partial \varepsilon}(\varepsilon(u))(v-u)\,d\Omega = (f, v-u) + \int_{\Gamma_F} F_i(v_i - u_i)\,d\Gamma, \quad \forall v \in U_{ad} \qquad (6.1.12)$$

If we set $v - u = \pm\phi \in D(\Omega)$, it follows that (6.1.1) holds in the sense of distributions on Ω, when σ is defined by (6.1.3). Then by applying the Green–Gauss theorem we find (cf. Sec. 5.1.2) that (6.1.5) holds as an equality in $H^{-1/2}(\Gamma)$. If w is nondifferentiable, then (6.1.11) is equivalent to

$$-l \in \partial(W(\varepsilon(u)) \qquad (6.1.13)$$

for $u \in U_{ad}$, where l is a linear continuous form on $[H^1(\Omega)]^3$ defined by

$$l(v) = (f, v) + \int_{\Gamma_F} F_i v_i\,d\Gamma. \qquad (6.1.14)$$

From (6.1.11) we obtain the following minimization problem.

Proposition 6.1.1. Every solution $u \in U_{ad}$ of (6.1.11), if any exists, satisfies the minimization problem

$$\Pi(u) = \min\{\Pi(v) \mid v \in U_{ad}\}, \qquad (6.1.15)$$

where Π is defined on $[H^1(\Omega)]^3$ by the relation

$$\Pi(v) = W(\varepsilon(v)) - l(v), \qquad (6.1.16)$$

and conversely.

Proof. An immediate result of (6.1.11), q.e.d.

Functional Π expresses the total potential energy of the body considered and consists of the strain energy and the potential energy of the external loads. It is, obviously, convex, l.s.c. and proper on $[H^1(\Omega)]^3$.

Proposition 6.1.2. Every solution $u \in U_{ad}$ of the B.V.P. (6.1.1–5), if any exists, satisfies the variational inequality (6.1.11) and the minimization problem (6.1.15).

Proof. The first statement of the proposition has already been proved in (6.1.7–10). The second statement follows from Prop. 6.1.1., q.e.d.

Further, the existence and uniqueness of the solution u of (6.1.11) or (6.1.15) is studied.

6.1.2 The Existence and Uniqueness of the Solution

Proposition 6.1.3. Let mes $\Gamma_U > 0$. Then the vectors $u \in U_{ad}$ which minimize the functional Π on U_{ad} form a nonempty, closed and convex set. If, moreover, w is strictly convex, the solution is unique.

Proof. In order to apply Prop. 2.2.1, we must show that

$$\Pi(v) \to \infty \quad \text{as} \quad \|v\|_1 \to \infty, \qquad v \in U_{ad}. \tag{6.1.17}$$

The assumptions on U imply (cf. Sec. 5.1.2) that there exists a vector $v_0 \in [H^1(\Omega)]^3$ such that $v_0 = U$ on Γ_U. Then using the transformation $\tilde{v} = v - v_0 \in V_0 = \{v | v \in [H^1(\Omega)]^3,\ v = 0 \text{ on } \Gamma_U\}$, the Schwarz inequality and the inequality (1.4.28), we see that

$$\int_\Omega \varepsilon_{ij}(v)\varepsilon_{ij}(v)\, d\Omega \geq c\|\tilde{v}\|_1^2 + c$$

$$+ 2\int_\Omega \varepsilon_{ij}(\tilde{v})\varepsilon_{ij}(v_0)\, d\Omega \geq c_1\|\tilde{v}\|_1^2 - c_2\|\tilde{v}\|_1 + c_3, \tag{6.1.18}$$

where c_i is a constant > 0 for $i = 1$ and ≥ 0 for $i = 2, 3$. From (6.1.6) and (6.1.18), and from the continuity of the form $l(v)$, it follows that

$$\Pi(v) \geq c(\|v\|_1^2 - a\|v\|_1 - b) \qquad a, b, c \text{ const}, c > 0, \tag{6.1.19}$$

and thus (6.1.17) is verified. For the rest of the proposition we argue as in Prop. 2.2.1, q.e.d.

Next, the case $\Gamma_U = \varnothing$ is studied. Then a rigid-body displacement is possible, $U_{ad} = [H^1(\Omega)]^3$, and Γ_F coincides with Γ. Of course, the given system of the forces $\{f_i\}$ and $\{F_i\}$ must be statically equivalent to zero, i.e.,

$$\int_\Omega f_i r_i\, d\Omega + \int_{\Gamma_F} F_i r_i\, d\Gamma = 0, \qquad \forall r \in \mathscr{R}, \tag{6.1.20}$$

where $\mathscr{R}$ denotes the space of the rigid-body displacements. (6.1.20) is a necessary condition for the existence of a solution. and results directly from (6.1.11) by setting $v = u \pm r$, where it is assumed that u is a solution of the problem. Let us introduce the quotient space $[H^1(\Omega)]^3/\mathscr{R}$, in which we define first the linear form $\bar{l}(\cdot)$ by

$$\bar{l}([v]) = l(v), \qquad \forall v \in [v], \tag{6.1.21}$$

and then the bilinear form $\int_\Omega \varepsilon_{ij}(\cdot)\varepsilon_{ij}(\cdot)\, d\Omega$ by

$$\int_\Omega \varepsilon_{ij}([v])\varepsilon_{ij}([u])\, d\Omega = \int_\Omega \varepsilon_{ij}(v)\varepsilon_{ij}(u)\, d\Omega, \quad \forall v \in [v], \quad u \in [u]. \quad (6.1.22)$$

From (4.1.36) it results that

$$\int_\Omega \varepsilon_{ij}([v])\varepsilon_{ij}([v])\, d\Omega \geq c \|[v]\|^2, \qquad c \text{ const} > 0, \qquad (6.1.23)$$

where $\|[(v)]\|$ is the norm in the space $[H^1(\Omega)]^3/\mathscr{R}$. Let us define on $[H^1(\Omega)]^3/\mathscr{R}$ a functional $\bar{\Pi}$ by setting

$$\bar{\Pi}([v]) = \Pi(v). \qquad (6.1.24)$$

This definition is justified by (6.1.20–22).

Proposition 6.1.4. Suppose that (6.1.20) holds. Then there exists a solution $[u] \in [H^1(\Omega)]^3/\mathscr{R}$ of the variational inequality (6.1.11). If, moreover, w is strictly convex, the solution $[u]$ is uniquely determined.

Proof. $\bar{\Pi}$ is a convex, l.s.c. functional on $[H^1(\Omega)]^3/\mathscr{R}$. Now we may show that

$$\bar{\Pi}([v]) \to \infty \quad \text{as} \quad \|[v]\| \to \infty. \qquad (6.1.25)$$

By applying (6.1.23), we can show that

$$\bar{\Pi}([v]) \geq c(\|[v]\|^2 - \|[v]\|), \qquad c \text{ const} > 0, \qquad (6.1.26)$$

which yields (6.1.25). As in Prop. 2.2.1, if w is strictly convex, the solution $[u]$ is unique, q.e.d.[1]

The foregoing results hold for a material obeying the law (3.3.86–87) and for an elastic hardening material, among others. As indicated in Sec. 3.3.2, the functional w is in these cases continuously differentiable, strictly convex and satisfies the left-hand side inequality in (6.1.6). Note that thus far we have used only the left-hand side inequality in (6.1.6). Accordingly, the existence of a unique displacement field and of a corresponding stress and strain field is ensured. For an elastic, perfectly locking material, functional w has the same properties as before, with the difference that it is only continuous. For such a material we can prove the existence of a unique displacement field.

[1] Props. (6.1.3) and (6.1.4) are proved in [176] [177] under the more general assumption that

$$w(\varepsilon) \geq c(\varepsilon_{ij}\varepsilon_{ij})^{p/2}, \ p > 1.$$

Note that an elastic, perfectly locking material does not fulfill the right-hand side inequality in (6.1.6).

6.1.3 Duality

A variational problem will now be formulated with respect to the stresses. To this end, the spaces H and Σ are introduced (see (1.4.32, 33)). Then we define the admissible set for the stresses

$$\Sigma_{ad} = \{\tau \mid \tau \in H,\ \tau_{ij,j} + f_i = 0 \text{ in } \Omega,\ T_i = F_i \text{ on } \Gamma_F\}, \qquad (6.1.27)$$

which is a closed, convex subset of Σ. Relation (6.1.3) can be written in its inverse form (3.3.58). Moreover, as can be verified by means of (2.5.3), the right-hand side of (6.1.6) implies the relation

$$c(\tau_{ij}\tau_{ij}) \leq w^c(\tau), \qquad c \text{ const} > 0. \qquad (6.1.28)$$

In order to formulate a variational problem with respect to the stresses, we define the functional

$$W^c(\tau) = \begin{cases} \displaystyle\int_\Omega w^c(\tau)\, d\Omega & \text{if } w^c(\tau) \in L^1(\Omega) \\ \infty & \text{otherwise} \end{cases} \qquad (6.1.29)$$

which is convex, l.s.c. and proper on $[L^2(\Omega)]^6$ and extends (6.1.3) to $[L^2(\Omega)]^6$ (Prop. 3.3.1). The following proposition holds.

Proposition 6.1.5. Suppose that a stress field $\sigma \in \Sigma_{ad}$ can be associated with a solution $u \in U_{ad}$ of the B.V.P. (6.1.1–5) fulfilling the material law (6.1.3). This stress field satisfies the variational inequality

$$W^c(\tau) - W^c(\sigma) \geq \langle U, T - S\rangle_{\Gamma_U}, \qquad \forall \tau \in \Sigma_{ad}. \qquad (6.1.30)$$

which is equivalent to the minimization problem

$$\Pi^c(\sigma) = \min\{\Pi^c(\tau) \mid \tau \in \Sigma_{ad}\}, \qquad (6.1.31)$$

where

$$\Pi^c(\tau) = W^c(\tau) - \langle U, T\rangle_{\Gamma_U}. \qquad (6.1.32)$$

Proof. Multiplying (6.1.2) by $\tau_{ij} - \sigma_{ij}$, $\tau, \sigma \in \Sigma_{ad}$, and applying the Green–Gauss theorem, we get the relation

$$\int_\Omega \varepsilon_{ij}(u)(\tau_{ij} - \sigma_{ij})\, d\Omega = \langle U, T - S\rangle_{\Gamma_U}, \qquad \forall \tau \in \Sigma_{ad}. \qquad (6.1.33)$$

(6.1.33), when combined with the variational inequality

$$W^c(\tau) - W^c(\sigma) \geq \int_\Omega \varepsilon_{ij}(\tau_{ij} - \sigma_{ij})\, d\Omega, \qquad \forall \tau \in [L^2(\Omega)]^6, \qquad (6.1.34)$$

yields the variational inequality (6.1.30), which clearly is equivalent to the minimization problem (6.1.31), q.e.d.

The functional Π^c expresses the complementary energy of the body, (6.1.31) the principle of minimum complementary energy for the material law considered, and (6.1.30) the principle of complementary virtual work. Since we encountor variations of the stresses, we can speak, as is usual in mechanics, of the dual problem of (6.1.15). It remains for us to investigate the relation between the solution of (6.1.15) (primal problem) and (6.1.31) (dual problem). First, we prove an existence result for the solution of (6.1.30) or (6.1.31).

Proposition 6.1.6. Suppose that $\Sigma_{ad} \neq \emptyset$. Then the problem (6.1.31) has a solution $\sigma \in \Sigma_{ad}$. This solution is unique if w^c is strictly convex.

Proof. From (6.1.28),

$$W^c(\tau) \geq c\|\tau\|_H^2, \qquad c \text{ const} > 0, \qquad \forall \tau \in H. \tag{6.1.35}$$

Hereafter the proof follows that of Prop. 5.3.2, q.e.d.

The functional $A(\cdot)$ in (3.2.47) has, for the problem considered in this section, the form

$$A(u) = -(f, u) - \int_{\Gamma_F} F_i u_i \, d\Gamma + I_{U_{ad}}(u). \tag{6.1.36}$$

It can be verified by the same method as in Sec. 5.3.2 that

$$A^c(-\Lambda'\sigma) = I_{\Sigma_{ad}}(\sigma) - \langle S, U \rangle_{\Gamma_U}. \tag{6.1.37}$$

Moreover, $B(\Lambda u)$ in (3.2.47) is given by

$$B(\Lambda u) = W(\varepsilon(u)), \tag{6.1.38}$$

and thus

$$B^c(\sigma) = W^c(\sigma). \tag{6.1.39}$$

Accordingly, the dual problem

$$\sup_{\Sigma} [-A^c(-\Lambda'\sigma) - B^c(\sigma)] \tag{6.1.40}$$

coincides with (6.1.31). The general duality theory of Sec. 3.2.2. allows the formulation of the following propositions.

Proposition 6.1.7. Assume that there exists $u_0 \in U_{ad}$ such that $W(\varepsilon(u_0))$ is finite and continuous at $\varepsilon(u_0)$. Then every solution u of the primal problem (6.1.15) and every solution σ of the dual problem (6.1.31) satisfy the relation

$$\Pi(u) + \Pi^c(\sigma) = 0 \tag{6.1.41}$$

and are linked by the extremality condition

$$W(\varepsilon(u)) + W^c(\sigma) = (\varepsilon(u), \sigma). \tag{6.1.42}$$

Proof. Direct application of Prop. 3.2.7. q.e.d.

Note that (6.1.42) is an extension of the pointwise material law (6.1.3) (cf., Props. 2.5.5 and 3.3.1.). (6.1.41) implies for $u \in U_{ad}$ and $\sigma \in \Sigma_{ad}$ the energy inequalities

$$\Pi(v) \geq \Pi(u) = -\Pi^c(\sigma) \geq -\Pi^c(\tau), \quad \forall v \in U_{ad}, \quad \forall \tau \in \Sigma_{ad}. \tag{6.1.43}$$

Proposition 6.1.8. The conditions

(i) u minimizes the potential energy Π on U_{ad}, σ minimizes the complementary energy Π^c on Σ_{ad}, and (6.1.41) holds, and
(ii) $u \in U_{ad}$ and $\sigma \in \Sigma_{ad}$ satisfy the extremality condition (6.1.42)

are equivalent to each other.

Proof. Results directly from Prop. 3.2.6., q.e.d.

The question also arises as to whether a strain field and a displacement field satisfying the material law and the kinematical conditions (6.1.2) and (6.1.4) can be associated with the solution σ of the dual problem. This question is answered affirmatively in [177] on the assumption that w^c is continuously differentiable. For a material obeying (3.3.83), questions on the existence of the solution are studied in [313] and [5] by considering the space of functions of bounded deformation. This functional setting has been adopted in the present book in Chapter 9 for the study of problems arising in the flow theory of plasticity.

6.2 Linear Elastic Material Law and Subdifferential Boundary Conditions

6.2.1 Formulation of the Boundary Conditions

In this section we assume that the body is linearly elastic, i.e., that

$$\sigma_{ij} = C_{ijhk}\varepsilon_{hk}, \tag{6.2.1}$$

where the elasticity tensor $C = \{C_{ijhk}\}$, $C_{ijhk} \in L^\infty(\Omega)$, $i, j, h, k = 1, 2, 3$, satisfies the symmetry and ellipticity relations (3.3.61a) and (3.3.61b). The boundary Γ of Ω consists of three open, mutually disjoint parts Γ_U, Γ_F and Γ_S, i.e., $\Gamma = \bar{\Gamma}_U \cup \bar{\Gamma}_F \cup \bar{\Gamma}_S$. We do not exclude the case $\Gamma_U = \varnothing$ and/or $\Gamma_F = \varnothing$, but Γ_S is always nonempty. On Γ_U the displacements and on Γ_F the surface forces are given, i.e., (6.1.4) and (6.1.5) hold. On Γ_S a subdifferential boundary condition of the form

$$-S \in \partial j(u) \tag{6.2.2}$$

holds pointwise, where j is a convex, l.s.c. and proper functional on $\mathbb{R}^3$. Thus, the classical formulation of the B.V.P. consists of the relations (6.1.1, 2), (6.2.1) and the boundary conditions (6.1.4, 5) and (6.2.2). Boundary condition (6.2.2) corresponds to various classes of inequality problems, as was illustrated in Sec. 3.3.1. In contrast to the previous section, here the boundary condition is responsible for the unilateral character of the problem and the resulting variational inequalities. Along the lines of this problem two others can also be considered. First, on Γ_S a relation of the form

$$-S_i \in \beta_i(u_i), \qquad i = 1, 2, 3 \tag{6.2.3}$$

can be assumed to hold, where $\beta_i\colon \mathbb{R} \to \mathbb{R}$ are maximal monotone operators. Second, the energy functional j can be assumed to be decomposable (cf., (3.3.155)), i.e., that the relations

$$-S_N \in \partial j_N(u_N), \qquad -S_T \in \partial j_T(u_T) \tag{6.2.4}$$

hold. As is apparent, the procedure which will be used for the boundary conditions (6.2.2) can also be adapted to (6.2.3) or (6.2.4) as well. On the assumption that all functions are appropriately smooth, we can, by means of the Green–Gauss theorem, write the relation

$$\int_\Omega \sigma_{ij}(\varepsilon_{ij}(v) - \varepsilon_{ij}(u))\, d\Omega = \int_\Omega f_i(v_i - u_i)\, d\Omega + \int_{\Gamma_F} F_i(v_i - u_i)\, d\Gamma + \int_{\Gamma_S} S_i(v_i - u_i)\, d\Gamma \tag{6.2.5}$$

for every v, such that on Γ_U, $v_i = U_i$.

For the same functional framework as in the previous section, we define the functional Φ on $[L^2(\Gamma))]^3$ by the relation

$$\Phi(v) = \begin{cases} \displaystyle\int_{\Gamma_S} j(u)\, d\Gamma & \text{if } \; j(u) \in L^1(\Gamma_S) \\ \infty & \text{otherwise} \end{cases}, \tag{6.2.6}$$

and as in Sec. 3.3.4 we consider its restriction to $[H^{1/2}(\Gamma)]^3$, again denoting the restriction by Φ. $\Phi_1\colon u \to \Phi(u|_\Gamma)$ is convex, l.s.c. and proper on $[H^1(\Omega)]^3$. Taking into account (6.2.2), (6.2.5) and (6.2.6), we are in a position to formulate the following problem. (We simply write $\Phi(u)$ instead of $\Phi(u|_\Gamma)$).

Problem 1. Find a displacement field $u \in U_{ad}$ such that

$$a(u, v-u) + \Phi(v) - \Phi(u) \geq (f, v-u) + \int_{\Gamma_F} F_i(v_i - u_i)\, d\Gamma, \qquad \forall v \in U_{ad} \quad (6.2.7)$$

where $a(u, v)$ is given by (5.1.10).

We can formulate the following additional problem.

Problem 2. Find $u \in U_{ad}$ such that

$$\Pi(u) = \min \{\Pi(v) \mid v \in U_{ad}\}, \quad (6.2.8)$$

where

$$\Pi(v) = \tfrac{1}{2}\, a(v, v) + \Phi(v) - (f, v) - \int_{\Gamma_F} F_i v_i \, d\Gamma. \quad (6.2.9)$$

Problem 3. Find $u \in U_{ad}$ and $\sigma \in \Sigma_{ad}$ such that (6.2.1) holds and

$$\Phi(u) + \Phi^c(-S) = -\langle S, u\rangle_{\Gamma_S}. \quad (6.2.10)$$

These problems are related by the following proposition:

Proposition 6.2.1

(a) Every solution of Problem 1 is also a solution of Problem 2, and conversely.
(b) If a displacement field $u \in U_{ad}$ and a stress field $\sigma \in \Sigma_{ad}$ constitute a solution of Problem 3, then u is a solution of Problems 1 and 2.
(c) If the displacement field $u \in U_{ad}$ satisfies Problem 1, then the corresponding stress field $\sigma(\varepsilon(u))$ belongs to Σ_{ad} and $\{u, \sigma\}$ is a solution to Problem 3.

Proof. Part (a) follows directly by applying Prop. 2.2.2. Part (b) results by combining (6.2.5), written in the functional setting adopted herein (cf., (1.4.37)), with the inequality

$$\Phi(v) - \Phi(u) \geq -\langle S, v-u\rangle_{\Gamma_S}, \qquad \forall v \in [H^{1/2}(\Gamma)]^3 \quad (6.2.11)$$

for $u_i \in H^{1/2}(\Gamma)$. We recall that (6.2.11) is equivalent to (6.2.10). Then (6.2.7) is obtained. For the proof of part (c), we take $u \in U_{ad}$ and we insert $v - u = \pm\phi \in [D(\Omega)]^3$ in (6.2.7). Thus (6.1.1) holds in the sense of $[\mathscr{D}'(\Omega)]^3$ and since $f_i \in L^2(\Omega)$, in the sense of $[L^2(\Omega)]^3$. By combining (1.4.37) with (6.2.7), we obtain the expression

$$\langle S - F, v-u\rangle_{\Gamma_F} + \Phi(v) - \Phi(u) + \langle S, v-u\rangle_{\Gamma_S} \geq 0, \qquad \forall v \in U_{ad} \quad (6.2.12)$$

If in (6.2.12) we set $v_i - u_i = \pm\phi_i \in H^{1/2}(\Gamma)$, with support in the int Γ_F, (6.1.5) results as an equality in $L^2(\Gamma_F)$. If we take into account the surjectivity of the trace mapping, (6.2.12) can be written for $u_i, v_i \in H^{1/2}(\Gamma)$. From this, (6.2.11) results, and, equivalently, (6.2.10), q.e.d.

Functional Π represents the potential energy of the body. The minimum problem (6.2.8) expresses the fact that at the position of equilibrium the potential energy achieves its minimum. The variational inequality (6.2.7) is the expression of the principle of virtual work in its inequality form. Some further propositions on the existence and uniqueness of the solution of the problem can be formulated. Again the cases mes $\Gamma_U > 0$ and $\Gamma_U = \varnothing$ will be examined separately.

6.2.2 Existence and Uniqueness Propositions

Proposition 6.2.2. Suppose that mes $\Gamma_U > 0$. Then Problem 1 has a unique solution.

Proof. In order to apply Prop. 2.2.1 we have to verify (2.2.3). We proceed as in Prop. 5.2.2. Substituting $\tilde{v} = v - w \in V_0$, where $w = U$ on Γ_U, we obtain

$$\Pi(v) = \Pi(\tilde{v} + w) \geq a(w, \tilde{v}) + \tfrac{1}{2}a(\tilde{v}, \tilde{v}) + \Phi(v) - (f, v) - \int_{\Gamma_F} F_i v_i \, d\Gamma. \quad (6.2.13)$$

According to the Hahn–Banach theorem, there exist constants $c_1, c_2 \geq 0$ such that (cf., Prop. 2.3.3)

$$\Phi(v) \geq -(c_1 \| v \|_{1/2} + c_2), \qquad \forall v \in [H^{1/2}(\Gamma)]^3. \quad (6.2.14)$$

Combining (6.2.13) and (6.2.14) with Korn's inequality (1.4.28), and observing the continuity of the linear forms, we find that

$$\Pi(v) \geq c(\| v \|_1^2 + \tilde{c}_1 \| v \|_1 + \tilde{c}_2), \quad (6.2.15)$$

with $c, \tilde{c}_i$ constants, $c > 0$ and $\tilde{c}_i \leq 0$. Accordingly,

$$\Pi(v) \to \infty \quad \text{as} \quad \| v \|_1 \to \infty \quad \text{for} \quad v \in U_{ad}, \quad (6.2.16)$$

and the existence of at least one solution is ensured. In order to show the uniqueness of the solution, let us assume that u_1 and u_2 are both solutions of Problem 1. Then from (6.2.7), by substituting $u = u_1$, $v = u_2$ and $u = u_2$, $v = u_1$ we obtain

$$a(u_1 - u_2, u_1 - u_2) \leq 0, \quad (6.2.17)$$

which yields $u_1 = u_2$, q.e.d.

For the case $\Gamma_U = \varnothing$ first a uniqueness result is proved.

Proposition 6.2.3. Let $\Gamma_U = \varnothing$. Assume that u_1 and u_2 are two displacement fields satisfying Problem 1, and σ_1 and σ_2 the corresponding stress fields. Then

$$u_1 - u_2 \in \mathcal{R}, \qquad \varepsilon(u_1) = \varepsilon(u_2), \qquad \sigma(u_1) = \sigma(u_2). \tag{6.2.18}$$

Proof. As in Prop. 6.2.2, the inequality (6.2.17) is obtained. Thus (cf., (5.2.14)) (5.2.19) holds, q.e.d.

Proposition 6.2.4. Let $\Gamma_U = \varnothing$ and assume that

$$\Phi = \Phi_1 + \Phi_2, \tag{6.2.19}$$

where Φ_1 and Φ_2 are proper, convex and l.s.c., and are such that $D(\Phi_1) \cap D(\Phi_2) \neq \varnothing$. Moreover, assume that for $a > 1$ and for every t positive

$$\Phi_1(0) = 0, \qquad \Phi_1(tv) \leq t^{\alpha}\Phi_1(v), \qquad \forall v \in [H^1(\Omega)]^3 \tag{6.2.20}$$

and that

$$\Phi_1(r) > 0, \qquad \forall r \in \mathcal{R}, \qquad r \neq 0. \tag{6.2.21}$$

Then (6.2.7) has at least one solution.

Proof. By Prop. 2.2.1 we must show that (2.2.3) holds. If this is not true, then a sequence $\{v_n\} \in [H^1(\Omega)]^3$ exists such that $\| v_n \|_1 \to \infty$ as $n \to \infty$ and

$$\Pi(v_n) \leq c, \qquad c \text{ const} > 0. \tag{6.2.22}$$

Dividing by $\| v_n \|_1^2$ and denoting (n being sufficiently large) by w_n the quotient $v_n/\| v_n \|_1$, we find that

$$\tfrac{1}{2}a(w_n, w_n) + \frac{\Phi(v_n)}{\| v_n \|_1^2} \leq c\left(\frac{1}{\| v_n \|_1} + \frac{1}{\| v_n \|_1^2}\right), \tag{6.2.23}$$

with c constant > 0. Applying (6.2.14) to Φ_1 and Φ_2, we obtain from (6.2.23)

$$a(w_n, w_n) \leq c'\left(\frac{1}{\| v_n \|_1} + \frac{1}{\| v_n \|_1^2}\right), \qquad c' \text{ const} > 0. \tag{6.2.24}$$

Since $\| w_n \|_1 = 1$, a subsequence $\{w_{n'}\}$ exists such that $w_{n'} \to w$ weakly in $[H^1(\Omega)]^3$ and strongly in $[L^2(\Omega)]^3$ as $n' \to \infty$. By means of (4.1.23), (6.2.24) implies that as $n' \to \infty$

$$0 \geq \liminf a(w_{n'}, w_{n'}) \geq a(w, w), \tag{6.2.25}$$

and therefore $w \in \mathcal{R}$. But (cf., (1.4.29))

$$a(w_{n'}, w_{n'}) + c_1 | w_{n'} |_2^2 \geq c_2, \qquad c_1, c_2 \text{ const} > 0, \tag{6.2.26}$$

and by taking the liminf we obtain

$$| w |_2 \geq c, \qquad c \text{ const} > 0. \tag{6.2.27}$$

In (6.2.23), if Φ is written according to (6.2.19), and if (6.2.14) is then applied to Φ_2, there follows by means of (6.2.20)

$$\Phi_1(w_n) \le \frac{c}{\| v_n \|_1^{\alpha-1}} \left(1 + \frac{1}{\| v_n \|_1}\right), \tag{6.2.28}$$

and therefore

$$\Phi_1(w) \le \liminf \Phi_1(w_n) \le 0. \tag{6.2.29}$$

The latter contradicts (6.2.21) because $w \in \mathscr{R}$ and $w \neq 0$, q.e.d.

Proposition 6.2.5. Let $\Gamma_U = \varnothing$ and suppose that

$$\Phi(v) = \Phi_0(v) - \langle h, v \rangle, \tag{6.2.30}$$

where $h \in [H^{-1/2}(\Gamma)]^3$ is given and Φ_0 is a proper, convex and l.s.c. functional such that

$$\begin{gathered} \Phi_0(0) = 0 \quad \text{and} \quad \Phi_0(tv) = t\Phi_0(v), \\ \forall t > 0, \qquad \forall v \in [H^1(\Omega)]^3. \end{gathered} \tag{6.2.31}$$

(a) The condition

$$\Phi_0(r) \ge (f, r) + \int_{\Gamma_F} F_i r_i \, d\Gamma + \langle h, r \rangle, \qquad \forall r \in \mathscr{R} \tag{6.2.32}$$

is a necessary condition and

(b) the condition

$$\Phi_0(r) > (f, r) + \int_{\Gamma_F} F_i r_i \, d\Gamma + \langle h, r \rangle \qquad \forall r \in \mathscr{R} \qquad r \neq 0 \tag{6.2.33}$$

is a sufficient condition for the existence of a solution of (6.2.7).

Proof. (a) Similar to that of (5.2.10) (cf., also (5.2.8, 9)).

(b) As in Prop. 6.2.4, we argue by contradiction. First, we prove that $w \in \mathscr{R}$ and that (6.2.27) holds. Then with the aid of (6.2.31) it is shown that

$$\Phi_0(w_{n'}) \le (f, w_{n'}) + \int_{\Gamma_F} F_i w_{n'i} \, d\Gamma + \langle h, w_{n'} \rangle + \frac{c}{\| v_{n'} \|_1} \tag{6.2.34}$$

where c is a constant > 0, and therefore, by taking the liminf,

$$\Phi_0(w) \le (f, w) + \int_{\Gamma_F} F_i w_i \, d\Gamma + \langle h, w \rangle, \qquad w \in \mathscr{R}, \qquad w \neq 0 \tag{6.2.35}$$

holds, which contradicts (6.2.33), q.e.d.

This last proposition can obviously be applied to tangential or normal friction boundary conditions (cf., Prop. 5.2.4), for the Signorini–Fichera boundary conditions, etc. By means of Prop. 4.1.6, general necessary and sufficient conditions can be formulated. To this end, let us define the functional l by (6.1.14). Φ_1 below is defined in Sec. 6.2.1.

Proposition 6.2.6. Suppose that $\Gamma_U = \varnothing$. For a solution of the variational inequality (6.2.7) to exist, it is necessary that

$$l|_{\mathscr{R}} \in \overline{R(\partial \Phi_{1u_0})} \qquad \forall u_0 \in D(\Phi_1), \tag{6.2.36}$$

If, moreover, a $u_0 \in D(\Phi_1)$ exists such that

$$l|_{\mathscr{R}} \in \operatorname{relint} R(\partial \Phi_{1u_0}), \tag{6.2.37}$$

then (6.2.7) admits at least one solution.

Proof. Assumptions (i)–(vi) of Prop. (4.1.6) can be easily verified by taking $p(v) = a(v, v)^{1/2}$ and by introducing a linear continuous monotone operator $A: [H^1(\Omega)]^3 \to ([H^1(\Omega)]^3)'$ (cf., Eq. (1.2.25)) such that

$$\langle Au, v \rangle = a(u, v), \qquad \forall u, v \in [H^1(\Omega)]^3. \tag{6.2.38}$$

Then, (6.2.36) and (6.2.37) result, q.e.d.

This general proposition, applied to the friction problem (Sec. 5.1.1), yields the necessary and sufficient conditions (5.2.10) and (5.2.12). Similarly, Prop. 6.2.5 is a special case of Prop. 6.2.6.

6.2.3 Duality

In order to derive the principle of complementary potential energy, we proceed as in Sec. 6.1.3 and define the functional

$$\Phi^c(-T) = \sup_{v_i \in H^{1/2}(\Gamma)} (\langle -T, v \rangle_{\Gamma_S} - \Phi(v)), \tag{6.2.39}$$

which is convex, l.s.c. and proper on $[H^{-1/2}(\Gamma)]^3$.

Proposition 6.2.7. Suppose that the displacement field u and the stress field σ satisfy Problem 3. Then $\sigma \in \Sigma_{ad}$ is a solution of the variational inequality

$$A(\sigma, \tau - \sigma) + \Phi^c(-T) - \Phi^c(-S) \geq \langle T - S, U \rangle_{\Gamma_U}, \qquad \forall \tau \in \Sigma_{ad}, \tag{6.2.40}$$

where A is given by (5.3.5). (6.2.40) is equivalent to the minimization problem

$$\Pi^c(\sigma) = \min \{\Pi^c(\tau) \mid \tau \in \Sigma_{ad}\} \tag{6.2.41}$$

with

$$\Pi^c(\tau) = \tfrac{1}{2}A(\tau, \tau) + \Phi^c(-T) - \langle T, U \rangle_{\Gamma_U}. \tag{6.2.42}$$

Proof. By multiplying (6.1.2) by $\tau_{ij} - \sigma_{ij}$ and applying (1.4.37), we derive for $\sigma \in \Sigma_{ad}$ the variational equality

$$\int_\Omega \varepsilon_{ij}(u)(\tau_{ij} - \sigma_{ij})\, d\Omega = \langle T - S, u\rangle_{\Gamma_U \cup \Gamma_S}, \qquad \forall \tau \in \Sigma_{ad}. \quad (6.2.43)$$

But due to (6.2.39),

$$\Phi^c(-T) - \Phi^c(-S) \geq -\langle T - S, u\rangle_{\Gamma_S}, \qquad \forall \tau \in \Sigma_{ad}. \quad (6.2.44)$$

From (6.2.43, 44) and (6.2.1), the variational inequality (6.2.40) results. Finally, Prop. 2.2.2 implies (6.2.41) and conversely, q.e.d.

The variational inequality (6.2.40) expresses the principle of complementary virtual work, as does (6.2.41) for the minimum of complementary energy at the position of equilibrium.

Proposition 6.2.8. If $\Sigma_{ad} \neq \varnothing$, there exists a uniquely determined stress field $\sigma \in \Sigma_{ad}$ satisfying (6.2.41).

Proof. In order to apply Prop. 2.2.1, (2.2.3) must be verified when Π^c is given by (6.2.42). According to the Hahn–Banach theorem, there exist constants $c_1, c_2 \geq 0$ such that

$$\Phi^c(-T) \geq -c_1 \| T \|_{-1/2} - c_2. \quad (6.2.45)$$

By means of (6.2.45) and the continuity of the trace mapping $\bar{\gamma}$ (Sec. 1.4.3), it is shown, as in Prop. 5.3.2, that

$$\begin{gathered} \Pi^c(\tau) \geq c_3 \| \tau \|_\Sigma^2 - c_4 \| \tau \|_\Sigma - c_5, \\ c_3 \text{ const} > 0, \qquad c_4, c_5 \text{ const} \geq 0, \end{gathered} \quad (6.2.46)$$

and thus (2.2.3) is verified. Because of the strict convexity of $A(\tau, \tau)$, the solution is unique, q.e.d.

We shall refer to (6.2.41) as problem 4. As a result of Prop. 6.2.1, $\Sigma_{ad} \neq \varnothing$ if Problem 1 admits a solution. We shall now show the duality relation between the problems of minimum potential and minimum complementary energy in the sense of the mathematical theory of duality.

Functional $A(\cdot)$ (cf., (3.2.47)) of the primal problem has the form

$$A(v) = \Phi(v) - (f, v) - \int_{\Gamma_F} F_i v_i\, d\Gamma + I_{U_{ad}}(u). \quad (6.2.47)$$

It follows that

$$A^c(-\Lambda'\sigma) = I_{\Sigma_{ad}}(\sigma) + \Phi^c(-T) - \langle S, U\rangle_{\Gamma_U}. \quad (6.2.48)$$

Additionally, (5.3.21) and (5.3.22) hold, and thus (6.2.41) is the dual problem of (6.2.8). From Props. 3.2.6 and 3.2.7 the following propositions result.

Proposition 6.2.9. Assume that $D(\Phi) \neq 0$. Then every solution u of the primal problem (6.2.8) and every solution σ of the dual problem (6.2.41) together satisfy the relation

$$\Pi(u) + \Pi^c(\sigma) = 0 \tag{6.2.49}$$

and are related by the extremality conditions

$$\sigma_{ij} = C_{ijhk}\, \varepsilon_{hk}(u) \tag{6.2.50}$$

and (6.2.10).

Proposition 6.2.10. The two statements

(i) u minimizes Π on U_{ad}, σ minimizes Π^c on Σ_{ad}, and (6.2.49) is satisfied, and

(ii) $u \in U_{ad}$ and $\sigma \in \Sigma_{ad}$ satisfy the extremality conditions (6.2.10) and (6.2.50)

are equivalent to each other.

The inequalities (6.1.43) clearly hold for this B.V.P. as well and allow the derivation of global error estimates.

6.3 Subdifferential Material Laws and Subdifferential Boundary Conditions. Minimum Propositions for Nonmonotone Laws

6.3.1 Formulation and Study of the Problem

The B.V.P. studied in this section has a classical formulation consisting of the relations (6.1.1–3), along with the boundary conditions (6.1.4), (6.1.5) and (6.2.2) holding on Γ_U, Γ_F and Γ_S respectively. Moreover, if (6.1.6) holds, then the functional framework of Sec. 6.1 can be employed. By means of the Green–Gauss theorem, after defining the functionals W and Φ by (6.1.10) and (6.2.6) respectively, we are led to the following variational formulation.

Problem 1. Find a displacement field $u \in U_{ad}$ such that

$$W(\varepsilon(v)) - W(\varepsilon(u)) + \Phi(v) - \Phi(u) \geq (f, v - u) + \int_{\Gamma_F} F_i(v_i - u_i)\, d\Gamma, \qquad \forall v \in U_{ad}. \tag{6.3.1}$$

This variational inequality can be considered as defining the physical problem; it is the expression of the principle of virtual displacements. From (6.3.1) a minimization problem results, as the following proposition states.

Proposition 6.3.1. Every solution $u \in U_{ad}$ of (6.3.1), if any exists, satisfies the minimization problem:

$$\Pi(u) = \min \{\Pi(v) \mid v \in U_{ad}\} \tag{6.3.2}$$

where Π is defined on $[H^1(\Omega)]^3$ by

$$\Pi(v) = W(\varepsilon(v)) + \Phi(v) - (f, v) - \int_{\Gamma_F} F_i v_i \, d\Gamma. \tag{6.3.3}$$

The converse is also true.

Proof. Follows by substituting $\Pi(v)$ in (6.3.1), q.e.d.

We shall refer to (6.3.2) as Problem 2. The B.V.P. can be also expressed in another form.

Problem 3. Find $u \in U_{ad}$ and $\sigma \in \Sigma_{ad}$ such that (6.1.42) and (6.2.10) hold.

Proposition 6.3.2. Suppose that a displacement field $u \in U_{ad}$ and a stress field $\sigma \in \Sigma_{ad}$ satisfy Problem 3. Then u satisfies Problems 1 and 2.

Proof. (6.1.42) and (6.2.10) are equivalent to two subdifferential relations which imply by definition two variational inequalities (cf., Prop. 2.5.5). These last, when combined with (1.4.37) yield the desired result, q.e.d.

The existence of the solution of (6.3.1) or (6.3.2) can be proved only for the case mes $\Gamma_U > 0$. The question of the existence of a solution for $\Gamma_U = \varnothing$ is still open.

Proposition 6.3.3. Suppose that mes $\Gamma_U > 0$. Then (6.3.1) has a convex closed set of solutions which reduces to one element if w is strictly convex.

Proof. From (6.1.6), the estimate (6.1.18) holds. Moreover, (6.2.14) is also valid. Thus we can easily verify that

$$\Pi(v) \geq c_1 \| v \|_1^2 - c_2 \| v \|_1 - c_3, \qquad c_1 \text{ const} > 0, \quad c_2, c_3 \text{ const} \geq 0. \tag{6.3.4}$$

Thus Prop. 2.2.1 can be applied, q.e.d.

As in the problems treated in the previous sections, a variational formulation can also be considered with respect to the stresses. This leads to the proposition of minimum complementary energy for the present problem. This minimum problem is dual to Problem 2, which expresses the minimum of the potential energy at the position of equilibrium. The functionals W^c

and Φ^c defined respectively by (6.1.29) and (6.2.39) are now considered and the following propositions proved.

Proposition 6.3.4. Suppose that $u \in U_{ad}$ and $\sigma \in \Sigma_{ad}$ satisfy Problem 3. Then σ is a solution of the variational inequality

$$W^c(\tau) - W^c(\sigma) + \Phi^c(-T) - \Phi^c(-S) \geq \langle T - S, U \rangle_{\Gamma_U}, \qquad \forall \tau \in \Sigma_{ad} \quad (6.3.5)$$

(6.3.5) is equivalent to the minimization problem

$$\Pi^c(\sigma) = \min \{\Pi^c(\tau) \mid \tau \in \Sigma_{ad}\} \quad (6.3.6)$$

with

$$\Pi^c(\tau) = W^c(\tau) + \Phi^c(-T) - \langle T, U \rangle_{\Gamma_U}. \quad (6.3.7)$$

Proof. (6.2.43) combined with (6.2.44) and

$$W^c(\tau) - W^c(\sigma) \geq (\varepsilon(u), \tau - \sigma), \qquad \forall \tau \in H \quad (6.3.8)$$

implies (6.3.5), which is equivalent to (6.3.6), q.e.d.

Proposition 6.3.5. Suppose that $\Sigma_{ad} \neq \emptyset$. Then (6.3.5) has at least one solution.

Proof. In order to verify (2.2.3), when Π^c is given by (6.3.7), we take into account (6.1.28) and (6.2.45). Thus (cf., Props. 6.1.6 and 6.2.8) the estimate (6.2.46) results, q.e.d.

To investigate the duality between the problems of minimum potential and complementary energy, we assume that $A(u)$ in (3.3.47) is given by (6.2.47) and $B(\Lambda u)$ by (6.1.38). As in the two foregoing sections, it can be verified that the dual problem of (6.3.2) is problem (6.3.6). As a direct application of the general theory of Sec. 3.2.2, the following propositions result.

Proposition 6.3.6. Assume that a $u_0 \in D(\Phi)$ exists such that $W(\varepsilon(u))$ is finite and continuous at $\varepsilon(u_0)$. Then for every solution u (resp. σ) of the primal (resp. dual) problem 6.3.2 (resp. 6.3.6), the equality

$$\Pi(u) + \Pi^c(\sigma) = 0 \quad (6.3.9)$$

holds, and u and σ satisfy the extremality conditions (6.1.42) and (6.2.10).

Proposition 6.3.7. The two statements

(i) u minimizes Π on U_{ad}, σ minimizes Π^c on Σ_{ad} and (6.3.9) holds, and

(ii) $u \in U_{ad}$ and $\sigma \in \Sigma_{ad}$ satisfy the extremality conditions (6.1.42) and (6.2.10)

are equivalent to each other.

The propositions of this section can be applied to various problems which are important for possible applications in the engineering sciences. We mention here, e.g., friction- and unilateral-contact problems for locking or elastoplastic materials (deformation theory).

6.3.2 Nonmonotone Laws

All the combinations of the material laws and boundary conditions depicted schematically in Fig. 6.3.1 allow the formulation of minimization problems [257]. Further, the respective potential and complementary energy expressions will be derived. By taking into account the variational inequalities, which by definition hold, because of the superpotential laws, for cases 1a–2b of Fig. 6.3.1 we are led to potential energies Π given by the expressions

$$\Pi(v) = W(\pm\varepsilon(v)) + \Phi(\pm v) - (f, v) - \int_{\Gamma_F} F_i v_i \, d\Gamma, \tag{6.3.10}$$

or for cases 3c–4d given by the expressions

$$\Pi(v) = W(\pm\varepsilon(v)) + \Phi(\pm v) + (f, v) + \int_{\Gamma_F} F_i v_i \, d\Gamma. \tag{6.3.11}$$

The respective complementary energy expressions are

$$\Pi^c(\tau) = W^c(\pm\tau) + \Phi^c(\mp T) - \langle T, U\rangle_{\Gamma_U} \tag{6.3.12}$$

and

$$\Pi^c(\tau) = W^c(\mp\tau) + \Phi^c(\pm T) + \langle T, U\rangle_{\Gamma_U}. \tag{6.3.13}$$

τ in W^c has in cases 1a–2b (resp. 3c–4d) the same (resp. the opposite) sign as ε in W. On the other hand, T in Φ^c has in cases 1a–2b (resp. 3c–4d) the opposite (resp. the same) sign as v in Φ. If the potential or the complementary energy is strictly convex, the uniqueness of the solution (if any exists) of the corresponding problems is ensured (cf., the uniqueness proof in Prop. 2.2.1). Finally, let us note the following on the existence of the solution of the formulated minimum problems. In order to apply Prop. 2.2.1, functional Π must be convex and l.s.c. on $[H^1(\Omega)]^3$. These two properties can be fulfilled for many combinations of material laws and boundary conditions, in addition to the nondecreasing ones. However, the verification of the assumptions of Prop. 2.2.1. requires a more exact knowledge of the form of the functionals W, Φ, etc., depending in each case on the nature of the problem under study.

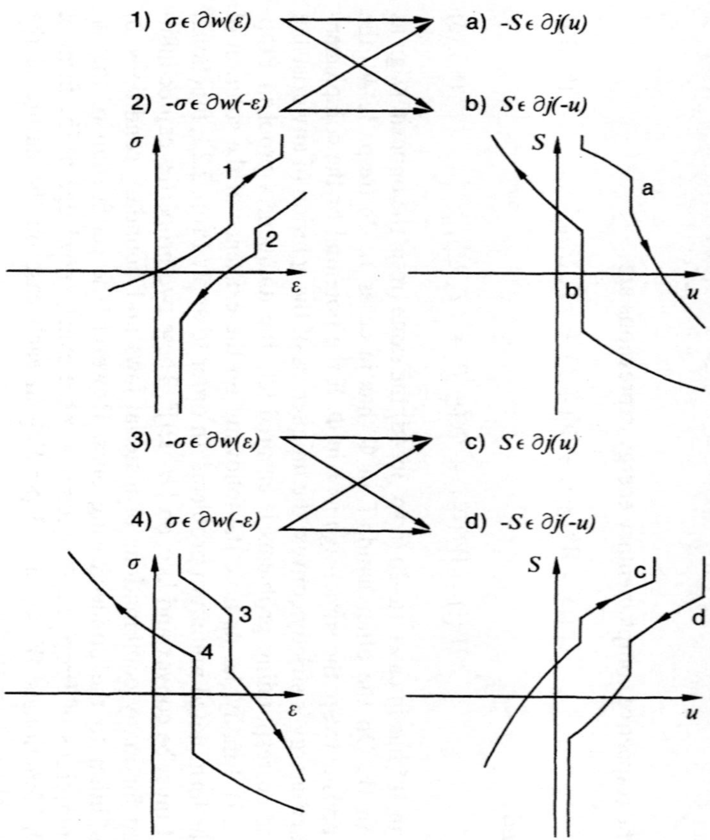

$$\left.\begin{array}{ll}(1\text{-a}) & W(\varepsilon(v)) - W(\varepsilon(u)) + \Phi(v) - \Phi(u) \geqslant \\ (1\text{-b}) & W(\varepsilon(v)) - W(\varepsilon(u)) + \Phi(-v) - \Phi(-u) \geqslant \\ (2\text{-a}) & W(-\varepsilon(v)) - W(-\varepsilon(u)) + \Phi(v) - \Phi(u) \geqslant \\ (2\text{-b}) & W(-\varepsilon(v)) - W(-\varepsilon(u)) + \Phi(-v) - \Phi(-u) \geqslant \end{array}\right\} \quad (f, v-u) + \int_{\Gamma_F} F_i(v_i - u_i)d\Gamma$$

$$\left.\begin{array}{ll}(3\text{-c}) & W(\varepsilon(v)) - W(\varepsilon(u)) + \Phi(v) - \Phi(u) \geqslant \\ (3\text{-d}) & W(\varepsilon(v)) - W(\varepsilon(u)) + \Phi(-v) - \Phi(-u) \geqslant \\ (4\text{-c}) & W(-\varepsilon(v)) - W(-\varepsilon(u)) + \Phi(v) - \Phi(u) \geqslant \\ (4\text{-d}) & W(-\varepsilon(v)) - W(-\varepsilon(u)) + \Phi(-v) - \Phi(-u) \geqslant \end{array}\right\} \quad -(f, v-u) - \int_{\Gamma_F} F_i(v_i - u_i)d\Gamma$$

Fig. 6.3.1. Possible path combinations for the formulation of variational inequalities.

6.4 The Corresponding Dynamic and Incremental Problems

On the assumption of small displacements, the dynamic problem corresponding to the static B.V.P. (6.1.1–5) is formulated by replacing (6.1.1) with (5.4.1) and by taking into account the initial conditions (5.4.9). f_i, F_i and U_i are time-dependent quantities, and U_i must be compatible with the initial conditions of the problem. The variational formulations given within this section are derived on the general assumption of appropriately smooth functions. The question of the existence of solutions of the formulated dynamic B.V.P.s is still open, with one exception for which an existence proof is possible. Note also that many of the variational problems formulated are not well-posed (cf., [295]).

Multiplying (5.4.1) by $v_i - u_i$ and integrating over Ω, we obtain the relation

$$\int_\Omega \sigma_{ij}(\varepsilon_{ij}(v) - \varepsilon_{ij}(u))\, d\Omega = \int_\Omega (f_i - \rho u_i'')(v_i - u_i)\, d\Omega + \int_{\Gamma_F} S_i(v_i - u_i)\, d\Gamma \quad (6.4.1)$$

for every v with $v_i = U_i$ on Γ_U, expressing d'Alembert's principle. From (6.4.1), (6.1.8) and (6.1.10), we are led to the variational inequality

$$\int_\Omega \rho u_i''(v_i - u_i)\, d\Omega + \Pi(v) - \Pi(u) \geq 0, \quad (6.4.2)$$

where Π is given by (6.1.16). The inequality (6.4.2) holds for every v with $v_i = U_i$ on Γ_U. The initial conditions (5.4.9) must also be fulfilled.

Similarly, it can be shown that the dynamic B.V.P. corresponding to the static problem studied in Sec. 6.2 (resp. 6.3) gives rise to a variational inequality similar to (6.4.2), the only difference being that the functional Π is given by (6.2.9) (resp. by (6.3.3)). For these two problems the following inequalities hold for every v with $v_i = U_i$ on Γ_U:

$$\int_\Omega \rho u_i''(v_i - u_i)\, d\Omega + a(u, v - u) + \Phi(v) - \Phi(u) - (f, v - u) - \int_{\Gamma_F} F_i(v_i - u_i)\, d\Gamma \geq 0 \quad (6.4.3)$$

and

$$\int_\Omega \rho u_i''(v_i - u_i)\, d\Omega + W(\varepsilon(v)) - W(\varepsilon(u)) + \Phi(v) - \Phi(u) - (f, v - u) - \int_{\Gamma_F} F_i(v_i - u_i)\, d\Gamma \geq 0. \quad (6.4.4)$$

These inequalities are formally derived and are expressions of d'Alembert's principle in its inequality form.

If, instead of (6.1.3) and (6.2.2), (3.3.93) and (3.3.11) hold (with the difference that because of the notation employed $\dot{\varepsilon}$ and $\partial u/\partial t$ are replaced by ε' and u') then the same procedure is followed as before by considering the velocity variations $v - u'$

instead of the displacement variations. Thus a variational inequality of the form

$$\int_{\Omega} \rho u_i''(v_i - u_i')\, d\Omega + \Pi(v) - \Pi(u') \geq 0 \tag{6.4.5}$$

results formally for every v with $v = U'$ on Γ_U. Here $\Pi(v)$ is given by (6.1.16), or (6.3.3) according to the type of problem studied. If the body obeys a linear elastic material law with subdifferential boundary conditions, the variational inequality describing its dynamic behavior reads

$$\int_{\Omega} \rho u_i''(v_i - u_i')\, d\Omega + a(u, v - u') + \Phi(v) - \Phi(u')$$

$$\geq (f, v - u') + \int_{\Gamma_F} F_i(v_i - u_i')\, d\Gamma \tag{6.4.6}$$

for every v with $v_i = U_i'$ on Γ_U. In order to investigate the uniqueness of the solution of (6.4.5) (if any exists), it is assumed that u_1 and u_2 are both solutions of (6.4.5). Then by inserting $v_i = u_{1i}'$, $u_i' = u_{2i}'$ and $v_i = u_{2i}'$, $u_i' = u_{1i}'$, we obtain

$$\int_{\Omega} \rho(u_{1i}'' - u_{2i}'')(u_{1i}' - u_{2i}')\, d\Omega \leq 0 \tag{6.4.7}$$

or

$$\frac{1}{2}\frac{d}{dt}\int_{\Omega} \rho(u_1' - u_2')^2\, d\Omega \leq 0 \tag{6.4.8}$$

From (6.4.8) and (5.4.9) it results that $u_1 = u_2$. For the uniqueness of the solution of (6.4.6) we argue as in Prop. 5.4.3. The solution u of (6.4.6) will be sought in the set

$$U_{ad} = \{v \mid v \in [H^1(\Omega)]^3,\ v = U' \text{ on } \Gamma_U\}. \tag{6.4.9}$$

The subspace V_0 (cf., (5.4.12)) is introduced, and by means of the substitutions $\tilde{v} = v - w$ and $\tilde{u} = u - w$, where w is defined as in Sec. 5.4, inequality (6.4.6) is transformed into the following

Problem 1. Find $\tilde{u}$: $[0, T] \to V_0$, with $\tilde{u}'(t) \in V_0$ and $\tilde{u}''(t) \in [L^2(\Omega)]^3$, which satisfies the variational inequality

$$(\rho\tilde{u}'', \tilde{v} - \tilde{u}') + a(\tilde{u}, \tilde{v} - \tilde{u}') + \Phi(\tilde{v} + w') - \Phi(\tilde{u}' + w')$$

$$\geq (\psi, \tilde{v} - \tilde{u}'), \qquad \forall \tilde{v} \in V_0 \tag{6.4.10}$$

and the initial conditions (5.4.14).

The functional ψ is again defined by (5.4.15), but with $C_N = 0$. In order to define the regularized problem, we assume that there exists a family of differentiable functionals Φ_ε, $\varepsilon > 0$, such that

(i)
$$\int_0^T \Phi_\varepsilon(\tilde{v} + w')\, dt \to \int_0^T \Phi(\tilde{v} + w')\, dt$$
$$\text{as } \varepsilon \to 0, \qquad \forall \tilde{v} \in L^2(0T, V_0), \tag{6.4.11}$$

(ii) there exists a bounded sequence $\{\bar{v}_\varepsilon\} \in V_0$ such that

$$\frac{d\Phi_\varepsilon}{dv}(\bar{v}_\varepsilon) = 0 \tag{6.4.12}$$

for every ε, and

(iii) if $v_\varepsilon \to v$ and $v'_\varepsilon \to v'$ in $L^\infty(0T, V_0)$ weakly$_*$ and

$$\int_0^T \Phi_\varepsilon(\tilde{v}'_\varepsilon + w')\, dt$$

is bounded, then

$$\liminf \int_0^T \Phi_\varepsilon(\tilde{v}'_\varepsilon + w')\, dt \geq \int_0^T \Phi(\tilde{v}' + w')\, dt. \tag{6.4.13}$$

The regularized problem has the form (5.4.16).

Proposition 6.4.1. Suppose that (5.4.17, 19, 20, 21, and 23) hold. Let

$$\Phi(\tilde{u}_1 + w'(0)) < \infty, \tag{6.4.14}$$

and let

$$v \to a(\tilde{u}_0, v) + \left\langle \frac{d\Phi_\varepsilon}{dv}(\tilde{u}_1 + w'(0)), v \right\rangle_{\Gamma_S} \tag{6.4.15}$$

be a continuous linear functional on V_0 for the $[L^2(\Omega)]^3$-topology. Then as $\varepsilon \to 0$ (u_ε satisfies the regularized problem),

$$\lim \tilde{u}_\varepsilon = \tilde{u} \quad \text{in} \quad L^\infty(0T, V_0) \quad \text{weakly}_*, \tag{6.4.16}$$

$$\lim \tilde{u}'_\varepsilon = \tilde{u}' \quad \text{in} \quad L^\infty(0T, V_0) \quad \text{weakly}_*, \tag{6.4.17}$$

$$\lim \tilde{u}''_\varepsilon = \tilde{u}'' \quad \text{in} \quad L^\infty(0T, [L^2(\Omega)]^3) \quad \text{weakly}_*. \tag{6.4.18}$$

The function $\tilde{u}$ is the solution of the variational inequality (6.4.10).

The proof is omitted here, since it follows the same steps as the proof of Props. 5.4.1 and 8.2.3, where a more complicated variational inequality arising in thermoelasticity is studied. Assumptions (6.4.11–13) are made in order to facilitate the proof. A more general proof which does not use these assumptions may be performed along the lines of [28]. Dynamic variational inequalities result as well for more complex relations between σ (resp. S) and ε or ε' (resp. u or u'), i.e., for the combinations of loading paths given in Fig. 6.3.1. Then for every path combination a different variational inequality results, having basically the same form as before, i.e., (6.4.2) or (6.4.5), but with Π given by (6.3.10) for the combinations 1a-2b and by (6.3.11) for 3c-4d. In these latter cases, minus signs appear before the inertia terms.

Chapter 7

Inequality Problems in the Theory of Thin Elastic Plates

This chapter is devoted to the study of certain classes of static and dynamic inequality problems pertaining to the theory of thin elastic plates. Here we consider the model proposed by von Kármán for plates undergoing large deflections relative to their thickness. Kirchhoff's theory for plates with small deflections constitutes a special case of von Kármán's theory. The latter permits a rational treatment of the problem of plate buckling and therefore is preferred in this chapter. The first section, in which some static unilateral problems are formulated and studied, is based on papers by G. Duvaut and J. L. Lions [87], J. Frančů [102], P. Hess [140], O. John [154], J. Naumann [225], M. Potier-Ferry [268], [269] among others. In the second section, to study the buckling problem, the corresponding eigenvalue problem is discussed. With reference to this problem, we give some general propositions concerning the eigenvalue problem for variational inequalities. This is a new area, only recently developed, and for this reason many questions are still open; we rely herein on the papers of A. F. Abeasis, J. P. Dias and A. Lopes-Pinto [1], M. Berger [22], M. Berger and P. Fife [23], A. Cimetière [43], [44], C. Do [71], [72], [73], [74], M. Kučera, J. Nečas and J. Souček [165], E. Miersemann [203], J. Naumann and H. Wenk [226], and M. Potier-Ferry [270], [271]. The third section concerns dynamic unilateral problems of von Kármán plates and is based on [88] and [248].

7.1 Static Unilateral Problems of von Kármán Plates

7.1.1 Generalities

The points of the plate are referred to a fixed, right-handed Cartesian coordinate system $0x_1x_2x_3$. The middle plane of the undeformed plate, assumed to have a constant thickness, coincides with the $0x_2x_3$-plane (cf., Fig. 3.3.7). The material points $x=(x_1, x_2, 0)$ of the undeformed plate constitute an open, bounded, connected subset $\Omega \subset \mathbb{R}^2$ with a regular[1] boundary Γ. We denote by $u=(u_1, u_2)$ the horizontal and by ζ the vertical displacements of the points $x \in \Omega$. The theory of von Kármán plates (cf., [172], and for a rigorous justification [42]), gives rise to the following differential equations:

$$\mathsf{K}\Delta\Delta\zeta - h(\sigma_{\alpha\beta}\zeta_{,\beta})_{,\alpha} = f_3 \quad \text{in} \quad \Omega; \tag{7.1.1}$$

$$\sigma_{\alpha\beta,\alpha} = 0 \quad \text{in} \quad \Omega; \tag{7.1.2}$$

$$\sigma_{\alpha\beta} = C_{\alpha\beta\gamma\delta}(\varepsilon_{\gamma\delta}(u) + \tfrac{1}{2}\zeta_{,\gamma}\zeta_{,\delta}) \quad \text{in} \quad \Omega; \tag{7.1.3}$$

and

$$\varepsilon_{\alpha\beta}(u) = \tfrac{1}{2}(u_{\alpha,\beta} + u_{\beta,\alpha}) \quad \text{in} \quad \Omega. \tag{7.1.4}$$

Here $\alpha, \beta, \gamma, \delta = 1, 2$, h is the plate thickness[2], $\Delta\Delta$ is the biharmonic operator, and $\mathsf{K} = Eh^3/12(1-\nu^2)$ is the bending rigidity of the plate with E the modulus of elasticity and ν the Poisson ratio. The plate is subjected to a distributed load $f = (0, 0, f_3)$, $f_3 = f_3(x)$, per unit area of the middle surface. $\sigma = \{\sigma_{\alpha\beta}\}$ (resp. $\varepsilon = \{\varepsilon_{\alpha\beta}\}$) denotes the stress (resp. strain) tensor in the plane of the plate, and $C = \{C_{\alpha\beta\gamma\delta}\}$ is the corresponding elasticity tensor. The components of C are assumed first to be elements of $L^\infty(\Omega)$ and second to satisfy the usual symmetry and ellipticity properties. From (7.1.1), assuming, as always, sufficiently smooth functions, multiplying by $z - \zeta$, integrating over Ω and applying the Green-Gauss theorem, we obtain the expression

$$\begin{aligned} \alpha(\zeta, z-\zeta) &+ \int_\Omega h\sigma_{\alpha\beta}\zeta_{,\alpha}(z-\zeta)_{,\beta}\, d\Omega \\ &= \int_\Gamma h\sigma_{\alpha\beta}\zeta_{,\alpha}n_\beta(z-\zeta)\, d\Gamma + \int_\Gamma \bar{Q}(\zeta)(z-\zeta)\, d\Gamma \\ &\quad - \int_\Gamma M(\zeta)\frac{\partial(z-\zeta)}{\partial n}\, d\Gamma + \int_\Omega f_3(z-\zeta)\, d\Omega, \end{aligned} \tag{7.1.5}$$

where $n = (n_1, n_2)$ denotes the outward normal unit vector on Γ and

$$\begin{aligned} \alpha(\zeta, z) = \mathsf{K}\int_\Omega &[(1-\nu)\zeta_{,\alpha\beta}z_{,\alpha\beta} \\ &+ \nu\Delta\zeta\Delta z]\, d\Omega \qquad 0 < \nu < \tfrac{1}{2} \end{aligned} \tag{7.1.6}$$

$$M(\zeta) = -\mathsf{K}[\nu\Delta\zeta + (1-\nu)(2n_1n_2\zeta_{,12} + n_1^2\zeta_{,11} + n_2^2\zeta_{,22})] \tag{7.1.7}$$

[1] It suffices to be Lipschitzian [225].

[2] With minor modifications, the case $h = h(x) > 0$ can also be treated.

and

$$\bar{Q}(\zeta) = -\mathsf{K}\left[\frac{\partial \Delta\zeta}{\partial n} + (1-\nu)\frac{\partial}{\partial \tau}\left[n_1 n_2(\zeta_{,22} - \zeta_{,11}) + (n_1^2 - n_2^2)\zeta_{,12}\right]\right]. \quad (7.1.8)$$

Here τ denotes the unit vector tangential to Γ such that n, τ and $0x_3$ form a right-handed system. In the same way, if we multiply Eqs. (7.1.2) by $v_1 - u_1$ and $v_2 - u_2$ respectively, add the results and then integrate over Ω, we get by the Green–Gauss theorem the relation

$$\int_\Omega \sigma_{\alpha\beta}\,\varepsilon_{\alpha\beta}(v - u)\,d\Omega = \int_\Gamma \sigma_{\alpha\beta}\,n_\beta(v_\alpha - u_\alpha)\,d\Gamma, \qquad \alpha, \beta = 1, 2. \quad (7.1.9)$$

(7.1.5) and (7.1.9) express the principle of virtual work for the bending and the stretching respectively of a free plate. M is the bending moment and $\bar{Q}$ the total shearing force on the plate boundary Γ (cf., Sec. 3.3.1). First we shall formulate a unilateral coercive B.V.P., and then a general unilateral semicoercive B.V.P.

7.1.2 Boundary Conditions and Corresponding Variational Formulations

Assume first that on Γ the boundary conditions

$$\zeta = 0, \quad (7.1.10)$$

$$u_1 = u_2 = 0 \quad (7.1.11)$$

and

$$M \in \beta\left(\frac{\partial\zeta}{\partial n}\right) = \partial j\left(\frac{\partial\zeta}{\partial n}\right) \quad (7.1.12)$$

hold (in the pointwise sense) with β given by (3.3.46). Relations (7.1.1–4) and (7.1.10–12) define the B.V.P. (classical formulation) which we shall study first. We can formulate analogous B.V.P.s by assuming similar relations between $-\bar{Q}$ and ζ on Γ or on a part of it.

The boundary conditions

$$M \in \beta_1\left(\frac{\partial\zeta}{\partial n}\right) = \partial j_1\left(\frac{\partial\zeta}{\partial n}\right) \quad \text{on} \quad \Gamma \quad (7.1.13)$$

$$-\bar{Q} \in \beta_2(\zeta) = \partial j_2(\zeta) \quad \text{on} \quad \Gamma \quad (7.1.14)$$

$$\sigma_{1\alpha} n_\alpha = \sigma_{2\alpha} n_\alpha = 0 \quad \text{on} \quad \Gamma \quad (7.1.15)$$

define the second—this time more general—B.V.P. to be studied. We assume, furthermore, that

$$f_3 = \bar{f}_3 + \bar{\bar{f}}_3, \quad (7.1.16)$$

where $\bar{\bar{f}}_3 \in L^2(\Omega)$ is a given load distribution, and that for $\Omega' \subset \Omega$, with $\bar{\Omega}' \cap \Gamma = \varnothing$,[3]

$$\begin{aligned} -\bar{f}_3 &\in \beta_3(\zeta) = \partial j_3(\zeta) \quad \text{in} \quad \Omega' \subset \Omega \\ \bar{f}_3 &= 0 \quad \text{in} \quad \Omega - \Omega'. \end{aligned} \tag{7.1.17}$$

The operators β_i are maximal monotone from $\mathbb{R}$ into $\mathbb{R}$, and the functionals j_i are convex, proper and l.s.c. on $\mathbb{R}$ according to Prop. 2.6.4. Relations (7.1.1–4) and (7.1.13–17) define the classical formulation of the second B.V.P. Generalizations of the conditions (7.1.13–17) are possible; for instance, we may assume that on $\Gamma_1 \subset \Gamma$, $M \in \beta_1'(\partial\zeta/\partial n)$, and on Γ_2 $(\Gamma = \bar{\Gamma}_1 \cup \bar{\Gamma}_2)$, $M \in \beta_2'(\partial\zeta/\partial n)$. The same could also be said for $\bar{Q}$. Then, only minor modifications are required in the procedure that follows.

By means of the notations

$$R(h, k) = \int_\Omega C_{\alpha\beta\gamma\delta} h_{\alpha\beta} k_{\gamma\delta} \, d\Omega, \qquad \alpha, \beta, \gamma, \delta = 1, 2 \tag{7.1.18}$$

and

$$P(\zeta, z) = \{\zeta_{,\alpha} z_{,\beta}\}, \qquad P(\zeta, \zeta) = P(\zeta), \tag{7.1.19}$$

where $h = \{h_{\alpha\beta}\}$ and $k = \{k_{\gamma\delta}\}$, (7.1.5) and (7.1.9) can be written in the forms

$$\begin{aligned} &\alpha(\zeta, z - \zeta) + hR(\varepsilon(u) + \tfrac{1}{2}P(\zeta), P(\zeta, z - \zeta)) \\ &\qquad \int_\Gamma h\, \sigma_{\alpha\beta} \zeta_{,\alpha} n_\beta (z - \zeta) \, d\Gamma + \int_\Gamma \bar{Q}(\zeta)(z - \zeta) \, d\Gamma \\ &\qquad - \int_\Gamma M(\zeta) \frac{\partial(z - \zeta)}{\partial n} \, d\Gamma + \int_\Omega f_3(z - \zeta) \, d\Omega \end{aligned} \tag{7.1.20}$$

and

$$R(\varepsilon(u) + \tfrac{1}{2}P(\zeta), \varepsilon(v - u)) = \int_\Gamma \sigma_{\alpha\beta} n_\beta (v_\alpha - u_\alpha) \, d\Gamma, \qquad \alpha, \beta = 1, 2. \tag{7.1.21}$$

For the first B.V.P., we assume that

$$v \in [\mathring{H}^1(\Omega)]^2 \quad \text{and} \quad z \in H^2(\Omega) \cap \mathring{H}^1(\Omega) \tag{7.1.22}$$

and we recall that in virtue of (1.3.52) $z \in C^0(\bar{\Omega})$, and therefore $z = 0$ holds at all points of Γ. Then we introduce the convex, closed subset

$$K = \left\{ z \mid z \in H^2(\Omega) \cap \mathring{H}^1(\Omega), \frac{\partial z}{\partial n} \geq 0 \text{ on } \Gamma \right\} \tag{7.1.23}$$

of $H^2(\Omega)$ and we assume that $f_3 \in L^2(\Omega)$. The inequality in (7.1.23) makes sense a.e. on Γ since $\partial z/\partial n \in H^{1/2}(\Gamma)$. It is easy to verify that $R(h, k)$ is a

[3] For a counterexample, see [269].

continuous, symmetric, coercive, bilinear form on $[L^2(\Omega)]^4$. By Sobolev's imbedding theorem the imbedding $H^2(\Omega) \subset W^{1,4}(\Omega)$ is compact (cf., (1.3.51)) and thus we may easily prove, by applying Hölder's inequality, that the function $P: [H^2(\Omega)]^2 \to [L^2(\Omega)]^4$ is completely continuous (cf., Sec. 1.4.3). From (7.1.20, 21), by taking into account both the variational inequality

$$M(\zeta)\left(\frac{\partial z}{\partial n} - \frac{\partial \zeta}{\partial n}\right) \leq 0 \tag{7.1.24}$$

holding on Γ for sufficiently regular functions and the remaining boundary conditions, we are led to the following variational formulation of the problem.

Problem 1. Find $u \in [\mathring{H}^1(\Omega)]^2$ and $\zeta \in K$ such that

$$\alpha(\zeta, z - \zeta) + hR(\varepsilon(u) + \tfrac{1}{2}P(\zeta), P(\zeta, z - \zeta)) \geq (f_3, z - \zeta), \qquad \forall z \in K \tag{7.1.25}$$

and

$$R(\varepsilon(u) + \tfrac{1}{2}P(\zeta), \varepsilon(v - u)) = 0 \qquad \forall v \in [\mathring{H}^1(\Omega)]^2. \tag{7.1.26}$$

It is interesting to discuss in what sense the variational solution u, ζ of (7.1.25–26) satisfies (7.1.1–4) and (7.1.10–12). From (7.1.25), by setting $z = \zeta \pm \phi \in K$, $\phi \in D(\Omega)$, we get (7.1.1) as an equality in $\mathscr{D}'(\Omega)$; the same applies to (7.1.2). In order to proceed, we are obliged to assume that $(\sigma_{\alpha\beta}\zeta_{,\beta})_{,\underline{\alpha}} \in L^2(\Omega)$. Then $\Delta\Delta\zeta \in L^2(\Omega)$, and thus we conclude (cf. [181]) for M, $\bar{Q}$ that

$$M(\zeta) \in H^{-1/2}(\Gamma), \qquad \bar{Q}(\zeta) \in H^{-3/2}(\Gamma) \tag{7.1.27}$$

and that for $\zeta \in H^2(\Omega)$ the equality

$$\alpha(\zeta, z) = \mathsf{K} \int_\Omega (\Delta^2\zeta) z \, d\Omega + \langle \bar{Q}(\zeta), z \rangle_{3/2} - \left\langle M(\zeta), \frac{\partial z}{\partial n} \right\rangle_{1/2}, \qquad \forall z \in H^2(\Omega) \tag{7.1.28}$$

holds. This is the expression of the Green–Gauss theorem for the plate problem. (7.1.28), combined with (7.1.1), (7.1.25), implies the inequality

$$\left\langle M(\zeta), \frac{\partial z}{\partial n} - \frac{\partial \zeta}{\partial n} \right\rangle_{1/2} \leq 0, \qquad \forall z \in K, \tag{7.1.29}$$

which is a weak formulation of the boundary condition (7.1.12). It is easy to show that if $M(\zeta) \in L^2(\Gamma)$, then (7.1.29) yields (7.1.12) a.e. on Γ.

In order to give the variational formulation of the second B.V.P., we define the functional Φ on $H^2(\Omega)$ by

$$\Phi(z) = \begin{cases} \displaystyle\int_\Gamma \left[j_1\left(\frac{\partial z}{\partial n}\right) + j_2(z) \right] d\Gamma + \int_{\Omega'} j_3(z)\, d\Omega \\ \qquad \text{if } j_1\left(\frac{\partial z}{\partial n}\right),\ j_2(z) \in L^1(\Gamma),\ j_3(z) \in L^1(\Omega) \\ \infty \qquad \text{otherwise.} \end{cases} \tag{7.1.30}$$

It is convex, l.s.c. and proper (cf., Prop. 3.3.1). We also assume that $D(\Phi)$ is nonempty. As in Chapter 6 the use of the variational inequalities equivalent to (7.1.13, 14, 17) leads us to the following variational formulation of the B.V.P.

Problem 2. Find $u \in [H^1(\Omega)]^2$ and $\zeta \in H^2(\Omega)$ to satisfy the variational inequality

$$\alpha(\zeta, z - \zeta) + hR(\varepsilon(u) + \tfrac{1}{2}P(\zeta), P(\zeta, z - \zeta))$$
$$+ \Phi(z) - \Phi(\zeta) \geq (\bar{f}_3, z - \zeta) \qquad \forall z \in H^2(\Omega) \tag{7.1.31}$$

and the variational equality

$$R(\varepsilon(u) + \tfrac{1}{2}P(\zeta), \varepsilon(v - u)) = 0, \qquad \forall v \in [H^1(\Omega)]^2. \tag{7.1.32}$$

7.1.3 The Existence of the Solution

We shall now study the existence of a solution (u, ζ) of the variational system (7.1.31, 32). Problem 1 is obviously a special case of Problem 2.

Proposition 7.1.1. To every deflection $\zeta \in H^2(\Omega)$ there corresponds, by means of (7.1.32), a plane displacement $u(\zeta) \in [H^1(\Omega)]^2$. The strain tensor $\varepsilon(u(\zeta))$: $H^2(\Omega) \to [L^2(\Omega)]^4$ is uniquely determined and is a completely continuous, quadratic function.

Proof. From Korn's inequality (1.4.28), $R(\varepsilon(u), \varepsilon(v))$ is a bilinear coercive form of the quotient space $[H^1(\Omega)]^2/\bar{\mathscr{R}}$, where $\bar{\mathscr{R}}$ denotes the space of plane rigid-body displacements, i.e.,

$$\bar{\mathscr{R}} = \{\bar{r} \mid \bar{r} \in [H^1(\Omega)]^2, \varepsilon(\bar{r}) = 0\}$$
$$= \{\bar{r} \mid \bar{r} \in [H^1(\Omega)]^2, \bar{r}_1 = a_1 + bx_2, \bar{r}_2 = a_2 - bx_1, a_1, a_2, b \in \mathbb{R}\}. \tag{7.1.33}$$

Because of this property, (7.1.32) has for every ζ (cf., the Lax–Milgram theorem in Sec. 1.2.4) a solution $u(\zeta)$, since the application $v \to R(\frac{1}{2}P(\zeta), \varepsilon(v))$ is linear and continuous on $[H^1(\Omega)]^2$, as is easily shown. $u(\zeta)$ is determined up to an element of $\bar{\mathscr{R}}$, and therefore $\varepsilon(u(\zeta))$ is unique. From (7.1.32), it also results that $\varepsilon(u(\zeta))$ is a linear continuous function of $P(\zeta)$ and thus it is a completely continuous quadratic function of ζ, q.e.d.

Further, the function $G: H^2(\Omega) \to [L^2(\Omega)]^4$ defined by

$$G(\zeta) = \varepsilon(u(\zeta)) + \tfrac{1}{2}P(\zeta) \tag{7.1.34}$$

is introduced. Obviously, G is completely continuous and quadratic and fulfills the equation

$$R(G(\zeta), \varepsilon(u(\zeta))) = 0. \tag{7.1.35}$$

Thus, Problem 2 takes the following form.

Problem 2. Find $\zeta \in H^2(\Omega)$ such that

$$\alpha(\zeta, z - \zeta) + hR(G(\zeta), P(\zeta, z - \zeta)) + \Phi(z) - \Phi(\zeta) \geq (\bar{f}_3 z - \zeta), \qquad \forall z \in H^2(\Omega). \tag{7.1.36}$$

We next introduce the finite dimensional space

$$\mathcal{Q} = \{q \mid q \in H^2(\Omega), \alpha(q, q) = 0\}. \tag{7.1.37}$$

This is the space of all polynomials of degree ≤ 1. (1.3.30) implies that the norm $\|\zeta\|_2$ on $H^2(\Omega)$ is equivalent to $p(\zeta) + \zeta|_2$, where $p(\zeta)$ is a seminorm in $H^2(\Omega)$ given by

$$p(\zeta) = \left(\sum_{\alpha, \beta} \int_\Omega \left(\frac{\partial^2 \zeta}{\partial x_\alpha \partial x_\beta} \right)^2 d\Omega \right)^{1/2} \tag{7.1.38}$$

For $q \in \mathcal{Q}$, $p(q) = 0$ and conversely. Let $(H^2(\Omega))'$ be the dual of $H^2(\Omega)$ and $\langle \cdot, \cdot \rangle$ the duality pairing. If $\mathcal{Q}'$ is the dual of $\mathcal{Q}$, and if $f|_{\mathcal{Q}}$ denotes an element of $\mathcal{Q}'$, then

$$\langle f, q \rangle = \langle f|_{\mathcal{Q}}, q \rangle, \qquad \forall f \in (H^2(\Omega))' \text{ and } \forall q \in \mathcal{Q}. \tag{7.1.39}$$

The following proposition gives a necessary and sufficient condition for the existence of at least one solution of Problem 2.

Proposition 7.1.2

(a) In order for (7.1.36) to have a solution it is necessary that

$$\bar{f}_3|_{\mathcal{Q}} \in \overline{R(\partial \Phi_\zeta)}, \qquad \forall \zeta \in D(\Phi). \tag{7.1.40}$$

(b) If a $\zeta_0 \in D(\Phi)$ exists such that

$$\bar{f}_3|_{\mathcal{Q}} \in \operatorname{rel\,int} R(\partial \Phi_{\zeta_0}), \tag{7.1.41}$$

then (7.1.36) has a solution.

Proof. A direct application of Prop. 4.1.6. First, we shall verify that for every $\zeta \in H^2(\Omega)$

$$\inf_{q \in \mathcal{Q}} |\zeta - q|_2 \leq cp(\zeta), \qquad c \text{ onst} > 0 \tag{7.1.42}$$

holds, or equivalently,

$$|\zeta - \Pi\zeta|_2 \leq cp(\zeta), \qquad c \text{ const} > 0, \tag{7.1.43}$$

where Π is the projection operator of $H^2(\Omega)$ onto $\mathcal{Q}$ with respect to the L_2-norm. Replacing ζ by $\zeta|\zeta - \Pi\zeta|^{-\frac{1}{2}}$, we must prove that a constant $c > 0$ exists such that $p(\zeta) > c$ for $|\zeta - \Pi\zeta|_2 = 1$. Suppose that this is not true. Then there exists a sequence $\{\zeta_n\}$ in $H^2(\Omega)$ such that

$$|\zeta_n - \Pi\zeta_n|_2 = 1 \quad \text{and} \quad p(\zeta_n) \to 0. \tag{7.1.44}$$

Obviously, $\bar{\zeta}_n = \zeta_n - \Pi\zeta_n \in \mathcal{Q}^\perp$, where $\mathcal{Q}^\perp$ is the orthogonal complement of $\mathcal{Q}$ for the L^2-norm. Because of (7.1.44),

$$|\bar{\zeta}_n|_2 = 1 \quad \text{and} \quad p(\bar{\zeta}_n) \to 0. \tag{7.1.45}$$

It results that $\{\bar{\zeta}_n\}$ is bounded in $H^2(\Omega)$, and therefore there exists a subsequence, again denoted by $\{\bar{\zeta}_n\}$, which converges weakly to $\bar{\zeta}$ in $H^2(\Omega)$. The compact imbedding $H^2(\Omega) \subset L^2(\Omega)$ implies that

$$\bar{\zeta}_n \to \bar{\zeta} \quad \text{strongly in} \quad L^2(\Omega), \tag{7.1.46}$$

and thus $|\bar{\zeta}|_2 = 1$. For $\bar{\zeta}_n \to \bar{\zeta}$ weakly in $H^2(\Omega)$,

$$p(\bar{\zeta}) \leq \liminf p(\bar{\zeta}_n) = 0. \tag{7.1.47}$$

Thus $p(\bar{\zeta}) = 0$, i.e., $\bar{\zeta} \in \mathcal{Q}$. On the other hand, $\bar{\zeta} \in \mathcal{Q}^\perp$, and therefore $\bar{\zeta} = 0$, which contradicts the assumption. Further let us introduce a linear continuous monotone operator A and a completely continuous operator C such that (cf., Eq. (1.2.25))

$$\alpha(\zeta, z) = \langle A\zeta, z\rangle \tag{7.1.48}$$

and

$$hR(G(\zeta), P(\zeta, z)) = \langle C(\zeta), z\rangle. \tag{7.1.49}$$

The pseudomonotonicity (Sec. 1.4.3) of $T = A + C$ is easily verified and is left as an exercise for the reader.

We shall also show that for every $\zeta \in H^2(\Omega)$ and $q \in \mathcal{Q}$

$$\langle T(\zeta), q\rangle = 0. \tag{7.1.50}$$

It suffices to show the same for the operator C. We may easily verify that

$$P(\zeta + q) = P(\zeta) + P(q) + P(\zeta, q) + P(q, \zeta). \tag{7.1.51}$$

$P(\zeta)$, $P(q)$ and $P(\zeta + q)$ define $G(\zeta)$, $G(q)$ and $G(\zeta + q)$ respectively by means of the equations

$$R(\varepsilon(u(\zeta)) + \tfrac{1}{2}P(\zeta), \varepsilon(v)) = 0, \qquad \forall v \in [H^1(\Omega)]^2, \tag{7.1.52}$$

$$R(\varepsilon(u(q)) + \tfrac{1}{2}P(q), \varepsilon(v)) = 0, \qquad \forall v \in [H^1(\Omega)]^2 \tag{7.1.53}$$

and

$$R(\varepsilon(u(\zeta + q)) + \tfrac{1}{2}P(\zeta + q), \varepsilon(v)) = 0, \qquad \forall v \in [H^1(\Omega)]^2. \tag{7.1.54}$$

The last equation, combined with (7.1.51–53), implies that

$$R(\varepsilon(\bar{u}) + \tfrac{1}{2}(P(\zeta, q) + P(q, \zeta)), \varepsilon(v)) = 0, \qquad \forall v \in [H^1(\Omega)]^2, \tag{7.1.55}$$

where $\varepsilon(\bar{u}) = \varepsilon(u(q + \zeta) - u(\zeta) - u(q))$. Let

$$G(\zeta, q) = \varepsilon(\bar{u}) + \tfrac{1}{2}(P(\zeta, q) + P(q, \zeta)). \tag{7.1.56}$$

It results that

$$G(\zeta + q) = G(\zeta) + G(q) + G(\zeta, q). \tag{7.1.57}$$

As in Prop. 7.1.1 (7.1.55) has a unique solution in $\bar{\mathscr{R}}^\perp = [H^1(\Omega)]^2 \ominus \bar{\mathscr{R}}$. We see that for $q = p_1 x_1 + p_2 x_2 + p_3$ the vector

$$\bar{u} = (-p_1\zeta + kx_2 + k', -p_2\zeta - kx_1 + \bar{k}') \in \bar{\mathscr{R}}^\perp, \tag{7.1.58}$$

where p_1, p_2, p_3, k, k' and $\bar{k}'$ are arbitrary reals, satisfies the equation

$$G(\zeta, q) = 0 \tag{7.1.59}$$

and therefore also (7.1.55). Thus (7.1.59) holds for every $\zeta \in H^2(\Omega)$ and every $q \in \mathscr{Q}$. Likewise, for every $q \in \mathscr{Q}$

$$G(q) = 0. \tag{7.1.60}$$

As a result, for every $\zeta \in H^2(\Omega)$ and every $q \in \mathscr{Q}$

$$\begin{aligned} \tfrac{1}{h}\langle C(\zeta), q\rangle = R(G(\zeta), R(\zeta, q)) = R(G(\zeta), \\ \varepsilon(v) + \tfrac{1}{2}(P(\zeta, q) + P(q, \zeta))) = R(G(\zeta), G(\zeta, q)) = 0. \end{aligned} \tag{7.1.61}$$

It remains to be shown that for ζ such that $p(\zeta)$ is sufficiently large

$$\langle T(\zeta), \zeta\rangle \geq c(p(\zeta))^a, \qquad c \text{ const} > 0, \qquad a \text{ const} > 1. \tag{7.1.62}$$

Indeed,

$$\langle T(\zeta), \zeta\rangle = \alpha(\zeta, \zeta) + hR(G(\zeta), 2G(\zeta)) \geq c(p(\zeta))^2. \tag{7.1.63}$$

All the assumptions of Prop. 4.1.6 are satisfied, and thus the proof of (7.1.40, 41) is completed, q.e.d.

Further, we shall give two applications of Prop. 7.1.2.

(i) First, assume that $\bar{Q} = 0$ and $\sigma_{\alpha\beta} n_\beta = 0$ $\alpha, \beta = 1, 2$ on Γ, and that (7.1.13) holds with β_1 given by (3.3.45). Moreover, let $f_3 = \bar{f}_3 \in L^2(\Omega)$. Then

$$\Phi(\zeta) = \int_\Gamma M_0 \left|\frac{\partial\zeta}{\partial n}\right| d\Gamma. \tag{7.1.64}$$

In order to obtain the necessary and sufficient conditions, we shall make use of Prop. 2.4.6 and give a characterization of relint $R(\partial\Phi(\zeta_0 + \cdot))$ (see also [296]). For $q = q_0 + q_1 x_1 + q_2 x_2 \in \mathscr{Q}$,

$$\frac{\partial q}{\partial n} = q_1 n_1 + q_2 n_2. \tag{7.1.65}$$

Then the subspaces $\mathscr{Q}_1 = \{q | q = q_0 \in \mathbb{R}\}$ and $\mathscr{Q}_2 = \{q | q = q_1 x_1 + q_2 x_2, q_1, q_2 \in \mathbb{R}\}$ are defined. Then $\bar{\bar{f}}_3$ is such that

$$(\bar{\bar{f}}_3, q) = 0, \qquad \forall q \in \mathscr{Q}_1, \tag{7.1.66}$$

because $\Phi(q) = 0$, $\forall q \in \mathscr{Q}_1$ and (cf., (2.4.27))

$$\Phi(q) - (\bar{\bar{f}}_3, q) \geq c \| q \|_2, \qquad \forall q \in \mathscr{Q}_2. \tag{7.1.67}$$

Accordingly, the sufficient condition (7.1.41) acquires the form

$$(\bar{\bar{f}}_3, q) < \int_\Gamma M_0 \left| \frac{\partial q}{\partial n} \right| d\Gamma \quad \forall q \in \mathscr{Q}_2, \quad q \neq 0, \text{ and } \quad (\bar{\bar{f}}_3, 1) = 0, \tag{7.1.68}$$

and the necessary condition (7.1.40) the form

$$(\bar{\bar{f}}_3, q) \leq \int_\Gamma M_0 \left| \frac{\partial q}{\partial n} \right| d\Gamma \quad \forall q \in \mathscr{Q}_2, \text{ and } \quad (\bar{\bar{f}}_3, 1) = 0, \tag{7.1.69}$$

(ii) Assume that on $\Gamma_1 \subset \Gamma$ $M = M_0 \in L^2(\Gamma_1)$, on $\Gamma - \Gamma_1$ $M = 0$, and on the whole of Γ $\sigma_{\alpha\beta} n_\alpha = 0$, $\alpha, \beta = 1, 2$ and $\bar{Q}(\zeta) \geq 0$, $\zeta \geq 0$, $\zeta\bar{Q}(\zeta) = 0$. Then the variational inequality (7.1.36) holds with $\Phi(\zeta) = \int_{\Gamma_1} M_0 (\partial\zeta/\partial n)\, d\Gamma + I_K(\zeta)$, where I_K is the indicator of the set $K = \{\zeta | \zeta \in H^2(\Omega), \zeta \geq 0 \text{ on } \Gamma\}$. $\Phi(\zeta)$ is invariant on $\mathscr{Q}_1 = \{0\}$. It results, then, from (2.4.27) that

$$(\bar{\bar{f}}_3, q) - \int_{\Gamma_1} M_0 \frac{\partial q}{\partial n} d\Gamma \leq c_1 - c_3 \| q \|_2, \qquad \forall q \in K \cap \mathscr{Q}, \tag{7.1.70}$$

where c_1 and c_3 are constants and $c_3 > 0$. Thus we obtain the sufficient condition (see also [154])

$$(\bar{\bar{f}}_3, q) - \int_{\Gamma_1} M_0 \frac{\partial q}{\partial n} d\Gamma < 0, \qquad \forall q \in K \cap \mathscr{Q} \qquad q \neq 0 \tag{7.1.71}$$

and the necessary condition

$$(\bar{\bar{f}}_3, q) - \int_{\Gamma_1} M_0 \frac{\partial q}{\partial n} d\Gamma \leq 0, \qquad \forall q \in K \cap Q. \tag{7.1.72}$$

For the *unilateral* problems of von Kármán plates, the variational inequality formulation is not equivalent to a minimum problem because of the geometric nonlinearity of the physical model. However, the following holds.

Proposition 7.1.3. The solution $\zeta \in H^2(\Omega)$ of the minimization problem

$$\Pi(\zeta) = \inf\{\Pi(z) \mid z \in H^2(\Omega)\}, \qquad (7.1.73)$$

where

$$\Pi(z) = \tfrac{1}{2}\alpha(z, z) + \frac{h}{2} R(G(\zeta), G(\zeta)) + \Phi(\zeta) - (\bar{f}_3, z), \qquad (7.1.74)$$

satisfies the variational inequality (7.1.36).

Proof. Applying (1.4.18) and (7.1.32), we see that $R(G(\zeta), P(\zeta, z))$ is the Gâteaux-differential of $\frac{1}{2}R(G(\zeta), G(\zeta))$ at ζ. We then proceed as in Prop. 2.2.2, q.e.d.

Generally speaking, the solution of the problems studied is not uniquely determined. This is also the case if classical (bilateral) boundary conditions hold [22]. For the question of the uniqueness, we refer the reader to Sec. 7.2. The variational inequality (7.1.36) is general and its study also includes the study of the coercive cases. Note that if, for instance, on $\Gamma_1 \subset \Gamma$ the plate is clamped, i.e., $\zeta = \partial\zeta/\partial n = 0$, then either this condition can be incorporated in Φ, or ζ and z must be regarded as elements of $Z = \{\zeta \mid \zeta \in H^2(\Omega),\ \zeta = 0 \text{ and } \partial\zeta/\partial n = 0 \text{ on } \Gamma_1\}$. Up to now, we have assumed that $\bar{f}_3 \in L^2(\Omega)$. With minor changes we may also take $\bar{\bar{f}} \in [C(\bar{\Omega})]'$, and thus the case of concentrated loads on a curve or at a point of Ω may be considered (see [227]).

Another useful comment concerns the case in which the plate boundary has corners. Then the plate exhibits an additional reaction at every corner [116] [236]; in order to avoid the corresponding additional deflection of the plate, we assume that this reaction is zero. If the plate lies on a rigid plane, the Signorini-Fichera problem can be formulated with respect to the aforementioned reactions [154]. Recall that this reaction results from the fact that the twisting moments along Γ are incorporated into the shearing forces and as a result give the total shearing force $\bar{Q}$ [116].

7.1.4 In-Plane Unilateral Boundary Conditions

In this section we give the variational formulations arising if (7.1.15) is replaced by a pointwise relation of the form

$$-S \in \partial j_4(u) \quad \text{on} \quad \Gamma, \qquad (7.1.75)$$

where $S = \{\sigma_{\alpha\beta} n_\beta\}$, $\alpha, \beta = 1, 2$, and j_4 is a convex, l.s.c. and proper functional on $\mathbb{R}^2$. (7.1.75) describes unilateral contact with respect to u, or a crack at the boundary, or in-plane friction conditions, etc. Moreover, assume for the sake of simplicity that $\zeta = 0$ on Γ. From (7.1.20, 21) we are led to the following variational formulation.

Problem 3. Find $u \in [H^1(\Omega)]^2$ and $\zeta \in H^2(\Omega) \cap \mathring{H}^1(\Omega)$, which satisfy the variational inequality (7.1.31) (but with $j_2 = 0$ in (7.1.30)) for every $z \in H^2(\Omega) \cap \mathring{H}^1(\Omega)$ and the variational inequality

$$R(\varepsilon(u) + \tfrac{1}{2}P(\zeta), \varepsilon(v - u)) + \bar{\Phi}(v) - \bar{\Phi}(u) \geq 0, \qquad \forall v \in [H^1(\Omega)]^2, \quad (7.1.76)$$

where

$$\bar{\Phi}(v) = \begin{cases} \displaystyle\int_\Gamma j_4(v)\, d\Gamma & \text{if } \; j_4(v) \in L^1(\Gamma) \\ \infty & \text{otherwise} \end{cases}. \qquad (7.1.77)$$

The material of this section can also be applied to the study of inequality problems arising in the Kirchhoff plate theory, i.e., the small-displacement theory. In this case, the bending is independent of the stretching of the plate, and therefore $R(\varepsilon(u)+\tfrac{1}{2}P(\zeta), P(\zeta, z)) = 0$. For the coercive problems the solution is uniquely determined, whereas for the semicoercive ones the only general result is that if ζ_1 and ζ_2 are two solutions then $\zeta_1 - \zeta_2 \in \mathcal{Q}$. In this context, see also [84]. The next section treats the problem of the buckling of a plate subjected to unilateral conditions.

7.2 The Unilateral Buckling Problem. Eigenvalue Problems For Variational Inequalities

7.2.1 Formulation of the Problem

There are many practical situations in which the study of the buckling of plates subjected to unilateral constraints is necessary, as for instance in the contact problems of sheet piles, tunnels and certain foundation structures. The small displacement theory is not adequate for precise description of buckling phenomena, and thus the von-Kármán theory must be applied. The buckling is caused by boundary loading in the plane of the plate, of the form

$$\sigma_{\alpha\beta} n_\beta = \lambda g_\alpha, \qquad \alpha = 1, 2 \quad \text{on} \quad \Gamma, \qquad (7.2.1)$$

where $g = g(x)$, $x \in \Gamma$, is a self-equilibrating compressive load distribution on Γ, i.e.,

$$\int_\Gamma g_\alpha \, d\Gamma = 0 \quad \text{and} \quad \int_\Gamma (g_1 x_2 - g_2 x_1)\, d\Gamma = 0, \qquad (7.2.2)$$

and λ is a real number measuring the magnitude of the load distribution. Let us consider, for instance, a plate with built-in edges resting without friction on a rigid support $\Omega' \subset \Omega$. When λ exceeds a certain value, called the critical value, the plate leaves its position of equilibrium and buckles.

Along the lines of Sec. 7.1 and for $g \in L^2(\Gamma)$, the following variational formulation is obtained for the problem defined by (7.1.1–4), (7.1.13, 14), (7.1.16, 17) with $\bar{\bar{f}}_3 = 0$, and (7.2.1, 2).

Problem 1. Find $\zeta \in H^2(\Omega)$, $u \in [H^1(\Omega)]^2$ and $\lambda \in \mathbb{R}$ which satisfy

$$\alpha(\zeta, z-\zeta) + \Phi(z) - \Phi(\zeta) + \int_\Omega h\sigma_{\alpha\beta}\zeta_{,\alpha}(z-\zeta)_{,\beta}\, d\Omega$$

$$\geq \lambda \int_\Gamma hg_\alpha \zeta_{,\alpha}(z-\zeta)\, d\Gamma, \qquad \forall z \in H^2(\Omega) \quad (7.2.3)$$

and

$$\int_\Omega \sigma_{\alpha\beta}\varepsilon_{\alpha\beta}(v)\, d\Omega = \lambda \int_\Gamma g_\alpha v_\alpha\, d\Gamma, \qquad \forall v \in [H^1(\Omega)]^2, \qquad (7.2.4)$$

where $\sigma = \{\sigma_{\alpha\beta}\}$ and $\varepsilon = \{\varepsilon_{\alpha\beta}\}$ are given by (7.1.3) and (7.1.4).

The triplet (ζ, u_1, u_2) is called the variational solution corresponding to λ. Since $\zeta = 0$ must be a solution for every λ, as is obvious from the nature of the mechanical problem, we assume that $0 \in \partial j_i(0)$, $i = 1, 2, 3$, in (7.1.13, 14, 17). In order to obtain a simpler form of the problem we introduce the auxiliary quantities $\sigma^0 = \{\sigma^0_{\alpha\beta}\}$, $\varepsilon^0 = \{\varepsilon^0_{\alpha\beta}\}$ and $u^0 = \{u^0_\alpha\}$, satisfying (7.1.2), (7.1.4), the stress-strain law

$$\sigma^0_{\alpha\beta} = C_{\alpha\beta\gamma\delta}\varepsilon^0_{\gamma\delta}(u^0) \qquad (7.2.5)$$

and the boundary conditions (7.2.1) for $\lambda = 1$. For the resulting plane elasticity problem there exists, due to (7.2.2) (cf., also [98] and Sec. 4.1.1), a unique solution $u^0 \in [H^1(\Omega)]^2 \ominus \bar{\mathcal{R}}$ such that

$$\int_\Omega \sigma^0_{\alpha\beta}\varepsilon_{\alpha\beta}(v)\, d\Omega = \int_\Gamma g_\alpha v_\alpha\, d\Gamma, \qquad \forall v \in [H^1(\Omega)]^2. \qquad (7.2.6)$$

Every other solution has the form $u^0 + \bar{r}$, $\bar{r} \in \bar{\mathcal{R}}$. Moreover, $\sigma^0_{\alpha\beta}(u^0) \in L^2(\Omega)$ and is uniquely determined.

Further, we consider the stress and strain tensors $\bar{\sigma} = \{\bar{\sigma}_{\alpha\beta}\}$ and $\bar{\varepsilon} = \{\bar{\varepsilon}_{\alpha\beta}\}$ and the displacement vector $\bar{u} = \{\bar{u}_\alpha\}$ satisfying (7.1.2–4) and the boundary conditions (7.1.15). Prop. 7.1.1 implies that $\bar{\varepsilon} = \bar{\varepsilon}(\zeta)$ is unique and is a completely continuous function from $H^2(\Omega)$ into $[L^2(\Omega)]^4$, and similarly for $\bar{\sigma} = \bar{\sigma}(\zeta)$. Then the following variational problem can be formulated.

Problem 2. Find $\zeta \in H^2(\Omega)$ and $\bar{u} \in [H^1(\Omega)]^2 \ominus \bar{\mathcal{R}}$ such that

$$\alpha(\zeta, z-\zeta) + \Phi(z) - \Phi(\zeta) + \int_\Omega h\bar{\sigma}_{\alpha\beta}\zeta_{,\alpha}(z-\zeta)_{,\beta}\, d\Omega$$

$$\geq -\lambda \int_\Omega h\sigma^0_{\alpha\beta}\zeta_{,\alpha}(z-\zeta)_{,\beta}\, d\Omega \qquad (7.2.7)$$

$$+ \lambda \int_\Gamma hg_\alpha\zeta_{,\alpha}(z-\zeta)\, d\Gamma, \qquad \forall z \in H^2(\Omega)$$

$$\int_\Omega \bar{\sigma}_{\alpha\beta}\varepsilon_{\alpha\beta}(v)\, d\Omega = 0, \qquad \forall v \in [H^1(\Omega)]^2 \ominus \bar{\mathcal{R}}, \qquad (7.2.8)$$

where $\bar{\sigma}_{\alpha\beta}$ and $\bar{\varepsilon}_{\alpha\beta}(\bar{u})$ satisfy (7.1.3) and (7.1.4).

Following [226], we call a solution $(\zeta, \bar{u}_1, \bar{u}_2)$ of Problem 2 a reduced variational solution. It is easily verified that for a given λ the triplet (ζ, u_1, u_2) is a solution to Problem 1 if and only if $(\zeta, \bar{u}_1, \bar{u}_2)$ is a reduced variational solution, where u and $\bar{u}$ are related by

$$u_\alpha = \bar{u}_\alpha + \lambda u_\alpha^0, \qquad \alpha = 1, 2 \tag{7.2.9}$$

and

$$\sigma_{\alpha\beta} = \bar{\sigma}_{\alpha\beta} + \lambda\sigma_{\alpha\beta}^0, \quad \alpha, \beta = 1, 2. \tag{7.2.10}$$

Note that for every λ, (0, 0, 0) is a solution of Problem 2. A nontrivial reduced variational solution of the form $(0, \bar{u}_1, \bar{u}_2)$ is not possible for every λ, since then (7.2.8) would imply that $\bar{u} = 0$ as well.

It results that any solution of the buckling problem which is not trivial, i.e., $\zeta \neq 0$, may be written as the sum of a nontrivial reduced variational solution $(\zeta, \bar{u}_1, \bar{u}_2)$ and a plane displacement $\lambda(0, u_1^0 + \bar{r}_1, u_2^0 + \bar{r}_2)$ (cf., (7.1.33)).

The relations (7.2.3, 4) or (7.2.7, 8) constitute an eigenvalue problem for variational inequalities. Here we will study only the coercive eigenvalue problem. We assume that the boundary condition (7.1.10) holds, and we introduce the space

$$Z = \{z \mid z \in H^2(\Omega),\ z = 0 \text{ on } \Gamma\}. \tag{7.2.11}$$

On Z, $\alpha(\zeta, \zeta)$ is coercive (cf., for instance [225]), and from the inequality $c_1 \|\zeta\|_2^2 \leq \alpha(\zeta, \zeta) \leq c_2 \|\zeta\|_2^2$ holding for every $\zeta \in Z$ (c_1, c_2 const > 0), we conclude that $(\alpha(\zeta, \zeta))^{1/2}$ is a norm on Z equivalent to $\|\zeta\|_2$. We denote by $(z, \zeta)_Z$ the scalar product $\alpha(z, \zeta)$ with which Z is henceforth considered to be equipped. The respective norm is denoted by $\|\cdot\|_Z$.

Let us define (cf., Sec. 1.2.4) a nonlinear, completely continuous mapping $\tilde{C}$: $Z \to Z$ by the relation

$$(\tilde{C}(\zeta), z)_Z = R(\varepsilon(u(\zeta)) + \tfrac{1}{2}P(\zeta), P(\zeta, z)), \qquad \forall \zeta, z \in Z, \tag{7.2.12}$$

and a linear mapping B: $Z \to Z$ such that

$$(B\zeta, z)_Z = -\int_\Omega h\sigma_{\alpha\beta}^0 \zeta_{,\alpha} z_{,\beta}\, d\Omega, \qquad \forall \zeta, z \in Z. \tag{7.2.13}$$

The triplet $(\zeta, \bar{u}_1, \bar{u}_2) \in Z \times ([H^1(\Omega)]^2 \ominus \mathscr{R})$ is a reduced variational solution if $\bar{u} = u(\zeta)$ (cf., Prop. 7.1.1) and

$$((I + \tilde{C})(\zeta), z - \zeta)_Z + \Phi(z) - \Phi(\zeta) \geq \lambda(B\zeta, z - \zeta)_Z, \qquad \forall z \in Z, \tag{7.2.14}$$

where I denotes the identity operator in Z. In (7.2.14) Φ is given by (7.1.30), but with $j_2 = 0$.

7.2.2 A General Proposition on the Existence of the Solution

From the previous section we arrive at the study of the following general eigenvalue problem.

Problem 1.′ Find $\lambda \in \mathbb{R}$ and $\zeta \in Z$, where Z is a Hilbert space so as to satisfy (7.2.14).

We make the following assumptions:

(i) the operator $\tilde{T} = I + \tilde{C}$ from Z into Z is pseudomonotone;
(ii) for every $z \in Z$, $z \neq 0$,

$$(\tilde{T}\zeta, \zeta)_Z > 0; \tag{7.2.15}$$

(iii) there exist two functionals $F: Z \to \mathbb{R}$ and $E: Z \to \mathbb{R}$ with $F(0) = E(0) = 0$ such that for all $z \in Z$ and for any $z_0 \in Z$ the graphs of $F(z)$ and $G(z)$ are tangential at z_0 (cf., Sec. 1.4.2) to the graphs of

$$\bar{F}(z) = F(z_0) + (\tilde{T}(z_0), z - z_0)_Z \tag{7.2.16}$$

and

$$\bar{E}(z) = E(z_0) + (Bz_0, z - z_0)_Z \tag{7.2.17}$$

respectively.
(iv) F is weakly l.s.c.;
(v) $F(z) \to \infty$ as $\|z\|_Z \to \infty$; (7.2.18)
(vi) B is a linear compact operator from Z into Z; and
(vii) Φ is a convex, l.s.c., proper functional on Z with the property that for each $\mu > 1$ a real number $\alpha(\mu) > 0$ can be determined such that

$$\Phi(\mu z) \leq \alpha(\mu)\Phi(z), \qquad \forall z \in Z. \tag{7.2.19}$$

Moreover, let

$$0 \in \partial\Phi(0) \tag{7.2.20}$$

and

$$\Phi(0) = 0. \tag{7.2.21}$$

Further, we denote by A_1 and A_2 the sets

$$A_1 = \{\zeta \mid \zeta \in Z, F(\zeta) + \Phi(\zeta) = \beta\} \tag{7.2.22}$$

$$A_2 = \{\zeta \mid \zeta \in Z, F(\zeta) + \Phi(\zeta) \leq \beta\}, \tag{7.2.23}$$

where $\beta \in \mathbb{R}$ and $\beta > 0$.

The following proposition holds [226].

Proposition 7.2.1. Assume that conditions (i)–(vii) above are satisfied and that a

$$\zeta_0 \in A_2 \quad \text{exists such that } E(\zeta_0) > 0. \tag{7.2.24}$$

Then there exists at least one pair $\{\zeta, \lambda\}$ with $\lambda \geq 0$ which is a solution of (7.2.14). Moreover, for $\zeta \in A_2$

$$E(\zeta) = \max \{E(z) \mid z \in A_2\}. \tag{7.2.25}$$

Proof. We apply here the well-known regularization procedure. Let us introduce the regularized functional Φ_ε on Z defined for $\varepsilon > 0$ by the relation

$$\Phi_\varepsilon(\zeta) = \min \left\{ \Phi(z) + \frac{1}{2\varepsilon} \| z - \zeta \|_Z^2 \mid z \in Z \right\}, \qquad \varepsilon > 0. \tag{7.2.26}$$

Φ_ε is convex and Fréchet-differentiable on Z and satisfies the relation $\Phi_\varepsilon(\zeta) \leq \Phi(\zeta)$, $\forall \zeta \in Z$. The Fréchet-derivative $d\Phi_\varepsilon/d\zeta$, which is monotone Lipschitzian, is equal to the Yosida-approximation $(\partial\Phi)_\varepsilon$ of $\partial\Phi$, i.e.,

$$\frac{d\Phi_\varepsilon}{d\zeta} = (\partial\Phi)_\varepsilon = \frac{1}{\varepsilon}(I - \mathfrak{J}_\varepsilon), \tag{7.2.27}$$

where $\mathfrak{J}_\varepsilon$ is the mapping $(I + \varepsilon\partial\Phi)^{-1}$. It also holds that

$$\Phi_\varepsilon(\zeta) = \Phi(\mathfrak{J}_\varepsilon(\zeta)) + \frac{1}{2\varepsilon} \| \zeta - \mathfrak{J}_\varepsilon(\zeta) \|_Z^2, \qquad \varepsilon > 0. \tag{7.2.28}$$

We denote by $A_{1\varepsilon}$ and $A_{2\varepsilon}$ the sets which are again defined by (7.2.22, 23), but now with Φ_ε replacing Φ. Clearly, $A_{2\varepsilon}$ is nonempty for every $\varepsilon > 0$. The weak lower semicontinuity of F implies that $A_{2\varepsilon}$ is weakly closed, and because of (7.2.18)

$$\| z \|_Z \leq c, \qquad \forall z \in A_{2\varepsilon} \qquad c \text{ const} > 0, \qquad \forall \varepsilon > 0. \tag{7.2.29}$$

Accordingly, a $\zeta_\varepsilon \in A_{2\varepsilon}$ exists such that

$$E(\zeta_\varepsilon) = \max \{E(z) \mid z \in A_{2\varepsilon}\}, \qquad \forall \varepsilon > 0. \tag{7.2.30}$$

The relation

$$E(\zeta_\varepsilon) \geq \max \{E(z) \mid z \in A_2\} = c > 0, \qquad \forall \varepsilon > 0 \tag{7.2.31}$$

results from (7.2.24) and from the inclusion $A_2 \subset A_{2\varepsilon}$. It is easily verified that for any subset $Z_0 \subset Z$, on which $E(z)$ is bounded from below by a positive constant, there exists another positive constant c such that $(Bz, z)_Z \geq c$ for every $z \in Z_0$. For $Z_0 = \{\zeta_\varepsilon \mid \varepsilon > 0\}$ it results that

$$(B\zeta_\varepsilon, \zeta_\varepsilon) \geq c > 0, \qquad \forall \varepsilon > 0. \tag{7.2.32}$$

All the constants used thus far are independent of ε. (7.2.32) and the continuity of $F + \Phi_\varepsilon$ imply that $\zeta_\varepsilon \in A_{1\varepsilon}$ and $\zeta_\varepsilon \neq 0$ for every $\varepsilon > 0$. Thus

$$E(\zeta_\varepsilon) = \max\{E(z) \mid z \in A_{2\varepsilon}\} = \max\{E(z) \mid z \in A_{1\varepsilon}\}. \qquad (7.2.33)$$

In addition, $(d\Phi_\varepsilon/d\zeta)(0) = 0$ for every $\varepsilon > 0$, and thus (7.2.15) implies that

$$\left(\tilde{T} + \frac{d\Phi_\varepsilon}{d\zeta}\right)(\zeta_\varepsilon) \neq 0, \qquad \forall \varepsilon > 0. \qquad (7.2.34)$$

Applying a theorem of Browder,[4] we conclude that there exists $\lambda_\varepsilon \in \mathbb{R}$ such that

$$\tilde{T}(\zeta_\varepsilon) + \frac{d\Phi_\varepsilon}{d\zeta}(\zeta_\varepsilon) = \lambda_\varepsilon B\zeta_\varepsilon, \qquad \forall \varepsilon > 0, \qquad (7.2.35)$$

or equivalently

$$(\tilde{T}(\zeta_\varepsilon), z - \zeta_\varepsilon)_Z + \Phi_\varepsilon(z) - \Phi_\varepsilon(\zeta_\varepsilon) \geq \lambda_\varepsilon (B\zeta_\varepsilon, z - \zeta_\varepsilon)_Z, \qquad \forall z \in Z. \quad (7.2.36)$$

Multiplying (7.2.35) by ζ_ε yields

$$\lambda_\varepsilon > 0, \qquad \forall \varepsilon > 0. \qquad (7.2.37)$$

For every $w \in Z$ and $\mu > 1$ we have

$$\Phi_\varepsilon(\mu z) \leq \frac{\mu^2}{2\varepsilon} \| z - w \|_Z^2 + \alpha(\mu)\Phi(w)$$

$$\leq \sup(\mu^2, \alpha(\mu)) \left\{ \frac{1}{2\varepsilon} \| z - w \|_Z^2 + \Phi(w) \right\}$$

Thus

$$\Phi_\varepsilon(\mu z) \leq \sup(\mu^2, \alpha(\mu))\, \Phi_\varepsilon(z). \qquad (7.2.38)$$

By setting $z = 2\zeta_\varepsilon$ in (7.2.36) and taking into account (7.2.29, 32) and the fact that $\tilde{T}$ is bounded, we find that

$$c\lambda_\varepsilon \leq (\tilde{T}(\zeta_\varepsilon), \zeta_\varepsilon)_Z + \sup\{4, \alpha(2)\}\Phi_\varepsilon(\zeta_\varepsilon) \leq c', \qquad \forall \varepsilon > 0, \qquad (7.2.39)$$

where c and c' are independent of ε. From the estimate

$$c > \Phi_\varepsilon(\zeta_\varepsilon) = \frac{1}{2\varepsilon} \| \zeta_\varepsilon - \mathfrak{J}_\varepsilon(\zeta_\varepsilon) \|_Z^2 + \Phi(\mathfrak{J}_\varepsilon(\zeta_\varepsilon)), \qquad c \text{ const} > 0 \quad (7.2.40)$$

we obtain as $\varepsilon \to 0$

$$\zeta_\varepsilon - \mathfrak{J}_\varepsilon(\zeta_\varepsilon) \to 0 \quad \text{strongly in} \quad Z. \qquad (7.2.41)$$

From (7.2.29, 37) and (7.2.39) it can be concluded that there exist subsequences, again denoted by $\{\zeta_\varepsilon\}$ and $\{\lambda_\varepsilon\}$, such that for $\varepsilon \to 0$

$$\zeta_\varepsilon \to \zeta \quad \text{weakly in} \quad Z \qquad (7.2.42)$$

[4] See [30] Theorem 4.

and

$$\lambda_\varepsilon \to \lambda. \tag{7.2.43}$$

It results that

$$\mathcal{J}_\varepsilon(\zeta_\varepsilon) \to \zeta \quad \text{weakly in} \quad Z. \tag{7.2.44}$$

Thus

$$F(\zeta) + \Phi(\zeta) \leq \liminf\,[F(\zeta_\varepsilon) + \Phi(\mathcal{J}_\varepsilon(\zeta_\varepsilon))] \leq \beta \tag{7.2.45}$$

and therefore $\zeta \in A_2$. Furthermore, it is found from (7.2.31) that

$$E(\zeta) = \max\,\{E(z) \,|\, z \in A_2\}. \tag{7.2.46}$$

It remains to be proved that the variational inequality (7.2.14) is satisfied by ζ and λ. Here the pseudomonotonicity of $\tilde{T}$ is used. First, we obtain from (7.2.36), by setting $z = \zeta$, the relation

$$(\tilde{T}(\zeta_\varepsilon), \zeta_\varepsilon - \zeta)_Z \leq \Phi(\zeta) - \Phi(\mathcal{J}_\varepsilon(\zeta_\varepsilon)) + \lambda_\varepsilon(B\zeta_\varepsilon, \zeta_\varepsilon - \zeta)_Z. \tag{7.2.47}$$

Finally, from (7.2.42-44) and because B is compact, we get

$$\limsup\,(\tilde{T}(\zeta_\varepsilon), \zeta_\varepsilon - \zeta)_Z \leq 0. \tag{7.2.48}$$

Since $\tilde{T}$ is pseudomonotone

$$\begin{aligned}(\tilde{T}(\zeta), \zeta - z)_Z &\leq \liminf\,(\tilde{T}(\zeta_\varepsilon), \zeta_\varepsilon - z)_Z \\ &\leq \Phi(z) - \Phi(\zeta) + \lambda(B(\zeta), z - \zeta)_Z, \qquad \forall z \in Z,\end{aligned} \tag{7.2.49}$$

q.e.d.

This proposition guarantees the existence of at least one "buckled" solution $\{\zeta, \lambda\}$. If, in addition, $\tilde{T}$ is continuous on Z and the application $\mu \to \Phi(\mu\zeta)$ is for $\mu \geq 0$ and for every $\zeta \in Z$ continuous and nondecreasing, then the solution ζ of (7.2.25) belongs to A_1.

The next proposition, stated here without proof, concerns the existence of higher eigenvalues. We make the following additional assumptions on $\tilde{T}$, B and Φ:

(i′) there exists a real-valued function $b = b(\gamma, \delta)$ on $[0, +\infty) \times [0, +\infty)$ which is bounded on bounded sets and is such that

$$\|\tilde{T}(\zeta) - \tilde{T}(z)\|_Z \leq b(\|z\|_Z, \|\zeta\|_Z)\,\|z - \zeta\|_Z, \forall z, \zeta \in Z; \tag{7.2.50}$$

(ii′)
$$(\tilde{T}(\zeta), \zeta)_Z \geq c\,\|\zeta\|_Z^2, \qquad \forall \zeta \in Z, \quad c \text{ const} > 0, \tag{7.2.51}$$

this assumption replacing (7.2.15);

(iii′)
$$\tilde{T}(-\zeta) = -\tilde{T}(\zeta), \qquad \forall \zeta \in Z; \tag{7.2.52}$$

(iv′)
$$(B\zeta, \zeta)_Z > 0, \qquad \forall \zeta \in Z, \qquad \zeta \neq 0; \tag{7.2.53}$$

(v′) $$0 \in \operatorname{int} D(\Phi),\ \Phi \text{ is continuous at } \zeta = 0, \tag{7.2.54}$$

$$\Phi(\mu\zeta) \geq \Phi(\zeta),\ \forall \mu > 1, \qquad \forall \zeta \in Z \tag{7.2.55}$$

and

$$\Phi(-\zeta) = \Phi(\zeta), \qquad \forall \zeta \in Z. \tag{7.2.56}$$

Proposition 7.2.2. Assume that $\tilde{T}$, B and Φ satisfy assumptions (i), (iii), (vi), (vii) and (i′)–(v′). Then there exists an infinite sequence $\{\zeta_n, \lambda_n\}$, $n = 1, 2, \ldots$, of solutions of (7.2.14) with the property that

$$\zeta_n \in A_2, \qquad \zeta_n \neq 0, \qquad \lambda_n \geq 0. \tag{7.2.57}$$

If $\{\zeta_n, \lambda_n\}$ is a solution of the eigenvalue problem, then $\{-\zeta_n, \lambda_n\}$ is also.

Possible applications of this proposition in mechanics are limited, because of the additional assumptions in comparison to Prop. 7.2.1. For the proof, which uses the results of the Lusternik-Schnirelmann theory, the reader is referred to [226]. It is worth noting here that Prop. 7.2.2 does not permit the drawing of conclusions on the existence of an infinite sequence of distinct solutions, i.e., that $u_{n'} \neq u_n$ for $n' \neq n$.

7.2.3 Application to the Buckling Problem

In order to give an application of the preceding propositions, let us consider a plate subjected to the boundary conditions

$$\zeta = 0 \quad \text{and} \quad M(\zeta) \in \beta\left(\frac{\partial \zeta}{\partial n}\right) \quad \text{on} \quad \Gamma, \tag{7.2.58}$$

where β is given by (3.3.45). On $\Omega' \subset \Omega(\bar{\Omega}' \cap \Gamma = \varnothing)$ the reaction $\bar{f}_3$ obeys a law of the form (7.1.17): we assume that $\bar{f}_3 = 0$ on $\Omega - \Omega'$ and

$$-\bar{f}_3 \in \beta(\zeta) = \begin{cases} [-f_0, +f_0] & \text{for } \zeta = 0 \\ f_0 & \text{for } \zeta > 0 \\ -f_0 & \text{for } \zeta < 0 \end{cases} \quad \text{on} \quad \Omega'. \tag{7.2.59}$$

Frictional effects are described by the second condition in (7.2.58) and by (7.2.59), the former regarding boundary rotation and the latter regarding plate deflection.

As previously for (7.1.73), it can be proved that $\tilde{C}(\zeta)$ is the Gâteaux-derivative of $F_1(\zeta) = \frac{1}{4}R(\varepsilon(u(\zeta)) + \frac{1}{2}P(\zeta), P(\zeta))$ at ζ. Actually, $\tilde{C}(\cdot)$ is continuous with respect to ζ, as we will see, and hence $\tilde{C}(\cdot)$ is a derivative in the sense of Fréchet. Indeed, the estimate

$$\begin{gathered} \| \tilde{C}(\zeta_1) - \tilde{C}(\zeta_2) \|_Z \leq c(\| \zeta_1 \|_Z^2 + \| \zeta_2 \|_Z^2) \| \zeta_1 - \zeta_2 \|_{1,4}, \\ c \text{ const} > 0, \qquad \forall \zeta_1, \zeta_2 \in Z \end{gathered} \tag{7.2.60}$$

holds.[5] We shall only outline the proof of (7.2.60) [226]. Korn's inequality (cf. Sec. 1.4.3) when applied to (7.1.32) implies, by means of the imbedding $H^1(\Omega) \subset W^{1,4}(\Omega)$, that $u(\zeta_1)$ and $u(\zeta_2)$ satisfy the estimate

$$\|u(\zeta_1) - u(\zeta_2)\|_1 \leq c(\|\zeta_1\|_Z + \|\zeta_2\|_Z)\, \|\zeta_1 - \zeta_2\|_{1,4}, \qquad c \text{ const} > 0, \quad \forall \zeta_1, \zeta_2 \in Z. \tag{7.2.61}$$

Applying (7.2.61), we obtain

$$|R(\varepsilon(u(\zeta_1)) + \tfrac{1}{2}P(\zeta_1), P(\zeta_1, z)) - R(\varepsilon(u(\zeta_2)) + \tfrac{1}{2}P(\zeta_2), P(\zeta_2, z))| \leq c(\|\zeta_1\|_Z^2 + \|\zeta_2\|_Z^2)\, \|\zeta_1 - \zeta_2\|_{1,4}\, \|z\|_Z, \qquad c \text{ const} > 0, \quad \forall \zeta_1, \zeta_2 \in Z, \tag{7.2.62}$$

from which (7.2.60) results. Accordingly, $\tilde{T} = I + \tilde{C}$ is the Fréchet-derivative of the functional

$$\tilde{F}(z) = \tfrac{1}{2}\|z\|_Z^2 + F_1(z). \tag{7.2.63}$$

The remaining assumptions on $\tilde{T}$ in Props. 7.2.1 and 7.2.2 are easily verified.

We make an additional assumption concerning the stresses σ^0: the boundary Γ_1 of every subdomain $\Omega_1 \subset \Omega$ of the plate is subjected to compressive forces, i.e., a.e. on Γ the inequality

$$\sigma_{ij}^0 \bar{n}_i \bar{n}_j \leq 0 \tag{7.2.64}$$

holds, where $\bar{n} = (\bar{n}_1, \bar{n}_2)$ denotes the unit vector normal to Γ_1 which is directed outwards with respect to Ω_1. It is readily seen that all the assumptions for B are verified. Accordingly, Props. 7.2.1 and 7.2.2 guarantee the existence of an infinite sequence of variational solutions $\{\lambda_n, \zeta_n, u_n\}$ with $\lambda_n \geq 0$ and $\zeta_n \neq 0$.

We may also easily verify that $\zeta = 0$ is the only solution if $\lambda \leq \lambda^*$, where λ^* is the smallest positive eigenvalue of B. To show this, let us put $z = \zeta/2$ (with $\zeta \neq 0$) in (7.2.14). By taking into account (7.2.53), the relation $(\tilde{C}\zeta, \zeta)_Z \geq 0$, $\forall \zeta \in Z$ and the form of Φ, we obtain

$$\lambda(B\zeta, \zeta)_Z > \|\zeta\|_Z^2, \qquad \zeta \neq 0. \tag{7.2.65}$$

(7.2.65) implies that $\lambda > 0$ and

$$\frac{1}{\lambda} < \sup_Z \frac{(B\zeta, \zeta)_Z}{\|\zeta\|_Z^2}. \tag{7.2.66}$$

[5] Suppose that $\zeta = 0$ on Γ, and let ζ_1 and ζ_2 be two solutions of (7.1.36). Then by adopting the notation of this section we obtain the inequality

$$(\tilde{T}(\zeta_1) - \tilde{T}(\zeta_2), \zeta_1 - \zeta_2)_Z \leq 0,$$

which, by means of (7.2.60) and the imbedding $H^2(\Omega) \subset W^{1,4}(\Omega)$, yields $\zeta_1 = \zeta_2$ if $\|\zeta_1\|_Z^2 + \|\zeta_2\|_Z^2$ is sufficiently small.

In addition, B is a linear, self-adjoint compact operator from Z into Z and, therefore, by applying the classical theory of eigenvalue problems we find that there exists an eigenvalue $\lambda \in \mathbb{R}$ defined by the relation $z = \lambda B z$, $z \neq 0$. The smallest positive eigenvalue λ^* of B is given by the Rayleigh quotient

$$\frac{1}{\lambda^*} = \sup_{Z} \frac{(Bz, z)_Z}{\| z \|_Z^2}. \tag{7.2.67}$$

Thus if $\zeta \neq 0$ then $\lambda^* < \lambda$, and accordingly, if $\lambda \leq \lambda^*$, the unique solution is $\zeta = 0$.

We state here without proof some conclusions from [72] on the bifurcation of the nonzero solutions from the trivial solution $\zeta = 0$. We denote by $\{\zeta_\beta, \lambda_\beta\}$ a nontrivial solution corresponding to $\beta > 0$ (cf., 7.2.25). As $\beta \to \infty$, $\|\zeta_\beta\|_Z \to \infty$ and $\lambda_\beta \to \infty$; as $\beta \to 0$, $\zeta_\beta \to 0$. Further, Z_0 denotes a closed linear subspace of Z with the property that $\Phi(z) = 0$ if and only if $z \in Z_0$. Two cases are distinguished:

(i) $Z_0 = \{0\}$. Let $\{\lambda_m, \zeta_m\}$ be a sequence of nontrivial solutions such that $\zeta_m \to 0$. Then

$$\lambda_m \to \infty. \tag{7.2.68}$$

(ii) $Z_0 \neq \{0\}$. Then as $\beta \to 0, \zeta_\beta \to 0$ and

$$\lambda_\beta \to \lambda^*, \tag{7.2.69}$$

where λ^* is the first positive eigenvalue of B.

As an application, we first consider a clamped plate subjected to the reaction-deflection law (7.1.17) on $\Omega' \subset \Omega$. It is assumed that $j_3(\zeta) = 0$ if and only if $\zeta = 0$; therefore $Z_0 = \{0\}$. It results that no solution bifurcates from the trivial solution. Thus $\lambda \to \infty$ is the only bifurcation value. To take another example, we consider a plate subjected to the boundary conditions (7.1.10) and (7.1.13), where $j_2(\partial\zeta/\partial n) = 0$ if and only if $\partial\zeta/\partial n = 0$. If, instead of (7.1.13), $\partial\zeta/\partial n = 0$ on Γ, i.e., if the plate is clamped, there will be a value λ_0 at which the solution bifurcates from the trivial solution $\zeta = 0$ [22]. λ_0 is thus the critical loading for the clamped plate. We see that λ_0 is a bifurcation value for the unilateral problem as well. Indeed, $Z_0 = \mathring{H}^2(\Omega)$ and therefore if $\beta \to 0$, $\zeta_\beta \to 0$ and $\lambda_\beta \to \lambda^* = \lambda^0$. We conclude that the previous subdifferential boundary conditions relating bending moments to rotations act, from the standpoint of bifurcation, like a clamping of the plate.

7.2.4 Extension of the Rayleigh-Quotient Rule to Unilateral Problems

As is well known, in eigenvalue problems of linear, self-adjoint, compact operators in a Hilbert space, the eigenvalues are calculated by means of Rayleigh-quotients. This has not yet been shown in the general case of variational inequality (7.2.14); only in the special case of the variational inequality for a plate resting on a rigid support does an analogous result hold. The plate is assumed to be clamped on $\Gamma_1 \subset \Gamma$ and simply supported on $\Gamma_2 = \Gamma - \Gamma_1$. Then

$$Z = \left\{ z \mid z \in H^2(\Omega),\ z = 0 \text{ on } \Gamma,\ \frac{\partial z}{\partial n} = 0 \text{ a.e. on } \Gamma_1 \right\}, \tag{7.2.70}$$

and in (7.2.14) Φ is the indicator I_K of the convex closed set K defined by

$$K = \{z \mid z \in Z,\ z \geq 0, \text{in } \Omega\}. \tag{7.2.71}$$

Then Problem 1 of Sec. 7.2.2 becomes: find $\zeta \in K$, $\lambda \in \mathbb{R}$, such that

$$((I + \tilde{C})(\zeta), z - \zeta)_Z - \lambda(B\zeta, z - \zeta)_Z \geq 0, \qquad \forall z \in K. \tag{7.2.72}$$

The following proposition holds for this eigenvalue problem [43].

Proposition 7.2.3. Let us define the numbers

$$\frac{1}{\lambda_1} = \sup_K \frac{(Bz, z)_Z}{\|z\|_Z^2}, \qquad \frac{1}{\lambda_1'} = \inf_K \frac{(Bz, z)_Z}{\|z\|_Z^2}. \tag{7.2.73}$$

When $\lambda_1' \leq \lambda \leq \lambda_1$, the variational inequality (7.2.72) admits only the trivial solution $\zeta = 0$, and conversely; when $\lambda > \lambda_1$ and $\lambda < \lambda_1'$, (7.2.72) has at least one nonzero solution ζ_λ such that

$$\lim_{\lambda \to \lambda_1 +} \|\zeta_\lambda\|_Z = 0. \qquad \lim_{\lambda \to \lambda_1' -} \|\zeta_\lambda\|_Z = 0. \tag{7.2.74}$$

Proof. Only the essential ideas of the proof are given here. A Galerkin basis is introduced and the cone K is approximated by a sequence of finite-dimensional, convex, closed cones K_m which for $m \to \infty$ tend to K and have as origin the zero point of Z. Let $\zeta_m \in K_m$ be a solution of the variational inequality

$$((I + \tilde{C})(\zeta_m), z - \zeta_m)_Z - \lambda(B\zeta_m, z - \zeta_m)_Z \geq 0, \qquad \forall z \in K_m. \tag{7.2.75}$$

The sequence of the solutions ζ_m is bounded and a subsequence can be selected converging strongly to the solution ζ of (7.2.72). By setting $z = 0$ and $z = 2\zeta$ in (7.2.72), it results that

$$((I + \tilde{C})(\zeta), \zeta)_Z - \lambda(B\zeta, \zeta)_Z = 0. \tag{7.2.76}$$

The boundary conditions on Γ ensure that $(\tilde{C}(\zeta), \zeta)_Z = 0$ implies $\zeta = 0$ [270]. If $\zeta \neq 0$, $(\tilde{C}(\zeta), \zeta)_Z > 0$ and thus by means of (7.2.76) and by taking into account the definitions (7.2.73) of λ_1 and λ_1', the first part of the proposition results. In order to prove (7.2.74), it is shown that for every solution ζ_λ of (7.2.72) and for every λ with $\lambda_1 \leq \lambda \leq 2\lambda_1$ there exists a constant $c > 0$ such that

$$\|\zeta_\lambda\|_Z^2 \leq c\left(\frac{\lambda}{\lambda_1} - 1\right), \tag{7.2.77}$$

and analogously for λ_1', q.e.d.

As is obvious from Prop. 7.2.3, λ_1 and λ_1' are respectively the positive and negative critical loads of the plate. The crucial point of the foregoing proof is that $(\tilde{C}(\zeta), \zeta)_Z =$

0 implies $\zeta = 0$. In [270], this question is discussed and a class of boundary conditions permitting this result is given.

Assume, further, that the plate lies on a linear elastic body $\Omega_1 \subset \mathbb{R}^3$, referred to the coordinate system $0x_1x_2x_3$ of the plate [44]. The displacement vector U of Ω_1 is assumed to belong to $[H^1(\Omega_1)]^3$. On a part Γ_1' of the boundary of Ω_1, $U_i = 0$, $i = 1, 2, 3$. The convex set K_1 of the admissible displacements now has the form (cf., (3.3.26); $\Gamma_1' = \Gamma_1 - \Omega$, Γ_1 is the boundary of Ω_1)

$$K_1 = \{(z, V) \,|\, z \in Z,\ V \in [H^1(\Omega_1)]^3,\ z - V_3 \geq 0 \text{ a.e. in } \Omega,\ V = 0 \text{ on } \Gamma_1'\}. \quad (7.2.78)$$

Then for $(\zeta, U) \in K_1$ the variational inequality

$$((I + \tilde{C})(\zeta), z - \zeta)_Z + \lambda(B\zeta, z - \zeta)_Z + \bar{a}(U, V - U) \geq 0, \qquad \forall (z, V) \in K_1 \quad (7.2.79)$$

holds, where $\bar{a}(U, V)$ is the bilinear form of the strain energy of Ω_1 (cf., 5.1.10). In this case the critical load is expressed by

$$\frac{1}{\lambda_1} = \sup_{K_1} \frac{(Bz, z)_Z}{\|z\|_Z + \bar{a}(V, V)}. \quad (7.2.80)$$

For $0 \leq \lambda \leq \lambda_1$, $\zeta = U_1 = U_2 = U_3 = 0$ is the only solution of (7.2.79). If $\lambda > \lambda_1$, there exists a solution with $\zeta \neq 0$.

An inherent weakness of the von Kármán plate theory is the assumed linear elasticity law between stresses and strains. A formal extension of the foregoing theory, based on a holonomic law of subdifferential nature between the in-plane stress and strain tensors, is straightforward; the corresponding eigenvalue problem can also be formulated. In this context, some results have already been obtained [74] [205] concerning the buckling of von Kármán plates where the material obeys the well-known law (3.3.83) regarding the stretching of the plate.

7.3 Dynamic Unilateral Problems of von Kármán Plates

7.3.1 Boundary Conditions and Variational Inequalities

Within the framework of Sec. 7.1.1, the dynamic behavior of von Kármán plates is governed by the following system of differential equations ($\alpha, \beta = 1, 2$):

$$\rho\zeta'' + \mathsf{K}\,\Delta\Delta\zeta - \mathsf{D}\,\Delta\zeta'' - h(\sigma_{\alpha\beta}\zeta_{,\beta})_{,\alpha} = f_3 \quad \text{in} \quad \Omega \times (0, T); \quad (7.3.1)$$

$$\rho u_\alpha'' - \sigma_{\alpha\beta,\beta} = f_\alpha \quad \text{in} \quad \Omega \times (0, T); \quad (7.3.2)$$

$$\sigma_{\alpha\beta} = C_{\alpha\beta\gamma\delta}(\varepsilon_{\gamma\delta}(u) + \tfrac{1}{2}\zeta_{,\gamma}\zeta_{,\delta}) \quad \text{in} \quad \Omega \times (0, T); \quad (7.3.3)$$

$$\varepsilon_{\alpha\beta}(u) = \tfrac{1}{2}(u_{\alpha,\beta} + u_{\beta,\alpha}) \quad \text{in} \quad \Omega \times (0, T). \quad (7.3.4)$$

This system originates formally from (7.1.1–4) by adding the inertia terms (the rotatory inertia terms corresponding to u_1 and u_2 have been neglected). Here t is the time variable, $[0, T]$ the time interval, ρ the plate density,

($\rho \in L^\infty(\Omega)$, ess inf $\rho(x) > 0$) and $f = (f_1, f_2, f_3)$, $f_i = f_i(x, t)$, the load vector of the plate. $\mathsf{D} = \rho h^3/12$ is the product of the density and the moment of inertia of a unit-width strip cut from the plate. For $t = 0$ the following initial conditions hold:

$$u = u_0; \tag{7.3.5}$$

$$u' = u_1; \tag{7.3.6}$$

$$\zeta = \zeta_0; \tag{7.3.7}$$

and

$$\zeta' = \zeta_1, \tag{7.3.8}$$

where u_0, u_1, ζ_0 and ζ_1 are functions of x. First, the relations analogous to (7.1.20–21) are formulated on the assumption of sufficiently regular functions. They read

$$\begin{aligned}
&\int_\Omega \rho\zeta''(z - \zeta')\,d\Omega + \bar{\alpha}(\zeta'', z - \zeta') + \alpha(\zeta, z - \zeta') \\
&\quad + \int_\Omega h\sigma_{\alpha\beta}\zeta_{,\alpha}(z - \zeta')_{,\beta}\,d\Omega \\
&= \int_\Gamma h\sigma_{\alpha\beta}\zeta_{,\alpha}n_\beta(z - \zeta')\,d\Gamma + \int_\Gamma \bar{Q}(\zeta)(z - \zeta')\,d\Gamma \\
&\quad - \int_\Gamma M(\zeta)\frac{\partial(z - \zeta')}{\partial n}\,d\Gamma \\
&\quad + \int_\Omega f_3(z - \zeta')\,d\Omega + \mathsf{D}\int_\Gamma \frac{\partial\zeta''}{\partial n}(z - \zeta')\,d\Gamma,
\end{aligned} \tag{7.3.9}$$

where $\alpha(\zeta, z)$, $M(\zeta)$ and $\bar{Q}(\zeta)$ are given by (7.1.6–8), and

$$\bar{\alpha}(\zeta'', z) = \mathsf{D}\int_\Omega \zeta''_{,\alpha} z_{,\alpha}\,d\Omega, \qquad \alpha = 1, 2. \tag{7.3.10}$$

Similarly, from (7.3.2) the relation

$$\begin{aligned}
&\int_\Omega \rho u''(v - u')\,d\Omega + \int_\Omega \sigma_{\alpha\beta}\,\varepsilon_{\alpha\beta}(v - u')\,d\Omega \\
&\qquad = \int_\Omega f_\alpha(v_\alpha - u'_\alpha)\,d\Omega + \int_\Gamma \sigma_{\alpha\beta}n_\beta(v_\alpha - u'_\alpha)\,d\Gamma.
\end{aligned} \tag{7.3.11}$$

is obtained. The following boundary conditions are assumed to hold pointwise on Γ for every $t \in [0, T]$:

$$M \in \beta_1\left(\frac{\partial\zeta'}{\partial n}\right) = \partial j_1\left(\frac{\partial\zeta'}{\partial n}\right); \tag{7.3.12}$$

$$-\bar{Q} \in \beta_2(\zeta') = \partial j_2(\zeta'); \tag{7.3.13}$$

and

$$\sigma_{1\alpha} n_\alpha = \sigma_{2\beta} n_\beta = 0. \tag{7.3.14}$$

Moreover, f_3 is split into $\bar{f}_3$ and $\bar{\bar{f}}_3$, where $\bar{\bar{f}}_3$ is given and

$$\begin{gathered} -\bar{f}_3 \in \beta_3(\zeta') = \partial j_3(\zeta') \text{ in } \Omega' \subset \Omega, \\ \bar{\Omega}' \cap \Gamma = \emptyset, \quad \bar{f}_3 = 0 \text{ in } \Omega - \Omega'. \end{gathered} \tag{7.3.15}$$

The β_i and j_i terms have the same properties as in Sec. 7.1.1. In order to give a variational formulation of the B.V.P. defined by (7.3.1–8) and (7.3.12–15) we assume that $\rho \in L^\infty(\Omega)$, $z \in H^2(\Omega)$, $v \in [H^1(\Omega)]^2$ and $f_i(t) \in L^2(\Omega)$, $i = 1, 2, 3$, for every $t \in [0, T]$, and we define the functional Φ on $H^2(\Omega)$ by (7.1.30). Taking into account (7.1.18, 19) and the variational inequalities resulting from (7.3.12, 13, 15), we are led to the following variational problem.

Problem 1. Find $u: [0, T] \to [H^1(\Omega)]^2$, $\zeta: [0, T] \to H^2(\Omega)$, with $u'(t) \in [H_1(\Omega)]^2$ and $\zeta'(t) \in H^2(\Omega)$, $u''(t) \in [L^2(\Omega)]^2$, $\zeta''(t) \in L^2(\Omega)$ such as to satisfy the variational system

$$\begin{aligned} &(\rho\, \zeta'', z - \zeta') + \alpha(\zeta, z - \zeta') + \bar{\alpha}(\zeta'', z - \zeta') \\ &\quad + hR(\varepsilon(u) + \tfrac{1}{2}P(\zeta), P(\zeta, z - \zeta')) \\ &\quad + \Phi(z) - \Phi(\zeta') \geq (\bar{\bar{f}}_3, z - \zeta') \\ &\quad + \mathrm{D} \cdot \int_\Gamma \frac{\partial \zeta''}{\partial n}(z - \zeta')\, d\Gamma, \qquad \forall z \in H^2(\Omega) \end{aligned} \tag{7.3.16}$$

$$\begin{aligned} &(\rho u'', v - u') + R(\varepsilon(u) + \tfrac{1}{2}P(\zeta), \varepsilon(v - u')) \\ &\qquad = (f_e, v - u'), \qquad \forall v \in [H^1(\Omega)]^2 \end{aligned} \tag{7.3.17}$$

and the initial conditions (7.3.5–8).

Here,

$$f_e = (f_1, f_2). \tag{7.3.18}$$

It is also possible to consider unilateral boundary conditions with respect to $S = \{S_a\} = \{\sigma_{\alpha\beta} n_\beta\}$ and u'; for instance, we can assume that on Γ (7.3.12) holds, as do the boundary conditions

$$-\bar{Q} \in \partial j_2(\zeta') \text{ on } \Gamma_1, \qquad \zeta = 0 \text{ on } \Gamma - \Gamma_1 \tag{7.3.19}$$

and

$$S = 0 \text{ on } \Gamma_1, \qquad -S \in \partial j_4(u') \text{ on } \Gamma - \Gamma_1. \tag{7.3.20}$$

In addition, let (7.3.15) hold as well. Here j_4 is again a convex, l.s.c. and proper functional on $\mathbb{R}^2$. Then the spaces $Z = \{z \mid z \in H^2(\Omega),\ z = 0$ on

$\Gamma - \Gamma_1\}$ and $V = \{v \mid v \in [H^1(\Omega)]^2\}$ are introduced, and the functionals Φ and $\bar{\Phi}$ are defined by (3.3.125) (cf., (7.1.30)). A pair (u, ζ) is a variational solution of this B.V.P. if and only if it satisfies the following problem.

Problem 2. Find u: $[0, T] \to V$ and ζ: $[0, T] \to Z$ with $u'(t) \in V$, $\zeta'(t) \in Z$ and $u''(t) \in [L^2(\Omega)]^2$, $\zeta''(t) \in L^2(\Omega)$, such as to satisfy

$$\begin{aligned}(\rho\zeta'', z - \zeta') &+ \alpha(\zeta, z - \zeta') + \bar{\alpha}(\zeta'', z - \zeta') \\ &+ hR(\varepsilon(u) + \tfrac{1}{2}P(\zeta), P(\zeta, z - \zeta')) \\ &+ \Phi(z) - \Phi(\zeta') \geq (\bar{f}_3, z - \zeta') \\ &+ D \int_\Gamma \frac{\partial \zeta''}{\partial n} (z - \zeta')\, d\Gamma, \qquad \forall z \in Z,\end{aligned} \tag{7.3.21}$$

$$\begin{aligned}(\rho u'', v - u') &+ R(\varepsilon(u) + \tfrac{1}{2}P(\zeta), \varepsilon(v - u')) \\ &+ \bar{\Phi}(v) - \bar{\Phi}(u') \geq (f_e, v - u'), \qquad \forall v \in V\end{aligned} \tag{7.3.22}$$

and the initial conditions (7.3.5–8).

7.3.2 Existence Proposition

The existence and uniqueness of the solution of the general unilateral B.V.P. formulated here have not yet been proved. Here, we shall study the variational system (7.3.16, 17), where (7.3.13) is replaced by the condition $\zeta = 0$ on Γ. We first formulate an abstract problem for which the existence and uniqueness proofs will be given; this problem describes large classes of unilateral B.V.P.s in the dynamic theory of von Kármán plates.

Let V, V_0, Z, Z_1 and $Z_0 = L^2(\Omega)$ be Hilbert spaces such that the inbeddings

$$V \subset V_0 = V'_0 \subset V', \tag{7.3.23}$$

and

$$Z \subset Z_1 \subset Z_0 = Z'_0 \subset Z'_1 \subset Z', \tag{7.3.24}$$

are dense. Moreover, let the injection $Z \subset Z_0$ be compact and the bilinear forms $\alpha(z, z)$ and $\bar{\alpha}(z, z)$ be symmetric, continuous and coercive on Z and Z_1 respectively. We define $R(h, k)$ as in Sec. 7.1.2 and assume that the applications ε, P and R have the properties:

(i) ε: $V \to [L^2(\Omega)]^4$ is linear and continuous; (7.3.25)

(ii) P: $Z \times Z \to [L^2(\Omega)]^4$ is bilinear, and completely continuous; (7.3.26)

(iii) $\| P(\zeta, \zeta) \|_{[L^2(\Omega)]^4} \leq c \| \zeta \|_Z \| \zeta \|_{Z_1}$, $\quad \forall \zeta \in Z$; and (7.3.27)

(iv) as $\zeta_i \to \zeta$ in $L^\infty(0T, Z)$ weakly$_*$ and $\zeta'_i \to \zeta'$ in $L^\infty(0T, Z_0)$ weakly$_*$

$$R(P(\zeta_i), P(\zeta_i, z)) \to R(P(\zeta), P(\zeta, z)) \text{ in } \mathscr{D}'(0T), \quad \forall z \in L^2(0T, Z). \quad (7.3.28)$$

We also assume that there exists a sequence of convex, differentiable functionals Φ_ε ($\varepsilon > 0$) on Z such that as $\varepsilon \to 0$

(i) $$\int_0^T \Phi_\varepsilon(z)\, dt \to \int_0^T \Phi(z)\, dt, \qquad \forall z \in L^2(0T, Z), \quad (7.3.29)$$

(ii) there exists a bounded sequence $\{\bar{\zeta}_\varepsilon\} \in Z$ such that

$$\frac{d\Phi_\varepsilon}{d\zeta}(\bar{\zeta}_\varepsilon) = 0, \qquad \forall \varepsilon > 0 \quad (7.3.30)$$

and

(iii) if $z_\varepsilon \to z$ and $z'_\varepsilon \to z'$ weakly$_*$ in $L^\infty(0T, Z)$ and $\int_0^T \Phi_\varepsilon(z'_\varepsilon)dt \le c$, then

$$\liminf \int_0^T \Phi_\varepsilon(z'_\varepsilon)\, dt \ge \int_0^T \Phi(z')\, dt. \quad (7.3.31)$$

We can now formulate the general problem.

Problem 3. Find u: $[0, T] \to V$ and ζ: $[0, T] \to Z$, with $u'(t) \in V$, $u''(t) \in V_0$ and $\zeta'(t) \in Z$, $\zeta''(t) \in Z_1$, which satisfy the variational system

$$(\rho\zeta'', z - \zeta') + \alpha(\zeta, z - \zeta') + \bar{\alpha}(\zeta'', z - \zeta') + hR(\varepsilon(u) + \tfrac{1}{2}P(\zeta), P(\zeta, z - \zeta'))$$
$$+ \Phi(z) - \Phi(\zeta') \ge (f_3, z - \zeta'), \qquad \forall z \in Z, \quad (7.3.32)$$

$$(\rho u'', v - u') + R(\varepsilon(u) + \tfrac{1}{2}P(\zeta),$$
$$\varepsilon(v - u')) = (f_e, v - u'), \qquad \forall v \in V \quad (7.3.33)$$

and the initial conditions (7.3.5–8). The following regularized problem corresponds to (7.3.32, 33).

Problem 3ε. Find u_ε: $[0, T] \to V$ and ζ_ε: $[0, T] \to Z$ with $u'_\varepsilon(t) \in V$, $u''_\varepsilon(t) \in V_0$ and $\zeta'_\varepsilon(t) \in Z$, $\zeta''_\varepsilon(t) \in Z_1$, which satisfy the equations

$$(\rho\zeta''_\varepsilon, z) + \alpha(\zeta_\varepsilon, z) + \bar{\alpha}(\zeta''_\varepsilon, z) + hR(\varepsilon(u_\varepsilon) + \tfrac{1}{2}P(\zeta_\varepsilon), P(\zeta_\varepsilon, z))$$
$$+ \left\langle \frac{d\Phi_\varepsilon}{d\zeta}(\zeta'_\varepsilon), z \right\rangle_Z = (f_3, z), \qquad \forall z \in Z \quad (7.3.34)$$

$$(\rho u''_\varepsilon, v) + R(\varepsilon(u_\varepsilon) + \tfrac{1}{2}P(\zeta_\varepsilon),$$
$$\varepsilon(v)) = (f_e, v), \qquad \forall v \in V \quad (7.3.35)$$

and the initial conditions

$$u_\varepsilon = u_0, \quad u'_\varepsilon = u_1, \qquad \zeta_\varepsilon = \zeta_0, \quad \zeta'_\varepsilon = \zeta_1. \quad (7.3.36)$$

The assumptions made, hold for a large class of problems related to the von Kármán plate theory. Let us consider, for instance, the boundary conditions $\zeta = 0$, (7.3.12) and (7.3.14), and let (7.3.15) hold in Ω. Then

$$Z = H^2(\Omega) \cap \mathring{H}^1(\Omega), \qquad Z_1 = \mathring{H}^1(\Omega), \qquad Z_0 = L^2(\Omega) \tag{7.3.37}$$

and

$$V = [H^1(\Omega)]^2, \qquad V_0 = [L^2(\Omega)]^2. \tag{7.3.38}$$

It can be seen that the assumptions on the spaces Z, Z_1, Z_0, V and V_0, and on the bilinear forms $\alpha(z, z)$, $\bar{\alpha}(z, z)$ and $R(h, k)$, are fulfilled as well as (7.3.25, 26). Next, we verify (7.3.27). By applying (1.2.24) for $\zeta_{,\alpha}$ with $\alpha_1 = 0$, $\alpha_2 = \frac{1}{2}$, $\alpha_3 = 1$, we find that

$$\|\zeta_{,\alpha}\|_{1/2} \leq |\zeta_{,\alpha}|_2^{1/2} \|\zeta_{,\alpha}\|_1^{1/2}, \qquad \forall \zeta_{,\alpha} \in H^1(\Omega). \tag{7.3.39}$$

This inequality together with (1.3.50) (for $n = 2$) yields

$$|\zeta_{,\alpha}|_4 \leq c \|\zeta_{,\alpha}\|_1^{1/2} |\zeta_{,\alpha}|_2^{1/2}, \qquad \forall \zeta_{,\alpha} \in H^1(\Omega), \qquad c \text{ const} > 0. \tag{7.3.40}$$

Thus, by means of Hölder's inequality, it results that

$$|\zeta_{,\alpha}\zeta_{,\beta}|_2 \leq c|\zeta_{,\alpha}|_4 |\zeta_{,\beta}|_4 \leq c \|\zeta_{,\alpha}\|_1^{1/2} \|\zeta_{,\beta}\|_1^{1/2} |\zeta_{,\alpha}|_2^{1/2} |\zeta_{,\beta}|_2^{1/2}$$
$$\leq c \|\zeta\|_2 \|\zeta\|_1, \qquad c \text{ const} > 0, \qquad \forall \zeta \in Z, \tag{7.3.41}$$

and so (7.3.27) is proved. In order to verify (7.3.28), we must show that

$$\int_0^T \int_\Omega C_{\alpha\beta\gamma\delta} \zeta_{i,\alpha} \zeta_{i,\beta} \zeta_{i,\gamma} z_{,\delta} \, d\Omega \, dt$$
$$\to \int_0^T \int_\Omega C_{\alpha\beta\gamma\delta} \zeta_{,\alpha} \zeta_{,\beta} \zeta_{,\gamma} z_{,\delta} \, d\Omega \, dt, \qquad \forall z \in L^2(0T, Z) \tag{7.3.42}$$

when $\zeta_i \to \zeta$ in $L^\infty(0T, Z)$ weakly$_*$ and $\zeta_i' \to \zeta'$ in $L^\infty(0T, Z_0)$ weakly$_*$. Because of these last assumptions, $\zeta_{i,\alpha} \to \zeta_{,\alpha}$ (resp. $\zeta'_{i,\alpha} \to \zeta'_{,\alpha}$) weakly$_*$ in $L^\infty(0T, H^1(\Omega))$ (resp. in $L^\infty(0T, H^{-1}(\Omega))$. The imbedding from $H^1(\Omega)$ into $L^p(\Omega)$ is compact for $1 \leq p < \infty$ (cf., (1.3.51)). Applying a compactness theorem of Lions,[6] we see that

$$\zeta_{i,\alpha} \to \zeta_{,\alpha} \quad \text{strongly in } L^p(\Omega \times (0T)) \text{ for every } p \text{ finite} > 1, \tag{7.3.43}$$

and thus (7.3.42) is verified. The following proposition holds.

Proposition 7.3.1. Suppose that

$$f_e, f_e' \in L^2(0T, V_0), (f_e(t), \bar{r}) = 0 \quad \forall \bar{r} \in \bar{\mathscr{R}} \quad \forall t \in (0, T), \tag{7.3.44}$$

$$f_3, f_3' \in L^2(0T, Z_0), \tag{7.3.45}$$

[6] See [182], p. 58, Theorem 5.1.

$$u_0, u_1 \in V, \tag{7.3.46}$$

$$\zeta_0, \zeta_1 \in Z, \tag{7.3.47}$$

$$\Phi(\zeta_1) < \infty, \tag{7.3.48}$$

the linear form

$$z \to F_1(z) = -\alpha(\zeta_0, z) - hR(\varepsilon(u_0) + \tfrac{1}{2}P(\zeta_0), P(\zeta_0, z)) - \left\langle \frac{d\Phi_\varepsilon}{d\zeta}(\zeta_1), z \right\rangle_Z, \quad \varepsilon > 0, \tag{7.3.49}$$

is continuous on Z for the Z_1-topology, and finally that the linear form

$$v \to F_2(v) = -R(\varepsilon(u_0) + \tfrac{1}{2}P(\zeta_0), \varepsilon(v)), \tag{7.3.50}$$

is continuous on V for the V_0-topology. Then a solution u_ε, ζ_ε of Problem 3ε exists. Moreover, a $\zeta \in L^\infty(0T, Z)$ and a $u \in L^\infty(0T, V_0)$ exist such that for $\varepsilon \to 0$

$$\zeta_\varepsilon \to \zeta, \quad \zeta'_\varepsilon \to \zeta' \quad \text{in} \quad L^\infty(0T, Z) \text{ weakly}_*, \tag{7.3.51}$$

$$\zeta''_\varepsilon \to \zeta'' \quad \text{in} \quad L^\infty(0T, Z_1) \text{ weakly}_*, \tag{7.3.52}$$

$$u_\varepsilon \to u, \quad u'_\varepsilon \to u', \quad u''_\varepsilon \to u'' \quad \text{in} \quad L^\infty(0T, V_0) \text{ weakly}_*, \tag{7.3.53}$$

and

$$\varepsilon(u_\varepsilon) \to \varepsilon(u), \quad \varepsilon(u'_\varepsilon) \to \varepsilon(u') \quad \text{in} \quad L^\infty(0T, [L^2(\Omega)]^4) \text{ weakly}_*. \tag{7.3.54}$$

Proof. (a) By means of the Galerkin method, it can be shown, as in Prop. 5.4.2, that the regularized problem admits a solution u_ε, ζ_ε such that $\zeta_\varepsilon, \zeta'_\varepsilon \in L^\infty(0T, Z)$, $\zeta''_\varepsilon \in L^\infty(0T, Z_1)$ and $u_\varepsilon, u'_\varepsilon, u''_\varepsilon \in L^\infty(0T, V_0)$.

(b) Let us put $z = \zeta'_\varepsilon$ and $z = \bar{\zeta}_\varepsilon$ into (7.3.34) and $u = u'_\varepsilon$ into (7.3.35). Then by using (7.3.30) we obtain the relations

$$\begin{aligned}(\rho\zeta''_\varepsilon, \zeta'_\varepsilon - \bar{\zeta}_\varepsilon) + \alpha(\zeta_\varepsilon, \zeta'_\varepsilon) + \bar{\alpha}(\zeta''_\varepsilon, \zeta'_\varepsilon - \bar{\zeta}_\varepsilon) + hR(\varepsilon(u_\varepsilon) \\ + \tfrac{1}{2}P(\zeta_\varepsilon), P(\zeta_\varepsilon, \zeta'_\varepsilon)) + \left\langle \frac{d\Phi_\varepsilon}{dz}(\zeta'_\varepsilon) - \frac{d\Phi_\varepsilon}{dz}(\bar{\zeta}_\varepsilon), \zeta'_\varepsilon - \bar{\zeta}_\varepsilon \right\rangle_Z \\ = (f_3, \zeta'_\varepsilon - \bar{\zeta}_\varepsilon) + \alpha(\zeta_\varepsilon, \bar{\zeta}_\varepsilon) + hR(\varepsilon(u_\varepsilon) + \tfrac{1}{2}P(\zeta_\varepsilon), P(\zeta_\varepsilon, \bar{\zeta}_\varepsilon))\end{aligned} \tag{7.3.55}$$

and

$$(\rho u''_\varepsilon, u'_\varepsilon) + R(\varepsilon(u_\varepsilon) + \tfrac{1}{2}P(\zeta_\varepsilon), \varepsilon(u'_\varepsilon)) = (f_\varepsilon, u'_\varepsilon). \tag{7.3.56}$$

By means of the identity

$$\begin{aligned}R(\varepsilon(u_\varepsilon) + \tfrac{1}{2}P(\zeta_\varepsilon), P(\zeta_\varepsilon, \zeta'_\varepsilon)) = \frac{1}{2}\frac{d}{dt} R(\varepsilon(u_\varepsilon) + \tfrac{1}{2}P(\zeta_\varepsilon), \varepsilon(u_\varepsilon) + \tfrac{1}{2}P(\zeta_\varepsilon)) \\ - R(\varepsilon(u_\varepsilon) + \tfrac{1}{2}P(\zeta_\varepsilon), \varepsilon(u'_\varepsilon)),\end{aligned} \tag{7.3.57}$$

there results from (7.3.55), through integration from 0 to t, the relation

$$\begin{aligned}
&\tfrac{1}{2}(\|\zeta_\varepsilon' - \bar{\zeta}_\varepsilon\|_{Z_0}^2 + c\|\zeta_\varepsilon\|_Z^2 + c\|\zeta_\varepsilon' - \bar{\zeta}_\varepsilon\|_{Z_1}^2 \\
&\quad + hR(\varepsilon(u_\varepsilon) + \tfrac{1}{2}P(\zeta_\varepsilon), \varepsilon(u_\varepsilon) + \tfrac{1}{2}P(\zeta_\varepsilon)) + h\|u_\varepsilon'\|_{V_0}^2) \\
&\quad \le \int_0^t [(f_3, \zeta_\varepsilon' - \bar{\zeta}_\varepsilon) + h(f_e, u_\varepsilon')]\, dt + \int_0^t c[\|\zeta_\varepsilon\|_Z \|\bar{\zeta}_\varepsilon\|_Z \qquad (7.3.58) \\
&\quad + \|\varepsilon(u_\varepsilon) + \tfrac{1}{2}P(\zeta_\varepsilon)\|_{[L^2(\Omega)]^4} \|P(\zeta_\varepsilon, \bar{\zeta}_\varepsilon)\|_{[L^2(\Omega)]^4}]\, dt \\
&\quad + \tfrac{1}{2}(\|\zeta_1 - \bar{\zeta}_\varepsilon\|_{Z_0}^2 + c\|\zeta_0\|_Z^2 + c\|\zeta_1 - \bar{\zeta}_\varepsilon\|_{Z_1} \\
&\quad + R(\varepsilon(u_0) + \tfrac{1}{2}P(\zeta_0), \varepsilon(u_0) + \tfrac{1}{2}P(\zeta_0)) + h\|u_1\|_{V_0}^2).
\end{aligned}$$

To derive (7.3.58), we have used the monotonicity of $d\Phi_\varepsilon/dz$, the continuity and the coerciveness of the bilinear forms and the continuity of the linear forms. From (7.3.58) the inequality

$$\begin{aligned}
&\|\zeta_\varepsilon\|_Z^2 + \|\zeta_\varepsilon' - \bar{\zeta}_\varepsilon\|_{Z_1}^2 + \|u_\varepsilon'\|_{V_0}^2 + \|\varepsilon(u_\varepsilon) + \tfrac{1}{2}P(\zeta_\varepsilon)\|_{[L^2(\Omega)]^4}^2 \\
&\quad \le c_1\left[\int_0^t (\|\zeta_\varepsilon\|_Z^2 + \|\zeta_\varepsilon' - \bar{\zeta}_\varepsilon\|_{Z_1}^2 + \|u_\varepsilon'\|_{V_0}^2 \qquad (7.3.59) \right. \\
&\quad \left. + \|\varepsilon(u_\varepsilon) + \tfrac{1}{2}P(\zeta_\varepsilon)\|_{[L^2(\Omega)]^4}^2)\, dt\right] + c_2, \qquad c_1, c_2 \text{ const} > 0
\end{aligned}$$

results. From (7.3.59), by applying Gronwall's inequality, we obtain

$$\begin{aligned}
&\|\zeta_\varepsilon\|_Z^2 \le c, \qquad \|\zeta_\varepsilon'\|_{Z_1} \le c, \qquad \|u_\varepsilon'\|_{V_0} \le c, \\
&\|\varepsilon(u_\varepsilon) + \tfrac{1}{2}P(\zeta_\varepsilon)\|_{[L^2(\Omega)]^4} \le c.
\end{aligned} \tag{7.3.60}$$

Accordingly, the sequences $\{\zeta_\varepsilon\}$, $\{\zeta_\varepsilon'\}$, $\{u_\varepsilon\}$, $\{u_\varepsilon'\}$ and $\{\varepsilon(u_\varepsilon)\}$ are bounded in the spaces $L^\infty(0T, Z)$, $L^\infty(0T, Z_1)$, $L^\infty(0T, V_0)$, $L^\infty(0T, V_0)$ and $L^\infty(0T, [L^2(\Omega)]^4)$ respectively.

(c) From (7.3.34, 35), by differentiating (cf., Prop. 5.4.1) with respect to t we obtain the system of variational equalities

$$\begin{aligned}
&(\rho\zeta_\varepsilon''', z) + \alpha(\zeta_\varepsilon', z) + \bar{\alpha}(\zeta_\varepsilon''', z) + hR(\varepsilon(u_\varepsilon') + P(\zeta_\varepsilon, \zeta_\varepsilon'), P(\zeta_\varepsilon, z)) \\
&+ hR(\varepsilon(u) + \tfrac{1}{2}P(\zeta_\varepsilon), P(\zeta_\varepsilon', z)) + \left\langle \left(\frac{d\Phi_\varepsilon}{dz}(\zeta_\varepsilon')\right)', z \right\rangle_Z = (f_3', z), \quad \forall z \in Z \qquad (7.3.61)
\end{aligned}$$

and

$$(\rho u_\varepsilon''', v) + R(\varepsilon(u_\varepsilon') + P(\zeta_\varepsilon', \zeta_\varepsilon), \varepsilon(v)) = (f_e', v), \qquad \forall v \in V. \tag{7.3.62}$$

Because of the monotonicity of $d\Phi_\varepsilon/dz$, we have

$$\left\langle \left(\frac{d\Phi_\varepsilon}{dz}(\zeta_\varepsilon')\right)', \zeta_\varepsilon'' \right\rangle_Z = \lim_{\delta t \to 0} \frac{\left\langle \dfrac{d\Phi_\varepsilon}{dz}(\zeta_\varepsilon'(t + \delta t)) - \dfrac{d\Phi_\varepsilon}{dz}(\zeta_\varepsilon'(t)), \zeta_\varepsilon'(t + \delta t) - \zeta_\varepsilon'(t) \right\rangle_Z}{(\delta t)^2} \ge 0. \tag{7.3.63}$$

Let us put $z=\zeta''_\varepsilon$ and $v=u''_\varepsilon$ into (7.3.61–62). Then the inequality

$$\frac{1}{2}\frac{d}{dt}F(t) \leq (f'_3, \zeta''_\varepsilon) + h(f'_e, u''_\varepsilon) + hR(\varepsilon(u'_\varepsilon) + P(\zeta_\varepsilon, \zeta'_\varepsilon), P(\zeta'_\varepsilon))$$

$$- hR\left(\varepsilon(u_\varepsilon) + \tfrac{1}{2}P(\zeta_\varepsilon), \frac{1}{2}\frac{d}{dt}P(\zeta'_\varepsilon)\right), \quad (7.3.64)$$

results with

$$F(t) = \int_\Omega \rho(\zeta''_\varepsilon)^2\, d\Omega + \alpha(\zeta'_\varepsilon, \zeta'_\varepsilon) + \bar{\alpha}(\zeta''_\varepsilon, \zeta''_\varepsilon)$$

$$+ hR(\varepsilon(u'_\varepsilon) + P(\zeta_\varepsilon, \zeta'_\varepsilon), \varepsilon(u'_\varepsilon) + P(\zeta_\varepsilon, \zeta'_\varepsilon)) + h\int_\Omega \rho(u''_\varepsilon)^2\, d\Omega. \quad (7.3.65)$$

Integrating (7.3.65) from 0 to t yields, by means of (7.3.27, 60), the inequality

$$F(t) \leq 2F(0) + c\int_0^t F(t)\, dt + c\int_0^t (\| f'_3 \|_{Z_1'} + \| f'_e \|_{V_0})\, dt$$

$$\leq 2F(0) + c\int_0^t F(t)\, dt + c. \quad (7.3.66)$$

The meaning of (7.3.49) and (7.3.50) is that there exist $z_1 \in Z'_1$ and $v_1 \in V_0$ respectively such that

$$F_1(z) = \langle z_1, z\rangle_{Z_1}, \qquad \forall z \in Z \quad (7.3.67)$$

and

$$F_2(v) = (v_1, v)_{V_0}, \qquad \forall v \in V. \quad (7.3.68)$$

By means of (7.3.67), the variational equality (7.3.34) yields

$$(\rho\zeta''_\varepsilon(0), z) + \bar{\alpha}(\zeta''_\varepsilon(0), z) = -\alpha(\zeta_0, z) - hR(\varepsilon(u_0) + \tfrac{1}{2}P(\zeta_0), P(\zeta_0, z))$$

$$- \left\langle \frac{d\Phi_\varepsilon}{dz}(\zeta_1), z\right\rangle_Z + (f_3(0), z) = \langle f_3(0) + z_1, z\rangle_Z \quad (7.3.69)$$

for $t = 0$, and thus

$$\| \zeta''_\varepsilon(0) \|_{Z_1} \leq c. \quad (7.3.70)$$

Similarly, for $t = 0$ and by means of (7.3.68), (7.3.35) implies the inequality

$$\| u''_\varepsilon(0) \|_{V_0} \leq c. \quad (7.3.71)$$

Accordingly,

$$F(0) \leq c, \quad (7.3.72)$$

and thus Gronwall's inequality can be applied to (7.3.66). It results that $F(t) \leq c$, or equivalently that

$$\|\zeta_\varepsilon'\|_Z \leq c, \qquad \|\zeta_\varepsilon''\|_{Z_1} \leq c, \qquad \|u_\varepsilon''\|_{V_0} \leq c,$$
$$\|\varepsilon(u_\varepsilon')\|_{[L^2(\Omega)]^4} \leq c. \tag{7.3.73}$$

Hence the sequences $\{\zeta_\varepsilon'\}$, $\{\zeta_\varepsilon''\}$, $\{u_\varepsilon''\}$ and $\{\varepsilon(u_\varepsilon')\}$ are bounded in the spaces $L^\infty(0T, Z)$, $L^\infty(0T, Z_1)$, $L^\infty(0T, V_0)$ and $L^\infty(0T, [L^2(\Omega)]^4)$ respectively.

(d) From (7.3.60) and (7.3.84), by considering subsequences if necessary, we obtain

$$\zeta_\varepsilon \to \zeta \quad \text{and} \quad \zeta_\varepsilon' \to \zeta' \quad \text{in} \quad L^\infty(0T, Z) \quad \text{weakly}_*,$$
$$\zeta_\varepsilon'' \to \zeta'' \quad \text{in} \quad L^\infty(0T, Z_1) \quad \text{weakly}_*,$$
$$u_\varepsilon \to u,\ u_\varepsilon' \to u',\ u_\varepsilon'' \to u'' \quad \text{in} \quad L^\infty(0T, V_0) \quad \text{weakly}_*$$

and

$$\varepsilon(u_\varepsilon) \to \varepsilon(u),\ \varepsilon(u_\varepsilon') \to \varepsilon(u') \quad \text{in} \quad L^\infty(0T, [L^2(\Omega)]^4) \quad \text{weakly}_*.$$

We may also verify that the initial conditions (7.3.5–8) are satisfied, q.e.d.

The next proposition completes the existence proof.

Proposition 7.3.2. The functions u, ζ satisfying (7.3.51–54) are solutions of Problem 3.

Proof. It is easily verified by taking the limit as $\varepsilon \to 0$ that u, ζ satisfy both (7.3.33) and the initial conditions (7.3.5–8). In (7.3.34) let us replace z with $z - \zeta_\varepsilon'$, where $z \in L^2(0T, Z)$. We find that

$$A_1 \geq A_2, \tag{7.3.74}$$

where

$$A_1 = \int_0^T [(\rho\zeta_\varepsilon'', z) + \alpha(\zeta_\varepsilon, z) + \bar{\alpha}(\zeta_\varepsilon'', z) + hR(\varepsilon(u_\varepsilon) + \tfrac{1}{2}P(\zeta_\varepsilon), P(\zeta_\varepsilon, z)) + \Phi_\varepsilon(z) - (f_3, z - \zeta_\varepsilon')]\, dt \tag{7.3.75}$$

and

$$A_2 = \int_0^T \frac{1}{2}\frac{d}{dt}\left[\int_\Omega \rho(\zeta_\varepsilon')^2\, d\Omega + \alpha(\zeta_\varepsilon, \zeta_\varepsilon) + \bar{\alpha}(\zeta_\varepsilon', \zeta_\varepsilon')\right] dt + \int_0^T hR(\varepsilon(u_\varepsilon) + \tfrac{1}{2}P(\zeta_\varepsilon), P(\zeta_\varepsilon, \zeta_\varepsilon'))\, dt + \int_0^T \Phi_\varepsilon(\zeta_\varepsilon')\, dt. \tag{7.3.76}$$

Let $z = z_1 \in L^2(0T, Z)$ such that $\int_0^T \Phi(z_1)\, dt < \infty$. Setting $z = z_1$, in (7.3.74) yields $\int_0^T \Phi_\varepsilon(\zeta_\varepsilon')\, dt \leq c$. By means of (7.3.31) we find that as $\varepsilon \to 0$

$$\liminf A_2 \geq \int_0^T [(\rho\zeta'', \zeta) + \alpha(\zeta, \zeta') + \bar{\alpha}(\zeta'', \zeta')$$
$$+ hR(\varepsilon(u) + \tfrac{1}{2}P(\zeta), P(\zeta, \zeta')) + \Phi(\zeta')]\, dt \quad (7.3.77)$$

and

$$\lim A_1 = \int_0^T [(\rho\zeta'', z) + \alpha(\zeta, z) + \bar{\alpha}(\zeta'', z)$$
$$+ hR(\varepsilon(u) + \tfrac{1}{2}P(\zeta), P(\zeta, z)) + \Phi(z) - (f_3, z - \zeta')]\, dt. \quad (7.3.78)$$

(7.3.74, 77, 78) imply that

$$\int_0^T [(\rho\zeta'', z - \zeta') + \alpha(\zeta, z - \zeta') + \bar{\alpha}(\zeta'', z - \zeta') + hR(\varepsilon(u) + \tfrac{1}{2}P(\zeta), P(\zeta, z - \zeta'))$$
$$+ \Phi(z) - \Phi(\zeta') - (f_3, z - \zeta')]\, dt \geq 0, \qquad \forall z \in L^2(0T, Z). \quad (7.3.79)$$

From (7.3.79), by the same procedure as in Prop. 5.4.1, the pointwise inequality (7.3.32) results, q.e.d.

After giving the definitive variational setting of the B.V.P. through Props. 7.3.1 and 7.3.2, we readily see in what sense the solution of the variational formulation satisfies the differential equations (7.3.1, 2) and the boundary conditions of the problem. For example, assume the boundary conditions $\zeta = 0$, (7.3.12) and (7.3.14), and let the reactions $\bar{f}_3$ on Ω' be zero. It is then easily verified, as in Sec. 5.4.2, that (7.3.1) and (7.3.2) are satisfied in the sense of distributions on Ω, as are the boundary conditions (7.3.12) and (7.3.14) in the sense of traces. Moreover, because of (7.3.51–53), the initial conditions (7.3.5–8) make sense.

Note that if the variational equality (7.3.33) is replaced by the variational inequality (7.3.22), then Proposition 7.3.1 holds, with minor modifications for this problem as well. However, it is not possible to prove that the limits u, ζ satisfy the system of the initial variational inequalities [248].

7.3.3 Uniqueness Proposition

Proposition 7.3.3. On the assumptions of Prop. 7.3.1, the variational system (7.3.32, 33) with the initial conditions (7.3.5–8) admits a unique solution.

Proof. Let u_1, ζ_1 and u_2, ζ_2 be two solutions of Problem 3, and let $u_1 - u_2 = w$, and $\zeta_1 - \zeta_2 = \eta$. Then (7.3.32) and (7.3.33) imply

$$-(\rho\eta'', \eta') - \alpha(\eta, \eta') - \bar{\alpha}(\eta'', \eta') \geq hR(\varepsilon(u_1) + \tfrac{1}{2}P(\zeta_1), P(\zeta_1, \eta'))$$
$$- hR(\varepsilon(u_2) + \tfrac{1}{2}P(\zeta_2), P(\zeta_2, \eta')) \quad (7.3.80)$$

and

$$(\rho w'', v) + R(\varepsilon(w) + \tfrac{1}{2}(P(\zeta_1) - P(\zeta_2)), \varepsilon(v)) = 0, \qquad \forall v \in V \tag{7.3.81}$$

respectively. By means of the relation

$$P(\zeta_1, \eta') = P(\zeta_1, \zeta_1') - P(\zeta_2, \zeta_2') - P(\eta, \zeta_2') \tag{7.3.82}$$

and the corresponding relation for $P(\zeta_2, \eta')$, (7.3.80) implies that

$$\frac{1}{2}\frac{d}{dt}\left[\int_\Omega \rho(\eta')^2\, d\Omega + \alpha(\eta, \eta) + \bar{\alpha}(\eta', \eta')\right] + hB \le 0, \tag{7.3.83}$$

where

$$\begin{aligned} B = \frac{1}{2}\frac{d}{dt}\, &[R(\varepsilon(w) + \tfrac{1}{2}(P(\zeta_1) - P(\zeta_2)), \varepsilon(w) \\ &+ \tfrac{1}{2}(P(\zeta_1) - P(\zeta_2))) + R(\varepsilon(u_2) + \tfrac{1}{2}P(\zeta_2), P(\eta))] \\ &- R(\varepsilon(w) + \tfrac{1}{2}(P(\zeta_1) - P(\zeta_2)), \varepsilon(w')) \\ &- R(\varepsilon(w) + \tfrac{1}{2}(P(\zeta_1) - P(\zeta_2)), P(\eta, \zeta_2')) \\ &- R(\varepsilon(u_2') + \tfrac{1}{2}P'(\zeta_2), \tfrac{1}{2}P(\eta)). \end{aligned} \tag{7.3.84}$$

For $v = w'$, (7.3.81) becomes

$$(\rho w'', w') + R(\varepsilon(w) + \tfrac{1}{2}(P(\zeta_1) - P(\zeta_2)), \varepsilon(w')) = 0. \tag{7.3.85}$$

(7.3.84), when combined with (7.3.85), implies that

$$\begin{aligned} B = \frac{1}{2}\frac{d}{dt}\Bigg[&R(\varepsilon(w) + \tfrac{1}{2}(P(\zeta_1) - P(\zeta_2)), \varepsilon(w) + \tfrac{1}{2}(P(\zeta_1) - P(\zeta_2))) \\ &+ R(\varepsilon(u_2) + \tfrac{1}{2}P(\zeta_2), P(\eta)) + \int_\Omega \rho(w')^2\, d\Omega\Bigg] \\ &- R(\varepsilon(w) + \tfrac{1}{2}(P(\zeta_1) - P(\zeta_2)), P(\eta, \zeta_2')) \\ &- R(\varepsilon(u_2') + \tfrac{1}{2}P'(\zeta_2), \tfrac{1}{2}P(\eta)). \end{aligned} \tag{7.3.86}$$

Further, we use the estimates

$$\begin{aligned} |R(\varepsilon(u_2) + \tfrac{1}{2}P(\zeta_2), P(\eta))| &\le c\|\varepsilon(u_2) \\ &+ \tfrac{1}{2}P(\zeta_2)\|_{[L^2(\Omega)]^4}\|P(\eta)\|_{[L^2(\Omega)]^4} \le c\|\eta\|_Z\|\eta\|_{Z_1} \\ &\le \tfrac{1}{2}\alpha(\eta, \eta) + c\|\eta\|_{Z_1}^2 \le \tfrac{1}{4}\alpha(\eta, \eta) + c\int_0^t \bar{\alpha}(\eta', \eta')\, dt, \end{aligned} \tag{7.3.87}$$

$$\int_0^t R(\varepsilon(w) + \tfrac{1}{2}(P(\zeta_1) - P(\zeta_2)), P(\eta, \zeta_2'))\, dt$$

$$\leq \int_0^t R(\varepsilon(w) + \tfrac{1}{2}(P(\zeta_1) - P(\zeta_2)), \varepsilon(w) \tag{7.3.88}$$

$$+ \tfrac{1}{2}(P(\zeta_1) - P(\zeta_2))) \, dt + c \int_0^t \| \eta \|_Z^2 \| \zeta_2' \|_Z^2 \, dt$$

and

$$\int_0^t R(\varepsilon(u_2') + \tfrac{1}{2}P'(\zeta_2), \tfrac{1}{2}P(\eta)) \, dt \leq c \int_0^t \| \eta \|_Z^2 \, dt. \tag{7.3.89}$$

Combining (7.3.86–89) with (7.3.83), we find

$$S(t) \leq c \int_0^t S(t) \, dt, \tag{7.3.90}$$

where

$$S(t) = \int_\Omega \rho(\eta')^2 \, d\Omega + \alpha(\eta, \eta) + \bar{\alpha}(\eta', \eta') + hR(\varepsilon(w) + \tfrac{1}{2}(P(\zeta_1)$$

$$- P(\zeta_2)), \varepsilon(w) + \tfrac{1}{2}(P(\zeta_1) - P(\zeta_2))) + h \int_\Omega \rho(w')^2 \, d\Omega. \tag{7.3.91}$$

From (7.3.90, 91), it results that $\eta = \eta' = w = w' = 0$, and thus the uniqueness of the solution is proved, q.e.d.

To close this section, we would like to mention that the dynamic analysis of unilateral von Kármán plates includes as a special case the corresponding dynamic unilateral problems of Kirchhoff plates. The terms involving the influence of plate stretching on bending are omitted, and the variational formulation of the problem is reduced to only one variational inequality. After some modification, Propositions 7.3.1–3 can be specialized for this case. If, additionally, the rotatory inertia $D\Delta\zeta''$ is disregarded, then a variational inequality of the form

$$(\rho\zeta'', z - \zeta') + \alpha(\zeta, z - \zeta') + \Phi(z) - \Phi(\zeta') \geq (f_3, z - \zeta') \qquad \forall z \in Z \tag{7.3.92}$$

is obtained. Many of the results proved in [28] may be directly applied to this inequality.

Chapter 8

Variational and Hemivariational Inequalities in Linear Thermoelasticity

This chapter deals with the study of certain unilateral B.V.P.s which are formulated in linear thermoelasticity. It is assumed that on the boundary of the body under consideration subdifferential relations hold, first between temperature and the heat flux vector and second between velocity and the stress vector. After deriving the corresponding variational inequalities, we prove some propositions on the existence and uniqueness of the solution. These two problems were first studied by G. Duvaut and J. L. Lions [85]. Finally, by the method of Sec. 4.1.3, some general variational inequalities are formulated. They stem from the assumption that the linear elasticity law and the well-known Fourier's law of heat conduction are replaced by nonlinear laws described by superpotentials. This chapter closes with the study of certain hemivariational inequalities.

8.1 B.V.P.s and Their Variational Formulations

8.1.1 Classical Formulations

The configuration of the thermoelastic medium is embedded in the Euclidean space $\mathbb{R}^3$. A typical point of $\mathbb{R}^3$ is denoted by x and its coordinates in reference to a fixed Cartesian coordinate system $0x_1x_2x_3$ are denoted by $(x_i\ i=1,2,3)$. The time variable t takes values over $[0,T]\subset\mathbb{R}$. We further denote by $u=u(x,t)$ the displacement of the material point x at time t with reference to the natural state of the body, which is characterized by zero stresses and constant absolute temperature $\theta_0>0$. The density at point x of the natural state is denoted by $\rho=\rho(x)$. The open, bounded, connected subset of $\mathbb{R}^3$ occupied by the body is denoted by Ω. The boundary Γ of Ω is assumed to be regular. The behavior of a linear thermoelastic body is governed by the following constitutive equations for the stress tensor $\sigma=\{\sigma_{ij}\}$, $i,j=1,2,3$, and the specific entropy deviation $\eta-\eta_0$ (η_0 is the specific entropy of the natural state)

$$\sigma_{ij} = t_{ij} - m_{ij}(\theta - \theta_0) = C_{ijhk}\varepsilon_{hk} - m_{ij}(\theta - \theta_0), \tag{8.1.1}$$

$$\eta - \eta_0 = \frac{1}{\theta_0} c_D(\theta - \theta_0) + \frac{1}{\rho} m_{ij}\varepsilon_{ij}. \tag{8.1.2}$$

Here $\theta = \theta(x, t)$ is the absolute temperature, and $\varepsilon = \{\varepsilon_{ij}\}$ the strain tensor which is related to the displacements by

$$\varepsilon_{ij}(u) = \tfrac{1}{2}(u_{i,j} + u_{j,i}) \tag{8.1.3}$$

for a small deformation theory. $C = \{C_{ijhk}\}$, $i, j, h, k = 1, 2, 3$, is the elasticity tensor satisfying the well-known symmetry and ellipticity conditions (3.3.61a, b). $m = \{m_{ij}\}$ is the symmetric tensor of thermal expansion, and $c_D = c_D(x) > 0$ is the specific heat at zero strain of the body. $C(x)$, $m(x)$ and $c_D(x)$ are referred to the natural state of the body. The equations of motion assuming small displacements and the law of conservation of energy read

$$\rho u_i'' = \sigma_{ij,j} + f_i, \tag{8.1.4}$$

and

$$\rho\theta_0\eta' = -q_{i,i} + Q, \tag{8.1.5}$$

where $f = \{f_i\}$, $f_i = f_i(x, t)$, is the volume force vector, $q = \{q_i\}$, $q_i = q_i(x, t)$, is the heat flux vector and $Q = Q(x, t)$ is the radiant heating per unit volume. To complete the list of equations of linear thermoelasticity, we introduce Fourier's law of heat conduction. It has the form

$$q_i = -k_{ij}\theta_{,j}, \tag{8.1.6}$$

where $k = \{k_{ij}\}$, $k_{ij} = k_{ij}(x)$ is the symmetric tensor of thermal conductivity. k refers to the natural state of the body and satisfies the condition

$$k_{ij}a_i a_j \geq c a_i a_i, \qquad \forall a = \{a_i\} \in \mathbb{R}^3, \qquad c \text{ const} > 0. \tag{8.1.7}$$

Assuming that all the quantities are sufficiently smooth for (8.1.1–6), we obtain the following system of differential equations (cf., also [65]):

$$\rho u_i'' = f_i + (C_{ijhk}\varepsilon_{hk})_{,j} - (m_{ij}(\theta - \theta_0))_{,j} \quad \text{in} \quad \Omega \times (0, T), \tag{8.1.8}$$

$$\rho c_D \theta' \_ (k_{ij}\theta_{,j})_{,i} + m_{ij}\theta_0\varepsilon_{ij}' = Q \quad \text{in} \quad \Omega \times (0, T). \tag{8.1.9}$$

These last two differential equations describe the linear thermoelastic behavior of a generally nonhomogeneous and nonisotropic body. For $t = 0$ we assume the *initial conditions*

$$u_i = u_{0i} \quad \text{in} \quad \Omega, \tag{8.1.10}$$

$$u_i' = u_{1i} \quad \text{in} \quad \Omega \tag{8.1.11}$$

and

$$\theta = \bar{\theta} \quad \text{in} \quad \Omega. \tag{8.1.12}$$

Here $u_0 = u_0(x)$, $u_1 = u_1(x)$ and $\bar{\theta} = \bar{\theta}(x)$ are given functions on Ω.

The first B.V.P. which we will study results if between the boundary temperature and the heat flux vector the relation

$$q_i n_i = -k_{ij}\theta_{,j} n_i \in \beta(\theta) \quad \text{on} \quad \Gamma \times (0, T), \tag{8.1.13}$$

holds. For the displacements, we assume simply that

$$u_i = 0 \quad \text{on} \quad \Gamma \times (0, T). \tag{8.1.14}$$

Here $n = \{n_i\}$ denotes, as usual, the unit normal to Γ directed towards the exterior of Ω, and β is a maximal monotone operator. The boundary condition (8.1.13) describes a broad class of semipermeability problems (cf. Sec. 3.3.1).

The second B.V.P. which we shall study results if instead of (8.1.13, 14) we consider the boundary conditions

$$\theta = \theta_0 \quad \text{on} \quad \Gamma \times (0, T), \tag{8.1.15}$$

$$u_i = U_i \quad \text{on} \quad \Gamma_U \times (0, T), \tag{8.1.16}$$

and

$$-S = \{-S_i\} = \{-\sigma_{ij} n_i\} \in \partial\psi(u') \quad \text{on} \quad \Gamma_S \times (0, T). \tag{8.1.17}$$

Here $\Gamma = \bar{\Gamma}_U \cup \bar{\Gamma}_S$, where Γ_U and Γ_S are nonempty, disjoint, open sets, $U_i = U_i(x, t)$ is a prescribed displacement vector on Γ_U, assumed to be compatible with (8.1.10, 11), and ψ is a convex, l.s.c. and proper functional on $\mathbb{R}^3$. A more general B.V.P. results by replacing the boundary condition (8.1.15) with (8.1.13); in this case, unilateral phenomena, both kinematical and thermal, are considered.

In order to facilitate further calculations, we introduce the following notation:

$$a(u, v) = \int_\Omega C_{ijhk}\varepsilon_{ij}(u)\varepsilon_{hk}(v)\, d\Omega; \tag{8.1.18}$$

$$(u, v) = \int_\Omega u_i v_i\, d\Omega, \tag{8.1.19}$$

$$M_1(\theta, v) = \int_\Omega (m_{ij}\theta)_{,j} v_i\, d\Omega; \tag{8.1.20}$$

$$M_2(u, \phi) = \int_\Omega m_{ij} u_{i,j}\phi\, d\Omega; \tag{8.1.21}$$

$$K(\theta, \phi) = \int_\Omega k_{ij}\theta_{,j}\phi_{,i}\, d\Omega; \tag{8.1.22}$$

$$(\theta, \phi) = \int_\Omega \theta\phi \, d\Omega. \tag{8.1.23}$$

Assuming that the variations $v - u'$ and $\phi - \theta$ are sufficiently smooth, then by multiplying (8.1.8) and (8.1.9) by $v - u'$ and $\phi - \theta$ respectively, integrating over Ω, and using the Green–Gauss theorem, we obtain the variational equalities

$$(\rho u'', v - u') + a(u, v - u') + M_1(\theta - \theta_0, v - u')$$
$$= (f, v - u') + \int_\Gamma t_{ij} n_j (v_i - u_i') \, d\Gamma \quad \text{in} \quad \Omega \times (0, T) \tag{8.1.24}$$

and

$$(\rho c_D \theta', \phi - \theta) + K(\theta, \phi - \theta) + M_2(\theta_0 u', \phi - \theta)$$
$$= (\mathcal{Q}, \phi - \theta) + \int_\Gamma k_{ij} \theta_{,j} n_i (\phi - \theta) \, d\Gamma \quad \text{in} \quad \Omega \times (0, T). \tag{8.1.25}$$

8.1.2 Variational Formulations

Next assume[1] that C_{ijhk}, k_{ij}, m_{ij}, ρ and c_D are elements of $L^\infty(\Omega)$, and that $f(t) \in [L^2(\Omega)]^3$ and $\mathcal{Q}(t) \in L^2(\Omega)$. Let us introduce the spaces $[\mathring{H}^1(\Omega)]^3$ for v, u' and $H^1(\Omega)$ for ϕ, θ. Since β in (8.1.13) is a maximal monotone operator on $\mathbb{R}$, a convex, l.s.c., proper functional φ can be determined up to an additive constant (cf., Prop. 2.6.4) such that $\beta = \partial\varphi$. A convex, l.s.c., proper functional Φ is then defined on $L^2(\Gamma)$ by the relation (cf., Prop. 3.3.1)

$$\Phi(\phi) = \begin{cases} \int_\Gamma \varphi(\phi) \, d\Gamma & \text{if} \quad \varphi(\phi) \in L^1(\Gamma) \\ \infty & \text{otherwise} \end{cases} \tag{8.1.26}$$

As in Sec. 3.3.1, the restriction of Φ to $H^{1/2}(\Gamma)$ is again denoted by Φ. Further, the variational equalities (8.1.24) and (8.1.25) are combined with the boundary conditions (8.1.13) and (8.1.14). (8.1.13), for $q_i n_i$, $\theta \in L^2(\Gamma)$ (or more regular), implies the variational inequality

$$\Phi(\phi) - \Phi(\theta) \geq \int_\Gamma - k_{ij} \theta_{,j} n_i (\phi - \theta) \, d\Gamma, \qquad \forall \phi \in L^2(\Gamma), \tag{8.1.27}$$

which, together with (8.1.24, 25), allows us to define the following variational problem.

Problem 1. Find functions u: $[0, T] \to [\mathring{H}^1(\Omega)]^3$ and θ: $[0, T] \to H^1(\Omega)$, with $u'(t) \in [\mathring{H}^1(\Omega)]^3$, $u''(t) \in [L^2(\Omega)]^3$, $\theta'(t) \in L^2(\Omega)$, which satisfy the variational equality

[1] Moreover, let ess inf $c_D(x) > 0$, ess inf $\rho(x) > 0$.

$$(\rho u'', v - u') + a(u, v - u') + M_1(\theta - \theta_0, v - u')$$
$$= (f, v - u'), \qquad \forall v \in [\mathring{H}^1(\Omega)]^3 \quad (8.1.28)$$

the variational inequality

$$\begin{aligned}(\rho c_D \theta', \phi - \theta) + K(\theta, \phi - \theta) + M_2(\theta_0 u', \phi - \theta)\\ + \Phi(\phi) - \Phi(\theta) \geq (\mathfrak{Q}, \phi - \theta), \qquad \forall \phi \in H^1(\Omega)\end{aligned} \quad (8.1.29)$$

and the initial conditions (8.1.10–12).

In Sec. 8.2 we shall give an existence and uniqueness result after assuming a more precise dependence of the data on time. Moreover, we will discuss there in what sense the solution of Problem 1 satisfies (8.1.8–14). Let us go on to consider the B.V.P. which is defined by the relations (8.1.8–12) and (8.1.15–17).

Let $v, u' \in [H^1(\Omega)]^3$ be such that $v = u' = U'(t)$ on Γ_U and $\phi, \theta \in H^1(\Omega)$ with $\phi = \theta = \theta_0$ on Γ. A convex, l.s.c. and proper functional Ψ is defined on $[L^2(\Gamma)]^3$ by the relation

$$\Psi(v) = \begin{cases} \displaystyle\int_{\Gamma_S} \psi(v)\, d\Gamma & \text{if } \psi(v) \in L^1(\Gamma). \\ \infty & \text{otherwise} \end{cases} \quad (8.1.30)$$

We keep the same notation for the restriction of Ψ to $[H^{1/2}(\Gamma)]^3$. Arguing as before, by using the given boundary conditions, we obtain from (8.1.24, 25) the relations

$$(\rho u'', v - u') + a(u, v - u') + M_1(\theta - \theta_0, v - u') + \Psi(v)$$
$$- \Psi(u') \geq (f, v - u'), \qquad \forall v \in [H^1(\Omega)]^3 \text{ with } v = U'(t) \text{ on } \Gamma_U \quad (8.1.31)$$

and

$$(\rho c_D \theta', \phi - \theta) + K(\theta, \phi - \theta) + M_2(\theta_0 u', \phi - \theta)$$
$$= (\mathfrak{Q}, \phi - \theta), \qquad \forall \phi \in H^1(\Omega), \text{ with } \phi = \theta_0 \text{ on } \Gamma. \quad (8.1.32)$$

Let us consider a function w: $[0, T] \to [H^1(\Omega)]^3$ such that $w(t) = U(t)$ on Γ_U and $U_i(t) \in \mathring{H}^{1/2}(\Gamma_U)$ (cf., Sec. 5.1.2), and make the substitutions $\tilde{u}' = u' - w'$, $\tilde{v} = v - w'$ (resp. $\tilde{\theta} = \theta - \theta_0$, $\tilde{\phi} = \phi - \phi_0$). For the sake of simplicity, we omit the tildes in the sequel. Thus we obtain from (8.1.31, 32) the variational system

$$(\rho u'', v - u') + a(u, v - u) + M_1(\theta, v - u') + \Psi(v + w')$$
$$- \Psi(u' + w') \geq (\bar{f}, v - u'), \qquad \forall v \in V_0, \quad (8.1.33)$$

$$(\rho c_D \theta', \phi - \theta) + K(\theta, \phi - \theta) + M_2(\theta_0 u', \phi - \theta)$$
$$= (\bar{\mathfrak{Q}}, \phi - \theta), \qquad \forall \phi \in \mathring{H}^1(\Omega), \quad (8.1.34)$$

where $V_0 = \{v \mid v \in [H^1(\Omega)]^3, v = 0 \text{ on } \Gamma_U\}$ and $\bar{f}$, $\bar{\mathfrak{Q}}$ are given by

$$(\bar{f}, v) = (f, v) - a(w, v) - (\rho w'', v) \tag{8.1.35}$$

and

$$(\bar{\mathfrak{Q}}, \phi) = (\mathfrak{Q}, \phi) - M_2(\theta_0 w', \phi). \tag{8.1.36}$$

The initial conditions (8.1.10–12) take the form

$$u_i = u_{0i} - w_i(0), \qquad u_i' = u_{1i} - w_i'(0), \qquad \theta = \bar{\theta} - \theta_0 . \tag{8.1.37}$$

Thus we may consider the following variational problem.

Problem 2. Find functions u: $[0, T] \to V_0$ and θ: $[0, T] \to \mathring{H}^1(\Omega)$, with $u'(t) \in V_0$, $u''(t) \in [L^2(\Omega)]^3$, $\theta'(t) \in L^2(\Omega)$, which satisfy (8.1.33, 34) and the initial conditions (8.1.37).

Further, the B.V.P. defined by (8.1.8–13) and (8.1.16–17) will be considered. Arguing as for the preceding problems, and by means of the substitutions used in Problem 2, we obtain the following variational formulation (retaining the symbols u, v instead of $\tilde{u}$, $\tilde{v}$).

Problem 3. Find u: $[0, T] \to V_0$ and θ: $[0, T] \to H^1(\Omega)$, with $u'(t) \in V_0$, $u''(t) \in [L^2(\Omega)]^3$, $\theta'(t) \in L^2(\Omega)$, which satisfy the variational inequalities

$$(\rho u'', v - u') + a(u, v - u') + M_1(\theta - \theta_0, v - u') +$$
$$+ \Psi(v + w') - \Psi(u' + w') \geq (\bar{f}, v - u') + \int_{\Gamma_S} m_{ij}(\theta - \theta_0)(v_i - u_i') n_j \, d\Gamma, \ \forall v \in V_0 \tag{8.1.38}$$

$$(\rho c_D \theta', \phi - \theta) + K(\theta, \phi - \theta) + M_2(\theta_0 u', \phi - \theta) + \Phi(\phi)$$
$$- \Phi(\theta) \geq (\bar{\mathfrak{Q}}, \phi - \theta), \qquad \forall \phi \in H^1(\Omega), \tag{8.1.39}$$

and the initial conditions (8.1.37) (the first two) and (8.1.12).

In the next section, the existence and uniqueness of the solution are discussed.

8.2 Existence and Uniqueness Propositions

8.2.1 Study of Problem 1

In order to give an existence proof for Problem 1, we assume that there exists a family of differentiable functionals Φ_ε such that

$$\text{(i)} \int_0^T \Phi_\varepsilon(\phi)\, dt \to \int_0^T \Phi(\phi)\, dt \text{ as } \varepsilon \to 0, \qquad \forall \phi \in L^2(0T, H^1(\Omega)), \tag{8.2.1}$$

(ii) there exists a bounded sequence $\{\bar{\phi}_\varepsilon\}$ in $H^1(\Omega)$ with the property that for every $\varepsilon > 0$

$$\frac{d\Phi_\varepsilon}{d\phi}(\bar{\phi}_\varepsilon) = 0, \tag{8.2.2}$$

and

(iii) if for $\varepsilon \to 0$, $\phi_\varepsilon \to \phi$, $\phi'_\varepsilon \to \phi'$ in $L^2(0\ T, H^1(\Omega))$ weakly and the integral $\int_0^T \Phi_\varepsilon(\phi_\varepsilon)\, dt$ is bounded from above, then

$$\liminf \int_0^T \Phi_\varepsilon(\phi_\varepsilon)\, dt \geq \int_0^T \Phi(\phi)\, dt. \tag{8.2.3}$$

By means of $\{\Phi_\varepsilon\}$, the regularized problem can now be formulated (cf., Sec. 5.4.3).

Problem 1ε. Find u_ε: $[0, T] \to [\mathring{H}^1(\Omega)]^3$, with $u'_\varepsilon(t) \in [\mathring{H}^1(\Omega)]^3$, $u''_\varepsilon(t) \in [L^2(\Omega)]^3$, and θ_ε: $[0, T] \to H^1(\Omega)$, with $\theta'_\varepsilon(t) \in L^2(\Omega)$, to solve the variational system

$$(\rho u''_\varepsilon, v) + a(u_\varepsilon, v) + M_1(\theta_\varepsilon - \theta_0, v) = (f, v), \qquad \forall v \in [\mathring{H}^1(\Omega)]^3, \tag{8.2.4}$$

$$(\rho\, c_D \theta'_\varepsilon, \phi) + K(\theta_\varepsilon, \phi) + M_2(\theta_0 u'_\varepsilon, \phi)$$

$$+ \left\langle \frac{d\Phi_\varepsilon}{d\theta}(\theta_\varepsilon), \phi \right\rangle_\Gamma = (\mathcal{Q}, \phi),\ \forall \phi \in H^1(\Omega), \tag{8.2.5}$$

with the initial conditions

$$u_\varepsilon = u_0, \qquad u'_\varepsilon = u_1, \qquad \theta_\varepsilon = \bar{\theta}. \tag{8.2.6}$$

For the definitive variational formulation of the B.V.P., let us assume that

$$f, f' \in L^2(0\ T, L^2(\Omega)), \tag{8.2.7}$$

$$\mathcal{Q}, \mathcal{Q}' \in L^2(0\ T, [L^2(\Omega)]^3), \tag{8.2.8}$$

$$u_0 \in [H^2(\Omega) \cap \mathring{H}^1(\Omega)]^3, \tag{8.2.9}$$

$$u_1 \in [\mathring{H}^1(\Omega)]^3 \tag{8.2.10}$$

and

$$\bar{\theta} \in H^1(\Omega), \qquad \Phi(\bar{\theta}) < \infty. \tag{8.2.11}$$

The following proposition can now be proved.

Proposition 8.2.1. Assume that (8.2.1–3) and (8.2.7–11) hold, that

$$m_{ij,\,i} \in L^\infty(\Omega), \qquad i, j = 1, 2, 3, \tag{8.2.12}$$

and that

$$F(\phi) = K(\bar{\theta}, \phi) + \left\langle \frac{d\Phi_\varepsilon(\bar{\theta})}{d\theta}, \phi \right\rangle_\Gamma + M_2(\theta_0 u_1, \phi) \tag{8.2.13}$$

is a continuous linear functional with respect to the $L^2(\Omega)$-topology on $H^1(\Omega)$. Then, as $\varepsilon \to 0$

$$u_\varepsilon \to u \text{ weakly}_* \text{ in } L^\infty(0\,T, [\mathring{H}^1(\Omega)]^3), \tag{8.2.14}$$

$$u'_\varepsilon \to u' \text{ weakly}_* \text{ in } L^\infty(0\,T, [\mathring{H}^1(\Omega)]^3), \tag{8.2.15}$$

$$u''_\varepsilon \to u'' \text{ weakly}_* \text{ in } L^\infty(0\,T, [L^2(\Omega)]^3) \tag{8.2.16}$$

and

$$\begin{aligned} \theta_\varepsilon \to \theta, \qquad \theta'_\varepsilon \to \theta' &\text{ weakly in } L^2(0\,T, \cdot H^1(\Omega)) \\ &\text{and weakly}_* \text{ in } L^\infty(0T, L^2(\Omega)) \end{aligned} \tag{8.2.17}$$

The pair $\{u, \theta\}$ thus obtained is a solution of Problem 1.

Proof. The regularized problem under the previous assumptions admits a solution $\{u_\varepsilon, \theta_\varepsilon\}$ such that

$$u_\varepsilon, u'_\varepsilon \in L^\infty(0T, [\mathring{H}^1(\Omega)]^3),$$
$$u''_\varepsilon \in L^2(0T, [L^2(\Omega)]^3),$$
$$\theta_\varepsilon, \theta'_\varepsilon \in L^2(0\,T, H^1(\Omega)) \cap L^\infty(0\,T, L^2(\Omega)).$$

This can be shown by the Galerkin approximation, as was done for Prop. 5.4.2. We put $v = u'_\varepsilon$ in (8.2.4) and $\phi = \theta_\varepsilon - \bar{\phi}_\varepsilon$ in (8.2.5) and we obtain

$$(\rho u''_\varepsilon, u'_\varepsilon) + a(u_\varepsilon, u'_\varepsilon) + M_1(\theta_\varepsilon - \theta_0, u'_\varepsilon) = (f, u'_\varepsilon) \tag{8.2.18}$$

and

$$\begin{aligned} &\left(\frac{\rho\, c_D\, \theta'_\varepsilon}{\theta_0}, \theta_\varepsilon - \bar{\phi}_\varepsilon\right) + \frac{1}{\theta_0} K(\theta_\varepsilon, \theta_\varepsilon - \bar{\phi}_\varepsilon) + M_2(u'_\varepsilon, \theta_\varepsilon - \bar{\phi}_\varepsilon) \\ &\qquad + \frac{1}{\theta_0} \left\langle \frac{d\Phi_\varepsilon}{d\theta}(\theta_\varepsilon), \theta_\varepsilon - \bar{\phi}_\varepsilon \right\rangle_\Gamma = \left(\frac{\mathcal{Q}}{\theta_0}, \theta_\varepsilon - \bar{\phi}_\varepsilon\right). \end{aligned} \tag{8.2.19}$$

By addition and by means of (8.2.2), the relation

$$\begin{aligned} &\frac{1}{2}\frac{d}{dt}\left[\int_\Omega \rho(u'_\varepsilon)^2\, d\Omega + a(u_\varepsilon, u_\varepsilon) + \int_\Omega \frac{\rho\, c_D}{\theta_0}(\theta_\varepsilon - \bar{\phi}_\varepsilon)^2\, d\Omega\right] \\ &\qquad + M_1(\theta_\varepsilon - \theta_0, u'_\varepsilon) + M_2(u'_\varepsilon, \theta_\varepsilon - \bar{\phi}_\varepsilon) + \frac{1}{\theta_0} K(\theta_\varepsilon, \theta_\varepsilon - \bar{\phi}_\varepsilon) \\ &\qquad + \frac{1}{\theta_0}\left\langle \frac{d\Phi_\varepsilon}{d\theta}(\theta_\varepsilon) - \frac{d\Phi_\varepsilon}{d\theta}(\bar{\phi}_\varepsilon), \theta_\varepsilon - \bar{\phi}_\varepsilon \right\rangle_\Gamma = (f, u'_\varepsilon) + \left(\frac{\mathcal{Q}}{\theta_0}, \theta_\varepsilon - \bar{\phi}_\varepsilon\right) \end{aligned} \tag{8.2.20}$$

results. But since $u'_\varepsilon(t) = 0$ on Γ, we find, by means of the Green–Gauss theorem, that

$$M_1(\theta_\varepsilon - \theta_0, u'_\varepsilon) + M_2(u'_\varepsilon, \theta_\varepsilon - \bar{\phi}_\varepsilon) = -M_1(\theta_0, u'_\varepsilon) - M_2(u'_\varepsilon, \bar{\phi}_\varepsilon). \tag{8.2.21}$$

Using the monotonicity inequality

$$\left\langle \frac{d\Phi_\varepsilon}{d\theta}(\theta_\varepsilon) - \frac{d\Phi_\varepsilon}{d\theta}(\bar{\phi}_\varepsilon), \theta_\varepsilon - \bar{\phi}_\varepsilon \right\rangle_\Gamma \geq 0, \tag{8.2.22}$$

the coerciveness of $a(v, v)$ and of $K(\phi, \phi) + |\phi|_2^2$ and the continuity of the linear forms, from (8.2.20, 21) we find

$$[|u_\varepsilon'|_2^2 + \|u_\varepsilon\|_1^2 + |\theta_\varepsilon - \bar{\phi}_\varepsilon|_2^2] + c\int_0^t \|\theta_\varepsilon\|_1^2\,dt$$

$$\leq c_1 \int_0^t (|u_\varepsilon'|_2^2 + \|u_\varepsilon\|_1^2 + |\theta_\varepsilon - \bar{\phi}_\varepsilon|_2^2)\,dt + c_2. \tag{8.2.23}$$

Here c, c_1, c_2 are various positive constants depending on the data of the problem (θ_0, ess $\inf_\Omega \rho$, ess $\sup_\Omega \rho$, etc.), and $\|\cdot\|_1$ in this section denotes the norm in both $[\mathring{H}^1(\Omega)]^3$ and $H^1(\Omega)$. (8.2.23) implies by Gronwall's inequality that

$$|u_\varepsilon'|_2 \leq c, \qquad \|u_\varepsilon\|_1 \leq c, \qquad |\theta_\varepsilon - \bar{\phi}_\varepsilon|_2 \leq c, \tag{8.2.24}$$

and subsequently, if we take $t = T$, that

$$\int_0^T \|\theta_\varepsilon\|_1^2\,dt \leq c. \tag{8.2.25}$$

We now put $t = 0$ in (8.2.4, 5). It results that

$$(\rho u_\varepsilon''(0), v) + a(u_0, v) + M_1(\bar{\theta} - \theta_0, v)$$

$$= (f(0), v), \qquad \forall v \in [\mathring{H}^1(\Omega)]^3, \tag{8.2.26}$$

from which

$$|u_\varepsilon''(0)|_2 \leq c. \tag{8.2.27}$$

Similarly, from (8.2.5)

$$(\rho c_D \theta_\varepsilon'(0), \phi) + K(\bar{\theta}, \phi) + M_2(\theta_0 u_1, \phi)$$

$$+ \left\langle \frac{d\Phi_\varepsilon}{d\theta}(\bar{\theta}), \phi \right\rangle_\Gamma = (\mathfrak{Q}(0), \phi), \qquad \forall \phi \in H^1(\Omega). \tag{8.2.28}$$

Due to (8.2.13), there exists $\tau \in L^2(\Omega)$ such that for every $\phi \in H^1(\Omega)$

$$F(\phi) = (\tau, \phi), \tag{8.2.29}$$

and thus it can be easily shown that

$$|\theta_\varepsilon'(0)|_2 \leq c. \tag{8.2.30}$$

Differentiating (8.2.4, 5) with respect to t (cf., the proof of Prop. 5.4.1), we obtain

$$(\rho u_\varepsilon''', v) + a(u_\varepsilon', v) + M_1(\theta_\varepsilon', v) = (f', v), \qquad \forall v \in [\mathring{H}^1(\Omega)]^3 \quad (8.2.31)$$

and

$$\left(\frac{\rho c_D \theta_\varepsilon''}{\theta_0}, \phi\right) + \frac{1}{\theta_0} K(\theta_\varepsilon', \phi) + M_2(u_\varepsilon'', \phi)$$
$$+ \frac{1}{\theta_0}\left\langle \left(\frac{d\Phi_\varepsilon}{d\theta}(\theta_\varepsilon)\right)', \phi \right\rangle_\Gamma = \left(\frac{\mathcal{Q}'}{\theta_0}, \phi\right), \qquad \forall \phi \in H^1(\Omega). \quad (8.2.32)$$

We set $v = u_\varepsilon''$ in (8.2.31) and $\phi = \theta_\varepsilon'$ in (8.2.32). Because of the monotonicity of $d\Phi_\varepsilon/d\theta$, the inequality

$$\left\langle \left(\frac{d\Phi_\varepsilon}{d\theta}(\theta_\varepsilon)\right)', \theta_\varepsilon' \right\rangle_\Gamma$$
$$= \lim_{\delta t \to 0} \left[\frac{\left\langle \frac{d\Phi_\varepsilon}{d\theta}(\theta_\varepsilon(t + \delta t)) - \frac{d\Phi_\varepsilon}{d\theta}(\theta_\varepsilon(t)), \theta_\varepsilon(t + \delta t) - \theta_\varepsilon(t) \right\rangle}{(\delta t)^2} \right] \geq 0 \quad (8.2.33)$$

is easily verified. Thus (8.2.31, 32) imply that

$$\frac{1}{2}\frac{d}{dt}\left[\int_\Omega \rho (u_\varepsilon'')^2\, d\Omega + a(u_\varepsilon', u_\varepsilon') + \int_\Omega \frac{\rho c_D}{\theta_0}(\theta_\varepsilon')^2\, d\Omega\right]$$
$$+ M_1(\theta_\varepsilon', u_\varepsilon'') + M_2(u_\varepsilon'', \theta_\varepsilon')$$
$$+ \frac{1}{\theta_0} K(\theta_\varepsilon', \theta_\varepsilon') \leq (f', u_\varepsilon'') + \left(\frac{\mathcal{Q}'}{\theta_0}, \theta_\varepsilon'\right). \quad (8.2.34)$$

From (8.2.34), by applying the Green–Gauss theorem to $M_1(\theta_\varepsilon', u_\varepsilon'') + M_2(u_\varepsilon'', \theta_\varepsilon')$ and by integrating from 0 to t, we deduce the relation

$$c_1[|u_\varepsilon''|_2^2 + \|u_\varepsilon'\|_1^2 + |\theta_\varepsilon'|_2^2] + \int_0^t \frac{1}{\theta_0} K(\theta_\varepsilon', \theta_\varepsilon')\, dt$$
$$\leq c_2 \int_0^t [|f'|_2 |u_\varepsilon''|_2 + \frac{1}{\theta_0}|\mathcal{Q}'|_2 |\theta_\varepsilon'|_2]\, dt \quad (8.2.35)$$
$$+ c_1(|u_\varepsilon''(0)|_2^2 + \|u_1\|_1^2 + |\theta_\varepsilon'(0)|_2^2),$$

where the c_i terms are positive constants. From (8.2.27) and (8.2.30), by using Gronwall's inequality and the positive semidefiniteness of $K(\phi, \phi)$, we obtain from (8.2.35) the relations

$$|u_\varepsilon''|_2 \leq c, \qquad \|u_\varepsilon'\|_1 < c, \qquad |\theta_\varepsilon'|_2 \leq c. \quad (8.2.36)$$

If we set $t = T$ in (8.2.35),

$$\int_0^T \|\theta_\varepsilon'\|_1^2\, dt \leq c \quad (8.2.37)$$

results. Because of (8.2.24, 25, 36, 37), we can extract subsequences again denoted by u_ε, u'_ε, u''_ε and θ_ε, θ'_ε, such that for $\varepsilon \to 0$

$$u_\varepsilon \to u, \quad u'_\varepsilon \to u' \quad (\text{resp. } u''_\varepsilon \to u'') \quad \text{weakly}_* \text{ in } \quad L^\infty(0T, [\mathring{H}^1(\Omega)]^3) \tag{8.2.38}$$

$$(\text{resp. } L^\infty(0\,T, [L^2(\Omega)]^3))$$

and

$$\theta_\varepsilon \to \theta, \qquad \theta'_\varepsilon \to \theta' \quad \text{weakly in} \quad L^2(0T, H^1(\Omega)) \quad \text{and} \quad \text{weakly}_* \text{ in} \quad L^\infty(0T, L^2(\Omega)). \tag{8.2.39}$$

It now remains to be proved that the obtained functions u, θ satisfy the variational inequalities (8.1.28) and (8.1.29) and the initial conditions. From (8.2.45) it follows that

$$(\rho u''_\varepsilon, v - u'_\varepsilon) + a(u_\varepsilon, v - u'_\varepsilon) + M_1(\theta_\varepsilon - \theta_0, v - u'_\varepsilon) = (f, v - u'_\varepsilon), \qquad \forall v \in \mathring{H}^1(\Omega) \tag{8.2.40}$$

and

$$\begin{aligned}(\rho c_D \theta'_\varepsilon, \phi - \theta_\varepsilon) &+ K(\theta_\varepsilon, \phi - \theta_\varepsilon) + M_2(\theta_0 u'_\varepsilon, \phi - \theta_\varepsilon) \\ &+ \Phi_\varepsilon(\phi) - \Phi_\varepsilon(\theta_\varepsilon) - (\mathcal{Q}, \phi - \theta_\varepsilon) = \Phi_\varepsilon(\phi) - \Phi_\varepsilon(\theta_\varepsilon) \\ &- \left\langle \frac{d\Phi_\varepsilon}{d\theta}(\theta_\varepsilon), \phi - \theta_\varepsilon \right\rangle_\Gamma \geq 0, \qquad \forall \phi \in H^1(\Omega)\end{aligned} \tag{8.2.41}$$

respectively. Let $\phi = \phi_1 \in L^2(0T, H^1(\Omega))$ such that $\int_0^T \Phi(\phi_1)\,dt < \infty$. Setting $\phi = \phi_1$ in (8.2.41), we conclude by means of (8.2.1) and the previously proven estimates that

$$\begin{aligned}\int_0^T \Phi_\varepsilon(\theta_\varepsilon)\,dt \leq \int_0^T &[(\rho c_D \theta'_\varepsilon, \phi_1 - \theta_\varepsilon) + K(\theta_\varepsilon, \phi_1 - \theta_\varepsilon) \\ &+ M_2(\theta_0 u'_\varepsilon, \phi_1 - \theta_\varepsilon) - (\mathcal{Q}, \phi_1 - \theta_\varepsilon) + \Phi_\varepsilon(\phi_1)]\,dt \leq c.\end{aligned} \tag{8.2.42}$$

Hence, according to (8.2.3), the relation

$$\liminf \int_0^T \Phi_\varepsilon(\theta_\varepsilon)\,dt \geq \int_0^T \Phi(\theta)\,dt \tag{8.2.43}$$

holds as $\varepsilon \to 0$.

Further, let us take $v = v(t)$ in (8.2.40) and $\phi = \phi(t)$ in (8.2.41), v and ϕ being arbitrary functions of $L^2(0\,T, [\mathring{H}^1(\Omega)]^3)$ and $L^2(0\,T, H^1(\Omega))$. From (8.2.40) it results that

$$\int_0^T [(\rho u''_\varepsilon, v) + a(u_\varepsilon, v) + M_1(\theta_\varepsilon - \theta_0, v) - (f, v - u'_\varepsilon)]\,dt$$

$$= \int_0^T [(\rho u_\varepsilon'', u_\varepsilon') + a(u_\varepsilon, u_\varepsilon') + M_1(\theta_\varepsilon - \theta_0, u_\varepsilon')]\, dt$$

$$= \int_\Omega \frac{\rho}{2} (u_\varepsilon'(T))^2\, d\Omega - \int_\Omega \frac{\rho}{2} (u_1)^2\, d\Omega$$

$$+ \int_0^T [a(u_\varepsilon, u_\varepsilon') + M_1(\theta_\varepsilon - \theta_0, u_\varepsilon')]\, dt. \tag{8.2.44}$$

For $\varepsilon \to 0$, noting that u and θ satisfy the initial conditions, we deduce the relation

$$\int_0^T [(\rho u'', v - u') + a(u, v - u') + M_1(\theta - \theta_0, v - u') - (f, v - u')]\, dt = 0, \qquad \forall v \in L^2(0T, \mathring{H}^1(\Omega)). \tag{8.2.45}$$

Analogously, from (8.2.41) the relation

$$\int_0^T [(\rho c_D \theta_\varepsilon', \phi) + K(\theta_\varepsilon, \phi) + M_2(\theta_0 u_\varepsilon', \phi) + \Phi_\varepsilon(\phi) - (\mathfrak{Q}, \phi - \theta_\varepsilon)]\, dt$$

$$\geq \int_0^T [(\rho c_D \theta_\varepsilon', \theta_\varepsilon) + K(\theta_\varepsilon, \theta_\varepsilon) + M_2(\theta_0 u_\varepsilon', \theta_\varepsilon) + \Phi_\varepsilon(\theta_\varepsilon)]\, dt \tag{8.2.46}$$

$$= \int_\Omega \frac{\rho}{2} c_D(\theta_\varepsilon(T))^2\, d\Omega - \int_\Omega \frac{\rho}{2} c_D(\bar{\theta})^2\, d\Omega$$

$$+ \int_0^T [K(\theta_\varepsilon, \theta_\varepsilon) + M_2(\theta_0 u_\varepsilon', \theta_\varepsilon) + \Phi_\varepsilon(\theta_\varepsilon)]\, dt$$

is obtained. We denote the right-hand side of (8.2.46) by G_ε. For $\varepsilon \to 0$ we can easily verify, by means of (8.2.43), that

$$\liminf G_\varepsilon \geq \int_0^T [(\rho c_D \theta', \theta') + K(\theta, \theta) + M_2(\theta_0 u', \theta) + \Phi(\theta)]\, dt. \tag{8.2.47}$$

Thus, by taking the limit in (8.2.46), we derive the variational inequality

$$\int_0^T [(\rho c_D \theta', \phi - \theta) + K(\theta, \phi - \theta) + M_2(\theta_0 u', \phi - \theta) + \Phi(\phi) - \Phi(\theta) - (\mathfrak{Q}, \phi - \theta)]dt \geq 0, \qquad \forall \phi \in L^2(0T, H^1(\Omega)). \tag{8.2.48}$$

From (8.2.45) and (8.2.48) we obtain the pointwise variational inequalities (8.1.28) and (8.1.29) by using the same procedure as in the proof of Prop. 5.4.1, q.e.d.

It now remains for us to discuss how the previously found functions u, θ solve the initial B.V.P. By taking $v - u' \in [D(\Omega)]^3$ in (8.1.28), we find

(8.1.8) in the sense of distributions on Ω, provided (8.1.3) is taken into account. Similarly, from (8.1.29) we obtain, for $\phi - \theta \in D(\Omega)$, relation (8.1.9) in the sense of distributions on Ω. Since $k_{ij}\theta_{,j}$ and $(k_{ij}\theta_{,j})_{,i} \in L_2(\Omega)$, $k_{ij}\theta_{,j}n_i$ can be uniquely defined as an element of $H^{-1/2}(\Gamma)$. Then the Green–Gauss theorem can be applied (cf., (1.4.37)). As in the previous chapters, it results that

$$-k_{ij}\theta_{,j}n_i \in \partial\Phi(\theta) \tag{8.2.49}$$

on $H^{1/2}(\Gamma) \times H^{-1/2}(\Gamma)$, which consititutes a weak form of the boundary condition (8.1.13). Since $u \in L^{\infty}(0T, [\mathring{H}^1(\Omega)]^3)$ and $u' \in L^{\infty}(0T, [\mathring{H}^1(\Omega)]^3)$, it follows (Sec. 1.4.1) that $u \in C(0T, [\mathring{H}^1(\Omega)]^3)$, and thus the initial condition (8.1.10) makes sense. We can proceed similarly for the other initial conditions (8.1.11, 12). Next, the uniqueness of the solution of Problem 1 is proved.

Proposition 8.2.2. The solution $\{u, \theta\}$ of Problem 1 is unique.

Proof. Assume that the pairs $\{\bar{u}, \bar{\theta}\}$ and $\{\bar{\bar{u}}, \bar{\bar{\theta}}\}$ are both solutions to Problem 1. Then from (8.1.28, 29), by substituting $v = \bar{u}'$, $u = \bar{\bar{u}}$, $\phi = \bar{\theta}$, $\theta = \bar{\bar{\theta}}$ and $v = \bar{\bar{u}}'$, $u = \bar{u}$, $\phi = \bar{\bar{\theta}}$, $\theta = \bar{\theta}$ and adding, we obtain

$$-(\rho(\bar{\bar{u}}'' - \bar{u}''), (\bar{\bar{u}}' - \bar{u}')) - a(\bar{u} - \bar{\bar{u}}, \bar{u}' - \bar{\bar{u}}') - M_1(\bar{\theta} - \bar{\bar{\theta}}, \bar{u}' - \bar{\bar{u}}') = 0,$$

$$-\left(\frac{\rho c_D}{\theta_0}(\bar{\theta}' - \bar{\bar{\theta}}'), \bar{\theta} - \bar{\bar{\theta}}\right) - \frac{1}{\theta_0}K(\bar{\theta} - \bar{\bar{\theta}}, \bar{\theta} - \bar{\bar{\theta}}) - M_2(\bar{u}' - \bar{\bar{u}}', \bar{\theta} - \bar{\bar{\theta}}) \geq 0. \tag{8.2.50}$$

It follows that

$$\frac{1}{2}\frac{d}{dt}\left[\int_\Omega \rho(\bar{\bar{u}}' - \bar{u}')^2\, d\Omega + a(\bar{u} - \bar{\bar{u}}, \bar{u} - \bar{\bar{u}}) + \int_\Omega \frac{\rho c_D}{\theta_0}(\bar{\theta} - \bar{\bar{\theta}})^2\, d\Omega\right]$$
$$+\frac{1}{\theta_0}K(\bar{\theta} - \bar{\bar{\theta}}, \bar{\theta} - \bar{\bar{\theta}}) \leq 0, \tag{8.2.51}$$

and thus

$$c(|\bar{u}' - \bar{\bar{u}}'|_2^2 + \|\bar{u} - \bar{\bar{u}}\|_1^2 + |\bar{\theta} - \bar{\bar{\theta}}|_2^2) + \frac{1}{\theta_0}\int_0^t K(\bar{\theta} - \bar{\bar{\theta}}, \bar{\theta} - \bar{\bar{\theta}})\, dt \leq 0 \tag{8.2.52}$$

From the last relation it results that $\bar{u} = \bar{\bar{u}}$, $\bar{\theta} = \bar{\bar{\theta}}$, q.e.d.

8.2.2 Study of Problem 2. Some Remarks on Related Problems

We assume that Ψ has the same properties as Φ (see (8.2.1–3)):

$$\text{(i')} \int_0^T \Psi_\varepsilon(v + w')\,dt \to \int_0^T \Psi(v + w')\,dt \text{ as } \varepsilon \to 0,\ \forall v \in L^2(0T, V_0); \tag{8.2.53}$$

(ii') there exists a bounded sequence $\{\bar{v}_\varepsilon\}$ in V_0 such that

$$\frac{d\Psi_\varepsilon}{dv}(\bar{v}_\varepsilon) = 0 \tag{8.2.54}$$

for every $\varepsilon > 0$; and

(iii') if $v_\varepsilon \to v$, $v'_\varepsilon \to v'$ in $L^\infty(0T, V_0)$ weakly$_*$, and if

$$\int_0^T \Psi_\varepsilon(v_\varepsilon + w')\,dt \text{ is bounded from above, then} \tag{8.2.55}$$

$$\liminf \int_0^T \Psi_\varepsilon(v'_\varepsilon + w')\,dt \geq \int_0^T \Psi(v' + w')\,dt. \tag{8.2.56}$$

By means of Ψ_ε we define the regularized problem.

Problem 2ε. Find a solution u_ε: $[0, T] \to V_0$, θ_ε: $[0, T] \to \mathring{H}^1(\Omega)$, with $u'_\varepsilon(t) \in V_0$, $u''_\varepsilon(t) \in [L^2(\Omega)]^3$ and $\theta'_\varepsilon(t) \in L^2(\Omega)$, of the variational system

$$(\rho u''_\varepsilon, v) + a(u_\varepsilon, v) + M_1(\theta_\varepsilon, v) + \left\langle \frac{d\Psi_\varepsilon}{dv}(u'_\varepsilon + w'), v \right\rangle_\Gamma = (\bar{f}, v), \qquad \forall v \in V_0, \tag{8.2.57}$$

$$(\rho c_D \theta'_\varepsilon, \phi) + K(\theta_\varepsilon, \phi) + M_2(\theta_0 u'_\varepsilon, \phi) = (\bar{\mathfrak{Q}}, \phi) \qquad \forall \phi \in \mathring{H}^1(\Omega) \tag{8.2.58}$$

such that for $t = 0$

$$u_\varepsilon = u_0 - w(0), \qquad u'_\varepsilon = u_1 - w'(0), \qquad \theta_\varepsilon = \bar{\theta} - \theta_0. \tag{8.2.59}$$

We assume that

$$f, f', f'' \in L^2(0T, [L^2(\Omega)]^3), \tag{8.2.60}$$

$$\mathfrak{Q}, \mathfrak{Q}' \in L^2(0T, H^1(\Omega)), \tag{8.2.61}$$

$$w, w', w'', w''', w'''' \in L^2(0T, [H^1(\Omega)]^3), \tag{8.2.62}$$

$$w(0) \in [H^2(\Omega)]^3, \tag{8.2.63}$$

$$u_0 \in [H^2(\Omega)]^3; \bar{\theta} \in H^1(\Omega), \tag{8.2.64}$$

and

$$u_1 \in [H^1(\Omega)]^3,\ \Psi(u_1) < \infty. \tag{8.2.65}$$

The following proposition can now be proved.

Proposition 8.2.3. Assume that (8.2.53–56), (8.2.12) and (8.2.60–65) hold and that

$$F_1(v) = a(u_0 - w(0), v) + M_1(\bar{\theta}, v) + \left\langle \frac{d\Psi_\varepsilon}{dv}(u_1), v \right\rangle_\Gamma \tag{8.2.66}$$

is a continuous linear functional for the $L^2(\Omega)$-topology on V_0. Then, as $\varepsilon \to 0$

$$u_\varepsilon \to u \quad \text{weakly}_* \text{ in } \quad L^\infty(0T, V_0), \tag{8.2.67}$$

$$u'_\varepsilon \to u' \quad \text{weakly}_* \text{ in } \quad L^\infty(0T, V_0), \tag{8.2.68}$$

$$u''_\varepsilon \to u'' \quad \text{weakly}_* \text{ in } \quad L^\infty(0T, [L^2(\Omega)]^3), \tag{8.2.69}$$

$$\begin{aligned} \theta_\varepsilon \to \theta \quad \text{and} \quad \theta'_\varepsilon \to \theta' \quad &\text{weakly in} \quad L^2(0T, \mathring{H}^1(\Omega)) \\ \text{and} \quad &\text{weakly}_* \text{ in} \quad L^\infty(0T, L^2(\Omega)), \end{aligned} \tag{8.2.70}$$

where $\{u_\varepsilon, \theta_\varepsilon\}$ solves Problem 2ε. The pair $\{u, \theta\}$ thus obtained satisfies Problem 2.

Proof. Problem 2ε corresponds to a system of nonlinear differential equations. We can show, as in Prop. 5.4.2, by means of a Galerkin approximation, that there exists a solution $\{u_\varepsilon, \theta_\varepsilon\}$ such that

$$u_\varepsilon, u'_\varepsilon \in L^\infty(0T, V_0), \qquad u''_\varepsilon \in L^\infty(0T, [L^2(\Omega)]^3)$$

and

$$\theta_\varepsilon, \theta'_\varepsilon \in L^2(0T, \mathring{H}^1(\Omega)) \cap L^\infty(0T, L^2(\Omega)).$$

Further, let us insert $v = u'_\varepsilon + w' - \bar{v}_\varepsilon$ into (8.2.57) (where $(d\Psi/dv)(\bar{v}_\varepsilon) = 0$ in accordance with (8.2.54)) and $\phi = \theta_\varepsilon$ in (8.2.58). Arguing as for (8.2.19), we obtain the estimations

$$|u'_\varepsilon|_2 \le c, \qquad \|u_\varepsilon\|_1 \le c, \qquad |\theta_\varepsilon|_2 \le c, \qquad \int_0^T \|\theta_\varepsilon\|_1^2 \, dt \le c. \tag{8.2.71}$$

Now we take $t = 0$ in the regularized problem.

From (8.2.57, 58), in view of the assumptions made above, and especially of (8.2.66), it follows that

$$|u''_\varepsilon(0)|_2 \le c \tag{8.2.72}$$

and

$$|\theta'_\varepsilon(0)|_2 \le c. \tag{8.2.73}$$

Differentiating (8.2.57) and (8.2.58) with respect to t and using the monotonicity of $d\Psi_\varepsilon/dv$, the coerciveness of $a(v, v)$ and $K(\phi, \phi) + |\phi|_2^2$ and the estimates (8.2.71–73), we deduce, through integration from 0 to t, the inequalities

$$|u_\varepsilon''|_2 \le c, \qquad \|u_\varepsilon'\|_1 \le c, \qquad |\theta_\varepsilon'|_2 \le c, \qquad \int_0^T \|\theta_\varepsilon'\|_1^2 \, dt \le c. \tag{8.2.74}$$

Because of (8.2.71) and (8.2.74) we can select subsequences, again denoted by $\{u_\varepsilon\}$, $\{u_\varepsilon'\}$, $\{u_\varepsilon''\}$, $\{\theta_\varepsilon\}$, $\{\theta_\varepsilon'\}$, such that

$$\begin{gathered} u_\varepsilon \to u,\ u_\varepsilon' \to u' \quad (\text{resp. } u_\varepsilon'' \to u'') \quad \text{weakly}_* \text{ in } \ L^\infty(0T, V_0) \\ (\text{resp. in } L^\infty(0T, [L^2(\Omega)]^3)) \end{gathered} \tag{8.2.75}$$

and

$$\begin{gathered} \theta_\varepsilon \to \theta,\ \theta_\varepsilon' \to \theta' \quad \text{weakly in } \ L^2(0T, \mathring{H}^1(\Omega)) \\ \text{and weakly}_* \text{ in } \ L^\infty(0T, L^2(\Omega)). \end{gathered} \tag{8.2.76}$$

Finally, it can be proved, analogously to Prop. 8.2.1, that the pair $\{u, \theta\}$ is a solution of Problem 2, q.e.d.

As previously in Problem 1, we can easily verify that the obtained functions u and θ satisfy (8.1.8, 9) in the sense of distributions on Ω. Then, by the standard procedure used in similar cases throughout this book, we obtain the variational inequality

$$\Psi(v + w') - \Psi(u' + w') \ge -\langle \sigma_{ij} n_j, v - u' \rangle_\Gamma, \qquad \forall v \in V_0. \tag{8.2.77}$$

It can be written equivalently in the form

$$-S = -\{\sigma_{ij} n_j\} \in \partial \Psi(u'), \tag{8.2.78}$$

on $[H^{1/2}(\Gamma)]^3 \times [H^{-1/2}(\Gamma)]^3$. (8.2.77) is a weak form of the boundary condition (8.1.17). Finally, arguing as in Problem 1, we see that the initial conditions make sense.

Proposition 8.2.4. The solution $\{u, \theta\}$ of Problem 2 is unique.

Proof. The same as for Prop. 8.2.2, q.e.d.

With the same method, we can also study a thermoelastic body for which the normal (resp. tangential) component S_N (resp. S_T) of the stress vector is given on Γ_S and a subdifferential boundary condition holds between S_T (resp. S_N) and the tangential (resp. normal) component v_T (resp. v_N) of the velocity, as happens, for example, in the case of friction boundary conditions. Finally, concerning the general case of Problem 3, which includes the two variational inequalities (8.1.38) and (8.1.39), to the best of the author's knowledge the existence proof remains an open question. We can again define a regularized problem, but we cannot obtain results analogous to (8.2.14–17).

Another family of variational inequalities of a rather technical nature results by replacing (8.1.13) with the relation

$$q_i n_i = -k_{ij}\theta_{,j} n_i \in \beta(\theta') \quad \text{on} \quad \Gamma \times (0, T). \tag{8.2.79}$$

Note that in the framework of linear thermoelasticity (8.2.79) cannot be justified. The respective variational formulation is obtainable, as is obvious, by replacing $\phi - \theta$ with $\phi - \theta'$ and $\Phi(\theta)$ with $\Phi(\theta')$ in (8.1.29) or in (8.1.39); the existence proof, however, remains an open problem.

8.3 Generalizations and Related Variational Inequalities

We shall close this chapter by deriving some general variational inequalities, on the assumption that the linear elasticity law, and/or Fourier's law, is replaced by a nonlinear law derived from superpotentials. Thus in place of (8.1.1) the law

$$\sigma = t - m(\theta - \theta_0) \quad \text{with} \quad t = \{t_{ij}\} \in \partial w(\varepsilon) \tag{8.3.1}$$

is considered, and in place of (8.1.6)

$$q \in -\partial R\,(\text{grad}\,\theta). \tag{8.3.2}$$

w and R are convex, l.s.c. and proper functionals on $\mathbb{R}^6$ and $\mathbb{R}^3$ respectively.

It is doubtful whether (8.3.1) and (8.3.2) are legitimate from the point of view of rational physical theory. For example, (8.3.2) is not compatible with the linear laws between fluxes and forces in thermodynamics. However, both (8.3.1) and (8.3.2) can be used in certain cases of strong physical nonlinearities, which often arise in technical problems, such as jumps, and which cannot be ignored after linearization.

We next consider a B.V.P. defined by (8.1.2–5) (8.3.1–2), the boundary conditions (8.1.14, 15) and the initial conditions (8.1.10–12). On the assumption that all functions involved are sufficiently smooth, multiplying (8.1.4) by $v - u$, where $v = 0$ on Γ, and using the Green–Gauss theorem, we obtain (formally) the relation

$$(\rho u'', v - u) + \int_\Omega \sigma_{ij}[\varepsilon_{ij}(v) - \varepsilon_{ij}(u)]\, d\Omega - (f, v - u) = \int_\Gamma \sigma_{ij} n_j (v_i - u_i)\, d\Gamma = 0. \tag{8.3.3}$$

Similarly, if we multiply (8.1.5) by $\phi - \theta$ (with $\phi = \theta_0$ on Γ), take into account (8.1.2) and again apply the Green–Gauss theorem, the result is

$$(\rho c_D \theta', \phi - \theta) + M_2(\theta_0 u', \phi - \theta) - \int_\Omega q_i (\phi - \theta)_{,i}\, d\Omega - (Q, \phi - \theta) = -\int_\Gamma q_i (\phi - \theta) n_i\, d\Gamma. \tag{8.3.4}$$

Assume that $f(t) \in [L^2(\Omega)]^3$, $Q(t) \in L^2(\Omega)$, $v \in [\mathring{H}^1(\Omega)]^3$, $\phi \in W_0 = \{\phi \mid \phi \in H^1(\Omega), \phi = \theta_0 \text{ on } \Gamma\}$. Then we can define a functional Y_1 (resp. Y_2) on $[\mathring{H}^1(\Omega)]^3$ (resp. on $H_1(\Omega)$) such that

$$Y_1(v) = \begin{cases} \int_\Omega w(\varepsilon(v))\, d\Omega & \text{if} \quad w(\varepsilon(v)) \in L^1(\Omega) \\ \infty & \text{otherwise} \end{cases} \tag{8.3.5}$$

resp.

$$Y_2(\phi) = \begin{cases} \int_\Omega R(\text{grad}\, \phi)\, d\Omega & \text{if} \quad R(\text{grad}\, \phi) \in L^1(\Omega) \\ \infty & \text{otherwise} \end{cases}. \tag{8.3.6}$$

From (8.3.3, 4) and by means of the subdifferential inequalities

$$Y_1(v) - Y_1(u) \geq \int_\Omega t_{ij}(\varepsilon_{ij}(v) - \varepsilon_{ij}(u))\, d\Omega, \qquad \forall v \in [\mathring{H}^1(\Omega)]^3 \tag{8.3.7}$$

and

$$Y_2(\phi) - Y_2(\theta) \geq -\int_\Omega q_i(\phi_{,i} - \theta_{,i})\, d\Omega, \qquad \forall \phi \in H^1(\Omega), \tag{8.3.8}$$

we are led to the variational inequalities

$$(\rho u'', v - u) + Y_1(v) - Y_1(u) + M_1(\theta - \theta_0, v - u) - (f, v - u) \geq 0, \qquad \forall v \in [\mathring{H}^1(\Omega)]^3 \tag{8.3.9}$$

and

$$(\rho c_D \theta', \phi - \theta) + Y_2(\phi) - Y_2(\theta) + M_2(\theta_0 u', \phi - \theta) - (\mathcal{Q}, \phi - \theta) \geq 0, \ \forall \phi \in W_0 . \tag{8.3.10}$$

Accordingly, the variational problem consists of seeking a solution u: $[0, T] \to [\mathring{H}^1(\Omega)]^3$, θ: $[0, T] \to W_0$ to (8.3.9–10), with $u'(t) \in [\mathring{H}^1(\Omega)]^3$, $u''(t) \in [L^2(\Omega)]^3$ and $\theta'(t) \in L^2(\Omega)$, assuming (8.1.10–12) as initial conditions. If in a linear thermoelastic body (8.1.17) is replaced by the holonomic boundary condition on Γ_S

$$-S \in \partial\psi(u), \tag{8.3.11}$$

it results formally that the B.V.P. defined by (8.1.8–13, 16) and (8.3.11) gives rise to the following variational problem: find u: $[0, T] \to U_{ad} = \{v | v \in [H^1(\Omega)]^3; v_i = U_i$ on $\Gamma_U\}$ and $\theta \in H^1(\Omega)$, with $u'(t) \in [H^1(\Omega)]^3$, $u''(t) \in [L^2(\Omega)]^3$ and $\theta'(t) \in L^2(\Omega)$, to satisfy the variational inequalities

$$\begin{aligned}(\rho u'', v - u) &+ a(u, v - u) + M_1(\theta - \theta_0, v - u) \\ &+ \Psi(v) - \Psi(u) \geq (f, v - u) + \int_{\Gamma_S} m_{ij}(\theta - \theta_0)(v_i - u_i)\, n_j d\Gamma, \qquad \forall v \in U_{ad}\end{aligned} \tag{8.3.12}$$

$$\begin{aligned}(\rho c_D \theta', \phi - \theta) &+ K(\theta, \phi - \theta) + M_2(\theta_0 u', \phi - \theta) \\ &+ \Phi(\phi) - \Phi(\theta) \geq (\mathcal{Q}, \phi - \theta), \quad \forall \phi \in H^1(\Omega)\end{aligned} \tag{8.3.13}$$

and the initial conditions (8.1.10–12).

Suppose, further, that instead of the linear elasticity law and Fourier's law relations (8.3.1, 2) hold. Then (8.3.12) and (8.3.13) are replaced respectively by the inequalities

$$(\rho u'', v - u) + Y_1(v) - Y_1(u) + M_1(\theta - \theta_0, v - u) + \Psi(v) - \Psi(u)$$

$$\geq (f, v - u) + \int_{\Gamma_S} m_{ij}(\theta - \theta_0)(v_i - u_i)n_j \, d\Gamma, \qquad \forall v \in U_{ad} \quad (8.3.14)$$

and

$$(\rho c_D \theta', \phi - \theta) + Y_2(\phi) - Y_2(\theta) + M_2(\theta_0 u', \phi - \theta)$$

$$+ \Phi(\phi) - \Phi(\theta) \geq (\mathcal{Q}, \phi - \theta), \qquad \forall \phi \in H^1(\Omega) \quad (8.3.15)$$

If we consider the material law (8.3.1) but with

$$t \in \partial w(\varepsilon'), \qquad (8.3.16)$$

and if in addition the boundary condition (8.1.17) holds, then variations $v - u'$ must be considered. The corresponding variational formulations result from (8.3.9, 12, 14) simply by changing u to u'.

8.4 Hemivariational Inequalities in Linear Thermoelasticity

8.4.1. Formulation of Certain General Problems

In Sec. 4.3 nonconvex superpotentials were defined through the notion of generalized gradient and by means of them several material laws and boundary conditions were studied. Here we shall assume that boundary conditions derived from a nonconvex superpotential hold, and we shall formulate the corresponding hemivariational inequalities. First we consider that (8.1.13) is replaced by the condition

$$q_i n_i = -k_{ij}\theta_{,j} n_i \in \bar{\partial}\tilde{\varphi}(\theta) \qquad \text{on} \qquad \Gamma \times (0, T) \qquad (8.4.1)$$

where $\tilde{\varphi}$ is an extended real-valued functional on $\mathbb{R}$. Analogously to the examples (ix) and (x) of Sec. 3.3.1, (8.4.1) may arise in a semipermeability problem or in a temperature control problem with a nonmonotone $\{q_i n_i, \theta\}$ diagram. It is also possible to consider a condition of the following form in the interior of the body: suppose that

$$\mathcal{Q} = \mathcal{Q}_1 + \mathcal{Q}_2; \qquad (8.4.2)$$

here $\mathcal{Q}_2$ is given and $\mathcal{Q}_1$ is related to θ by the condition

$$-\mathcal{Q}_1 \in \bar{\partial}\tilde{j}(\theta) \qquad \text{in} \qquad \Omega \times (0, T), \qquad (8.4.3)$$

where $\tilde{j}$ is an extended real-valued functional on $\mathbb{R}$. Similarly to (8.4.1), (8.4.3) may describe semipermeability and temperature control conditions of nonmonotone nature in the interior of the body. Functions $\tilde{\varphi}$ and $\tilde{j}$ are generally nonconvex and nondifferentiable. If they are convex, (8.4.1, 3) become subdifferential relations and correspond to monotone $\{q_i n_i, \theta\}$ and

$\{-\mathfrak{Q}_1, \theta\}$ diagrams (cf. [84]). Also the boundary condition (8.1.17) may be replaced by a more general condition of the form

$$-S \in \bar{\partial}\tilde{\psi}(u') \quad \text{on} \quad \Gamma \times (0, T) \tag{8.4.4}$$

where $\tilde{\psi}$ is an extended real-valued functional on $\mathbb{R}^3$.

Let us now consider the B.V.P. defined by (8.1.8–12) and the boundary conditions (8.1.14) and (8.4.1). Note that for $\tilde{\varphi}(\theta)$ finite the inequality

$$\tilde{\varphi}^{\uparrow}(\theta, \phi-\theta) \geq q_i n_i(\phi-\theta) \qquad \forall \phi \in \mathbb{R} \tag{8.4.5}$$

holds by definition (cf. (4.3.13)). Combining (8.4.5) with (8.1.24, 25) we are led to a problem similar to Problem 1 of Sec. 8.1.2, but now with the variational inequality (8.1.29) replaced by the hemivariational inequality

$$(\rho\, c_D \theta', \phi-\theta) + K(\theta, \phi-\theta) + M_2(\theta_0 u', \phi-\theta) + {}_\Gamma\!\int \tilde{\varphi}^{\uparrow}(\theta, \phi-\theta)\, d\Gamma \geq$$
$$\geq (\mathfrak{Q}, \phi-\theta) \qquad \forall \phi \in H^1(\Omega). \tag{8.4.6}$$

(8.4.6) is formulated on the assumption that ${}_\Gamma\!\int \tilde{\varphi}^{\uparrow}(\theta, \phi-\theta)\, d\Gamma$ is defined (finitely). We may proceed in the same way if on Γ the classical boundary conditions (8.1.14, 15) hold, but in $\Omega \times (0, T)$ (8.4.2, 3) must be satisfied. In this case the following problem may be formulated, if ${}_\Omega\!\int \tilde{j}^{\uparrow}(\theta, \phi-\theta)\, d\Omega$ exists and is finite: find u: $[0, T] \to [\mathring{H}^1(\Omega)]^3$ and θ: $[0, T] \to W_0$ with $u'(t) \in [\mathring{H}^1(\Omega)]^3$, $u''(t) \in [L^2(\Omega)]^3$, $\theta'(t) \in L^2(\Omega)$ which satisfy (8.1.28),

$$(\rho\, c_D \theta', \phi-\theta) + K(\theta, \phi-\theta) + M_2(\theta_0 u', \phi-\theta) + \int_\Omega \tilde{j}^{\uparrow}(\theta, \phi-\theta)\, d\Omega \geq$$
$$\geq (\mathfrak{Q}_2, \phi-\theta) \qquad \forall \phi \in W_0 \tag{8.4.7}$$

and the initial conditions (8.1.10–12).

If on Γ the boundary conditions (8.4.1) and (8.4.4) hold the following problem may be posed provided that ${}_\Gamma\!\int \tilde{\varphi}^{\uparrow}(\theta, \phi-\theta)\, d\Gamma$ and ${}_\Gamma\!\int \tilde{\psi}^{\uparrow}(u', v-u')\, d\Gamma$ are finitely defined: find u: $[0, T] \to [H^1(\Omega)]^3$ and θ: $[0, T] \to H^1(\Omega)$ with $u'(t) \in [H^1(\Omega)]^3$, $u''(t) \in L^2(\Omega)$ and $\theta'(t) \in L^2(\Omega)$ to satisfy

$$(\rho u'', v-u') + a(u, v-u') + M_1(\theta-\theta_0, v-u') + \int_\Gamma \tilde{\psi}^{\uparrow}(u', v-u')\, d\Gamma \geq$$
$$\geq (f, v-u) + \int_\Gamma m_{ij}(\theta-\theta_0)\,(v_i-u_i')\, n_j d\Gamma \qquad \forall v \in [H^1(\Omega)]^3 \tag{8.4.8}$$

$$(\rho\, c_D \theta', \phi-\theta) + K(\theta, \phi-\theta) + M_2(\theta_0 u', \phi-\theta) + \int_\Gamma \tilde{\varphi}^{\uparrow}(\theta, \phi-\theta)\, d\Gamma \geq$$
$$\geq (\mathfrak{Q}, \phi-\theta) \qquad \forall \phi \in H^1(\Omega) \tag{8.4.9}$$

and the initial conditions (8.1.10–12).

If in all these problems $\tilde{\varphi}$, $\tilde{j}$ and $\tilde{\psi}$ are lipschitzian and $\bar{\partial}$-regular, we can define functions $\tilde{\Phi}$, $\tilde{J}$ and $\tilde{\Psi}$ by setting

$$\tilde{\Phi}(\phi)=\int_\Gamma \tilde{\varphi}(\phi)\,d\Gamma,\quad \tilde{J}(\phi)=\int_\Omega \tilde{j}(\phi)\,d\Omega,\quad \tilde{\Psi}(v)=\int_\Gamma \tilde{\psi}(v)\,d\Gamma. \tag{8.4.10}$$

In this case (cf. Prop. 4.3.1) the terms causing the unilateral character of the variational expressions (8.4.6–9), i.e. $\int_\Gamma \tilde{\varphi}^\uparrow(\theta,\phi-\theta)\,d\Gamma$, $\int_\Omega \tilde{j}^\uparrow(\theta,\phi-\theta)\,d\Omega$ etc., could be replaced by $\tilde{\Phi}^\uparrow(\theta,\phi-\theta)$, $\tilde{J}^\uparrow(\theta,\phi-\theta)$ etc. It is easy to verify that (8.4.6) implies (8.1.9) in the sense of distributions on Ω and the boundary condition (8.4.1) in the weak form

$$\tilde{\Phi}^\uparrow(\theta,\phi-\theta)\ge\langle -k_{ij}\theta_{,j}n_i,\phi-\theta\rangle_\Gamma \qquad \forall\phi\in H^1(\Omega). \tag{8.4.11}$$

We may also easily derive the resulting variational forms if in (8.3.1) and/or (8.3.2) w and R are nonconvex superpotentials.

In the convex case, to every maximal monotone function $\bar{\beta}$: $\mathbb{R}\to\mathscr{P}(\mathbb{R})$ there corresponds a convex superpotential f such that (2.6.6) holds. In the nonconvex case an analogous result can be obtained.

Suppose that β: $\mathbb{R}\to\mathbb{R}$ is such that its restriction $\beta|_I$ is in $L^\infty(I)$ for any bounded interval I. For any $\rho>0$ and $\xi\in\mathbb{R}$ let us define

$$\check{\beta}_\rho(\xi)=\operatorname*{ess\,inf}_{|\xi_1-\xi|<\rho}\beta(\xi_1)\quad\text{and}\quad \hat{\beta}_\rho(\xi)=\operatorname*{ess\,sup}_{|\xi_1-\xi|<\rho}\beta(\xi_1), \tag{8.4.12}$$

and let

$$\check{\beta}(\xi)=\lim_{\rho\to 0}\check{\beta}_\rho(\xi)\quad\text{and}\quad \hat{\beta}(\xi)=\lim_{\rho\to 0}\hat{\beta}_\rho(\xi). \tag{8.4.13}$$

$\check{\beta}$ and $\hat{\beta}$ exist, due to the monotonicity properties of $\check{\beta}_\rho$ and $\hat{\beta}_\rho$ when they are considered as functions of ρ; $\check{\beta}$ and $\hat{\beta}$ are l.s.c. and u.s.c. respectively. Then the multivalued function $\tilde{\beta}$ defined by

$$\tilde{\beta}(\xi)=[\check{\beta}(\xi),\hat{\beta}(\xi)] \tag{8.4.14}$$

is introduced [347]. For instance, if β has a finite jump at ξ and $\beta(\xi_+)>\beta(\xi_-)$ (resp. $\beta(\xi_+)<\beta(\xi_-)$),

$$\tilde{\beta}(\xi)=[\beta(\xi_-),\beta(\xi_+)]\ \text{(resp. }[\beta(\xi_+),\beta(\xi_-)]) \tag{8.4.15}$$

since $\hat{\beta}(\xi)=\sup\{\beta(\xi_+),\beta(\xi_-)\}$ and $\check{\beta}(\xi)=\inf\{\beta(\xi_+),\beta(\xi_-)\}$. A locally Lipschitz function $\tilde{j}$ is defined by the relation (cf. (4.3.11))

$$\tilde{j}^0(\xi,z)=\limsup_{\substack{\mu\to 0+\\ h\to 0}}\frac{1}{\mu}\int_{\xi+h}^{\xi+h+\mu z}\beta(\xi_1)\,d\xi_1, \tag{8.4.16}$$

and $\bar{\partial}\tilde{j}(\xi)\subset\tilde{\beta}(\xi)$ [37]. If, moreover, $\beta(\xi_\pm)$ exist for each $\xi\in\mathbb{R}$, then

$$\bar{\partial}\tilde{j}(\xi)=\tilde{\beta}(\xi). \tag{8.4.17}$$

In the next section we shall give a method based on a regularization procedure for proving the existence of the solution of a hemivariational inequality. The method is illustrated with respect to a simple static hemivariational inequality resulting from (8.4.7) by assuming that no interaction exists between elastic deformations and thermal phenomena and, for the sake of simplicity, that $\theta_0 = 0$ on Γ.

8.4.2 An Existence Result for a Hemivariational Inequality. — A Model Problem

In this section the following problem is considered.

Problem 1. For $\mathfrak{Q}_2 \in L^2(\Omega)$ find $\theta \in \mathring{H}^1(\Omega)$ which satisfies the inequality

$$K(\theta, \phi - \theta) + \int_\Omega \tilde{j}^0(\theta, \phi - \theta)\, d\Omega \geq (\mathfrak{Q}_2, \phi - \theta) \qquad \forall \phi \in \mathring{H}^1(\Omega). \quad (8.4.18)$$

Here $\tilde{j}$ fullfils (8.4.17) and $\tilde{\beta}$ is given by (8.4.14). With respect to (8.4.18) the following regularization method is employed. Let $p \in D(-1, +1)$ such that $p \geq 0$ and $\int_{-\infty}^{+\infty} p(\xi)\, d\xi = 1$. We consider the function $p_\varepsilon(\xi) = \frac{1}{\varepsilon} p\left(\frac{\xi}{\varepsilon}\right)$, $\varepsilon > 0$ and the convolution

$$\beta_\varepsilon = p_\varepsilon * \beta. \quad (8.4.19)$$

$\beta_\varepsilon \in C^\infty(\mathbb{R})$ is a regularized form of β and leads to the following problem.

Problem 1ε. For $\mathfrak{Q}_2 \in L^2(\Omega)$ find $\theta_\varepsilon \in \mathring{H}^1(\Omega)$ such that

$$K(\theta_\varepsilon, \phi) + (\beta_\varepsilon(\theta_\varepsilon), \phi) = (\mathfrak{Q}_2, \phi) \qquad \forall \phi \in \mathring{H}^1(\Omega). \quad (8.4.20)$$

The following proposition holds.

Proposition 8.4.1. Suppose that for some ξ

$$\operatorname*{ess\,sup}_{(-\infty, -\xi)} \beta(\xi) \leq 0 \leq \operatorname*{ess\,inf}_{(\xi, \infty)} \beta(\xi). \quad (8.4.21)$$

Then Problem 1 has at least one solution.

Proof. Let us consider a Galerkin basis and let V_n be the corresponding n-dimensional subspace. We choose ($\mathring{H}^1(\Omega) \cap L^\infty(\Omega)$ is dense in $\mathring{H}^1(\Omega)$) $V_n \subset \mathring{H}^1(\Omega) \cap L^\infty(\Omega)$ and we consider the following problem: find $\theta_{\varepsilon n} \in V_n$ such that

$$K(\theta_{\varepsilon n}, \phi) + (\beta_\varepsilon(\theta_{\varepsilon n}), \phi) = (\mathfrak{Q}_2, \phi) \qquad \forall \phi \in V_n. \quad (8.4.22)$$

This finite-dimensional problem has a solution as we may easily show by applying Brouwer's fixed point theorem.[2] To this end (8.4.22) is written in the form

$$(T(\theta_{\varepsilon n}), \phi) = 0 \qquad \forall \phi \in V_n \tag{8.4.23}$$

where $T: V_n \to V_n'$. Two positive numbers ρ_1 and ρ_2 can be determined such that (cf. (8.4.21))

$$\begin{aligned} \beta_\varepsilon(\xi) &\geq 0 \quad \text{if} \quad \xi > \rho_1 \\ \beta_\varepsilon(\xi) &\leq 0 \quad \text{if} \quad \xi < -\rho_1, \quad \text{and} \\ -\rho_2 \leq \beta_\varepsilon(\xi) &\leq \rho_2 \quad \text{if} \quad |\xi| \leq \rho_1 . \end{aligned} \tag{8.4.24}$$

It results that

$$(\beta_\varepsilon(\theta_{\varepsilon n}), \theta_{\varepsilon n}) = \int_{|\theta_{\varepsilon n}(x)| > \rho_1} \beta_\varepsilon(\theta_{\varepsilon n})\, \theta_{\varepsilon n} d\Omega + \int_{|\theta_{\varepsilon n}(x)| \leq \rho_1} \beta_\varepsilon(\theta_{\varepsilon n})\, \theta_{\varepsilon n} d\Omega \geq$$

$$\geq -\rho_1 \rho_2 |\Omega| \tag{8.4.25}$$

and since $K(\phi, \phi)$ is coercive on $\mathring{H}^1(\Omega)$, it can be shown that

$$(T(\theta_{\varepsilon n}), \theta_{\varepsilon n}) \geq c_1 \|\theta_{\varepsilon n}\|_1^2 - c_2 \|\theta_{\varepsilon n}\|_1 - c_3 \quad c_1, c_2, c_3 \text{ const} > 0. \tag{8.4.26}$$

Thus, an $r > 0$ can be determined such that $(T(\theta_{\varepsilon n}), \theta_{\varepsilon n}) \geq 0$ for $\|\theta_{\varepsilon n}\|_1 = r$; accordingly there exists $\theta_{\varepsilon n}$, with $\|\theta_{\varepsilon n}\|_1 \leq r$, such that $T(\theta_{\varepsilon n}) = 0$, i.e. (8.4.22) admits a solution. Since the sequence $\{\theta_{\varepsilon n}\}$ is bounded in $\mathring{H}^1(\Omega)$ we may extract a subsequence, again denoted as $\{\theta_{\varepsilon n}\}$, such that as $\varepsilon \to 0$, $n \to \infty$

$$\theta_{\varepsilon n} \to \theta \quad \text{weakly in} \quad \mathring{H}^1(\Omega) \tag{8.4.27}$$

and (cf. (1.3.53))

$$\theta_{\varepsilon n} \to \theta \quad \text{strongly in} \quad L^2(\Omega). \tag{8.4.28}$$

It results that

$$\theta_{\varepsilon n} \to \theta \quad \text{a.e. in } \Omega. \tag{8.4.29}$$

Further we will show that $\{\beta_\varepsilon(\theta_{\varepsilon n})\}$ is weakly relatively compact in $L^1(\Omega)$. This is equivalent to the assertion that for each $\mu > 0$, there exists $\delta > 0$ such that, if $\omega \subset \Omega$ with $|\omega| < \delta$, then $\int_\omega |\beta_\varepsilon(\theta_{\varepsilon n})| d\Omega \leq \mu$ (cf. [91] p. 239). To prove this we use the following estimate for $\int_\Omega |\theta_{\varepsilon n} \beta_\varepsilon(\theta_{\varepsilon n})| d\Omega$ [347]:

[2] This theorem states that if A is a bounded, convex and open subset of $\mathbb{R}^n$ and the function $T: \bar{A} \to \bar{A}$ is continuous on $\bar{A}$, then T has at least one fixed point in $\bar{A}$ [153]. It can be shown, by means of this theorem, that if $T: \mathbb{R}^n \to \mathbb{R}^n$ is continuous such that for $r > 0$ $(T(x))_i x_i \geq 0$ $\forall x \in \mathbb{R}^n$ with $|x| = r$, then an x_0 with $|x_0| \leq r$ exists such that $T(x_0) = 0$ (see [182] p. 53).

$$\int_\Omega |\theta_{\varepsilon n}\beta_\varepsilon(\theta_{\varepsilon n})|d\Omega = \int_{|\theta_{\varepsilon n}(x)|>\rho_1} \theta_{\varepsilon n}\beta_\varepsilon(\theta_{\varepsilon n})\, d\Omega +$$

$$+\int_{|\theta_{\varepsilon n}(x)|\le\rho_1} |\theta_{\varepsilon n}\beta_\varepsilon(\theta_{\varepsilon n})|d\Omega \le (\beta_\varepsilon(\theta_{\varepsilon n}), \theta_{\varepsilon n}) +$$

$$+2\int_{|\theta_{\varepsilon n}(x)|\le\rho_1} |\theta_{\varepsilon n}\beta_\varepsilon(\theta_{\varepsilon n})|d\Omega = (\mathfrak{Q}_2, \theta_{\varepsilon n}) - K(\theta_{\varepsilon n}, \theta_{\varepsilon n}) +$$

$$+2\int_{|\theta_{\varepsilon n}(x)|\le\rho_1} |\theta_{\varepsilon n}\beta_\varepsilon(\theta_{\varepsilon n})|d\Omega \le c + 2\rho_1\rho_2|\Omega|. \tag{8.4.30}$$

We then choose α so that $c+2\rho_1\rho_2|\Omega| < \frac{\alpha\mu}{2}$ for all n, ε, and δ so that

$$\operatorname*{ess\,sup}_{|\xi|\le\alpha+1} |\beta(\xi)| \le \frac{\mu}{2\delta} \tag{8.4.31}$$

and we find that for $|\omega| < \delta$

$$\int_\omega |\beta_\varepsilon(\theta_{\varepsilon n})|d\Omega \le \frac{1}{\alpha}\int_\omega |\theta_{\varepsilon n}\beta_\varepsilon(\theta_{\varepsilon n})|d\Omega + \int_\omega \sup_{|\theta_{\varepsilon n}(x)|\le\alpha} |\beta_\varepsilon(\theta_{\varepsilon n})|d\Omega \le$$

$$\le \frac{1}{\alpha}\int_\Omega |\theta_{\varepsilon n}\beta_\varepsilon(\theta_{\varepsilon n})|d\Omega + \delta \operatorname*{ess\,sup}_{|\xi|\le\alpha+1} |\beta(\xi)| \le \mu. \tag{8.4.32}$$

Thus the weak relative compactness of $\{\beta_\varepsilon(\theta_{\varepsilon n})\}$ in $L^1(\Omega)$ is shown, and therefore, as $\varepsilon\to 0$, $n\to\infty$ a subsequence exists such that

$$\beta_\varepsilon(\theta_{\varepsilon n}) \to \chi \quad \text{weakly in } L^1(\Omega). \tag{8.4.33}$$

Taking into account that $V_n \subset L^\infty(\Omega) \cap \mathring{H}^1(\Omega)$ we can pass to the limit and using the density property of the Galerkin approximation we obtain

$$K(\theta, \phi) + (\chi, \phi) = (\mathfrak{Q}_2, \phi) \qquad \forall\phi\in\mathring{H}^1(\Omega), \tag{8.4.34}$$

where $\chi \in L^1(\Omega) \cap H^{-1}(\Omega)$. To complete the proof it will be shown that

$$\chi \in \partial\tilde{j}(\theta) \quad \text{a.e. in} \quad \Omega. \tag{8.4.35}$$

(8.4.29) implies that for any $\gamma>0$ we may choose $\omega\subset\Omega$, with $|\omega|<\gamma$, so that

$$\theta_{\varepsilon n} \to \theta \quad \text{uniformly in} \quad \Omega-\omega \tag{8.4.36}$$

where $\theta\in L^\infty(\Omega-\omega)$. Accordingly, for any $\rho>0$ an $n_0 > \frac{2}{\rho}$ and an $\varepsilon_0 < \frac{\rho}{2}$

can be determined such that for $n > n_0$, $\varepsilon < \varepsilon_0$ and for every $x \in \Omega - \omega$

$$|\theta_{\varepsilon n}(x) - \theta(x)| < \frac{\rho}{2}. \tag{8.4.37}$$

But

$$\beta_\varepsilon(\xi) = \int_{-\varepsilon}^{+\varepsilon} \beta(\xi - t) p_\varepsilon(t)\, dt \leq \operatorname*{ess\,sup}_{|t| < \varepsilon} \beta(\xi - t) \tag{8.4.38}$$

and analogously

$$\operatorname*{ess\,inf}_{|t| < \varepsilon} \beta(\xi - t) \leq \beta_\varepsilon(\xi). \tag{8.4.39}$$

Thus

$$\operatorname*{ess\,inf}_{|\xi - \theta_{\varepsilon n}(x)| < \varepsilon} \beta(\xi) \leq \beta_\varepsilon(\theta_{\varepsilon n}(x)) \leq \operatorname*{ess\,sup}_{|\xi - \theta_{\varepsilon n}(x)| < \varepsilon} \beta(\xi). \tag{8.4.40}$$

For $\varepsilon < \varepsilon_0$ and $n > n_0$ (8.4.37) and (8.4.40) imply the inequality

$$\breve{\beta}_\rho(\theta(x)) = \operatorname*{ess\,inf}_{|\xi - \theta(x)| < \rho} \beta(\xi) \leq \beta_\varepsilon(\theta_{\varepsilon n}(x)) \leq \operatorname*{ess\,sup}_{|\xi - \theta(x)| < \rho} \beta(\xi) = \hat{\beta}_\rho(\theta(x)). \tag{8.4.41}$$

For $e \geq 0$ a.e. in $\Omega - \omega$, $e \in L^\infty(\Omega - \omega)$,

$$\int_{\Omega - \omega} \breve{\beta}_\rho(\theta)\, e d\Omega \leq \int_{\Omega - \omega} \beta_\varepsilon(\theta_{\varepsilon n})\, e d\Omega \leq \int_{\Omega - \omega} \hat{\beta}_\rho(\theta)\, e d\Omega, \tag{8.4.42}$$

which in the limit $\varepsilon \to 0$, $n \to \infty$ yields

$$\int_{\Omega - \omega} \breve{\beta}_\rho(\theta)\, e d\Omega \leq \int_{\Omega - \omega} \chi e d\Omega \leq \int_{\Omega - \omega} \hat{\beta}_\rho(\theta)\, e d\Omega. \tag{8.4.43}$$

From (8.4.43), and because $e \geq 0$ is arbitrary, we find by taking the limit as $\rho \to 0$ (Lebesgue's theorem) that

$$\breve{\beta}(\theta) \leq \chi \leq \hat{\beta}(\theta) \quad \text{a.e. in } \Omega - \omega. \tag{8.4.44}$$

If we choose γ as small as we wish, we obtain (8.4.35) from (8.4.44) and (8.4.14, 17) q.e.d.

The crucial point of the proof is the weak relative compactness of $\{\beta_\varepsilon(\theta_{\varepsilon n})\}$. Note the difference from analogous proofs for variational inequalities, where, due to the convexity, the monotonicity argument is employed.

Chapter 9

Variational Inequalities in the Theory of Plasticity and Viscoplasticity

In this chapter, we first study B.V.P.s arising in the framework of the geometrically linear theory of elastic viscoplastic and elastic perfectly plastic solids, and second B.V.P.s concerning rigid viscoplastic flows. We formulate the problems in terms of variational inequalities, and then discuss the existence and uniqueness of their solutions.

In preparing this chapter, several sources have been consulted, in particular, C. Johnson [155], H. Matthies, G. Strang and E. Christiansen [200], R. Temam and G. Strang [313], P. M. Suquet [305] [306] [307], G. Duvaut [89], G. Duvaut and J. Lions [82] [84] [86], G. Anzellotti and M. Giaquinta [5], and P. P. Mosolov and V. P. Miasnikov [217].

9.1 Elastic Viscoplastic Materials

9.1.1 Formulation of the Dynamic Problem, Existence and Uniqueness of the Solution

In this section we will study the dynamic behavior of an elastic viscoplastic body. We assume that the body occupies in its undeformed state an open, bounded, connected subset Ω of $\mathbb{R}^3$ having a regular boundary Γ. As usual, the points $x=\{x_i\}\in\Omega$ are referred to a fixed Cartesian coordinate system. If we assume small displacements and small displacement gradients, the behavior of an elastic viscoplastic body with constant density ρ is governed by the equations

$$\rho u_i'' = \sigma_{ij,j} + f_i \quad \text{in} \quad \Omega \times (0, T), \tag{9.1.1}$$

$$\varepsilon_{ij} = \tfrac{1}{2}(u_{i,j} + u_{j,i}) \quad \text{in} \quad \Omega \times (0, T) \tag{9.1.2}$$

and

$$\varepsilon_{ij}' = \varepsilon_{ij}'^{E} + \varepsilon_{ij}'^{VP} \quad \text{in} \quad \Omega \times (0, T), \tag{9.1.3}$$

where (see (3.3.113))

$$\varepsilon'^{E} = c\sigma', \qquad c = \{c_{ijhk}\} \tag{9.1.4}$$

and

$$\varepsilon'^{VP}(x,t) \in \partial J_\mu(x, \sigma(x,t)),\ \mu > 0, \quad x = \{x_i\} \in \Omega,\ t \in [0,T]. \tag{9.1.5}$$

In the above relations the prime denotes the partial derivative with respect to time. $\sigma = \{\sigma_{ij}\}$ (resp. $\varepsilon = \{\varepsilon_{ij}\}$) is the stress (resp. strain) tensor, $u = \{u_i\}$ is the displacement vector and $f = \{f_i\}, f_i = f_i(x, t)$, is the volume force vector. Hooke's tensor $c = \{c_{ijhk}\}$ with $c_{ijhk} \in L^\infty(\Omega), i, j, h, k = 1, 2, 3$, which relates the elastic strain rate ε'^E with the stress rate σ', fulfills the symmetry and ellipticity properties (3.3.61a, b). The superpotential relation (9.1.5) which provides the viscoplastic strain rate tensor ε'^{VP}, is a slight generalization of (3.3.113). Indeed, the superpotential J_μ is assumed to be a function of x, thus allowing consideration of the nonhomogeneity of the material. J_μ is for every x a convex differentiable function of σ. We assume that the boundary Γ consists of two nonempty parts Γ_U and Γ_F. On Γ_U (resp. Γ_F) the displacements (resp. the forces) are taken to be equal to zero, i.e.

$$u_i = 0 \quad \text{on} \quad \Gamma_U \times (0, T) \tag{9.1.6}$$

and

$$S_i = \sigma_{ij} n_j = 0 \quad \text{on} \quad \Gamma_F \times (0, T). \tag{9.1.7}$$

For $t = 0$ the initial conditions

$$u_i = u_{0i} \quad \text{in} \quad \Omega, \tag{9.1.8}$$

and

$$u_i' = u_{1i} \quad \text{in} \quad \Omega \tag{9.1.9}$$

$$\sigma_{ij} = \sigma_{0ij} \quad \text{in} \quad \Omega \tag{9.1.10}$$

hold, where u_0, u_1, σ_0, are functions of x only and satisfy on Γ the boundary conditions (9.1.6, 7). Moreover, let σ_0 satisfy the equilibrium equation

$$\sigma_{0ij,j} = 0, \tag{9.1.11}$$

since we assume that $f_i(0) = 0$ at $t = 0$. Suppose that at $t = 0$ the body behaves elastically and that initial strains can be present, i.e.,

$$c\sigma_0 \neq \varepsilon(u_0). \tag{9.1.12}$$

For the stress tensors we introduce the space H (see (1.4.32)) and let

$$(\sigma, \tau) = \int_\Omega \sigma_{ij}\tau_{ij}\, d\Omega, \qquad \|\sigma\|^2 = \int_\Omega \sigma_{ij}\sigma_{ij}\, d\Omega \tag{9.1.13}$$

be the inner product and the norm in H. Further, we introduce the sets

$$\Sigma_0 = \{\tau \mid \tau \in H, \quad \tau_{ij,j} \in L^2(\Omega), \quad \tau_{ij} n_j = 0 \quad \text{on} \quad \Gamma_F, i, j = 1, 2, 3\} \tag{9.1.14}$$

and

$$U_0 = \{u \mid u \in [L^2(\Omega)]^3, \quad \varepsilon_{ij}(u) \in H, \\ u_i = 0 \quad \text{on} \quad \Gamma_U, i, j = 1, 2, 3\}. \tag{9.1.15}$$

For every $u \in V_0 = \{u \mid u \in [H^1(\Omega)]^3$, with $u = 0$ on $\Gamma_U\}$, and for every $\tau \in \Sigma_0$, the Green–Gauss relation

$$\int_\Omega \tau_{ij}\varepsilon_{ij}(u)\, d\Omega + \int_\Omega u_i \tau_{ij,j}\, d\Omega = 0 \tag{9.1.16}$$

holds. The dynamic problem for elastic-viscoplastic bodies takes the following form.

Problem 1. Find σ: $[0, T] \to \Sigma_0$ and u: $[0, T] \to U_0$ to satisfy the equations (9.1.1–5) and the initial conditions (9.1.8–10). It is assumed that $f \in W^{1,2}(0T, [L^2(\Omega)]^3)$, $u_0, u_1 \in V_0$ and $\sigma_0 \in H$.

The following proposition will now be proved.

Proposition 9.1.1. Problem 1 has a unique solution (σ, u) with the properties:

$$\sigma \in AC^{1,\infty}(0T, H), \tag{9.1.17}$$

$$\sigma' \in L^\infty(0T, H), \tag{9.1.18}$$

$$u \in AC^{2,\infty}(0T, [L^2(\Omega)]^3), \tag{9.1.19}$$

$$u' \in AC^{1,\infty}(0T, [L^2(\Omega)]^3), \tag{9.1.20}$$

$$u'' \in L^\infty(0T, [L^2(\Omega)]^3). \tag{9.1.21}$$

Proof. (a) We discretize Problem 1 with respect to t. We divide the time interval $[0, T]$ into N subintervals and let $k = T/N$, $t^{(n)} = nk$, $n = 0, 1, \ldots, N$. We use the notations $v = u'$

$$u^{(n)} = u(t^{(n)}), \qquad v^{(n)} = v(t^{(n)}), \qquad \sigma^{(n)} = \sigma(t^{(n)}),$$

$$f^{(n)} = f(t^{(n)}), \qquad c\sigma^{(n)} = \{c_{ijhk}\sigma_{hk}^{(n)}\}, \qquad D\sigma^{(n)} = \{\sigma_{ij,j}^{(n)}\},$$

$$D\tau = \{\tau_{ij,j}\} \text{ and } (D\sigma^{(n)}, w) = \int_\Omega \sigma_{ij,j}^{(n)} w_i\, d\Omega.$$

We then seek by induction $\sigma^{(n+1)} \in \Sigma_0$ and $v^{(n+1)} \in V_0$ such that

$$c(\sigma^{(n+1)} - \sigma^{(n)}) + k\frac{dJ_\mu}{d\sigma}(\sigma^{(n+1)}) = k\varepsilon(v^{(n+1)}), \tag{9.1.22}$$

$$\rho v^{(n+1)} - \rho v^{(n)} = kD\sigma^{(n+1)} + kf^{(n+1)} \tag{9.1.23}$$

and

$$\sigma^{(0)} = \sigma_0, \qquad v^{(0)} = u_1. \tag{9.1.24}$$

(b) We now consider for $\sigma_\eta^{(n+1)} \in \Sigma_0$ and $v_\eta^{(n+1)} \in V_0$ the variational system (η-regularized form of the problem)

$$\left(c(\sigma_\eta^{(n+1)} - \sigma_\eta^{(n)}) + k\,\frac{dJ_\mu}{d\sigma}(\sigma_\eta^{(n+1)}),\ \tau\right) + \eta(D\sigma_\eta^{(n+1)}, D\tau)$$
$$= k(\varepsilon(v_\eta^{(n+1)}), \tau), \qquad \forall \tau \in \Sigma_0, \tag{9.1.25}$$

$$(\rho(v_\eta^{(n+1)\cdot} - \rho v_\eta^{(n)}) - kD\sigma_\eta^{(n+1)} - kf^{(n+1)}, w)$$
$$+ \eta(\varepsilon(v_\eta^{(n+1)}), \varepsilon(w)) = 0, \qquad \forall w \in V_0, \tag{9.1.26}$$

which has for every $\eta > 0$ a solution[1] on $\Sigma_0 \times V_0$.

(c) By setting $\tau = \sigma_\eta^{(n+1)}$ in (9.1.25) and $w = v_\eta^{(n+1)}$ in (9.1.26), we obtain

$$(c\sigma_\eta^{(n+1)}, \sigma_\eta^{(n+1)}) + \eta\,|D\sigma_\eta^{(n+1)}|_2^2 + k\left(\frac{dJ_\mu}{d\sigma}(\sigma_\eta^{(n+1)}), \sigma_\eta^{(n+1)}\right)$$
$$+ \rho\,|v_\eta^{(n+1)}|_2^2 + \eta\,\|\varepsilon(v_\eta^{(n+1)})\|^2 = (\rho v_\eta^{(n)} + kf^{(n+1)}, v_\eta^{(n+1)}) \tag{9.1.27}$$
$$+ (c\sigma_\eta^{(n)}, \sigma_\eta^{(n+1)}).$$

By means of the monotonicity of $\sigma_\eta^{(n+1)} \to (dJ_\mu/d\sigma)\,(\sigma_\eta^{(n+1)})$ and the inequality

$$\int_\Omega [J_\mu(\sigma_\eta^{(n+1)}) - J_\mu(0)]\, d\Omega \geq \left(\frac{dJ_\mu}{d\sigma}(\sigma_\eta^{(n+1)}), \sigma_\eta^{(n+1)}\right), \tag{9.1.28}$$

we obtain

$$\|\sigma_\eta^{(n+1)}\| \leq c, \qquad |v_\eta^{(n+1)}|_2 \leq c, \qquad \int_\Omega J_\mu(\sigma_\eta^{(n+1)})\, d\Omega \leq c, \tag{9.1.29}$$

where c is a constant > 0. It results that

$$\left\|\frac{dJ_\mu}{d\sigma}(\sigma_\eta^{(n+1)})\right\| \leq c \tag{9.1.30}$$

[1] This can be proved either by a saddle-point method or by means of the a priori estimates which follow. Let us show here the first method. A solution, if any exists, is a saddle-point of the Lagrangian

$$L(\tau, w) = \tfrac{1}{2}(c\tau, \tau) + k\Phi(\tau) - (c\sigma^{(n)}, \tau) - k(\varepsilon(w), \tau)$$
$$- \tfrac{1}{2}\rho\,|w|_2^2 + ((\rho v^{(n)} + kf^{(n+1)}), w) + \frac{\eta}{2}\,|D\tau|_2^2 - \eta\,\|\varepsilon(w)\|^2$$

on $\Sigma_0 \times V_0$, where $\Phi(\tau) = \int_\Omega J_\mu(x, \tau(x))\, d\Omega$ and $\|\cdot\|$ denotes the norm in H and $|\cdot|_2$ in $[L^2(\Omega)]^3$. From a proposition of the saddle-point theory (see [91] Prop. 2.2, Chap. VI), it results that L has a saddle-point $(\sigma^{(n+1)}, v^{(n+1)}) \in \Sigma_0 \times V_0$ which satisfies (9.1.25, 26).

and

$$|\eta^{1/2}D\sigma_\eta^{(n+1)}|_2 \leq c, \qquad \|\eta^{1/2}\varepsilon(v_\eta^{(n+1)})\| \leq c. \tag{9.1.31}$$

Thus there exist $\sigma^{(n+1)} \in H$, $v^{(n+1)} \in [L^2(\Omega)]^3$ and $\psi \in H$ such that as $\eta \to 0$

$$\sigma_\eta^{(n+1)} \to \sigma^{(n+1)} \quad \text{weakly in} \quad H, \tag{9.1.32}$$

$$v_\eta^{(n+1)} \to v^{(n+1)} \quad \text{weakly in} \quad [L^2(\Omega)]^3 \tag{9.1.33}$$

and

$$\frac{dJ_\mu}{d\sigma}(\sigma_\eta^{(n+1)}) \to \psi \quad \text{weakly in} \quad H. \tag{9.1.34}$$

Further, we take the limit in (9.1.25, 26) as $\eta \to 0$, noting that

$$\lim_{\eta \to 0} \eta\, \varepsilon(v_\eta^{(n+1)}) = 0 \qquad \text{in} \quad H. \tag{9.1.35}$$

Indeed,

$$\eta(\varepsilon(v_\eta^{(n+1)})) = \eta^{1/2}(\varepsilon_\eta^{(n+1)})(\eta^{1/2}\,\|\,\varepsilon(v_\eta^{(n+1)})\,\|) \tag{9.1.36}$$

with

$$\varepsilon_\eta^{(n+1)} = \frac{\varepsilon(v_\eta^{(n+1)})}{\|\,\varepsilon(v_\eta^{(n+1)})\,\|},$$

and $\{\varepsilon_\eta^{(n+1)}\}$ is bounded. Hence (9.1.36) along with the second of (9.1.31) implies (9.1.35). It results from (9.1.25, 26) that as $\eta \to 0$, $\sigma^{(n+1)}$, $v^{(n+1)}$ and ψ satisfy the equations

$$c(\sigma^{(n+1)} - \sigma^{(n)}) + k\psi = k\varepsilon(v^{(n+1)}) \tag{9.1.37}$$

and

$$\rho(v^{(n+1)} - v^{(n)}) - kD\sigma^{(n+1)} - kf^{(n+1)} = 0 \tag{9.1.38}$$

in the sense of $\mathscr{D}'(\Omega)$. Thus $\varepsilon(v^{(n+1)}) \in H$ and $D\sigma^{(n+1)} \in [L^2(\Omega)]^3$. The Green–Gauss theorem now yields

$$(v_\eta^{(n+1)}, D\tau) + (\varepsilon(v_\eta^{(n+1)}), \tau) = 0, \qquad \forall \tau \in \Sigma_0, \tag{9.1.39}$$

from which as $\eta \to 0$ it results that

$$(v^{(n+1)}, D\tau) + (\varepsilon(v^{(n+1)}), \tau) = 0, \qquad \forall \tau \in \Sigma_0, \tag{9.1.40}$$

which shows that $v^{(n+1)} = 0$ a.e. on Γ_U, and thus $v^{(n+1)} \in V_0$. Arguing similarly, we find that $\sigma^{(n+1)} \in \Sigma_0$.

Finally, by the classical monotonicity argument (cf., the proof of (5.4.77)) we find that

$$\psi = \frac{dJ_\mu}{d\sigma}(\sigma^{(n+1)}). \tag{9.1.41}$$

Thus we have proved the existence of a solution of the time-discretized problem.

(d) In order to examine the limit of the solution $(\sigma^{(n+1)}, v^{(n+1)})$ as the number N of subdivisions of the time interval $[0, T]$ tends to ∞ we consider the functions $\sigma_N \in L^\infty(0T, H)$ and $v_N \in L^\infty(0T, [L^2(\Omega)]^3)$, which vary linearly as functions of time within each time inverval, i.e.

$$\sigma_N(t) = \sigma^{(n)} + \frac{(t - t^{(n)})}{k}(\sigma^{(n+1)} - \sigma^{(n)}) \quad \text{for.} \ t \in [t^{(n)}, t^{(n+1)}] \quad (9.1.42)$$

and

$$v_N(t) = v^{(n)} + \frac{(t - t^{(n)})}{k}(v^{(n+1)} - v^{(n)}) \quad \text{for} \quad t \in [t^{(n)}, t^{(n+1)}]. \quad (9.1.43)$$

As we have proved, $\sigma^{(n+1)}$ and $v^{(n+1)}$ satisfy the variational equalities

$$\left(c(\sigma^{(n+1)} - \sigma^{(n)}) + k\frac{dJ_\mu}{d\sigma}(\sigma^{(n+1)}), \tau\right)$$
$$= k(\varepsilon(v^{(n+1)}), \tau), \qquad \forall \tau \in \Sigma_0 \quad (9.1.44)$$

and

$$(\rho(v^{(n+1)} - v^{(n)}) - kD\sigma^{(n+1)} - kf^{(n+1)}, w) = 0, \qquad \forall w \in V_0. \quad (9.1.45)$$

We set $\tau = \sigma^{(n+1)}$ and $w = v^{(n+1)}$; after summation, when n takes the values $1, 2, \ldots, m-1$ it results easily that

$$\tfrac{1}{2}(c\sigma^{(m)}, \sigma^{(m)}) + \tfrac{1}{2}\rho\,|v^{(m)}|_2^2 \le c, \quad (9.1.46)$$

from which it follows that

$$\|\sigma_N\| \le c \quad \text{and} \quad |v_N|_2 \le c. \quad (9.1.47)$$

Further, (9.1.44) is formulated for $t = nk$ and $t = (n+1)k$, and then the two resulting variational equalities are subtracted. The same is done for (9.1.45). We find then, that

$$(c(\sigma^{(n+2)} - 2\sigma^{(n+1)} + \sigma^{(n)}) + k\left(\frac{dJ_\mu}{d\sigma}(\sigma^{(n+2)}) - \frac{dJ_\mu}{d\sigma}(\sigma^{(n+1)})\right)$$
$$- k\varepsilon(v^{(n+2)} - v^{(n+1)}), \tau) = 0, \qquad \forall \tau \in \Sigma_0, \quad (9.1.48)$$

and

$$(\rho(v^{(n+2)} - 2v^{(n+1)} + v^{(n)}) - k(D\sigma^{(n+2)}$$
$$+ f^{(n+2)} - D\sigma^{(n+1)} - f^{(n+1)}), w) = 0, \qquad \forall w \in V_0. \quad (9.1.49)$$

We now set $\tau = \sigma^{(n+2)} - \sigma^{(n+1)}$ and $w = v^{(n+2)} - v^{(n+1)}$ and use the relations

$$\begin{aligned}(c\sigma^{(n+2)} - 2c\sigma^{(n+1)} + c\sigma^{(n)}, \sigma^{(n+2)} - \sigma^{(n+1)}) \\ = \tfrac{1}{2}(c(\sigma^{(n+2)} - \sigma^{(n+1)}), (\sigma^{(n+2)} - \sigma^{(n+1)})) \\ - \tfrac{1}{2}(c(\sigma^{(n+1)} - \sigma^{(n)}), (\sigma^{(n+1)} - \sigma^{(n)})) \\ + \tfrac{1}{2}(c(\sigma^{(n+2)} - 2\sigma^{(n+1)} + \sigma^{(n)}), (\sigma^{(n+2)} - 2\sigma^{(n+1)} + \sigma^{(n)}))\end{aligned} \tag{9.1.50}$$

and

$$\begin{aligned}(\rho v^{(n+2)} - 2\rho v^{(n+1)} + \rho v^{(n)}, v^{(n+2)} - v^{(n+1)}) \\ = \tfrac{1}{2}\rho\,|v^{(n+2)} - v^{(n+1)}|_2^2 - \tfrac{1}{2}\rho|v^{(n+1)} - v^{(n)}|_2^2 \\ + \tfrac{1}{2}\rho\,|v^{(n+2)} - 2v^{(n+1)} + v^{(n)}|_2^2\,.\end{aligned} \tag{9.1.51}$$

Dividing by k^2 and summing over n $(0 \leq n \leq m-1)$ the resulting equations from (9.1.50–51), we obtain the relation

$$\begin{aligned}\left(c\left(\frac{\sigma^{(m+1)} - \sigma^{(m)}}{k}\right), \frac{\sigma^{(m+1)} - \sigma^{(m)}}{k}\right) + \left|\frac{v^{(m+1)} - v^{(m)}}{k}\right|_2^2 \\ \leq c_1 + c_2 \sum_{n=0}^{m-1} k \left\|\frac{\sigma^{(n+1)} - \sigma^{(n)}}{k}\right\|^2.\end{aligned} \tag{9.1.52}$$

By applying the well-known Gronwall's lemma in its discrete form, we find that

$$\|\sigma_N'\| \leq c, \qquad |v_N'|_2 \leq c, \tag{9.1.53}$$

i.e., $\{\sigma_N'\}$ is a bounded sequence in $L^\infty(0T, H)$ and $\{v_N'\}$ is a bounded sequence in $L^\infty(0T, [L^2(\Omega)]^3)$. From the derived estimates and from (9.1.44, 45), we see that the sequence

$$\left\{\left(\frac{dJ_\mu}{d\sigma}(\sigma^{(n+1)}), \sigma^{(n+1)}\right)\right\} \quad \text{is bounded in} \quad L^\infty(0T)\,. \tag{9.1.54}$$

The continuity of $dJ_\mu/d\sigma$ implies that

$$\left\{\frac{dJ_\mu}{d\sigma}(\sigma_N)\right\} \quad \text{is a bounded sequence in} \quad L^\infty(0T, H). \tag{9.1.55}$$

Accordingly, we may extract subsequences, again denoted by $\{\sigma_N\}$ and $\{v_N\}$, such that

$$v_N \to v \quad \text{and} \quad v_N' \to v' \quad \text{weakly}_* \text{ in} \quad L^\infty(0T, [L^2(\Omega)]^3), \tag{9.1.56}$$

$$\sigma_N \to \sigma \quad \text{and } \sigma_N' \to \sigma' \quad \text{weakly}_* \text{ in} \quad L^\infty(0T, H), \tag{9.1.57}$$

$$\frac{dJ_\mu}{d\sigma}(\sigma_N) \to \bar{\psi} \quad \text{weakly}_* \text{ in} \quad L^\infty(0T, H). \tag{9.1.58}$$

Thus there exist $\sigma \in W^{1,\infty}(0T, H)$, $v \in W^{1,\infty}(0T, [L^2(\Omega)]^3)$ and $\bar{\psi} \in L^\infty(0T, H)$ which satisfy the relations

$$c_{ijhk}\sigma'_{hk} + \bar{\psi}_{ij} = \varepsilon_{ij}(v) \tag{9.1.59}$$

and

$$\rho v'_i - \sigma_{ij,j} = f_i\,. \tag{9.1.60}$$

From (9.1.42–45) we find that

$$\|\varepsilon(v_N)\| \le c \quad \text{and} \quad |D\sigma_N|_2 \le c. \tag{9.1.61}$$

We may also verify that σ_N and v_N fulfill the boundary conditions on Γ_F and on Γ_U respectively.

It results that $\sigma \in L^\infty(0T, \Sigma_0)$ and $v \in L^\infty(0T, V_0)$ (cf., the proof that $v^{(n+1)} \in V_0$ and $\sigma^{(n+1)} \in \Sigma_0$). σ and v also satisfy the initial conditions, as can be readily verified. It now remains to be shown that

$$\bar{\psi} = \frac{dJ_\mu}{d\sigma}(\sigma). \tag{9.1.62}$$

This is proved by using the monotonicity argument, i.e., we prove the inequality

$$\int_0^T \left(\frac{dJ_\mu}{d\sigma}(\tau) - \bar{\psi}, \tau - \sigma\right) dt \ge 0, \qquad \forall \tau \in L^\infty(0T, H). \tag{9.1.63}$$

(9.1.63) results by starting from the respective discrete inequality and passing to the limit as $N \to \infty$. From (9.1.56, 57) and the continuity theorem (cf., Sec. 1.4.1), (9.1.19–21) result.

(e) Next the uniqueness of the solution is proved.

Let (σ_1, u_1) and (σ_2, u_2) be two pairs of solutions of Problem 1. From (9.1.59, 60) and (9.1.62), we obtain for every $\tau \in \Sigma_0$ and $w \in V_0$

$$(c(\sigma'_1 - \sigma'_2), \tau) + (\rho(v'_1 - v'_2), w)$$
$$+ \left(\frac{dJ_\mu}{d\sigma}(\sigma_1) - \frac{dJ_\mu}{d\sigma}(\sigma_2), \tau\right) + (v_1 - v_2, D\tau) + (\sigma_1 - \sigma_2, \varepsilon(w)) = 0. \tag{9.1.64}$$

We now set in (9.1.64) $\tau = \sigma_1 - \sigma_2$ and $w = v_1 - v_2$. Due to (9.1.16) and the monotonicity of $(dJ_\mu/d\sigma)(\sigma)$, we find that

$$(c(\sigma'_1 - \sigma'_2), \sigma_1 - \sigma_2) + (\rho(v'_1 - v'_2), v_1 - v_2) \le 0. \tag{9.1.65}$$

At $t = 0$

$$\sigma_1(0) - \sigma_2(0) = 0 \quad \text{and} \quad v_1(0) - v_2(0) = 0. \tag{9.1.66}$$

(9.1.65, 66) imply that $\sigma_1 - \sigma_2 = 0$, $v_1 - v_2 = 0$, i.e., the uniqueness of the solution, q.e.d.

Proposition 9.1.1 holds under more general conditions.

(A) First we may assume that the boundary conditions on Γ_U and on Γ_F are nonhomogeneous, i.e.,

$$u_i = U_i \quad \text{on} \quad \Gamma_U \times (0, T) \tag{9.1.67}$$

and

$$S_i = \sigma_{ij} n_j = F_i \quad \text{on} \quad \Gamma_F \times (0, T). \tag{9.1.68}$$

In this case, we introduce the new unknown functions $\bar{v} = v - v^0$ and $\bar{\sigma} = \sigma - \sigma^0$, where v^0 and σ^0 satisfy the boundary conditions (9.1.67, 68) and the initial conditions (9.1.8–10). $\bar{v}$ and $\bar{\sigma}$ satisfy homogeneous boundary and initial conditions as well as the field equations

$$c_{ijhk} \bar{\sigma}'_{hk} + \left[\frac{dJ_\mu}{d\sigma}(\bar{\sigma} + \sigma^0)\right]_{ij} - \varepsilon_{ij}(\bar{v}) = g_{ij} \quad \text{in} \quad \Omega \times (0, T) \tag{9.1.69}$$

and

$$\rho \bar{v}'_i - \bar{\sigma}_{ij,j} = h_i \quad \text{in} \quad \Omega \times (0, T), \tag{9.1.70}$$

where

$$g_{ij} = \varepsilon_{ij}(v^0) - c_{ijhk} \sigma^{0\,\prime}_{hk}, \qquad h_i = f_i - \rho (v_i^0)' + \sigma^0_{ij,j}. \tag{9.1.71}$$

Concerning the regularity of g and h we may assume, e.g., that $g, g' \in L^2(0T, H)$ and $h, h' \in L^2(0T, [L^2(\Omega)]^3)$. We may reasonably consider v^0 and σ^0 to be the solution of the corresponding purely elastic problem with appropriate regularity.

(B) We may also assume that, instead of (9.1.5),

$$\varepsilon'^{VP}(x, t) \in \partial J(x, t, \sigma(x, t)) \tag{9.1.72}$$

holds, where J is a convex, nonnegative and finite function of σ for every $t \in [0, T]$, $x \in \Omega$, which is measurable in x and t for each σ. Thus (see Sec. 3.3.4 and [287]) J is a normal convex integrand. Suppose that (4.3.48) holds for the pair $\{\Phi, J\}$ as an equality, and that Φ defined for $\tau \in H$ by

$$\Phi(t, \tau) = \int_\Omega J(x, t, \tau(x, t))\, d\Omega \tag{9.1.73}$$

is for a.e. t a convex, l.s.c. and finite functional. Let the domain of definition of Φ in H be denoted $\tilde{H}$. Φ is the extension of J to $\tilde{H}$. Obviously, for $J = J_\mu$, $\tilde{H} = H$. We assume that $\tilde{H}$ endowed with the norm $\|\cdot\|_{\tilde{H}}$ is a reflexive Banach space imbedded into H. Moreover, let $\tilde{D} = \{\tau \mid \tau = \{\tau_{ij}\},\ \tau_{ij} = \tau_{ji},\ \tau_{ij} \in D(\Omega)\}$ be dense in $\tilde{H}$ for its strong topology and let $\tilde{L}^q \subset \tilde{H}$ (imbedding), $q \leq 3$, where $\tilde{L}^q = \{\tau \mid \tau \in \{\tau_{ij}\},\ \tau_{ij} \in L^q(\Omega),\ \tau_{ij} = \tau_{ji}\}$. By $\tilde{H}'$ we denote the dual of $\tilde{H}$ for the duality pairing (σ, τ) (cf., (9.1.13)). Further, we define the sets

$$\Sigma_{ad} = \{\tau \mid \tau \in \tilde{H},\ \tau_{ij,j} \in L^3(\Omega),\ \tau_{ij} n_j = F_i \quad \text{on} \quad \Gamma_F\}, \tag{9.1.74}$$

$$U_{ad} = \{u \mid u \in [L^{3/2}(\Omega)]^3,\ \varepsilon_{ij} \in \tilde{H}',\ u_i = U_i \quad \text{on} \quad \Gamma_U\}, \tag{9.1.75}$$

$$V_{ad} = \{v \mid v \in [L^{3/2}(\Omega)]^3,\ \varepsilon_{ij} \in \tilde{H}',\ v_i = U'_i \quad \text{on} \quad \Gamma_U\} \tag{9.1.76}$$

and

$$\tilde{U}_0 = \{u \mid u \in [L^{3/2}(\Omega)]^3,\ \varepsilon_{ij} \in \tilde{H}',\ u_i = 0 \quad \text{on} \quad \Gamma_U\}. \tag{9.1.77}$$

Obviously, (9.1.16) holds for every $v \in \tilde{U}_0$ and every $\tau \in \tilde{\Sigma}_0 = \{\tau \mid \tau \in \tilde{H},\ \tau_{ij,j} \in L^3(\Omega),\ \tau_{ij}n_j = 0 \text{ on } \Gamma_F\}$, and since $\tilde{L}^q \subset \tilde{H} \subset H$, the boundary conditions are well defined.[2] We make the following assumptions for Φ:

(i) Every sequence $\{\sigma_n\} \in \tilde{H}$ bounded in H and such that $\Phi(t, \sigma_n) < c$, is also bounded in $\tilde{H}$;

(ii) $d\Phi/d\sigma$ is a continuous operator from $\tilde{H}$ into $\tilde{H}'$ in the sense that there exists a function $r \in C^0(\mathbb{R}, \mathbb{R})$ with $r(0) = 0$ such that

$$\left\| \frac{d\Phi}{d\sigma}(\tau_1) - \frac{d\Phi}{d\sigma}(\tau_2) \right\|_{\tilde{H}'} \leq r(\| \tau_1 - \tau_2 \|_{\tilde{H}}); \tag{9.1.78}$$

(iii) With respect to t, for $\tau \in H$, $\Phi(t, \tau)$ and $(d\Phi/d\tau)(t, \tau)$ satisfy the condition

$$\left\| \frac{d\Phi(t_1, \tau)}{d\tau} - \frac{d\Phi(t_2, \tau)}{d\tau} \right\|_{\tilde{H}'} + |\Phi(t_1, \tau) - \Phi(t_2, \tau)|$$
$$\leq c_1 |t_1 - t_2| (\Phi(t_2, \tau) + c_2). \tag{9.1.79}$$

(C) Finally, we may assume that the elasticity tensor is time dependent, i.e.,

$$c_{ijhk} \in W^{2,\infty}(0T, L^\infty(\Omega)), \qquad i, j, h, k = 1, 2, 3, \tag{9.1.80}$$

and satisfies, besides the symmetry and ellipticity properties, the inequality

$$\left\| \frac{\partial^\alpha c(t)\tau}{\partial t^\alpha} \right\| \leq c \| \tau \|, \qquad \forall \tau \in H, \qquad c \text{ const} > 0 \tag{9.1.81}$$

for $\alpha = 0, 1, 2$.

The dependence of the elasticity tensor c and/or of the superpotential of dissipation J on t permits consideration of more complicated effects, e.g., the dependence of c and J on temperature, which is taken to be a known function of time. Assumptions (i), (ii) and (iii) are easily satisfied on Φ; for instance, if (3.3.110) holds and g and μ are Lipschitz functions of t, then (9.1.78) and (9.1.79) are fulfilled. For the formulation of the B.V.P., any combination of conditions (A), (B) and (C) can be considered. Let us briefly discuss the most general case, that of a problem subjected to all three conditions.

We have the following B.V.P.

Problem 2. Find σ: $[0, T] \to \Sigma_{ad}$ and u: $[0, T] \to U_{ad}$ such as to satisfy (9.1.1) (with $f \in W^{1,2}(0T, [L^3(\Omega)]^3)$),

$$(c_{ijhk}(t)\sigma_{hk})' + \left(\frac{dJ(t, \sigma)}{d\sigma} \right)_{ij} = \varepsilon_{ij}(u') \tag{9.1.82}$$

and the initial conditions (9.1.8–10), where $u_0 \in U_{ad}$, $u_1 \in V_{ad}$ and $\sigma_0 \in \Sigma_{ad}$.

[2] We may assume, e.g., that $F_i \in L^2(0T, L^2(\Gamma_F))$; $U_{ad} \subset W^{1,3/2}(\Omega)$ since $X = \{u_i | u_i \in L^p(\Omega),\ \varepsilon_{ij}(u) \in L^p(\Omega)\ i, j = 1, 2, 3\} = W^{1,p}(\Omega),\ p > 1$ [306]. The trace theorem for $W^{1,3/2}(\Omega)$ supplies the possible space for U_i.

This problem can be treated by the same method as in Problem 1, and the following proposition can then be proved.

Proposition 9.1.2. Problem 2 has a unique solution with the properties

$$\sigma \in AC^{1,\infty}(0T, H) \cap L^\infty(0T, \tilde{H}), \tag{9.1.83}$$

$$\sigma' \in L^\infty(0T, H), \tag{9.1.84}$$

$$u \in AC^{2,\infty}(0T, [L^2(\Omega)]^3), \tag{9.1.85}$$

$$v \in AC^{1,\infty}(0T, [L^2(\Omega)]^3) \tag{9.1.86}$$

and

$$v' \in L^\infty(0T, [L^2(\Omega)]^3). \tag{9.1.87}$$

Proof. The proof follows the pattern of that of Prop. 9.1.1 (see also [307]). Here we sketch only the differences from the previous case, leaving the complete proof as an exercise for the reader. The problem is homogenized by introducing the functions σ^0 and u^0, and a new functional $\bar{\Phi}$ is defined by setting $\bar{\Phi}(t, \bar{\sigma}) = \Phi(t, \bar{\sigma} + \bar{\sigma}^0)$. The η-regularized form of the time-discrete problem has a slight difference from (9.1.26): here the term $\eta\langle \mathcal{M}\varepsilon(v_\eta^{(n+1)}), \varepsilon(w)\rangle$, is used instead of $\eta(\varepsilon(v_\eta^{(n+1)}), \varepsilon(w))$. The operator $\mathcal{M}$ between $\tilde{H}'$ and $\tilde{H}$ is such that for $\varepsilon \in \tilde{H}'$

$$\langle \mathcal{M}\varepsilon, \varepsilon\rangle = \|\varepsilon\|_{\tilde{H}'}^2 = \|\mathcal{M}\varepsilon\|_{\tilde{H}}^2 \tag{9.1.88}$$

and

$$\mathcal{M}(\alpha\varepsilon) = \alpha\mathcal{M}(\varepsilon), \qquad \alpha \in \mathbb{R}. \tag{9.1.89}$$

$\mathcal{M}$ is called the duality mapping between the two normed spaces $\tilde{H}$ and $\tilde{H}'$. Moreover, within each time interval, say $t \in [t^{(n)}, t^{(n+1)}]$, we define a convex functional Φ_N by setting $\Phi_N(t, \tau) = \bar{\Phi}(t_n, \tau)$.

9.1.2. The Quasi-static Problem

In this section we shall study the quasi-static problem corresponding to Problem 1. It has the following form.

Problem 3. Find u: $[0, T] \to U_0$ and σ: $[0, T] \to \Sigma_0$ such as to satisfy

$$c_{ijhk}\sigma'_{hk} + \left(\frac{dJ_\mu}{d\sigma}(\sigma)\right)_{ij} = \varepsilon_{ij}(u'), \tag{9.1.90}$$

$$\sigma_{ij,j} + f_i = 0 \tag{9.1.91}$$

and the initial conditions (9.1.8, 10).

First we shall prove the existence and uniqueness of the solution, and then we shall investigate the behavior of the solution of the dynamic problem as $\rho \to 0$. We shall show that in the absence of initial strains at $t = 0$, i.e., if (9.1.12) holds as an equality, the solution of Problem 1 tends to the solution of Problem 3. The following propositions hold.

Proposition 9.1.3. The quasi-static problem has a unique solution (σ, u) with the properties

$$\sigma \in L^\infty(0T, H), \tag{9.1.92}$$

$$\sigma' \in L^\infty(0T, H) \tag{9.1.93}$$

and

$$v \in L^\infty(0T, [L^2(\Omega)]^3). \tag{9.1.94}$$

Proof. The time-discretization procedure will be used as in Prop. 9.1.1. We introduce the sets

$$\Sigma(t) = \{\tau \mid \tau \in \Sigma_0, \tau_{ij,j} + f_i = 0 \quad \text{in} \quad \Omega\}, \tag{9.1.95}$$

$$\bar{\Sigma}_0 = \{\tau \mid \tau \in \Sigma_0, \tau_{ij,j} = 0 \quad \text{in} \quad \Omega\}, \tag{9.1.96}$$

and we define a time-discrete form of the problem: when $(\sigma^{(n)}, u^{(n)})$ are given, find $\sigma^{(n+1)} \in \Sigma_0$ and $u^{(n+1)} \in U_0$ such that

$$c(\sigma^{(n+1)} - \sigma^{(n)}) + k\frac{dJ_\mu}{d\sigma}(\sigma^{(n+1)}) = \varepsilon(u^{(n+1)}) - \varepsilon(u^n), \tag{9.1.97}$$

$$D\sigma^{(n+1)} + f^{(n+1)} = 0. \tag{9.1.98}$$

There exists a unique (cf., Prop. 2.2.1) $\sigma^{(n+1)}$ which minimizes over $\Sigma(t^{(n+1)})$ the functional

$$F(\tau) = \tfrac{1}{2}(c\tau, \tau) - (c\sigma^{(n)}, \tau) + k\Phi_\mu(\tau) + (\varepsilon(u^{(n)}), \tau). \tag{9.1.99}$$

It is verified (see Prop. 2.2.2) that

$$e^{(n+1)} = c\sigma^{(n+1)} - c\sigma^{(n)} + k\frac{dJ_\mu}{d\sigma}(\sigma^{(n+1)}) + \varepsilon(u^{(n)}) \tag{9.1.100}$$

satisfies the variational inequality

$$(e^{(n+1)}, \tau - \sigma^{(n+1)}) \geq 0, \qquad \forall \tau \in \Sigma(t^{(n+1)}), \tag{9.1.101}$$

and therefore

$$(e^{(n+1)}, \tau) = 0, \qquad \forall \tau \in \bar{\Sigma}_0. \tag{9.1.102}$$

Accordingly,

$$e^{(n+1)} \in (\bar{\Sigma}_0)^\perp. \tag{9.1.103}$$

Let

$$H_0 = \{e \mid e = \{e_{ij}\}, e_{ij} = e_{ji}, e_{ij} \in L^2(\Omega), \quad \text{there exists} \\ u \in U_0 \quad \text{such that} \quad e_{ij} = \varepsilon_{ij}(u)\}. \tag{9.1.104}$$

We shall prove that

$$(\bar{\Sigma}_0)^{\perp} = H_0. \tag{9.1.105}$$

It is obvious that $H_0 \subset (\bar{\Sigma}_0)^{\perp}$; we shall show that $(\bar{\Sigma}_0)^{\perp} \subset H_0$, or that $H_0^{\perp} \subset \bar{\Sigma}_0$. Let $\tau \in H_0^{\perp}$. Then

$$(\varepsilon(u), \tau) = 0, \qquad \forall u \in U_0. \tag{9.1.106}$$

The Green-Gauss theorem, with (9.1.106), implies that

$$\int_{\Omega} \tau_{ij,j} u_i d\Omega = \langle \bar{\gamma}(\tau), \gamma(u) \rangle \qquad \forall u \in U_0, \tag{9.1.107}$$

and thus for $u \in \bar{U}_0 = \{u | u \in U_0, u = 0 \text{ on } \Gamma\}$

$$\int_{\Omega} \tau_{ij,j} u_i d\Omega = 0, \qquad \forall u \in \bar{U}_0 = [\mathring{H}^1(\Omega)]^3. \tag{9.1.108}$$

Applying again the Green–Gauss theorem, we find that

$$\langle \bar{\gamma}(\tau), \gamma(u) \rangle = 0, \qquad \forall u \in U_0. \tag{9.1.109}$$

From (9.1.108, 109), we conclude that $\tau \in \bar{\Sigma}_0$, and thus the proof of (9.1.105) is completed. As a result, $e^{(n+1)}$ is the deformation tensor of a displacement field $u^{(n+1)} \in U_0$. The remaining steps of the proof are the same as for that of Prop. 9.1.1, q.e.d.

Proposition 9.1.4. Suppose that (9.1.12) holds as an equality. Let us denote by (σ_ρ, u_ρ) the solution of the dynamic problem (Problem 1) for a given ρ. Then as $\rho \to 0$,

$$\sigma_\rho \to \sigma \quad \text{in} \quad C^0(0T, H) \quad \text{strongly}, \tag{9.1.110}$$

$$\sigma'_\rho \to \sigma' \quad \text{in} \quad L^\infty(0T, H) \quad \text{weakly}_*, \tag{9.1.111}$$

$$u_\rho \to u \quad \text{in} \quad L^\infty(0T, [L^2(\Omega)]^3) \quad \text{weakly}_* \tag{9.1.112}$$

and

$$u'_\rho = v_\rho \to v = u' \quad \text{in} \quad L^\infty(0T, [L^2(\Omega)]^3) \quad \text{weakly}_*, \tag{9.1.113}$$

Proof. Since the proof follows the general pattern encountered in all the regularization procedures for dynamic variational inequalities, here we shall only outline the steps of the proof and the results of each step. The variational form of Problem 1 reads: find σ: $[0, T] \to \Sigma_0$ and u: $[0, T] \to U_0$ such as to satisfy the initial conditions (9.1.8–10) and the variational equalities

$$\left(c\sigma' + \frac{dJ_\mu}{d\sigma}(\sigma) - \varepsilon(v), \tau\right) = 0, \qquad \forall \tau \in \Sigma_0 \tag{9.1.114}$$

and

$$(\rho v' - D\sigma - f, w) = 0, \qquad \forall w \in U_0. \tag{9.1.115}$$

First we take σ_ρ and v_ρ as test functions in (9.1.114–115). By integration from 0 to t and by Gronwall's lemma, it results that

$$\|\sigma_\rho\| \le c, \qquad |\rho^{1/2}v_\rho|_2 \le c, \qquad \int_0^t \int_\Omega J_\mu(\sigma_\rho)\,d\Omega\,dt \le c. \tag{9.1.116}$$

Let us now consider the variational equalities[3]

$$\left(c\,\sigma'' + \left(\frac{dJ_\mu}{d\sigma}(\sigma)\right)' - \varepsilon(v'), \tau\right) = 0, \qquad \forall \tau \in \Sigma_0 \tag{9.1.117}$$

and

$$(\rho v'' - D\sigma' - f', w) = 0, \qquad \forall w \in U_0, \tag{9.1.118}$$

in which we set $\tau = \sigma'_\rho$ and $w = v'_\rho$. After integration, by means of the monotonocity of $dJ_\mu/d\sigma$ and Gronwall's Lemma, and using the estimates

$$\|\sigma'(0)\| < c, \qquad \rho\,|v'(0)|_2 \le c, \tag{9.1.119}$$

we obtain

$$\|\sigma'_\rho\| \le c, \qquad |\rho^{1/2}v'_\rho|_2 \le c. \tag{9.1.120}$$

From the foregoing estimates, we obtain

$$\left\|\frac{dJ_\mu}{d\sigma}(\sigma_\rho)\right\| \le c, \qquad \|\varepsilon(v_\rho)\| \le c, \qquad |D\sigma_\rho|_2 \le c. \tag{9.1.121}$$

By passing to subsequences if necessary, we find that there exist $\sigma \in L^\infty(0T, \Sigma_0)$, $v \in L^\infty(0T, U_0)$ and $\tilde{\psi} \in L^\infty(0T, H)$ such that as $\rho \to 0$

$$\sigma_\rho \to \sigma \quad \text{in} \quad L^\infty(0T, H) \quad \text{weakly}_*, \tag{9.1.122}$$

$$\sigma'_\rho \to \sigma' \quad \text{in} \quad L^\infty(0T, H) \quad \text{weakly}_*, \tag{9.1.123}$$

$$u_\rho \to u \quad \text{and} \quad v_\rho \to v \quad \text{in} \quad L^\infty(0T, [L^2(\Omega)]^3) \quad \text{weakly}_* \tag{9.1.124}$$

and

$$\frac{dJ_\mu}{d\sigma}(\sigma_\rho) \to \tilde{\psi} \quad \text{in} \quad L^\infty(0T, H) \quad \text{weakly}_*. \tag{9.1.125}$$

In order to show the strong convergence of σ_ρ to σ in $L^\infty(0T, H)$, we proceed as follows. From (9.1.114) and (9.1.115), using the monotonocity of $dJ_\mu/d\sigma$, we obtain the inequality

$$(c\sigma'_{\rho_1} - c\sigma'_{\rho_2}, \sigma_{\rho_1} - \sigma_{\rho_2}) + \rho_1(v'_{\rho_1}, v_{\rho_1} - v_{\rho_2}) - \rho_2(v'_{\rho_2}, v_{\rho_1} - v_{\rho_2}) \ge 0. \tag{9.1.126}$$

[3] As was remarked in Sec. 5.4.2, it would be more precise if instead of the time derivatives v' and σ' the time difference quotients $[v(t+\delta t) - v(t)]/\delta t$ and $[\sigma(t+\delta t) - \sigma(t)]/\delta t$ were taken.

Integrating (9.1.126) from 0 to t and applying Gronwall's Lemma, we establish the relation

$$\| \sigma_{\rho_1}(t) - \sigma_{\rho_2}(t) \| \leq c_1 \sqrt{| \rho_1 - \rho_2 |}\, e^{c_2 t}, \qquad c_1, c_2 \text{ const} > 0, \quad (9.1.127)$$

which proves that $\{\sigma_\rho\}$ is a Cauchy sequence in $L^\infty[0T, H]$, and thus the strong convergence in (9.1.110) results. By taking the limit in (9.1.114, 115) as $\rho \to 0$ and using the monotonicity argument to show that $\tilde{\psi} = dJ_\mu/d\sigma$, we prove that (σ, u) is a solution of the quasi-static problem, q.e.d.

If the nonhomogeneous boundary conditions (9.1.67, 68) hold, as well as the general law (9.1.72), and if the elasticity tensor depends on time, the last two propositions still hold, with certain modifications. Assuming, more precisely, that conditions (A), (B) and (C) of Sec. 9.1.1 hold, we can prove the following proposition for the quasi-static problem (called here Problem 4) corresponding to Problem 2.

Proposition 9.1.5. Problem 4 has a unique solution (σ, u) such that

$$\sigma \in C^0(0T, H) \cap L^\infty(0T, \tilde{H}), \quad (9.1.128)$$

$$\sigma' \in L^\infty(0T, H) \quad (9.1.129)$$

and

$$v \in L^\infty(0T, [L^2(\Omega)]^3). \quad (9.1.130)$$

Moreover, if (σ_ρ, u_ρ) denotes the solution of Problem 2 for a given ρ, and (9.1.12) is verified as an equality, then as $\rho \to 0$

$$\sigma_\rho \to \sigma \quad \text{in} \quad C^0(0T, H) \quad \text{strongly and in} \quad L^\infty(0T, \tilde{H}) \quad \text{weakly}_*, \quad (9.1.131)$$

and (9.1.111–113) hold.

Proof. After the homogenization of the boundary conditions, the proof results by slightly modifying the proofs of Props. 9.1.3 and 9.1.4, q.e.d.

9.2 Elastic Perfectly Plastic Materials

9.2.1 Formulation of the Quasi-static Problem

In this section, we consider elastoplastic solids obeying the Prandtl–Reuss law, and we study the resulting B.V.P. We assume quasi-static behavior for the body under consideration, which corresponds to a slow loading process. Variational principles for elastic-perfectly plastic solids have already been formulated by Hill, Hodge, Prager, *et al*; for an account of this subject, the reader is referred to the article by Koiter [163]. For the case of work-hardening and work-softening behavior, we should also mention the variational formulations by Maier and the Italian school.[4] Certain mathematical questions on the problem have been studied by Duvaut and Lions [84], who applied a regularization procedure to discuss the properties of the stress field. For the same problem, Moreau [214] developed a geometric theory concerning a dynamic problem associated with a moving convex set. The properties of the velocity field have been studied by Johnson [155], who used for the proof a time-discretization procedure like the one we have applied in the previous section. The basic difficulty of the study of a Prandtl–Reuss material is the irregularity of the strain-rate tensor $\varepsilon(v)$, since, as has been experimentally verified, the velocity field presents discontinuities. For this reason we assume, following Suquet [306] [307], that the velocities belong to the space $BD(\Omega)$ (cf., Sec. 1.3.5). This space has been studied in [200] [304] [305] [306] and [314], and constitutes the appropriate functional framework for the study of Prandtl–Reuss materials.

The solution of the quasi-static B.V.P. for an elastic perfectly plastic body obeying (9.1.6, 7) will be approximated by the solution of the B.V.P. studied in Prop. 9.1.5. To this end, the superpotential law (cf., (3.3.98))

$$\dot{\varepsilon}^p(x, t) \in \partial I_K(x, \sigma(x, t)), \tag{9.2.1}$$

where K is a convex closed set which may depend on x and t, will be approximated by material laws of the form (9.1.5), or more generally by material laws of the form (9.1.72), which depend on a parameter ε tending to zero. We also assume that the elasticity tensor is independent of t. Let us introduce the set

$$P(t) = \{\tau \mid \tau \in H,\ \tau(x) \in K(x, t) \text{ a.e. in } \Omega\}. \tag{9.2.2}$$

We now formulate the following problem for Σ_0 given in (9.1.14) but with $D\tau \in [L^3(\Omega)]^3$.

Problem 5. Find σ: $[0, T] \to P(t) \cap \Sigma_0$ and v: $[0, T] \to BD(\Omega)$ such that for $f \in W^{1,\infty}(0T, [L^3(\Omega)]^3)$ and $\sigma_0 \in P(0) \cap \Sigma_0$, with $D\sigma_0 = -f(0) = 0$ at $t = 0$

$$((c\sigma)', \tau - \sigma) + \int_\Omega v_i(\tau_{ij,j} - \sigma_{ij,j})\, d\Omega \geq 0, \qquad \forall \tau \in P(t) \cap \Sigma_0, \tag{9.2.3}$$

$$\sigma_{ij,j} + f_i = 0 \tag{9.2.4}$$

and for $t = 0$

$$\sigma = \sigma_0. \tag{9.2.5}$$

This problem constitutes a weak formulation of the quasi-static elastoplastic problem and results by formally applying the Green-Gauss theorem.

[4] See, e.g., [33] [57] [76] [186–189] [192–194].

For the study of Problem 5, we make the following assumptions, in addition to those used for the proof of Prop. 9.1.3:

(i) (Safe load assumption) there exists a stress field $\tilde{\sigma}=\{\tilde{\sigma}_{ij}\}$, such that: $\tilde{\sigma}\in\Sigma_0$, $\tilde{\sigma}(0)=\sigma_0$,

$$\tilde{\sigma}_{ij}\in W^{1,\infty}(0T, L^\infty(\Omega)) \quad D\tilde{\sigma}+f=0 \quad \text{a.e. on} \quad [0, T], \tag{9.2.6}$$

with the property that for every $\tau=\{\tau_{ij}\}$, $\tau\in\tilde{L}^\infty=\{\tau|\tau=\{\tau_{ij}\}$, $\tau_{ij}=\tau_{ji}$, $\tau_{ij}\in L^\infty(\Omega)\}$ with

$$\|\tau\|_{\tilde{L}^\infty}<q\,, \ \text{one has} \quad \tilde{\sigma}+\tau\in P(t) \ \text{a.e. on} \quad [0, T]; \tag{9.2.7}$$

(ii) $0\in \operatorname{int} P(t)\ \forall t\in[0, T]$. $\bigcap_t P(t)$contains a neighborhood of 0.

(iii) there exists a sequence of finite superpotentials $\Phi_\varepsilon(t, \tau)$ defined on H such that the assumptions in (B) of Sec. 9.1.1 are satisfied, as are the assumptions

$$\Phi_\varepsilon(t, \tau)=0 \quad \text{a.e. on } [0, T]\ \forall\tau\in P(t) \tag{9.2.8}$$

and, if a constant α exists such that ${}^T\!\int_0 \Phi_\varepsilon(t, \tau_\varepsilon)\,dt<\alpha$ and

$$\tau_\varepsilon\to\tau \text{ weakly}_* \text{ in } L^\infty(0T, H), \text{ then} \quad \tau\in P(t) \quad \text{a.e. on } [0, T]; \tag{9.2.9}$$

and

(iv) the elasticity tensor satisfies (3.3.61a) and (3.3.61b), the standard symmetry and ellipticity properties.

For instance, Φ_ε can be equal to $\int_\Omega J_\mu\, d\Omega$ in which case $\varepsilon=\mu$. The solution of the corresponding quasi-static problem will be denoted by $(\sigma_\varepsilon, v_\varepsilon)$.

9.2.2 Existence and Uniqueness Propositions

In this section, we study the existence and uniqueness of the solution of Problem 5.

Proposition 9.2.1. Let assumptions (i)–(iv) of the previous section hold. Then Problem 5 admits a solution (σ, v) such that

$$\sigma\in L^\infty(0T, H) \tag{9.2.10}$$

$$\sigma'\in L^2(0T, H) \tag{9.2.11}$$

and

$$v\in L^2_w(0T, BD(\Omega)). \tag{9.2.12}$$

Moreover, the solution (σ, v) can be approximated by the solution $(\sigma_\varepsilon, v_\varepsilon)$ as follows:

$$\sigma_\varepsilon \to \sigma \quad \text{in} \quad L^\infty(0T, H) \quad \text{weakly}_*, \tag{9.2.13}$$

$$\sigma'_\varepsilon \to \sigma' \quad \text{in} \quad L^2(0T, H) \quad \text{weakly}, \tag{9.2.14}$$

and

$$v_\varepsilon \to v \quad \text{in} \quad L^2_w(0T, BD(\Omega)) \quad \text{weakly}_*. \tag{9.2.15}$$

Proof. Let $(\sigma_\varepsilon, v_\varepsilon)$ be the solution of the problem

$$(c_{ijhk}\sigma_{\varepsilon hk})' + \left[\frac{d\Phi_\varepsilon}{d\sigma}(t, \sigma_\varepsilon)\right]_{ij} = \varepsilon_{ij}(v_\varepsilon), \tag{9.2.16}$$

$$\sigma_{\varepsilon ij,j} + f_i = 0, \tag{9.2.17}$$

$$\sigma_\varepsilon(0) = \sigma_0, \tag{9.2.18}$$

which by Prop. 9.1.5 exists and is unique[5]. We consider the variational equality formulation of (9.2.16) (cf., 9.1.114) and we consider the function $\bar{\sigma}_\varepsilon = \sigma_\varepsilon - \tilde{\sigma}$. In this variational equality we set $\tau = \bar{\sigma}_\varepsilon$ and we obtain $D_{\bar{\sigma}_\varepsilon} = \{0\}$

$$((c\sigma_\varepsilon)', \bar{\sigma}_\varepsilon) + \left(\frac{d\Phi_\varepsilon}{d\sigma}(t, \sigma_\varepsilon), \bar{\sigma}_\varepsilon\right) = (\varepsilon(v_\varepsilon), \bar{\sigma}_\varepsilon) + (v_\varepsilon, D\bar{\sigma}_\varepsilon). \tag{9.2.19}$$

Using the relation

$$(\varepsilon(v_\varepsilon), \bar{\sigma}_\varepsilon) + (v_\varepsilon, D\bar{\sigma}_\varepsilon) = 0 \tag{9.2.20}$$

by integration from 0 to t we obtain the inequality

$$c\,\|\bar{\sigma}_\varepsilon(t)\|^2 + \int_0^t \left(\frac{d\Phi_\varepsilon}{d\sigma}(t, \sigma_\varepsilon), \bar{\sigma}_\varepsilon\right) dt \le c \int_0^t \|c\tilde{\sigma}'\| \; \|\bar{\sigma}_\varepsilon\|\, dt. \tag{9.2.21}$$

From Gronwall's lemma and the assumptions on Φ_ε, it follows that

$$\|\bar{\sigma}_\varepsilon\| \le c, \tag{9.2.22}$$

and then that

$$\int_0^T \left(\frac{d\Phi_\varepsilon}{d\sigma}(t, \sigma_\varepsilon), \bar{\sigma}_\varepsilon\right) dt \le c. \tag{9.2.23}$$

We use $\Phi_\varepsilon(t, \sigma_\varepsilon) \ge 0$ and the inequality

$$\Phi_\varepsilon(t, \tilde{\sigma}) - \Phi_\varepsilon(t, \sigma_\varepsilon) \ge \left(\frac{d\Phi_\varepsilon}{d\sigma}(t, \sigma_\varepsilon), \tilde{\sigma} - \sigma_\varepsilon\right) = -\left(\frac{d\Phi_\varepsilon}{d\sigma}(t, \sigma_\varepsilon), \bar{\sigma}_\varepsilon\right), \tag{9.2.24}$$

[5] In Problem 5 an initial condition for v is not needed. $(\sigma_\varepsilon, v_\varepsilon)$ is obtained by assuming that *homogeneous* boundary conditions hold on Γ_U and on Γ_F and that (9.1.8) holds with u_0 arbitrary. Note the difference between (9.2.16) and (9.1.22), (9.1.90), etc (cf., also Sec. 3.3.4). The problem of the optimal assumptions on the dependance of K on t is still open. The proof of prop. 9.2.1 is also valid with minor modifications if $\{c_{ijhk}\}$ is dependent on t and (9.1.80, 81) hold.

to prove the estimate

$$\int_0^T \Phi_\varepsilon(t, \sigma_\varepsilon)\, dt \leq c. \tag{9.2.25}$$

If $\tau = \{\tau_{ij}\}$, $\tau_{ij} = \tau_{ji} \in L^\infty(0T, L^\infty(\Omega))$ and $\|\tau\|_{\tilde{L}^\infty} < q$ a.e. on $[0, T]$, then $\tau + \tilde{\sigma} \in P(t)$ a.e. on $[0, T]$. Using the inequality

$$\int_0^T \left(\frac{d\Phi_\varepsilon}{d\sigma}(t, \sigma_\varepsilon), \tau - \bar{\sigma}_\varepsilon\right) dt \leq \int_0^T (\Phi_\varepsilon(\tilde{\sigma} + \tau) - \Phi_\varepsilon(\sigma_\varepsilon))\, dt, \tag{9.2.26}$$

we deduce that for every $\tau \in L^\infty(0T, \tilde{L}^\infty)$ with $\|\tau\|_{\tilde{L}^\infty} < q$ a.e. on $[0, T]$, the estimate

$$\int_0^T \left(\frac{d\Phi_\varepsilon}{d\sigma}(t, \sigma_\varepsilon), \tau\right) dt \leq c \tag{9.2.27}$$

holds. For the derivation of (9.2.27), we have used (9.2.23, 24) and assumptions (i) and (iii). From all the above estimates it results that

$\{\sigma_\varepsilon\}$ is bounded in $L^\infty(0T, H)$ and $\left\{\frac{d\Phi_\varepsilon}{d\sigma}(t, \sigma_\varepsilon)\right\}$ is bounded[6] in $L^1(0T, \tilde{L}^1)$.

Let us now take $\bar{\sigma}'_\varepsilon$ as a test function. We obtain, then,

$$((c\bar{\sigma}_\varepsilon)', \bar{\sigma}'_\varepsilon) + \left(\frac{d\Phi_\varepsilon}{d\sigma}(t, \sigma_\varepsilon), \bar{\sigma}'_\varepsilon\right) = (\varepsilon(v_\varepsilon), \bar{\sigma}'_\varepsilon) - ((c\tilde{\sigma})', \bar{\sigma}'_\varepsilon). \tag{9.2.28}$$

Using the relation

$$(\varepsilon(v_\varepsilon), \bar{\sigma}'_\varepsilon) = 0 \tag{9.2.29}$$

and

$$\frac{d\Phi_\varepsilon}{dt}(t, \sigma_\varepsilon) = \frac{\partial\Phi_\varepsilon}{\partial t}(t, \sigma_\varepsilon) + \left(\frac{d\Phi_\varepsilon}{d\sigma}(t, \sigma_\varepsilon), \sigma'_\varepsilon\right), \tag{9.2.30}$$

by integrating from 0 to T we find that

$$\begin{aligned} c\int_0^T \|\bar{\sigma}_\varepsilon\|^2 dt &+ \Phi_\varepsilon(T, \sigma_\varepsilon(T)) - \Phi_\varepsilon(0, \sigma_0) \\ &\leq \int_0^T \|c\tilde{\sigma}'\|\, \|\bar{\sigma}'_\varepsilon(t)\|\, dt \\ &+ \int_0^T \left|\frac{\partial\Phi_\varepsilon}{\partial t}(t, \sigma_\varepsilon)\right| dt + \int_0^T \left(\frac{d\Phi_\varepsilon}{d\sigma}(t, \sigma_\varepsilon), \tilde{\sigma}'\right) dt. \end{aligned} \tag{9.2.31}$$

[6] We apply the following proposition for $p = \infty$ (cf., [2] p. 39): for each $g \in L^{p'}(\Omega)$, where $1 \leq p \leq \infty$ and p' is the conjugate exponent, we can define $\tilde{g} \in [L^p(\Omega)]^*$ through $\tilde{g}(f) = \int_\Omega g(x) f(x)\, d\Omega, f \in L^p(\Omega)$. Then $\|\tilde{g}\|_{[L^p(\Omega)]'} = \|g\|_{L^{p'}(\Omega)}$.

Because of (9.2.27) and (9.1.79), we may conclude from (9.2.31) that

$$\int_0^T \| \bar{\sigma}'_\varepsilon \|^2 \, dt \leq c, \tag{9.2.32}$$

i.e., $\{\bar{\sigma}'_\varepsilon\}$ (or $\{\sigma'_\varepsilon\}$) is bounded in $L^2(0T, H)$. In order to prove that $\{((d\Phi_\varepsilon/d\sigma)(t, \sigma_\varepsilon), \bar{\sigma}_\varepsilon)\}$ is bounded in $L_2(0, T)$, we take $\bar{\sigma}_\varepsilon$ as a test function. Then (9.2.19) yields

$$\left(\frac{d\Phi_\varepsilon}{d\sigma}(t, \sigma_\varepsilon), \bar{\sigma}_\varepsilon\right) = -((c\sigma_\varepsilon)', \bar{\sigma}_\varepsilon) \tag{9.2.33}$$

from which, since $\{\sigma'_\varepsilon\}$ is bounded in $L^2(0T, H)$ and $\{\sigma_\varepsilon\}$ in $L^\infty(0T, H)$, the result follows. By means of the same method as before (cf., the proof that $\{d\Phi_\varepsilon/d\sigma\}$ is bounded in $L^1(0T, \tilde{L}^1)$), we may show that $\{(d\Phi_\varepsilon/d\sigma)(t, \sigma_\varepsilon)\}$ is bounded in $L^2(0T, \tilde{L}^1)$. (9.2.16) implies that $\{\varepsilon(v_\varepsilon)\}$ is bounded in $L^2(0T, \tilde{L}^1)$. From (1.4.49), by taking $p(u) = \int_{\Gamma_U} |u| \, d\Gamma$ (Γ_U must imply: $p(u)$ is a norm on $\mathscr{R}$) we find that for every $v \in BD(\Omega)$ with $v = 0$ on Γ_U the inequality

$$\| v \|_{BD(\Omega)} \leq c \sum_{i,j=1}^{3} \| \varepsilon_{ij}(v) \|_{\mathscr{M}^1(\Omega)} \tag{9.2.34}$$

holds. This inequality implies – because if $\{\varepsilon(v_\varepsilon)\}$ is bounded in $L^2(0T, \tilde{L}^1)$, then it is also bounded in $L^2(0T, \tilde{\mathscr{M}}^1)$ ($\tilde{\mathscr{M}}^1$ is defined analogously to $\tilde{L}^1$) – that

$$\{v_\varepsilon\} \quad \text{is bounded in} \quad L^2(0T, BD(\Omega)). \tag{9.2.35}$$

From (9.2.22) and (9.2.32), by considering subsequences we find that

$$\sigma_\varepsilon \to \sigma \quad \text{in} \quad L^\infty(0T, H) \quad \text{weakly}_*, \tag{9.2.36}$$

$$\sigma'_\varepsilon \to \sigma' \quad \text{in} \quad L^2(0T, H) \quad \text{weakly} \tag{9.2.37}$$

and

$$v_\varepsilon \to v \quad \text{in} \quad L^2_w(0T, BD(\Omega)) \quad \text{weakly}_*. \tag{9.2.38}$$

It remains to be proved that (σ, v) is a solution of the plasticity B.V.P. From (9.2.36, 37) we immediately see that (9.2.4) and (9.2.5) are satisfied. Moreover, due to (9.2.36), and by means of (9.2.25) and (9.2.9) we conclude that $\sigma(t) \in P(t)$ for a.e. t. In order to verify (9.2.3), we multiply (9.2.16) by $\tau - \sigma_\varepsilon$, with $\tau(t) \in P(t) \cap \Sigma_0$, and integrate over $[0, T]$. As a result,

$$\int_0^T ((c\sigma_\varepsilon)', \tau - \sigma_\varepsilon) \, dt + \int_0^T \left(\frac{d\Phi_\varepsilon}{d\sigma}(t, \sigma_\varepsilon), \tau - \sigma_\varepsilon\right) dt + \int_0^T (v_\varepsilon, D\tau - D\sigma_\varepsilon) \, dt = 0. \tag{9.2.39}$$

We may now verify, since $BD(\Omega) \subset [L^{3/2}(\Omega)]^3$, that

$$\lim_{\varepsilon\to 0} \int_0^T \left[\int_\Omega v_{\varepsilon i}(\tau_{ij,j} - \sigma_{\varepsilon ij,j})\, d\Omega \right] dt = \int_0^T \left[\int_\Omega v_i(\tau_{ij,j} - \sigma_{ij,j})\, d\Omega \right] dt, \quad (9.2.40)$$

$$\limsup_{\varepsilon\to 0} \int_0^T ((c\sigma_\varepsilon)', \tau - \sigma_\varepsilon)\, dt = \int_0^T ((c\sigma)', \tau - \sigma)\, dt \quad (9.2.41)$$

and

$$\left(\frac{d\Phi_\varepsilon}{d\sigma}(t, \sigma_\varepsilon), \tau - \sigma_\varepsilon \right) \le \Phi_\varepsilon(t, \tau) - \Phi_\varepsilon(t, \sigma_\varepsilon) \le 0 \quad (9.2.42)$$

From (9.2.38–42) we obtain (9.2.3), q.e.d.

Proposition 9.2.2. The stress field is uniquely determined.

Proof. If (σ_1, v_1) and (σ_2, v_2) are two solutions of Problem 5, then from (9.2.3)

$$(c(\sigma_1' - \sigma_2'), \sigma_1 - \sigma_2) \le 0, \quad (9.2.43)$$

from which, by integrating from 0 to t, it results by means of (3.3.61a, b) that

$$\|\sigma_1(t) - \sigma_2(t)\|^2 \le c\|\sigma_1(0) - \sigma_2(0)\|^2. \quad (9.2.44)$$

This inequality and the fact that $\sigma_1(0) = \sigma_2(0)$ imply the uniqueness of the stress field, q.e.d.

The use of the space $BD(\Omega)$ for the velocities allows us to consider velocity discontinuities with respect to a surface S lying in Ω. They are distinguished either as tangential discontinuities corresponding to slip phenomena or as normal discontinuities. However, if the yield criterion depends only on the stress deviator, as for instance in metals, only slip discontinuities may occur. To show this, we shall apply for a surface S in Ω with normal $n = \{n_i\}$, the properties given in Sec. 1.4.3. If S divides Ω into Ω_+ and Ω_-, then from (1.4.47) and (1.4.48) we get the relation

$$\int_S (\gamma_+ v - \gamma_- v)_i \tau_{ij} n_j\, dS = \int_S \tau_{ij}\, d\varepsilon_{ij}(u), \qquad \forall \tau \in \tilde{D}(\bar{\Omega}), \quad (9.2.45)$$

which explains mathematically the origin of the velocity jump $(\gamma_+ v - \gamma_- v)$ along S. Since the yield criterion depends only on the stress deviator; we may set in (9.2.3) $\tau_{ij} = \sigma_{ij} + p\delta_{ij}$, where $p \in \mathring{H}^1(\Omega)$. The Green-Gauss theorem implies that for every $p \in \mathring{H}^1(\Omega)$

$$((c\sigma)', \tau - \sigma) + \int_\Omega v_i(\tau_{ij,j} - \sigma_{ij,j})\, d\Omega$$

$$= \int_\Omega (c_{iihk}\sigma_{hk})' p\, d\Omega + \int_\Omega v_i p_{,i}\, d\Omega = 0. \quad (9.2.46)$$

Accordingly,

$$\operatorname{div} v = v_{i,i} = (c_{iihk}\sigma_{hk})' \quad \text{in} \quad L^2(\Omega), \quad (9.2.47)$$

and thus div $v = \varepsilon'_{ii}$ has zero measure on S. By setting $\tau_{ij} = p\delta_{ij}$ in (9.2.45), we find that

$$(\gamma_+ v)_i n_i = (\gamma_- v)_i n_i \quad \text{in} \quad L^1(S), \quad (9.2.48)$$

i.e., only slip discontinuities occur. This result agrees with experimental data which confirm the appearance of strain concentrations in thin bands. Such bands can be considered as discontinuity surfaces.

Normal discontinuities appear in the case of yield criteria depending on the complete stress tensor, as happens for instance in several types of soils. For these discontinuities there are several theories explaining their origin; the reader may refer to the respective literature (see, e.g., [306]).

The discontinuities of the velocity field oblige us to be content with the weak formulation (9.2.3–5) of the elastoplasticity B.V.P. A stronger formulation, for instance as we have done in the viscoplastic case, would lead to continuous velocity fields. It is obvious from the above considerations that the boundary Γ of the body may be a slip line, and for this reason the boundary condition on Γ_U is satisfied by the external trace of the velocity. Therefore a plasticity B.V.P. should be formulated more strictly not on Ω as usual, but at least on $\bar{\Omega}$. We may thus define the field of admissible velocities by introducing the space $BD(\mathbb{R}^3)$ and by defining v_1 to be equivalent to v_2 if

$$v_1|_\Omega = v_2|_\Omega \quad \text{in} \quad [L^1(\Omega)]^3 \quad \text{and} \quad \varepsilon_{ij}(v_1)|_{\bar{\Omega}} = \varepsilon_{ij}(v_2)|_{\bar{\Omega}} \quad \text{in} \quad \mathscr{M}^1(\bar{\Omega}). \quad (9.2.49)$$

Hence $\gamma_+ v_1 = \gamma_+ v_2$ on Γ, and an equivalence relation denoted by P is introduced leading to the quotient space $BD(\bar{\Omega}) = BD(\mathbb{R}^3)/P$. Then the set of kinematically admissible velocities is

$$BD_{ad}(\bar{\Omega}) = \{v \mid v \in BD(\bar{\Omega}),\ \gamma_+ v = 0 \quad \text{on} \quad \Gamma_U\}. \quad (9.2.50)$$

It can be verified that the solution of Problem 5 is an element of $BD_{ad}(\bar{\Omega})$ [307].

As has been observed in [314], concerning the functional setting the problem being considered here reveals certain similarities to the minimal surface problem, where the space of functions of bounded variation ($BV(\Omega)$)

is used. The methods applied to the study of the latter problem, especially regarding fulfillment of the boundary conditions, have, in a sense, been "transferred" to the plasticity problem.

9.3 Rigid Viscoplastic Flow Problems

9.3.1 Classical Formulation of the General Dynamic Problem

In this section we will study inequality problems formulated in the framework of rigid-viscoplastic flows (see Sec. 3.3.3). The influence of thermal phenomena is neglected; we are interested only in the velocity and stress fields. We assume, for the sake of simplicity, that the density ρ of the fluid is constant, i.e., that

$$\rho(x, t) = \rho_0 = c^{st}. \tag{9.3.1}$$

Let Ω be an open, bounded,[7] connected subset of $\mathbb{R}^3$ and Γ its boundary. We seek a velocity field $v = \{v_i\}$ and a stress field $\sigma = \{\sigma_{ij}\}$, $i, j = 1, 2, 3$ such as to satisfy in $\Omega \times (0, T)$ the equation of motion

$$\rho\gamma_i = \sigma_{ij,j} + f_i \tag{9.3.2}$$

with $\gamma_i = \partial v_i/\partial t + v_{i,j} v_j$ and $f_i = f_i(x, t)$, the equation of incompressibility

$$\operatorname{div} v = 0 \tag{9.3.3}$$

and the material law (3.3.110). On Γ the boundary condition

$$v_i = 0 \tag{9.3.4}$$

is considered for every $t \in (0, T)$ and at $t = 0$ the initial condition

$$v_i(x, 0) = v_{0i}, \quad \text{with} \quad v_{0i} = v_{0i}(x) \tag{9.3.5}$$

holds. The viscosity coefficient μ and the yield limit g in (3.3.110) are assumed here to be positive constants. (A generalization for $\mu = \mu(x, t)$, $g = g(x, t)$ is also possible). Let us introduce the notation

$$D_{II}(v) = \tfrac{1}{2} D_{ij}(v) D_{ij}(v), \tag{9.3.6}$$

$$\Phi(v) = 2 \int_\Omega g(D_{II}(v))^{1/2}\, d\Omega \tag{9.3.7}$$

and

$$a(v, w) = 2 \int_\Omega \mu D_{ij}(v) D_{ij}(w)\, d\Omega, \qquad (f, v) = \int_\Omega f_i v_i\, d\Omega. \tag{9.3.8}$$

[7] The assumption "Ω bounded" is abandoned if we study the flow of a Bingham fluid around a fixed obstacle, in which case, it is assumed that as $x \to \infty$ $v(x, t) \to \{c, 0, 0\}$ where c is a constant, and a proposition similar to Prop. 9.3.1 may be proved [84].

If u is a sufficiently smooth solution of (9.3.1–5) and (3.3.110), it satisfies for every $t \in (0, T)$ the variational inequality

$$a(u, v-u) + \Phi(v) - \Phi(u) \geq (f - \rho\gamma, v-u), \qquad \forall v \in V_{ad}, \tag{9.3.9}$$

the condition of incompressibility and the boundary and initial conditions. The velocity fields v fulfilling these three conditions constitute the set V_{ad} in the velocity space, which remains to be chosen. The proof of (9.3.9) is straightforward if we multiply (9.3.2) by $v_i - u_i$, sum with respect to i and formally apply the Green–Gauss theorem. There results the equality

$$\int_\Omega \sigma_{ij}(u)[D_{ij}(v) - D_{ij}(u)]\, d\Omega = (f - \rho\gamma, v-u), \qquad \forall v \in V_{ad}, \tag{9.3.10}$$

which, combined with (3.3.112), yields (9.3.9) by means of the well-known Schwarz inequality. Conversely, if Φ is differentiable at u, i.e., if $D_{II}(u) \neq 0$, then we can show formally that for $u \in V_{ad}$ (9.3.9) is equivalent to the initial problem. Indeed, by forming grad $\Phi(u)$ we obtain from (9.3.9) the relation

$$\int_\Omega \left(\left(g\frac{D_{ij}(u)}{(D_{II}(u))^{1/2}} + 2\mu(D_{ij}(u)\right)_{,i} + f_i - \rho\gamma_i\right) v_i\, d\Omega = 0, \qquad \forall v \in V_{ad}, \tag{9.3.11}$$

from which we conclude, since div $v = 0$, that there exists a scalar function p such that the term in parentheses in (9.3.11) is equal to grad p. This completes the proof of our assertion.

9.3.2 The Functional Framework and Existence Propositions

Let us consider generally that $\Omega \subset \mathbb{R}^n$. We denote by $\mathscr{W}$ the space

$$\mathscr{W} = \{\phi \mid \phi \in [D(\Omega)]^n, \operatorname{div} \phi = 0 \text{ in } \Omega\}, \qquad n \geq 2. \tag{9.3.12}$$

The closure of $\mathscr{W}$ in $[H^\alpha(\Omega)]^n$ is the Hilbert space W_α. We denote by $[\cdot, \cdot]_\alpha$ and $|\cdot|_\alpha$ the scalar product and the norm on W_α respectively. Obviously,

$$\begin{aligned}&[v, w]_{n/2} = ((v_i, w_i))_{H^{n/2}(\Omega)},\\ &[v, w]_0 = ((v_i, w_i))_{H^0(\Omega)} = (v_i, w_i), \qquad |v|_0 = |v|_2.\end{aligned} \tag{9.3.13}$$

Note that $v \in W_{n/2}$ implies

$$v_{i,j} \in L^n(\Omega), \quad i, j = 1, 2, 3. \tag{9.3.14}$$

Indeed, for $v \in W_{n/2}$, $v_{i,j} \in H^{n/2-1}$, and thus (9.3.14) holds (see (1.3.50)). If $W'_{n/2}$ denotes the dual of $W_{n/2}$ for the duality pairing (v, w) then it can be easily shown that the spaces $\{W'_\alpha, W_\alpha\}$, $\alpha \geq 1$, have the properties of a Hilbert scale of spaces with $W_0 = W'_0$ as the middle element of the scale.

Moreover, the injection from W_α into W_0 is compact, since the injection $W_1 \subset W_0$ is compact.

Further, we introduce the notation

$$E(u, v, w) = \int_\Omega \rho u_i v_{j,i} w_j \, d\Omega. \tag{9.3.15}$$

For E the equality

$$E(u, v, w) + E(u, w, v) = 0, \qquad \forall u, v, w \in \mathscr{W} \tag{9.3.16}$$

holds, which can be appropriately extended by density. Due to the inequality[8]

$$\begin{aligned} |E(u, v, w)| &\leq c|u|_p|w|_p \sum_{ij} |v_{i,j}|_n \\ &\leq c|u|_p|w|_p \| v \|_{n/2} \quad \text{for } \frac{2}{p} + \frac{1}{n} = 1, \end{aligned} \tag{9.3.17}$$

which results from (9.3.14) and from (9.3.15) by applying Hölder's inequality, $(u, v, w) \to E(u, v, w)$ is continuous on $L^p \times W_{n/2} \times L^p (L^p = [L^p(\Omega)]^n)$. In terms of the velocities, and using the relations

$$E(u, u, u) = 0 \quad \text{and} \quad E(u, u, v) = -E(u, v, u), \tag{9.3.18}$$

(9.3.9) acquires the form

$$\begin{aligned} (\rho u', v - u) + a(u, v - u) - E(u, v, u) \\ + \Phi(v) - \Phi(u) \geq (f, v - u), \qquad \forall v \in V_{ad}. \end{aligned} \tag{9.3.19}$$

We shall distinguish the two-dimensional ($n = 2$) from the three-dimensional ($n = 3$) flow problem, where a weaker variational form of the problem is studied. Integrating (9.3.19) from 0 to T and assuming that v is a function of t, we get the inequality

$$\begin{aligned} \int_0^T ((\rho v', v - u) + a(u, v - u) - E(u, v, u) \\ + \Phi(v) - \Phi(u) - (f, v - u)) \, dt \\ + \frac{\rho}{2} |v(0) - u(0)|^2 \geq 0 \forall v \in \tilde{\mathscr{W}}, \end{aligned} \tag{9.3.20}$$

whose solution u will be sought in a set $\tilde{\mathscr{W}}$, which we now define

Proposition 9.3.1. (a) Let $\Omega \subset \mathbb{R}^3$ (or $\mathbb{R}^n$ with $n \geq 3$),

$$f \in L^2(0T, W_1') \tag{9.3.21}$$

and

$$v_0 \in W_0. \tag{9.3.22}$$

[8] $u \in W_{n/2} \subset W_1$ implies by (1.3.48) that $u_i \in L^q(\Omega)$: $1 \leq q \leq 2n/n - 2$ if $n > 2$, $1 \leq q < \infty$ if $n = 2$.

We denote by $\tilde{\mathscr{W}}$ the set

$$\tilde{\mathscr{W}} = \{v | v \in L^2(0T, W_{n/2}), v' \in L^2(0T, W_0), v(0) = v_0\}. \tag{9.3.23}$$

Then there exists a u such that[9]

$$u \in L^2(0T, W_1) \cap L^\infty(0T, W_0), \tag{9.3.24}$$

$$u' \in L^2(0T, W'_{n/2}), \tag{9.3.25}$$

which satisfies the variational inequality

$$\int_0^T [(\rho v', v - u) + a(u, v - u) + E(u, u, v - u)$$

$$+ \Phi(v) - \Phi(u) - (f, v - u)] \, dt \geq 0, \forall v \in \tilde{\mathscr{W}} \tag{9.3.26}$$

and the initial condition (9.3.5).

(b) Let $\Omega \subset \mathbb{R}^2$ and suppose that (9.3.21, 22) hold. Then there exists a unique u such that

$$u \in L^2(0T, W_1) \cap C(0T, W_0), \qquad u' \in L^2(0T, W'_1), \tag{9.3.27}$$

which satisfies a.e. on $[0, T]$ the variational inequality

$$(\rho u', v - u) + a(u, v - u) + E(u, u, v - u)$$

$$+ \Phi(v) - \Phi(u) \geq (f, v - u), \qquad \forall v \in W_1 \tag{9.3.28}$$

and the initial condition (9.3.5).

Proof. (a) For simplicity, we assume that $v_0 = 0$. (If $v_0 \neq 0$, we can make the transformation $\tilde{v} = v - v_0$, $\tilde{u} = u - u_0$). We will use the finite difference method employed in Prop. 9.1.1. Again, let $k = T/N$, N integer > 0, $t^{(\nu)} = \nu k$, $\nu = 0, 1, 2, \ldots, N$, $\sigma^{(\nu)} = \sigma(t^{(\nu)})$, $u^{(\nu)} = u(t^{(\nu)})$, $\Phi^{(\nu)} = \Phi(u(t^{(\nu)}))$ and $f^{(\nu)} = f(t^{(\nu)})$.

Furthermore, functional Φ is regularized; let

$$\Phi_\varepsilon(v) = \frac{2g}{1 + \varepsilon} \int_\Omega (\sqrt{D_{II}(v)})^{1+\varepsilon} \, d\Omega. \tag{9.3.29}$$

Then

$$\left(\frac{d\Phi_\varepsilon(u)}{dv}, v\right) = g \int_\Omega (\sqrt{D_{II}(u)})^{\varepsilon - 1} D_{ij}(u) D_{ij}(v) \, d\Omega, \tag{9.3.30}$$

and for Φ_ε relations analogous to (5.1.20, 21) can be verified. We now pose the following problem: find $u_\varepsilon^{(\nu+1)} \in W_{n/2}$ such that $u_\varepsilon^{(0)} = 0$ and, for $\nu = 0, 1, \ldots, N$,

[9] Then (see [182] p. 73) $u \in L^4(0T, [L^p(\Omega)]^n)$, where $1/p + 1/2n = 1/2$. By the theory of interpolation spaces [181], u is a.e. continuous from $[0, T] \longrightarrow W'_{n/4-1/2}$.

$$\left(\rho\,\frac{u_\varepsilon^{(\nu+1)}-u_\varepsilon^{(\nu)}}{k},\,v\right)+a(u_\varepsilon^{(\nu+1)},v)+E(u_\varepsilon^{(\nu+1)},u_\varepsilon^{(\nu+1)},v)$$
$$+\left(\frac{d\Phi_\varepsilon}{dv}(u_\varepsilon^{(\nu+1)}),\,v\right)-(f^{(\nu+1)},v)=0,\qquad \forall v\in W_{n/2}.\quad (9.3.31)$$

To study this problem, we must perform another regularization by means of the viscosity term $\eta[u_\varepsilon^{(\nu+1)}, v]_{n/2}$, $\eta > 0$, where η as well as ε tends to zero. Then the variational equality

$$\left(\rho\left(\frac{u_{\varepsilon\eta}^{(\nu+1)}-u_{\varepsilon\eta}^{(\nu)}}{k}\right),\,v\right)+\eta[u_{\varepsilon\eta}^{(\nu+1)},v]_{n/2}$$
$$+\,a(u_{\varepsilon\eta}^{(\nu+1)},v)+E(u_{\varepsilon\eta}^{(\nu+1)},u_{\varepsilon\eta}^{(\nu+1)},v)\qquad (9.3.32)$$
$$+\left(\frac{d\Phi_\varepsilon(u_{\varepsilon\eta}^{(\nu+1)})}{du},\,v\right)-(f^{(\nu+1)},v)=0,\qquad \forall v\in W_{n/2}$$

is considered. The proof is divided into two main steps. First, we prove the existence of a solution $u_{\varepsilon\eta}^{(\nu+1)} \in W_{n/2}$ of (9.3.32) and we let k tend to zero. In the second step, ε and η tend to zero and $u_{\varepsilon\eta}$ tends to a solution of (9.3.26). Let $w_1, w_2, \ldots,$ be a special base of eigenfunctions in $W_{n/2}$ defined by [182]

$$[w_j, v]_{n/2}=\lambda_j(w_j, v),\qquad \forall v\in W_{n/2},\qquad |w_j|_2=1.\qquad (9.3.33)$$

The w_j's are the eigenfunctions of the canonical isometry B between $W_{n/2}$ and $W'_{n/2}$. Due to the compactness properties of the spaces $W_{n/2}$, this eigenvalue problem admits a sequence of nonzero solutions w_j corresponding to a sequence of eigenvalues $\lambda_j > 0$ (cf. Sec. 7.2.4). We denote by $u_m = u_{\varepsilon\eta m}^{(\nu+1)} \in W_{n/2}^{(m)}$ (the finite-dimensional space spanned by w_j, $1 \le j \le m$) the solution of the problem

$$\left(\rho\left(\frac{u_m-u_{\varepsilon\eta}^{(\nu)}}{k}\right),\,w_j\right)+\eta[u_m,w_j]_{n/2}+a(u_m,w_j)$$
$$+\,E(u_m,u_m,w_j)+\left(\frac{d\Phi_\varepsilon(u_m)}{du},\,w_j\right)-(f^{(\nu+1)},w_j)=0,\qquad 1\le j\le m.\quad (9.3.34)$$

Using the monotonicity of $d\Phi_\varepsilon/du$ and the coerciveness of $a(u_m, u_m)$ on W_1 (the latter due to (9.3.4)), we conclude from (9.3.34) that

$$|u_m|_1\le c,\qquad |u_m|_2\le c,\qquad \eta^{1/2}\,|u_m|_{n/2}\le c\qquad (9.3.35)$$

independently of m, ε, η, k. We shall show that $v \to E(u_m, u_m, v)$ is a continuous linear functional on $W_{n/2}$. We consider the inequality

$$|v|_p\le c|v|_2^{1/2}\,\|v\|_1^{1/2},\qquad \forall v\in \mathring{H}^1(\Omega),\qquad (9.3.36)$$

which holds for $1/p + 1/2n = 1/2$. Indeed, (1.2.24), applied for $a_1 = 0$, $a_2 = 1/2$, $a_3 = 1$, and the imbedding of $H^{1/2}(\Omega)$ into $L^p(\Omega)$ (Sec. 1.3.4) imply (9.3.36).

From (9.3.17) and (9.3.36) we obtain

$$|E(u, v, w)| \leq c\,|\dot{u}|_1^{1/2}|u|_2^{1/2}\,|w|_1^{1/2}|w|_2^{1/2}\,|v|_{n/2}. \qquad (9.3.37)$$

Since $E(u_m, u_m, v) = -E(u_m, v, u_m)$, (9.3.37) implies that $v \rightarrow E(u_m, u_m, v)$ is a continuous linear functional on $W_{n/2}$, because of (9.3.35). Thus there exists $h(u_m) \in W'_{n/2}$ such that

$$E(u_m, u_m, v) = (h(u_m), v), \quad \forall v \in W_{n/2} \quad \text{and} \quad |h(u_m)|_{W'_{n/2}} \leq c|u_m|_p^2. \quad (9.3.38)$$

The function $v \rightarrow a(u, v)$ is continuous on W_1, and by means of (1.2.25) we can write that

$$a(u, v) = (Au, v), \qquad (9.3.39)$$

where $A: W_1 \rightarrow W'_1$ is linear and continuous. Thus (9.3.34) acquires the form

$$\left(\rho\frac{(u_m - u_{\varepsilon\eta}^{(v)})}{k} + \eta Bu_m + Au_m + h(u_m)\right.$$
$$\left. + \frac{d\Phi_\varepsilon}{du}(u_m) - f_i^{v+1} w_j\right) = 0, \qquad 1 \leq j \leq m. \quad (9.3.40)$$

The existence of a solution $u_m \in W_{n/2}^{(m)}$ of (9.3.40) may be easily proved by applying the fixed-point theorem of Brouwer.[10]

From (9.3.35) and (9.3.40), it results by considering subsequences that as $m \rightarrow \infty$

$$u_m \rightarrow u_{\varepsilon\eta}^{(v+1)} \quad \text{weakly in} \quad W_1, \qquad (9.3.41)$$

$$u_{mi}u_{mj} \rightarrow k_{ij}^{(v+1)} \quad \text{weakly in} \quad L^{p/2}(\Omega) \qquad (9.3.42)$$

and

$$\frac{d\Phi_\varepsilon}{du}(u_m) \rightarrow \psi^{(v+1)} \quad \text{weakly in} \quad W'_1. \qquad (9.3.43)$$

(9.3.42) is obtained by applying Hölder's inequality to $|u_{mi}u_{mj}|_{p/2}$ and taking into account (9.3.36, 35). This implies that

$$|u_{mi}u_{mj}|_{p/2} \leq c. \qquad (9.3.44)$$

[10] See Sec. 8.4.2. In our case, we put $u_m = \sum_{i=1}^m \xi_i w_i \in W_{n/2}^{(m)}$ and

$$(T\xi)_i = \left(\left(\rho\frac{u_m}{k} + \eta Bu_m + Au_m + \frac{d\Phi_\varepsilon(u_m)}{du}\right.\right.$$
$$\left.\left. - \left(f^{(v+1)} - h_m + \rho\frac{u_{\varepsilon\eta}^{(v)}}{k}\right)\right), w_i\right) \text{ with } \quad h_m = h(u_m).$$

Due to the compactness of the imbedding of W_1 in W_0,

$$u_m \to u_{\varepsilon\eta}^{(\nu+1)} \quad \text{strongly in} \quad W_0, \tag{9.3.45}$$

and by a well-known theorem of analysis (see, e.g., [302]), there exists a subsequence converging a.e. in Ω. It results that

$$u_{mi}\, u_{mj} \to u_{\varepsilon\eta i}^{(\nu+1)}\, u_{\varepsilon\eta j}^{(\nu+1)} \quad \text{in} \quad \mathscr{D}'(\Omega), \tag{9.3.46}$$

and thus $k_{ij} = u_{\varepsilon\eta i}^{(\nu+1)} u_{\varepsilon\eta j}^{(\nu+1)}$. Accordingly,

$$E(u_m, u_m, v) \to E(u_{\varepsilon\eta}^{(\nu+1)}, u_{\varepsilon\eta}^{(\nu+1)}, v), \qquad \forall v \in W_{n/2}. \tag{9.3.47}$$

From (9.3.34), since the basis $\{w_j\}$ is complete in $W_{n/2}$, there follows the inequality

$$\left(\rho\left(\frac{u_{\varepsilon\eta}^{(\nu+1)} - u_{\varepsilon\eta}^{(\nu)}}{k}\right), v\right) + \eta[u_{\varepsilon\eta}^{(\nu+1)}, v]_{n/2} + a(u_{\varepsilon\eta}^{(\nu+1)}, v)$$
$$+ E(u_{\varepsilon\eta}^{(\nu+1)}, u_{\varepsilon\eta}^{(\nu+1)}, v) + (\psi^{(\nu+1)}, v) - (f^{(\nu+1)}, v) = 0, \qquad \forall v \in W_{n/2}. \tag{9.3.48}$$

By the well-known monotonicity argument, we verify that

$$\psi^{(\nu+1)} = \frac{d\Phi_\varepsilon}{dv}(u_{\varepsilon\eta}^{(\nu+1)}). \tag{9.3.49}$$

Thus we have proved the existence of $u_{\varepsilon\eta}^{(\nu+1)}$. Further, we let $k \to 0$. From (9.3.32), by setting $v = u_{\varepsilon\eta}^{(\nu+1)}$ we obtain

$$\left(\rho\,\frac{(u_{\varepsilon\eta}^{(\nu+1)} - u_{\varepsilon\eta}^{(\nu)})}{k}, u_{\varepsilon\eta}^{(\nu+1)}\right) + \eta\,|\,u_{\varepsilon\eta}^{(\nu+1)}\,|_{n/2}^2 + c\,|\,u_{\varepsilon\eta}^{(\nu+1)}\,|_1^2$$
$$\leq c\,|\,f^{(\nu+1)}\,|_{W_1'}\,|\,u_{\varepsilon\eta}^{(\nu+1)}\,|_1. \tag{9.3.50}$$

Here we have used the monotonicity of $d\Phi_\varepsilon(v)/dv$, the coerciveness of $a(v,v)$ on W_1 and the fact that $E(u, u, u) = 0$. Note that

$$(\rho(u_{\varepsilon\eta}^{(\nu+1)} - u_{\varepsilon\eta}^{(\nu)}), u_{\varepsilon\eta}^{(\nu+1)}) = \tfrac{1}{2}\rho\,|\,u_{\varepsilon\eta}^{(\nu+1)}\,|_2^2$$
$$- \tfrac{1}{2}\rho\,|\,u_{\varepsilon\eta}^{(\nu)}\,|_2^2 + \tfrac{1}{2}\rho\,|\,u_{\varepsilon\eta}^{(\nu+1)} - u_{\varepsilon\eta}^{(\nu)}\,|_2^2. \tag{9.3.51}$$

Summing the inequalities (9.3.50) over ν, by means of (9.3.51) for $\nu = 0, 1 \ldots, M-1$ and for $\alpha > 0$ we get the inequality

$$\tfrac{1}{2}\rho\,|\,u_{\varepsilon\eta}^M\,|_2^2 + \sum_{\nu=0}^{M-1} [\tfrac{1}{2}\rho\,|\,u_{\varepsilon\eta}^{(\nu+1)} - u_{\varepsilon\eta}^{(\nu)}\,|_2^2$$
$$+ \eta k\,|\,u_{\varepsilon\eta}^{(\nu+1)}\,|_{n/2}^2 + ck\,|\,u_{\varepsilon\eta}^{(\nu+1)}\,|_1^2] \tag{9.3.52}$$
$$\leq c \sum_{\nu=0}^{M-1}\left(\alpha k\,|\,u_{\varepsilon\eta}^{(\nu+1)}\,|_1^2 + \frac{k}{\alpha}\,|\,f^{(\nu+1)}\,|_{W_1'}^2\right),$$

from which it follows that there exist constants c independent of ε, η, k such that

$$\max_M |u_{\varepsilon\eta}^M|_2^2 \leq c, \tag{9.3.53}$$

$$\sum_{\nu=0}^{M-1} k\eta\, |u_{\varepsilon\eta}^{(\nu+1)}|_{n/2}^2 \leq c \tag{9.3.54}$$

and

$$\sum_{\nu=0}^{M-1} k\, |u_{\varepsilon\eta}^{(\nu+1)}|_1^2 \leq c. \tag{9.3.55}$$

Moreover, from (9.3.30) we obtain

$$\sum_{\nu=0}^{M-1} k \left|\frac{d\Phi_\varepsilon(u_{\varepsilon\eta}^{(\nu+1)})}{dv}\right|_{W_1'}^2 \leq c, \tag{9.3.56}$$

and, from (9.3.55) and (9.3.38)

$$\sum_{\nu=0}^{M-1} k\, |h(u_{\varepsilon\eta}^{(\nu+1)})|_{W_{n/2}'}^2 \leq c. \tag{9.3.57}$$

From (9.3.32) and the foregoing estimates, we find that

$$\sum_{\nu=0}^{M-1} k \left|\frac{u_{\varepsilon\eta}^{(\nu+1)} - u_{\varepsilon\eta}^{(\nu)}}{k}\right|_{W_{n/2}'}^2 \leq c. \tag{9.3.58}$$

Next, we extend $u_{\varepsilon\eta}^{(\nu+1)}$ to be defined on $[0, T]$ by assuming a linear interpolation between the time moments $t^{(\nu)}$, i.e., we define

$$u_{\varepsilon\eta}^{(N)} = u_{\varepsilon\eta}^{(\nu)} + \frac{(t - t^{(\nu)})(u_{\varepsilon\eta}^{(\nu+1)} - u_{\varepsilon\eta}^{(\nu)})}{t^{(\nu+1)} - t^{(\nu)}} \quad \text{if} \quad t \in [t^{(\nu)}, t^{(\nu+1)}]. \tag{9.3.59}$$

The foregoing estimations lead us to the assertions that, independently of ε, η, k,

$$u_{\varepsilon\eta}^{(N)} \quad \text{remains in a bounded set in} \quad L^\infty(0T, W_0) \cap L^2(0T, W_1), \tag{9.3.60}$$

$$\eta^{1/2} u_{\varepsilon\eta}^{(N)} \text{ in a bounded set in } \quad L^2(0T, W_{n/2}), \tag{9.3.61}$$

$$\frac{d\Phi_\varepsilon(u_{\varepsilon\eta}^{(N)})}{dv} \text{ in a bounded set in } \quad L^2(0T, W'_{n/2}) \tag{9.3.62}$$

and

$$(u_{\varepsilon\eta}^{(N)})' \text{ in a bounded set in } \quad L^2(0T, W'_{n/2}). \tag{9.3.63}$$

Accordingly, as $k \to 0$, there exist $u_{\varepsilon\eta} \in L^\infty(0T, W_0) \cap L^2(0T, W_1)$,

$$\eta^{1/2} u_{\varepsilon\eta} \in L^2(0T, W_{n/2}), \qquad \psi \in L^2(0T, W'_{n/2}) \quad \text{and} \quad u'_{\varepsilon\eta} \in L^2(0T, W'_{n/2})$$

such that

$$u_{\varepsilon\eta}^{(N)} \to u_{\varepsilon\eta} \quad \text{weakly}_* \text{ in} \quad L^\infty(0T, W_0) \text{ and weakly in } L^2(0T, W_1), \tag{9.3.64}$$

$$\eta^{1/2} u_{\varepsilon\eta}^{(N)} \to \eta^{1/2} u_{\varepsilon\eta} \quad \text{weakly in} \quad L^2(0T, W_{n/2}), \tag{9.3.65}$$

$$\frac{d\Phi_\varepsilon(u_{\varepsilon\eta}^{(N)})}{dv} \to \psi \quad \text{weakly in} \quad L^2(0T, W'_{n/2}) \tag{9.3.66}$$

and

$$(u_{\varepsilon\eta}^{(N)})' \to u'_{\varepsilon\eta} \quad \text{weakly in} \quad L^2(0T, W'_{n/2}). \tag{9.3.67}$$

By the monotonicity argument, we can verify that

$$\psi = \frac{d\Phi_\varepsilon(u_{\varepsilon\eta})}{dv}. \tag{9.3.68}$$

Thus $u_{\varepsilon\eta}$ satisfies the variational equality

$$(\rho u'_{\varepsilon\eta}, v) + \eta[u_{\varepsilon\eta}, v]_{n/2} + a(u_{\varepsilon\eta}, v) + E(u_{\varepsilon\eta}, u_{\varepsilon\eta}, v) + \left(\frac{d\Phi_\varepsilon}{dv}(u_{\varepsilon\eta}), v\right) - (f, v) = 0, \qquad \forall v \in W_{n/2} \tag{9.3.69}$$

and the initial condition

$$u_{\varepsilon\eta}(0) = 0. \tag{9.3.70}$$

In the next step, we let ε and η tend to zero, and we prove that the limit of $u_{\varepsilon\eta}$, denoted by u, has the properties (9.3.24, 25) and fulfills the variational inequality (9.3.26). For a fixed v in $\mathscr{W}$, we consider the expression

$$D_{\varepsilon\eta} = \int_0^T [(\rho v', v - u_{\varepsilon\eta}) + a(u_{\varepsilon\eta}, v - u_{\varepsilon\eta}) + E(u_{\varepsilon\eta}, u_{\varepsilon\eta}, v - u_{\varepsilon\eta}) + \eta[u_{\varepsilon\eta}, v - u_{\varepsilon\eta}]_{n/2} + \Phi_\varepsilon(v) - \Phi_\varepsilon(u_{\varepsilon\eta}) - (f, v - u_{\varepsilon\eta})]\, dt, \tag{9.3.71}$$

which by means of (9.3.69) becomes

$$D_{\varepsilon\eta} = \int_0^T (\rho(v' - u'_{\varepsilon\eta}), v - u_{\varepsilon\eta})\, dt + \int_0^T \left[\Phi_\varepsilon(v) - \Phi_\varepsilon(u_{\varepsilon\eta}) - \left(\frac{d\Phi_\varepsilon}{dv}, v - u_{\varepsilon\eta}\right)\right] dt = \tfrac{1}{2}\rho\,|v(T) - u_{\varepsilon\eta}(T)|^2 + \int_0^T \left[\Phi_\varepsilon(v) - \Phi_\varepsilon(u_{\varepsilon\eta}) - \left(\frac{d\Phi_\varepsilon}{dv}, v - u_{\varepsilon\eta}\right)\right] dt \geq 0. \tag{9.3.72}$$

It results from (9.3.71) that

$$\int_0^T [(\rho v', v - u_{\varepsilon\eta}) + a(u_{\varepsilon\eta}, v) + E(u_{\varepsilon\eta}, u_{\varepsilon\eta}, v) + \eta[u_{\varepsilon\eta}, v]_{n/2} + \Phi_\varepsilon(v) - (f, v - u_{\varepsilon\eta})]\, dt \geq \int_0^T [a(u_{\varepsilon\eta}, u_{\varepsilon\eta}) + \Phi_\varepsilon(u_{\varepsilon\eta})]\, dt. \tag{9.3.73}$$

Due to (9.3.60–63), we can determine a subsequence $\{u_{\varepsilon\eta}\}$ such that as $\varepsilon \to 0$ and $\eta \to 0$

$$u_{\varepsilon\eta} \to u \quad \text{weakly}_* \text{ in } \; L^\infty(0T, W_0) \quad \text{and weakly in } \; L^2(0T, W_1) \tag{9.3.74}$$

and

$$u'_{\varepsilon\eta} \to u' \quad \text{weakly in } \; L^2(0T, W'_{n/2}). \tag{9.3.75}$$

From (9.3.73), we obtain

$$\int_0^T [(\rho v', v - u) + a(u, v) + E(u, u, v) + \Phi(v) - (f, v - u)]\, dt \geq \liminf \int_0^T [a(u_{\varepsilon\eta}, u_{\varepsilon\eta}) + \Phi_\varepsilon(u_{\varepsilon\eta})]\, dt. \tag{9.3.76}$$

For $E(u_{\varepsilon\eta}, u_{\varepsilon\eta}, v)$, the procedure used for the proof of (9.3.47) is again applied. The inequality

$$\liminf \int_0^T [a(u_{\varepsilon\eta}, u_{\varepsilon\eta}) + \Phi_\varepsilon(u_{\varepsilon\eta})]\, dt \geq \int_0^T [a(u, u) + \Phi(u)]\, dt \tag{9.3.77}$$

also holds, since $v \to \int_0^T a(v, v)\, dt$ is l.s.c. on $L^2(0T, W_1)$ and

$$\liminf \int_0^T \Phi_\varepsilon(u_{\varepsilon\eta})\, dt \geq \int_0^T \Phi(u)\, dt. \tag{9.3.78}$$

Indeed [84],

$$\int_0^T \Phi(u)\, dt \leq 2\left[\int_0^T \left(\int_\Omega (g\sqrt{D_{II}(u)})^{1+\varepsilon}\, d\Omega\right) dt\right]^{1/(1+\varepsilon)} \left[\int_0^T \left(\int_\Omega d\Omega\right) dt\right]^{\varepsilon/(1+\varepsilon)} \tag{9.3.79}$$

from which we find that

$$(1+\varepsilon)\,(2g)^\varepsilon \int_0^T \Phi_\varepsilon(u_{\varepsilon\eta})\, dt \geq (|T| \operatorname{mes} \Omega)^{-\varepsilon} \left[\int_0^T \Phi(u_{\varepsilon\eta})\, dt\right]^{1+\varepsilon}. \tag{9.3.80}$$

Thus

$$\liminf \int_0^T \Phi_\varepsilon(u_{\varepsilon\eta})\,dt \ge \liminf \int_0^T \Phi(u_{\varepsilon\eta})\,dt \ge \int_0^T \Phi(u)\,dt, \quad (9.3.81)$$

because $v \to \int_0^T \Phi(v)\,dt$ is continuous and convex on $L^2(0T, W_1)$. Therefore $\int_0^T \Phi(v)\,dt$ is l.s.c. for the weak topology of $L^2(0T, W_1)$.

From (9.3.72–75) and (9.3.77), u is a solution of (9.3.26) and satisfies the initial condition (9.3.5), q.e.d.

(b) For the case $\Omega \subset \mathbb{R}^2$, the introduction of the viscosity term $\eta[u, v]_1$ is immaterial, since $W_{n/2} = W_1$ and the coerciveness of the bilinear form $a(v, v)$ on W_1 is sufficient for the existence proof. By the method followed in (a), we can show the existence of a u_ε satisfying the variational equality

$$(\rho u'_\varepsilon, v - u_\varepsilon) + a(u_\varepsilon, v - u_\varepsilon) + E(u_\varepsilon, u_\varepsilon, v) + \left(\frac{d\Phi_\varepsilon}{dv}(u_\varepsilon), v - u_\varepsilon\right) - (f, v - u_\varepsilon) = 0, \qquad \forall v \in W_1 \quad (9.3.82)$$

and the initial condition (9.3.5).

Moreover, it can be shown (cf. 9.3.74, 75) that

$$u_\varepsilon \to u \quad \text{weakly in} \quad L^2(0T, W_1) \quad \text{and weakly}_* \text{ in} \quad L^\infty(0T, W_0), \quad (9.3.83)$$

$$u'_\varepsilon \to u' \quad \text{weakly in} \quad L^2(0T, W'_1) \quad (9.3.84)$$

and

$$u_\varepsilon(T) \to u(T) \quad \text{weakly in} \quad W_0. \quad (9.3.85)$$

The fact that $L^2(0T, W_1)$ and $L^2(0T, W'_1)$ are dual to each other enables us to prove that u satisfies the variational inequality (9.3.28). Let us define for $v \in L^2(0T, W_1)$ the quantity

$$D_\varepsilon = \int_0^T [(\rho u'_\varepsilon, v - u_\varepsilon) + a(u_\varepsilon, v - u_\varepsilon) + E(u_\varepsilon, u_\varepsilon, v - u_\varepsilon) + \Phi_\varepsilon(v) - \Phi_\varepsilon(u_\varepsilon) - (f, v - u_\varepsilon)]\,dt, \quad (9.3.86)$$

which, combined with (9.3.82), implies that

$$D_\varepsilon = \int_0^T \left[\Phi_\varepsilon(v) - \Phi_\varepsilon(u_\varepsilon) - \left(\frac{d\Phi_\varepsilon(u_\varepsilon)}{dv}, v - u_\varepsilon\right)\right] dt \ge 0. \quad (9.3.87)$$

Thus

$$\int_0^T [(\rho u'_\varepsilon, v) + a(u_\varepsilon, v) + E(u_\varepsilon, u_\varepsilon, v) + \Phi_\varepsilon(v) - (f, v - u_\varepsilon)]\,dt \ge \int_0^T [(\rho u'_\varepsilon, u_\varepsilon) + a(u_\varepsilon, u_\varepsilon) + \Phi_\varepsilon(u_\varepsilon)]\,dt \quad (9.3.88)$$

$$= \tfrac{1}{2}\rho\,|u_\varepsilon(T)|_2^2 - \tfrac{1}{2}\rho|v_0|_2^2 + \int_0^T [a(u_\varepsilon, u_\varepsilon) + \Phi_\varepsilon(u_\varepsilon)]\,dt.$$

From (9.3.88), by taking the lim inf of both sides we have the relation (cf. the proof of (5.4.59))

$$\int_0^T [(\rho u', v-u) + a(u, v-u) + E(u, u, v)$$
$$+ \Phi(v) - \Phi(u) - (f, v-u)]\, dt \geq 0, \qquad \forall v \in L^2(0T, W_1). \quad (9.3.89)$$

From (9.3.89) the pointwise inequality (9.3.28) results (cf. (5.4.60–62)). Moreover, due to (9.3.83, 84), u satisfies the initial condition (9.3.5).

In order to show the uniqueness of u, we argue by contradiction. Let u_1 and u_2 be two functions satisfying (9.3.27, 28). By setting in (9.3.28) $v = u_1$, $u = u_2$ and $v = u_2, u = u_1$, we obtain by addition the inequality

$$-(\rho z', z) - a(z, z) - E(u_1, u_1, z) + E(u_2, u_2, z) \geq 0, \quad (9.3.90)$$

where $z = u_1 - u_2$. From (9.3.90) we obtain

$$\frac{1}{2}\frac{d}{dt}\rho |z|_2^2 + \alpha |z|_1^2 \leq E(z, z, u_1), \qquad \alpha \text{ const} > 0. \quad (9.3.91)$$

We can determine constants $c_1, c_2 > 0$ such that (cf., (9.3.37))

$$|E(z, z, u_1)| \leq c_1 |z|_1 |z|_2 |u_1|_1 \leq \alpha |z|_1^2 + c_2 |u_1|_1^2 |z|_2^2. \quad (9.3.92)$$

From (9.3.91, 92) we get

$$\frac{1}{2}\frac{d}{dt}\rho |z|_2^2 \leq c_2 |u_1|_1^2 |z|_2^2, \quad (9.3.93)$$

and since $|u_1|_1^2 \in L^2(0T)$ we conclude from (9.3.93) through integration from 0 to t that $z = u_1 - u_2 = 0$, q.e.d.

For the three-dimensional flow problem, the uniqueness of the solution remains to be proved. It is easy to show that for $n = 2$ a solution of (9.3.26) satisfies (9.3.28), and is therefore unique. Indeed, (9.3.26) is also true for every $v \in \mathscr{W}_1 = \{v | v \in L^2(0T, W_1),\ v' \in L^2(0T, W_1'),\ v(0) = 0\}$ since all the functionals in (9.3.26) are continuous on $\mathscr{W}$ (when $\mathscr{W}$ is endowed with the topology induced by $\mathscr{W}_1$) and $\mathscr{W}$ is dense in $\mathscr{W}_1$. Noting that $u \in \mathscr{W}_1$, we set $v = (1-\lambda)u + \lambda v_1$ in (9.3.26) with $0 < \lambda < 1$ and $v_1 \in \mathscr{W}_1$. Dividing by λ and letting $\lambda \to 0_+$, we obtain (cf., the proof of Prop. 2.2.2) the variational inequality (9.3.89) for every $v \in \mathscr{W}_1$. Because this last space is dense in $L^2(0T, W_1)$, u satisfies (9.3.89) and thus (9.3.28) as well [84].

In [84] a proof of Prop. 9.3.1 without time discretization is given. A comparison of the proof of Prop. 9.3.1 with the existence proof for the dynamic variational inequality of Sec. 5.4.2 shows that here we have not used an estimate for the time derivative of the velocity u' at $t = 0$, and therefore Prop. 9.3.1 does not involve f'. However, assuming in the two-dimensional flow problem that $f, f' \in L^2(0T, W_1')$, $f(0) \in W_0$, $v_0 \in W_1 \cap$

$[H^2(\Omega)]^2$, and using an estimate for $u'(0)$, which results from the fact that $H^2(\Omega)$ is imbedded into $L^\infty(\Omega)$, we can show that $u \in L^2(0T, W_1)$ and $u' \in L^2(0T, W_1) \cap L^\infty(0T, W_0)$ [84].

9.3.3 The Relation Between Velocity and Stress Fields

In Sec. 9.3.1 we have established the formal equivalence of (9.3.9) with the initial problem. For the two-dimensional flow problem it is possible to prove this equivalence rigorously, thus exhibiting as well the relation existing between the velocity and the stress fields of the fluid.

Proposition 9.3.2. Suppose that $n = 2$ and

$$f \in L^2(0T, W_1'), \qquad v_0 \in W_0. \tag{9.3.94}$$

Let $u = \{u_i\}$, $\varphi = \{\varphi_{ij}\}$, p be such that

$$u \in L^2(0T, W_1), \qquad u' \in L^2(0T, W_1'), \tag{9.3.95}$$

$$u(0) = v_0, \tag{9.3.96}$$

$$\varphi_{ij} = \varphi_{ji}, \qquad \varphi_{ii} = 0, \qquad \varphi_{ij} \in L^\infty(\Omega \times (0, T)) \qquad i, j = 1, 2, 3, \tag{9.3.97}$$

$$\tfrac{1}{2}\varphi_{ij}\varphi_{ij} \leq g^2 \quad \text{a.e. in} \quad \Omega \times (0, T), \tag{9.3.98}$$

$$\varphi_{ij} D_{ij}(u) = \sqrt{2}\, g(D_{ij}(u) D_{ij}(u))^{1/2} \quad \text{a.e. in} \quad \Omega \times (0, T)). \tag{9.3.99}$$

$$p \in \mathscr{D}'(\Omega \times (0, T)) \tag{9.3.100}$$

and

$$\rho u_i' - \mu\, \Delta u_i + \rho u_{i,j} u_j - \varphi_{ij,j} = f_i - p_{,i} \text{ in } \mathscr{D}'(\Omega \times (0, T)). \tag{9.3.101}$$

Then u is the solution of (9.3.28) and (9.3.5). Conversely, if u satisfies (9.3.28) and (9.3.5), there exist $\varphi = \{\varphi_{ij}\}$ and p which satisfy (9.3.97–101).

Proof. The proof of the first part is obvous. For the converse, we shall give two proofs: a direct one using the Hahn–Banach theorem and another using the results of the duality theory. Let

$$\Sigma = \{\sigma \mid \sigma = \{\sigma_{ij}\}, \sigma_{ij} = \sigma_{ji}, \sigma_{ij} \in L^1(\Omega \times (0, T))\} \tag{9.3.102}$$

with the norm

$$\|\sigma\|_\Sigma = \int_0^T \int_\Omega (\sigma_{ij}\sigma_{ij})^{1/2}\, d\Omega\, dt \tag{9.3.103}$$

and let Σ' be its dual space. We denote by $\Lambda: v \to \{D_{ij}(v)\}$ the linear mapping from $L^2(0T, W_1)$ into Σ. But $v \to \int_0^T \Phi(v)\, dt$ is continuous on $L^2(0T, W_1)$, and since

$$\int_0^T \Phi(v)\, dt = g\sqrt{2}\, \|\Lambda(v)\|_\Sigma, \tag{9.3.104}$$

it results that Λ is continuous as well. We denote by B the functional

$$B(\Lambda(v)) = g\sqrt{2}\,\|\Lambda(v)\|_\Sigma \tag{9.3.105}$$

on Σ. We write the variational inequality (9.3.28) in the equivalent form

$$\int_0^T [(R(u), v - u) + \Phi(v) - \Phi(u)]\, dt \geq 0, \qquad \forall v \in L^2(0T, W_1), \tag{9.3.106}$$

where (cf., (9.3.38, 39))

$$R(u) = \rho u' - \mu\,\Delta u + h(u) - f \in L^2(0T, W_1'). \tag{9.3.107}$$

In the direct method of proof, we use the fact that (9.3.106) is equivalent to the problem: find $u \in L^2(0T, W_1)$ such that

$$\left|\int_0^T (R(u), v)\, dt\right| \leq \int_0^T \Phi(v)\, dt, \qquad \forall v \in L^2(0T, W_1) \tag{9.3.108}$$

and

$$\int_0^T (R(u), u)\, dt + \int_0^T \Phi(u)\, dt = 0. \tag{9.3.109}$$

This is easily proved using the special structure of Φ by the same method which was applied to the proof of Prop. 5.2.3. From (9.3.104, 108) the inequality

$$\left|\int_0^T (R(u), v)\, dt\right| \leq g\sqrt{2}\,\|\Lambda(v)\|_\Sigma \tag{9.3.110}$$

results. Thus by the Hahn–Banach theorem (see Sec. (1.2.1)), there exists $\varphi \in \Sigma'$, i.e., $\varphi_{ij} \in L^\infty(\Omega \times (0T))$ such that

$$\int_0^T (R(u), v)\, dt = -\int_0^T \int_\Omega \varphi_{ij} D_{ij}(v)\, d\Omega\, dt \tag{9.3.111}$$

and

$$\frac{\|\varphi\|_{\Sigma'}}{g\sqrt{2}} \leq 1. \tag{9.3.112}$$

Because of the incompressibility relation $D_{ii}(v) = 0$, in (9.3.111) we may take $\varphi_{ii} = 0$. From (9.3.109, 111), the equality

$$\int_0^T \int_\Omega \varphi_{ij} D_{ij}(u)\, d\Omega\, dt = \int_0^T \int_\Omega g\sqrt{2}(D_{ij}(u)D_{ij}(u))^{1/2}\, d\Omega\, dt \tag{9.3.113}$$

is obtained. From (9.3.111), by applying the Green–Gauss theorem, (9.3.100, 101) result,[11] as do the remaining relations (9.3.98, 99) from (9.3.112, 113). This completes the direct proof first given in [84].

For the second proof, we remark that the variational inequality (9.3.106) is not equivalent to a minimization problem. However, if R is considered as known, (9.3.106) becomes an optimization problem, and the results of the duality theory can be applied. It is easy to verify that the primal problem then has the form (cf., Sec. 3.2.2)

$$\inf_{L^2(0T, W_1)} (B(\Lambda v) + A(v)), \qquad A(v) = \int_0^T (R(u), v)\, dt, \tag{9.3.114}$$

and that u is a solution of it. Due to (3.2.48), the dual problem takes the form

$$\sup_{\Sigma'} [-B^c(\varphi) - A^c(-\Lambda'\varphi)] = \inf_{\Sigma'} [I_{\tilde{A}}(\varphi) + I_{\tilde{B}}(\varphi)], \tag{9.3.115}$$

where $q_{ii} = 0$ and $I_{\tilde{A}}$ and $I_{\tilde{B}}$ denote the indicators of the sets

$$\tilde{A} = \left\{\varphi \mid \varphi \in \Sigma', \frac{\|\varphi\|_{\Sigma'}}{g\sqrt{2}} \le 1, \right\}, \tag{9.3.116}$$

and

$$\tilde{B} = \{\varphi \mid \varphi \in \Sigma', \quad \varphi \text{ satisfies (9.3.101)}\}, \tag{9.3.117}$$

as can be easily verified by direct computation of the conjugate functionals. Prop. 3.2.7 ensures that a solution of the dual problem exists if R is given, and that (3.2.43) and (3.2.45) hold. The infimum in (9.3.114) and the supremum in (9.3.115) are zero, due to (9.3.109), and thus (9.3.97–101) are satisfied. One of the extremality conditions is (9.3.113), from which (9.3.99) results, q.e.d.

Function $\varphi = \{\varphi_{ij}\}$ expresses physically the components of a stress tensor and p the fluid pressure, so that

$$\sigma_{ij} = -p\, \delta_{ij} + 2\mu\, D_{ij}(u) + \varphi_{ij} \tag{9.3.118}$$

is the stress tensor of the fluid. We see that (9.3.97–99) are equivalent to the material law (3.3.110), while (9.3.101) corresponds to (9.3.2).

From the behavior of Bingham fluids, the behavior of Newtonian fluids is obtained if $g \to 0_+$. We denote by u_g the velocity field corresponding to

[11] We make use here of the following proposition: if $S \in [\mathscr{D}'(\Omega \times (0, T))]^n$ and $(S, \phi) = 0$ in $[\mathscr{D}'(0, T)]^n$, $\forall \phi \in \mathscr{W}$, then $S = -\operatorname{grad} p$ with $p \in \mathscr{D}'(\Omega \times (0, T))$.

g. For a two-dimensional flow problem, it may be shown (cf. [84]) that on the assumptions of Prop. 9.3.1 as $g \to 0_+$

$$u_g \to u \quad \text{weakly in} \quad L^2(0T, W_1), \tag{9.3.119}$$

$$u'_g \to u' \quad \text{weakly in} \quad L^2(0T, W'_1). \tag{9.3.120}$$

Here u is the solution of the Navier-Stokes equation, i.e. of the variational problem: find $u \in L^2(0T, W_1)$, $u' \in L^2(0T, W'_1)$, $u(0) = u_0$ such that

$$(\rho u'(t), v) + a(u(t), v) + E(u(t), u(t), v) = (f(t), v), \qquad \forall v \in W_1. \tag{9.3.121}$$

We shall close this section by giving a proposition on the stationary flow problem of a rigid viscoplastic material. In this case we seek a $u \in W_1$ which satisfies the variational inequality

$$a(u, v - u) + E(u, u, v) + \Phi(v) - \Phi(u) \geq (f, v - u), \qquad \forall v \in W_1, \tag{9.3.122}$$

both for the two-dimensional and three-dimensional flow problems.

Proposition 9.3.3. If $f \in W'_1$, there exists a solution $u \in W_1$ of (9.3.122).

The proof is similar to that of the first part of Prop. 9.3.1 and therefore is omitted here.

9.4 Other Problems on Bingham Fluids

9.4.1 Laminar Flow in a Cylindrical Pipe

We consider a cylindrical pipe with generators parallel to the $0x_3$-axis of the Cartesian system $0x_1x_2x_3$. The cross section of the cylinder is denoted by Σ and $\Omega = \Sigma \times [0, l]$, where l is a given length. We shall study the flow under a pressure drop, i.e.,

$$p = 0 \quad \text{at} \quad x = 0, \qquad p = -p_0 l, \qquad p_0 \geq 0 \quad \text{at} \quad x = l, \tag{9.4.1}$$

where p_0 is the pressure drop per unit length. If Γ is the boundary of Σ, we assume that

$$v = 0 \quad \text{on} \quad \Gamma \times [0, l]. \tag{9.4.2}$$

Moreover, let (9.3.1–3), (3.3.110) and (9.3.5) hold. We restrict ourselves here to the class of laminar flows. Then the velocity has only one component parallel to $0x_3$, which depends only on x_1 and x_2, i.e. $v = (0, 0, v_3(t, x_1, x_2))$. It is easy to verify that $D_{31} = D_{13} = \frac{1}{2}v_{3,1}$ and $D_{23} = D_{32} = \frac{1}{2}v_{3,2}$ are the only nonzero components of the tensor D. The same also happens for the corresponding components of the stress deviator. On the assumption that $f = 0$, the equation of motion (9.3.2) becomes

$$p_{,1} = 0, \qquad p_{,2} = 0, \qquad \rho v_3' - (\sigma_{31,1}^D + \sigma_{32,2}^D) = p_{,3}. \tag{9.4.3}$$

But since the right-hand side of (9.4.3) depends on x_3, whereas the left-hand side depends on x_1 and x_2, it results that $p_{,3}$ is constant. Therefore, from (9.4.1) it follows that

$$p(x_3) = -p_0 x_3. \tag{9.4.4}$$

We shall seek a solution u in the space $\mathring{H}^1(\Sigma)$ of the problem considered. We thus formulate the following problem: find $u\colon [0, T] \to \mathring{H}^1(\Sigma)$ such as to satisfy the variational inequality

$$(\rho u', v - u) + \mu \bar{a}(u, v - u) + g\bar{\Phi}(v) - g\bar{\Phi}(u) \geq (p_0, v - u), \qquad \forall v \in \mathring{H}^1(\Sigma). \tag{9.4.5}$$

and the initial condition

$$v_3(0, x_1, x_2) = u_0(x_1, x_2). \tag{9.4.6}$$

In (9.4.5) we use the notations

$$\bar{a}(u, v) = \int_\Sigma u_{,i} v_{,i}\, d\Sigma, \qquad (p_0, v) = \int_\Sigma p_0 v\, d\Sigma \tag{9.4.7}$$

and

$$\bar{\Phi}(v) = \int_\Sigma |\operatorname{grad} v|\, d\Sigma. \tag{9.4.8}$$

The respective stationary problem has the form: find $u \in \mathring{H}^1(\Sigma)$ such that

$$\mu\bar{a}(u, v - u) + g\bar{\Phi}(v) - g\bar{\Phi}(u) \geq (p_0, v - u), \qquad \forall v \in \mathring{H}^1(\Sigma). \tag{9.4.9}$$

The following proposition holds, even on the nonrealistic assumption that p_0 is not a constant.

Proposition 9.4.1. (a) Suppose that

$$p_0 \in L^2(\Sigma). \tag{9.4.10}$$

Then (9.4.9) has a unique solution which minimizes on $\mathring{H}^1(\Sigma)$ the functional

$$\Pi(v) = \tfrac{1}{2}\mu\bar{a}(v, v) + g\,\bar{\Phi}(v) - (p_0, v). \tag{9.4.11}$$

(b) If

$$p_0 \in L^2((0, T) \times \Sigma) \quad \text{and} \quad u_0 \in L^2(\Sigma), \tag{9.4.12}$$

then there exists a unique u such that

$$u \in L^2(0T, \mathring{H}^1(\Sigma)) \cap C^0(0T, L^2(\Sigma)) \tag{9.4.13}$$

and

$$u' \in L^2(0T, H^{-1}(\Sigma)), \tag{9.4.14}$$

which satisfies (9.4.5, 6).

Proof. (a) Since $a(v, v)$ is coercive on $\mathring{H}^1(\Sigma)$, (2.2.3) implies that $\Pi(v)$ admits a unique minimum on $\mathring{H}^1(\Sigma)$ at u. By Prop. 2.2.2, the variational inequality (9.4.9) is satisfied, q.e.d.

(b) A special case of Prop. 9.3.1, q.e.d.[12]

Due to the viscous character of the material, a pressure p is needed to initiate the flow. In [119] it is proved that if p_0 is a constant >0, then for the steady flow problem

(i) $u = 0$ if $g \geq p_0/A$ and (9.4.15)

(ii) $u \neq 0$ if $g < p_0/A$, where (9.4.16)

$$A = \inf\{\bar{\Phi}(v) \mid v \in \mathring{H}^1(\Sigma), |v|_{L^1(\Sigma)} = 1\}. \tag{9.4.17}$$

We leave the proof of (9.4.15, 16), which follows the same pattern as that of Prop. 7.2.1 (cf., also (7.2.65–67)), as an exercise for the reader. An analogous result can be obtained for the unsteady flow problem as well. Assume that p_0 and u_0 are elements of $L^2(\Sigma)$. We denote by $|\cdot|_2$ the norm $|\cdot|_{L^2(\Sigma)}$. If

$$|p_0|_2 \leq gA', \qquad A' = \inf\{\bar{\Phi}(v) | v \in \mathring{H}^1(\Sigma), |v|_2 = 1\}, \tag{9.4.18}$$

then

$$u(t) = 0 \quad \text{if} \quad t \geq T_c = \frac{\rho_0}{\lambda_0 \mu} \ln\left[1 + \frac{\lambda_0 \mu |u_0|_2}{A'g - |p_0|_2}\right], \tag{9.4.19}$$

where $u(t)$ denotes the solution of (9.4.5) and λ_0 is the smallest eigenvalue of the operator $-\Delta$ in $\mathring{H}^1(\Sigma)$.

In order to show (9.4.19), we put $v = 0$ and $v = 2u$ in (9.4.5), and by means of (9.4.18) we obtain a.e. on $[0, T]$

$$\frac{\rho_0}{2}\frac{d}{dt}|u(t)|_2^2 + \lambda_0 \mu |u(t)|_2^2 + (gA' - |p_0|_2)|u(t)|_2 \leq 0. \tag{9.4.20}$$

If $u(t) \neq 0$ for a.e. $t \geq 0$, (9.4.20) implies by integration that

$$|u(t)|_2 + \bar{A}' \leq (|u_0|_2 + \bar{A}')\, e^{-\frac{\mu\lambda_0 t}{\rho_0}}, \qquad \forall t \geq 0,\ \bar{A}' = \frac{gA' - |p_0|_2}{\mu\lambda_0}. \tag{9.4.21}$$

(9.4.21) cannot be true if t is sufficiently large; actually, if $t \geq T_c$, where T_c is given in (9.4.19). Thus for such a t, $u(t) = 0$, i.e., the flow stops. This is physically reasonable, considering the viscosity of the material.

[12] The reader may obtain the same result directly by regularization.

9.4.2 Heat Transfer in Rigid Viscoplastic Flows

In order to study the thermomechanical problem for a Bingham fluid, we consider a viscosity μ depending on the temperature $\theta = \theta(x, t)$. The fluid's natural state, to which the temperature is referred, is characterized by zero stresses and zero absolute temperature. This dependence is made more precise by assuming that $\theta \to \mu(\theta)$ is continuously differentiable, and that for $\theta \in \mathbb{R}, 0 < \mu_0 \le \mu(\theta) \le \mu_1 < \infty$. Besides (9.3.1–5) and the material law (3.3.110), the velocity and temperature fields satisfy the equation of energy balance (see (AIII.1)), and the heat flux vector $q = \{q_i\}$ and the temperature θ obey Fourier's law (8.1.6), where $k_{ij} \in L^\infty(\Omega)$. Between the specific internal energy ε and θ, we assume a linear relation of the form

$$\varepsilon = \gamma\theta, \qquad \gamma \text{ const} > 0. \tag{9.4.22}$$

For the temperature, the boundary condition

$$\theta = 0 \quad \text{on} \quad \Gamma \times (0, T) \tag{9.4.23}$$

is introduced, as well as the initial condition

$$\theta(x, 0) = \theta_0(x) \quad \text{at} \quad t = 0. \tag{9.4.24}$$

Further, let us denote by $X(\theta, u)$ the expression

$$X(\theta, u) = 2\mu(\theta)D_{II}(u) + g(D_{II}(v))^{1/2}, \tag{9.4.25}$$

and let $\bar{k}_{ij} = k_{ij}/\gamma$ and $\bar{\mathcal{Q}} = \mathcal{Q}/\gamma$. The energy balance equation then becomes

$$\frac{d\theta}{dt} = \frac{1}{\gamma} X(\theta, u) + (\bar{k}_{ij}\, \theta_{,i})_{,j} + \bar{\mathcal{Q}}. \tag{9.4.26}$$

Further, the notation

$$a(\theta, u, v) = 2\int_\Omega \mu(\theta) D_{ij}(u) D_{ij}(v)\, d\Omega \tag{9.4.27}$$

is used. Note that

$$a(\theta, u, u) + \Phi(u) = 2\int_\Omega X(\theta, u)\, d\Omega. \tag{9.4.28}$$

Let us consider the two-dimensional flow problem. As in Sec. 9.3.1, we seek a velocity field $u \in W_1$ which a.e. on $[0, T]$ satisfies (9.3.5) and the variational inequality

$$(\rho u', v - u) + a(\theta, u, v - u) + E(u, u, v - u) + \Phi(v) - \Phi(u) \ge (f, v - u), \qquad \forall v \in W_1. \tag{9.4.29}$$

We also introduce the space

$$Y = \{\phi \mid \phi, \phi_{,i}, \phi_{,ij}, \phi' \in L^{\alpha'}(\Omega \times (0, T)), \phi(x, T) = 0, \phi = 0 \quad \text{on} \quad \Gamma \times (0, T)\}, \tag{9.4.30}$$

where $1/\alpha + 1/\alpha' = 1$. Multiplying (9.4.26) by $\phi \in Y$ and integrating, we obtain the relation

$$\int_0^T \int_\Omega \left(\frac{d\theta}{dt} - (\bar{k}_{ij}\theta_{,i})_{,j}\right)\phi\, d\Omega\, dt = \int_0^T \int_\Omega \left(\frac{1}{\gamma}X(\theta, u)\phi + \bar{\mathcal{Q}}\phi\right) d\Omega\, dt. \tag{9.4.31}$$

In order to simplify the mathematical problem, we assume that $\theta_{,j}u_j \simeq 0$ and therefore $d\theta/dt \simeq \partial\theta/\partial t = \theta'$. From (9.4.31) we obtain, by integrating by parts, the variational equality

$$-\int_0^T \int_\Omega \theta\left(\frac{\partial \phi}{\partial t} + (\bar{k}_{ij}\phi_{,i})_{,j}\right) d\Omega\, dt$$

$$= \int_0^T \int_\Omega [\frac{1}{\gamma} X(\theta, u)\phi + \bar{\Omega}\phi] d\Omega dt + (\theta_0, \phi(x, 0)), \qquad \forall \phi \in Y. \quad (9.4.32)$$

The following variational problem is now considered [86]: find $u \in W_1$ and $\theta \in L^\alpha(\Omega \times (0, T))$ such as to satisfy the variational inequality (9.4.29), with the initial condition (9.3.5), and the variational equality (9.4.32).

The existence of the solution of this problem is given by the following proposition.

Proposition 9.4.2. Suppose that $1 < \alpha < \frac{3}{2}$

$$f \in L^2(0T, W_1), \qquad \Omega \in L^1(\Omega \times (0, T)), u_0 \in W_0, \text{ and } \theta_0 \in L^1(\Omega). \quad (9.4.33)$$

Then there exist u and θ where

$$u \in L^2(0T, W_1), \qquad u' \in L^2(0T, W'_1), \qquad \theta \in L^\alpha(\Omega \times (0, T)), \qquad (9.4.34)$$

which satisfy (9.4.29, 32) and (9.3.5).

Proof. Here we only sketch the proof. Let us consider (9.4.29) with θ in $a(\theta, u, v - u)$ replaced by $y \in L^\alpha(\Omega \times (0, T))$. If y is assumed to be given then Prop. 9.3.1 implies that (9.4.29) admits a unique solution $u(y) \in L^2(0T, W_1)$. The variational equality (9.4.32), when $X(\theta, u)$ is replaced by $X(y, u(y))$, defines θ as a function of y. By the Schauder fixed-point theorem, it may be shown that the application $y \to \theta(y)$ has a fixed point θ_0 which is a solution of the problem. For the complete proof, see [86].

The questions of the uniqueness of the solution in the two-dimensional flow problem as well as of the existence and uniqueness of the solution in the three-dimensional flow problem are still open. However, a proposition analogous to Prop. 9.3.2 can be proved.

Proposition 9.4.3. Suppose that (9.4.33) holds and let $u = \{u_i\}$, $\theta, \mathfrak{s} = \{\mathfrak{s}_{ij}\}$ and $\mathfrak{p}$ be such as to satisfy (9.4.34), (9.3.96–100) and (9.4.24), as well as (9.4.32) and

$$\rho u'_i - (\mu(\theta)u_{i,j})_{,j} + \rho u_{i,j}u_j - \mathfrak{s}_{ij,j} = f_i - \mathfrak{p}_{,i}, \qquad i = 1,2. \qquad (9.4.35)$$

Then (u, θ) is a solution of the problem. Conversely, if (u, θ) is a solution of (9.4.29), (9.3.5) and (9.4.32), then there exist $\mathfrak{s} = \{\mathfrak{s}_{ij}\}$ and $\mathfrak{p}$ which satisfy (9.3.97–100) and (9.4.35).

The proof is similar to that of Prop. 9.3.2 and therefore is omitted here.

9.4.3 Variational Inequalities in the Case of the General Law $\sigma \in \partial w(D)$

In this section, we study problems resulting from the general material law (3.3.90), where w is nonnegative and $w(0) = 0$. This law constitutes a generalization of the law of Bingham fluids. Obviously, when (9.3.10) is combined with (3.3.90), the inequality

$$W(v) - W(u) \geq (f - \rho\gamma, v - u), \qquad \forall v \in V_{ad} \tag{9.4.36}$$

results, where

$$W(v) = \begin{cases} \int_\Omega w(D(v))\,d\Omega, & \text{if} \quad w(D(v)) \in L^1(\Omega) \\ \infty & \text{otherwise.} \end{cases} \tag{9.4.37}$$

This problem can be studied on the assumption that

$$W(v) = \tfrac{1}{2}a(v, v) + \Phi(v), \tag{9.4.38}$$

where $a(v, v)$ is a bilinear, continuous, coercive form on W_1 and Φ is a convex, l.s.c. and proper functional on W_1. In this case, we obtain variational inequalities of the form (9.3.26) and (9.3.28). Assumption (9.4.38) arises if, e.g., $w(D)$ is the sum of $\bar{w}(D)$ and $\bar{\bar{w}}(D)$, where $\bar{w}(D)$ is a quadratic, positive definite form on $\mathbb{R}^6$ and $\bar{\bar{w}}(D)$ is a convex, l.s.c., proper functional on $\mathbb{R}^6$. It is assumed that Φ can be approximated by a sequence of convex Fréchet-differentiable functionals Φ_ε on W_1 such that

$$\text{(i)} \int_0^T \Phi_\varepsilon(v)\,dt \to \int_0^T \Phi(v)\,dt, \qquad \forall v \in L^2(0T, W_1), \tag{9.4.39}$$

$$\text{(ii)}\ d\Phi_\varepsilon(0)/dv = 0, \qquad \forall \varepsilon > 0, \tag{9.4.40}$$

(iii) for any sequence $\{v_\varepsilon\}$ such that $v_\varepsilon \to v$ in $L^2(0T, W_1)$ weakly as $\varepsilon \to 0$ and $\int_0^T \Phi_\varepsilon(v_\varepsilon)\,dt$ is bounded,

$$\liminf \int_0^T \Phi_\varepsilon(v_\varepsilon)\,dt \geq \int_0^T \Phi(v)\,dt. \tag{9.4.41}$$

Then the following proposition holds.

Proposition 9.4.4. With assumptions (i)–(iii), Prop. 9.3.1 is also valid for the general material law (3.3.90) where W is given by (9.4.38).

Proof. The same as that of Prop. 9.3.1 (cf., also the proof of Prop. 8.2.1) and therefore not given here.

If the decomposition (9.4.38) is not possible, we can again obtain the variational inequalities (9.3.26) and (9.3.28), but with $a(u, v-u) + \Phi(v) - \Phi(u)$ replaced by $W(v) - W(u)$. In this case, the questions of existence and uniqueness are still open.

The corresponding stationary problems strongly resemble the problem studied in Sec. 6.1. We consider here the stationary laminar flow problem in a cylindrical pipe under a given pressure gradient. It is assumed that

$$w(D) \geq c(D_{ij} D_{ij}), \qquad c \text{ const} > 0. \tag{9.4.42}$$

We then seek a $u \in \mathring{H}^1(\Sigma)$ such that

$$W(v) - W(u) \geq (p_0, v - u), \qquad \forall v \in \mathring{H}^1(\Sigma), \tag{9.4.43}$$

where W is given by (9.4.37) with the integral extended over Σ. This variational inequality is of the same nature as (6.1.11) and is equivalent to the minimization problem

$$\Pi(u) = \min\{\Pi(v) | v \in \mathring{H}^1(\Sigma)\}, \tag{9.4.44}$$

where

$$\Pi(v) = W(v) - (p_0, v). \tag{9.4.45}$$

The following proposition holds.

Proposition 9.4.5. There exists a solution $u \in \mathring{H}^1(\Sigma)$ of the variational inequality (9.4.43). If w is strictly convex, the solution is unique.

The proof is the same as that of Prop. 6.1.3 and therefore is not given here.

For the problem studied here, propositions analogous to those of Sec. 6.1 can be proved. If instead of (9.4.42) a relation of the form

$$w(D) \geq c(D_{ij} D_{ij})^{p/2}, \qquad c \text{ const} > 0, \qquad 1 < p < \infty, \tag{9.4.46}$$

holds, we may formulate (9.4.43) for $u, v \in \mathring{W}^{1,p}(\Sigma)$. In this case, the existence of a solution can be proved, as can certain duality propositions. In this context, see [177], where the problem of Sec. 6.1 is studied in L^p-spaces.

Part 3
Numerical Applications

Chapter 10

The Numerical Treatment of Static Inequality Problems

The aim of the two final chapters is to illustrate certain applications in the engineering sciences of the theory which has been developed up to this point. This chapter deals with the numerical solution of some static inequality problems.

First, the problem of frictional unilateral contact will be treated for a material obeying both Hooke's law and the general subdifferential holonomic law (3.3.57). For the latter law, the problem of torsion of cylindrical or prismatic bars will be studied. In the last section of the chapter a linear analysis approach to some classes of inequality problems is given.

In all the applications, we have to solve a quadratic or, generally, a nonlinear optimization problem. At this point we should note that, since Fichera's mathematical study [97] of the unilateral contact problem of a linear-elastic body with a rigid support, many authors have applied quadratic programming algorithms as a tool for the numerical treatment of inequality problems in applied mechanics and structural analysis, at the same time that the authors referred to in the previous chapters have been working on the mathematical aspects of these problems. Mention could be made here that G. Maier [184], [190], [191], [185], G. Dupuis and A. Probst [80], G. Nitsiotas [235], T. F. Conry and A. Seireg [56], J. J. Kalker and Y. van Randen [157], A. Sayegh [299], and F. D. Fischer [100], among others, early on recognized the need for a new numerical method—the use of optimization algorithms—for the treatment of inequality problems. Indeed, static unilateral problems are formulated as variational inequalities and as constrained minimization problems which are not equivalent to a system of differential equations, as is true with static bilateral problems. In the latter case, after discretization we obtain a system of algebraic equations, whereas in the former, a quadratic or a nonlinear programming problem is obtained depending on the degree of nonlinearity of the energy functions involved. In certain problems, also, derivative-free nonlinear programming algorithms must be used, if the energy function to be minimized is nondifferentiable. It is worth noting that we cannot treat all unilateral problems by means of the same optimization algorithm. The choice of the algorithm depends strongly on the very nature of the problem, i.e., on the function to be minimized and the subsidiary conditions. Because of this fact and the numerical difficulties resulting from the use of an optimization algorithm, research on the numerical treatment of inequality problems is still in progress. No attempt is made here to give a complete description of the optimization algorithms used. The reader can find the related theories in various treatises on quadratic and nonlinear programming (see, e.g., [143] [166]). For the numerical calculations the continuous problems are discretized by the finite element method. This chapter presumes a knowledge of this method (see, e.g., [101] [67] [345]).

All the above hold for convex energy functions. Nonconvex energy functions give rise to hemivariational inequalities, which after regularization and discretization lead to nonlinear differential (resp. algebraic) equations for dynamic (resp. static) problems [348].

A fundamental question in every approximation scheme is the proof of convergence and the derivation of error estimates for the approximate solution. The discussion of this question, and of all others resulting from the discretization process, is beyond the scope of the present book. Moreover, it should be pointed out here that a mathematically sound approximation theory has until now been developed only for some restricted classes of variational inequalities, which offer few possibilities of application and by no means cover all the problems studied in the previous chapters. On this subject the reader may refer to [120], [121], [131], [354], [355] and to the literature given there.

10.1 Unilateral Contact and Friction Problems

10.1.1 Discrete Forms of the Problems of Minimum Potential and Complementary Energy

In the various branches of engineering science, one frequently encounters structures which are in contact with a support but without being perfectly bonded to it. The possibility that the structure will lose contact with the support, and/or the possibility of slippage, is not excluded, and thus we speak of unilateral and/or frictional contact between structure and support. The friction B.V.P. has been discussed for a linear elastic body in Chapter 5, and the unilateral contact problem results as a special case of Sec. 6.2. Here the combined problem of frictional unilateral contact is numerically treated, first for a linear-elastic body and then for a body obeying the law (3.3.57). In this problem the regions where the structure remains in contact with the support and the regions where sliding occurs are not known a priori and hence must be determined.

First, the unilateral contact and the friction problems are considered separately. Along the lines of Sec. 6.2.1, we consider on a part Γ_S of the boundary of the structure the boundary condition (3.3.7) (resp. (3.3.10)), where j_N (resp. j_T) is given by (3.3.20) (resp. (3.3.32)). (3.3.7) (resp. (3.3.10)) and

$$\mathbf{S}_T = \mathbf{C}_T \quad \text{on} \quad \Gamma_S \tag{10.1.1}$$

resp.

$$\mathbf{S}_N = \mathbf{C}_N \quad \text{on} \quad \Gamma_S, \tag{10.1.2}$$

where $\mathbf{C}_T$ (resp. $\mathbf{C}_N$) is given, complete the boundary conditions for the unilateral contact (resp. friction) problem. Instead of (10.1.1) (resp. 10.1.2), other boundary conditions on $\mathbf{u}_T$, $\mathbf{S}_T$ (resp. $\mathbf{u}_N$, $\mathbf{S}_N$) could be introduced, as for instance $\mathbf{u}_T = \mathbf{0}$ (resp. $\mathbf{u}_N = \mathbf{0}$). Further, the body is divided into m elements whose stress and strain vectors are denoted by $\mathbf{s}_i$ and $\mathbf{e}_i$, $i = 1, \ldots, m$. In order to formulate discrete variational principles in a unified manner, the stress and strain state is defined in every element by means of natural generalized stress and strain components [7] [8] [9]. The generalized strains are not affected by rigid-body motion, and the generalized stresses are self-equilibrated in every element.[1] This approach has proved valuable in the derivation of discrete variational principles in the theory of plasticity (cf. [190]).

The configuration of the structure is described by the nodal displacement vector $\mathbf{u}$ with reference to a fixed Cartesian coordinate system. On a part Γ_U (resp. Γ_F) of the boundary, the given nodal displacement (resp. force) vector is denoted by $\mathbf{U}$ (resp. $\mathbf{F}$). The components of $\mathbf{U}$ are interpreted as initial strains in additional rigid bars located at the nodes of Γ_U in appropriate directions. $\mathbf{U}$ is incorporated in the initial strain vector $\mathbf{e}_0$, and $\mathbf{F}$ in the load vector $\mathbf{p}$ which consists of the vector $\bar{\mathbf{p}}$ of external loads and of the vector $\mathbf{S}$ of the reactions on the nodes of Γ_S, i.e., $\mathbf{p} = [\bar{\mathbf{p}}^T, \mathbf{S}^T]^T$. On the assumption of geometrically linear theory, the discretized unilateral contact problem consists of the following relations:

$$\mathbf{G}\mathbf{s} = \mathbf{p}; \tag{10.1.3}$$

$$\mathbf{e} = \mathbf{G}^T\mathbf{u}; \tag{10.1.4}$$

$$\mathbf{e} = \mathbf{e}_0 + \mathbf{F}_0\mathbf{s}; \tag{10.1.5}$$

$$-\mathbf{S}_N \in \mathbf{A}_N(\mathbf{u}_N) \quad \text{on} \quad \Gamma_S; \tag{10.1.6}$$

and

$$\mathbf{S}_T = \mathbf{C}_T \quad \text{on} \quad \Gamma_S. \tag{10.1.7}$$

[1] See also [276] where the vectors $\mathbf{s}_i$ and $\mathbf{e}_i$ of pin-jointed bars, beams, triangular and rectangular elements (in-plane forces), plate elements, solid tetrahedra, etc. are given.

Here $\mathbf{G}$ denotes the equilibrium matrix, $\mathbf{G}^T$ is its transpose, $\mathbf{F}_0$ is the natural flexibility matrix, $\mathbf{S}_N = [\mathbf{S}_{N_1}^T, \ldots, \mathbf{S}_{N_n}^T]$, $\mathbf{u}_N^T = [\mathbf{u}_{N_1}^T \cdots \mathbf{u}_{N_n}^T]$, with n the number of nodes of Γ_S, and $\mathbf{A}_N(\mathbf{u}_N) = [(\partial j_{N_1}(\mathbf{u}_{N_1}))^T, \ldots, (\partial j_{N_n}(\mathbf{u}_{N_n}))^T]^T$. For the friction problem the relations (10.1.6–7) must be replaced by (3.3.10) and (10.1.2) written for the discretized structure in the form

$$-\mathbf{S}_T \in \mathbf{A}_T(\mathbf{u}_T), \qquad \mathbf{S}_N = \mathbf{C}_N, \tag{10.1.8}$$

with $\mathbf{A}_T(\mathbf{u}_T) = [(\partial j_{T_1}(\mathbf{u}_{T_1}))^T, \ldots, (\partial j_{T_n}(\mathbf{u}_{T_n}))^T]^T$. As subsidiary conditions, the equations relating $\mathbf{u}_N$, $\mathbf{u}_T$ and $\mathbf{S}_N$, $\mathbf{S}_T$ with $\mathbf{u}$ and $\mathbf{S}$ respectively must be considered. We write these equations in the form

$$\mathbf{S}_N = \mathbf{Q}\mathbf{s}, \qquad \mathbf{S}_T = \mathbf{R}\mathbf{s}, \qquad \mathbf{u}_N = \mathbf{N}\mathbf{u}, \qquad \mathbf{u}_T = \mathbf{T}\mathbf{u} \tag{10.1.9}$$

where $\mathbf{Q}$, $\mathbf{R}$, $\mathbf{N}$ and $\mathbf{T}$ are appropriate matrices. Along the lines of the above discretization, we can verify[2] that for the discrete unilateral contact problem the proposition of minimum potential energy (cf., (6.2.8)) is written in the form

$$\begin{aligned} \Pi_1(\mathbf{u}) = \min\{\Pi_1(\mathbf{v}) | j_{N_i}(\mathbf{v}_{N_i}) < \infty,\\ \forall i \in \Gamma_S, \mathbf{v}_N = \mathbf{N}\mathbf{v}, \mathbf{v}_T = \mathbf{T}\mathbf{v}\}. \end{aligned} \tag{10.1.10}$$

Here

$$\Pi_1(\mathbf{v}) = \tfrac{1}{2}\mathbf{v}^T\mathbf{K}\mathbf{v} + \Phi(\mathbf{v}_N) - \mathbf{f}^T\mathbf{v} - \mathbf{C}_T^T\mathbf{v}_T, \tag{10.1.11}$$

where $\mathbf{K} = \mathbf{G}\mathbf{F}_0^{-1}\mathbf{G}^T$ is the stiffness matrix of the structure and $\Phi(\mathbf{v}_N)$ is the potential energy of the support given by

$$\Phi(\mathbf{v}_N) = \sum_{i=1}^{n} j_{N_i}(\mathbf{v}_{N_i}) \tag{10.1.12}$$

and

$$\mathbf{f} = \mathbf{f}_1 + \mathbf{f}_0 = \begin{bmatrix} \bar{\mathbf{p}} \\ \mathbf{0} \end{bmatrix} + \mathbf{f}_0, \qquad \mathbf{f}_0 = \mathbf{G}\mathbf{F}_0^{-1}\mathbf{e}_0. \tag{10.1.13}$$

Note that (10.1.10) holds for all displacement models (e.g., the generalized coordinate form of displacement model or the interpolation displacement model). Then $\mathbf{K}$ and $\mathbf{f}_0$ corresponding to this model (cf., eg., [67] [345]) will be taken into account in (10.1.11). The proposition of minimum complementary energy (cf., (6.2.41)) takes the form[3]

$$\begin{aligned} \Pi_1^c(\mathbf{s}) = \min\{\Pi_1^c(\mathbf{t}) | \bar{\mathbf{G}}\mathbf{t} = \bar{\mathbf{p}}, \mathbf{T}_T = \mathbf{C}_T, j_{N_i}^c(-\mathbf{T}_{N_i}) < \infty,\\ \forall i \in \Gamma_S, \mathbf{T}_N = \mathbf{Q}\mathbf{t}, \mathbf{T}_T = \mathbf{R}\mathbf{t}\}, \end{aligned} \tag{10.1.14}$$

[2] either by transcription of (6.2.8) into a discrete form, or directly from (10.1.3–7) by using the identity $\mathbf{p}^T(\mathbf{u}^* - \mathbf{u}) = \mathbf{s}^T(\mathbf{e}^* - \mathbf{e})$ (expression of the Green–Gauss theorem for a discretized free structure) which holds for every $\mathbf{e}^*$, $\mathbf{u}^*$ satisfying (10.1.4).

[3] For the direct derivation of (10.1.15), we could use the identity $\mathbf{e}^T(\mathbf{s}^0 - \mathbf{s}) = \mathbf{u}^T(\mathbf{p}^0 - \mathbf{p})$ holding for every $\mathbf{s}^0$, $\mathbf{p}^0$ which satisfy the equilibrium conditions.

where $\mathbf{G} = [\bar{\mathbf{G}}^T | \bar{\bar{\mathbf{G}}}^T]^T$ and

$$\Pi_1^c(\mathbf{t}) = \tfrac{1}{2}\mathbf{t}^T\mathbf{F}_0\,\mathbf{t} + \mathbf{\Phi}^c(-\mathbf{T}_N) + \mathbf{e}_0^T\,\mathbf{t}. \tag{10.1.15}$$

In the expression of the complementary energy, $\mathbf{\Phi}^c$ is the energy part due to the support conditions on Γ_S and is given by

$$\mathbf{\Phi}^c(-\mathbf{T}_N) = \sum_{i=1}^{n} j_{N_i}^c(-\mathbf{T}_{N_i}). \tag{10.1.16}$$

The subsidiary conditions $j_{N_i}(v_{N_i}) < \infty$, and $j_{N_i}^c(-T_{N_i}) < \infty$ are equivalent to $v_{N_i} \le 0$ and $T_{N_i} \le 0$ respectively, if the Signorini–Fichera conditions (3.3.22) hold on Γ_S, i.e. if the support is rigid. In this case the terms $\mathbf{\Phi}(\mathbf{u}_N)$ and $\mathbf{\Phi}^c(-\mathbf{T}_N)$ do not appear in (10.1.10) and (10.1.14) respectively. For the friction B.V.P., the proposition of minimum potential energy (see. (5.2.1)) takes, for the discretized structure, the form

$$\Pi_2(\mathbf{u}) = \min\,\{\Pi_2(\mathbf{v})\,|\,\mathbf{v}_N = \mathbf{N}\mathbf{v},\ \mathbf{v}_T = \mathbf{T}\mathbf{v}\}, \tag{10.1.17}$$

with

$$\Pi_2(\mathbf{v}) = \tfrac{1}{2}\mathbf{v}^T\mathbf{K}\mathbf{v} + \sum_{i=1}^{n} \mu\,|C_{N_i}|\,|v_{T_i}| - \mathbf{f}^T\mathbf{v} - \mathbf{C}_N^T\mathbf{v}_N. \tag{10.1.18}$$

Similarly, the proposition of minimum complementary energy (cf., (5.3.8)) is written in the form

$$\begin{gathered}\Pi_2^c(\mathbf{s}) = \min\,\{\Pi_2^c(\mathbf{t})\,|\,\bar{\mathbf{G}}\mathbf{t} = \bar{\mathbf{p}},\ \mathbf{T}_N = \mathbf{C}_N,\\ |T_{T_i}| \le \mu |C_{N_i}|\ \forall i \in \Gamma_S,\ \mathbf{T}_N = \mathbf{Q}\mathbf{t},\ \mathbf{T}_T = \mathbf{R}\mathbf{t}\},\end{gathered} \tag{10.1.19}$$

where

$$\Pi_2^c(\mathbf{t}) = \tfrac{1}{2}\mathbf{t}^T\mathbf{F}_0\,\mathbf{t} + \mathbf{e}_0^T\,\mathbf{t}. \tag{10.1.20}$$

If for every element a subdifferential law holds instead of Hooke's law, i.e., if

$$\mathbf{s}_i \in \partial w_i(\mathbf{e}_i - \mathbf{e}_{0i}), \tag{10.1.21}$$

where w_i is a convex, l.s.c. and proper functional, then the propositions of minimum potential and complementary energy again hold (cf., Sec. 6.3). Then the potential (resp. complementary) energy $\tilde{\Pi}_1$ (resp. $\tilde{\Pi}_1^c$) for the unilateral contact problem is

$$\tilde{\Pi}_1(\mathbf{v}) = W(\mathbf{v}) + \mathbf{\Phi}(\mathbf{v}_N) - \mathbf{f}_1^T\mathbf{v} - \mathbf{C}_T\mathbf{v}_T, \tag{10.1.22}$$

resp.

$$\tilde{\Pi}_1^c(\mathbf{t}) = W^c(\mathbf{t}) + \mathbf{\Phi}^c(-\mathbf{T}_N) + \mathbf{e}_0^T\,\mathbf{t}, \tag{10.1.23}$$

where W and W^c are

$$W(\mathbf{v}) = \sum_{i=1}^{m} w_i(\mathbf{e}_i(\mathbf{v}) - \mathbf{e}_{0i}) \tag{10.1.24}$$

and

$$W^c(\mathbf{t}) = \sum_{i=1}^{m} w_i^c(\mathbf{t}_i). \tag{10.1.25}$$

The solution $\mathbf{u}$ (resp. $\mathbf{s}$) of the minimization problems is sought in the same admissible sets as in problem (10.1.10) (resp. (10.1.14)), but must also fulfill the relation $w_i(\mathbf{e}_i(\mathbf{v}) - \mathbf{e}_{0i}) < \infty$ (resp. $w_i^c(\mathbf{t}_i) < \infty$) for $i = 1, \ldots, m$. For friction boundary conditions, the potential energy becomes

$$\tilde{\Pi}_2(\mathbf{v}) = W(\mathbf{v}) + \sum_{i=1}^{n} \mu |\mathbf{C}_{N_i}| |v_{T_i}| - \mathbf{f}_1^T \mathbf{v} - \mathbf{C}_N^T \mathbf{v}_N. \tag{10.1.26}$$

The solution of the minimum problem must satisfy the subsidiary conditions of (10.1.17) and also the inequalities $w_i(\mathbf{e}_i(\mathbf{v}) - \mathbf{e}_i^0) < \infty$ $i = 1, \ldots m$. The complementary energy reads

$$\tilde{\Pi}_2^c(\mathbf{t}) = W^c(\mathbf{t}) + \mathbf{e}_0^T \mathbf{t}. \tag{10.1.27}$$

Its minimum is constrained by the subsidiary conditions of (10.1.19) and also by the inequalities $w_i^c(\mathbf{t}_i) < \infty$, $i = 1, \ldots, m$. The boundary conditions describing both the unilateral contact and the friction phenomena have the form (see Sec. 3.3.1):

$$\begin{aligned} &\text{if} \quad u_N < 0, \quad \text{then} \quad S_N = 0 \quad \text{and} \quad S_T = 0; \\ &\quad \text{if} \quad u_N \geq 0, \quad \text{then} \quad -S_N \in \beta_1(u_N), \end{aligned} \tag{10.1.28}$$

and

$$\begin{aligned} &\text{if} \quad |S_T| < \mu |S_N|, \quad \text{then} \quad u_T = 0, \quad \text{whereas if} \\ &|S_T| = \mu |S_N|, \ \text{there exists} \ \lambda \geq 0 \ \text{such that} \ u_T = -\lambda S_T. \end{aligned} \tag{10.1.29}$$

These conditions stipulate that at those points where no contact exists between the body and the support there is no reaction force, whereas at the points of contact the normal reaction is $S_N \in -\beta_1(u_N)$ and frictional forces exist in the tangential direction. For this problem, it is not possible to formulate a variational inequality with respect to the displacements or the stresses, since S_N is not given in (10.1.29). Here, a heuristic algorithm is proposed for the numerical calculation [241], [249]. In the first step we assume that $\mathbf{S}_T^{(1)} = \mathbf{0}$ on Γ_S, and by solving the unilateral contact problem we obtain a normal reaction $\mathbf{S}_N$ on Γ_S. In the second step the friction problem is solved for $\mathbf{C}_N = \mathbf{S}_N^{(1)}$ and a value of $\mathbf{S}_T$, say $\mathbf{S}_T^{(2)}$, is obtained. Again the unilateral contact problem is solved for $\mathbf{C}_T = \mathbf{S}_T^{(2)}$, and so on, until the differences $\mathbf{S}_N^{(a+1)} - \mathbf{S}_N^{(a)}$ and $\mathbf{S}_T^{(a+1)} - \mathbf{S}_T^{(a)}$ become sufficiently small. In this context, see also [90] [232].

10.1.2 Applications

As a first application, the contact problem of the linear-elastic plate of Fig. 10.1.1 is examined. The plate lies on a rigid support, and so the Signorini–Fichera boundary conditions must be satisfied. The problem of

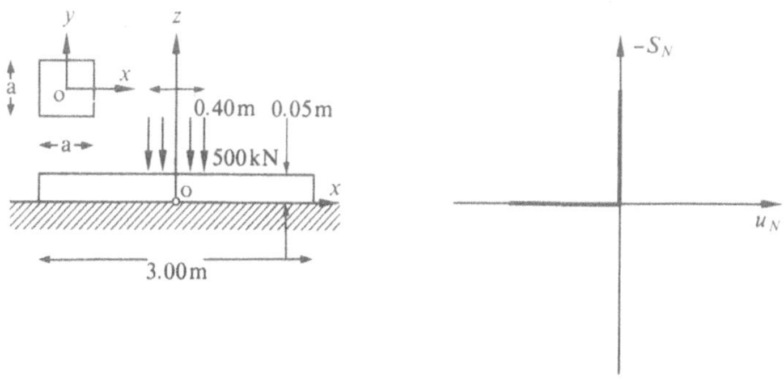

Fig. 10.1.1. The Signorini-Fichera problem for a plate ($E = 4.5 \times 10^3$ kN/cm^2, $\nu = 0.20$, $C_T = 0$).

minimum potential energy gives rise to a quadratic optimization problem, which has been solved by the algorithm of Hildreth and d'Esopo (see [166]). This algorithm is suitable for the minimization of strictly convex quadratic functions with equality and inequality subsidiary conditions and constitutes an iterative method, which is actually a modification of the classical Gauss–Seidel algorithm for the solution of linear systems of equations (Fig. 10.1.2).

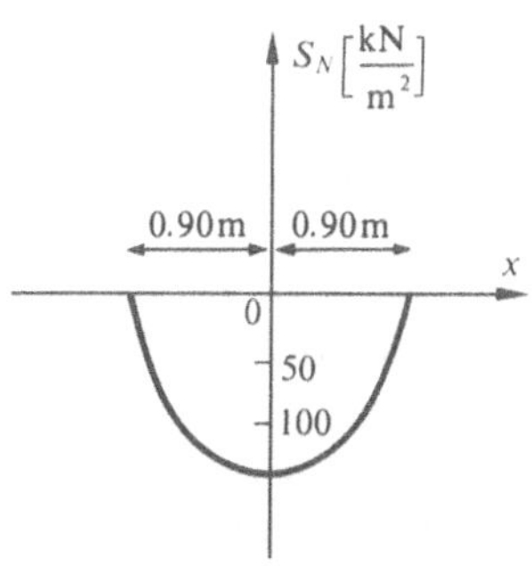

Fig. 10.1.2. Contact stress diagram.

As a second application, we solve the contact problem of the nonlinear elastic frame of Fig. 10.1.3a [249]. The beam elements have a bilinear stress-strain diagram and the foundation beam is in frictional unilateral contact with the soil, whose influence on the structure is given by the $S_N - u_N$ diagram of Fig. 10.1.4. The numerical calculations are performed by the "flexible tolerance" minimization algorithm of Paviani and Himmel-

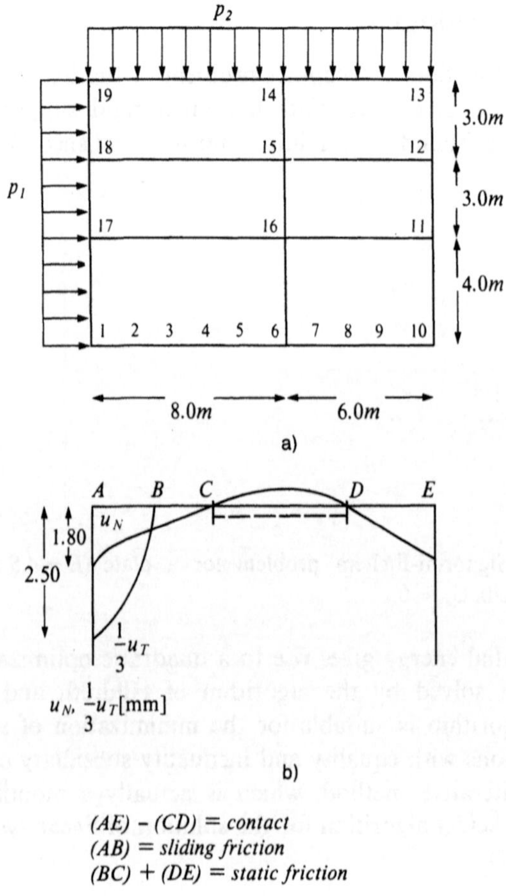

Fig. 10.1.3. The frictional unilateral contact problem for a frame: $p_2 = 2p_1 = 2\times 10^4$ N/m; $E = 2.05\times 10^3$ kN/cm² if $\varepsilon \leq 5^0/_{00}$, $E = 1.80\times 10^3$ kN/cm² if $\varepsilon > 5^0/_{00}$; beams and columns $F = 0.40\times 0.45$ m²; foundation beam $I = 10^{-2}$ m⁴, $F = 2$ m², $\mu = 0.50$.

blau [143]. The appearance of nondifferentiable terms in the potential energy leads to the use of a derivative-free minimization algorithm. The main feature of this algorithm is that the value of the objective function is improved by means of both feasible and near-feasible approximations, (i.e., approximations which do not belong to the feasible domain). There is also the advantage that computation time is not spent on strict feasibility requirements. In Fig. 10.1.3b the diagrams of the normal and tangential displacements are depicted.

The present section could be seen as a contribution to the solution, by means of variational methods, of the contact problems in which slippage and separation may occur between the contacting surfaces. Only convex potentials have been considered here (cf., Sec. 4.3). There is extensive

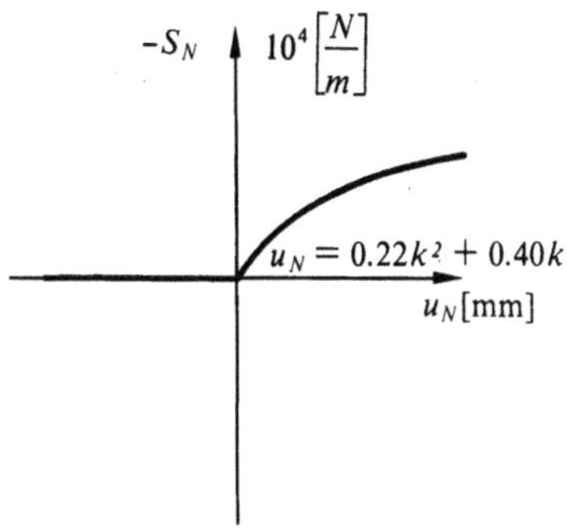

(1) (*l*)	$10^{-4} M$ (2) kNm	$10^{-4} M$ (3) kNm	$10^{-4} M$ (4) kNm
1 - 17	-83.38 -37.28	-70.63 -20.60	-50.03 -31.39
17 - 18	-48.07 0.00	-39.24 0.00	-44.14 - 1.96
18 - 19	81.42 90.25	74.56 93.20	78.48 93.20
6 - 16	-69.65 -68.67	-70.65 -63.76	-73.58 -68.67
10 - 11	-48.07 - 8.83	-44.14 -17.66	-34.34 -19.62

Fig. 10.1.4. The $S_N - u_N$ diagram and a comparison table for different contact assumptions: (1) beam elements, (2) frictional unilateral contact, (3) unilateral contact, (4) classical method.

literature on the treatment of various classes of contact problems using the classical finite element method. In these cases, first the contact conditions are appropriately modeled by means of bond elements, and then an incremental or a trial-and-error method is applied. For an account of this subject the reader may consult, among others, [139] [114].

10.2 Torsion of Cylindrical or Prismatic Bars With Convex Strain-Energy Density

10.2.1 Formulation of the Problem

In this section, we study the torsion problem of cylindrical or prismatic bars obeying the general superpotential material law (3.3.57). The theory to be developed follows the pattern of Sec. 6.1 when the procedure is specialized for the torsion problem. Concerning the classes of materials described by the superpotential law considered here, the reader may refer to Sec. 3.3.2.

Let Ω be an open, bounded, simply-connected subset of $\mathbb{R}^3$ occupied by the bar in its undeformed state. The bar is referred to a fixed Cartesian coordinate system $0x_1x_2x_3$. We denote by Σ the cross section of the bar parallel to the $0x_1x_2$-plane and by Γ_Σ its boundary, assumed to be regular.

Let Γ_l be the lateral surface of the bar and Γ_0 (resp. Γ_h) the cross sections corresponding to $x_3=0$ (resp. $x_3=h$), h being the length of the bar (Fig. 10.2.1).

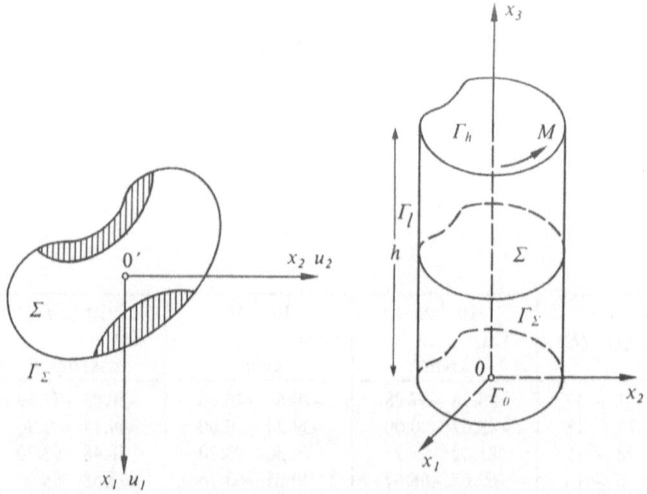

Fig. 10.2.1. The torsion problem.

The bar is twisted under the action of a moment M applied on the upper end Γ_h of the bar while its lower end remains fixed. In order to simplify the problem, we assume that free warping of the cross section in the x_3-direction is not prevented. The torsion B.V.P. takes the following form: find for a given angle of twist α a stress field $\sigma=\{\sigma_{ij}\}$, a strain field $\varepsilon=\{\varepsilon_{ij}\}$ and a displacement field $u=\{u_i\}$, $i,j=1, 2, 3$, satisfying in Ω the relations (6.1.2), (6.1.3) and

$$\sigma_{ij,j}=0 \tag{10.2.1}$$

and the boundary conditions

$$\sigma_{i1}n_1+\sigma_{i2}n_2=0, \qquad i=1,2,3 \quad \text{on} \quad \Gamma_l, \tag{10.2.2}$$

$$\sigma_{33}=0 \quad \text{on} \quad \Gamma_0\cup\Gamma_h, \tag{10.2.3}$$

$$u_1=u_2=0 \quad \text{on} \quad \Gamma_0 \tag{10.2.4}$$

and

$$u_1=-\alpha x_2, \qquad u_2=\alpha x_1 \quad \text{on} \quad \Gamma_h, \tag{10.2.5}$$

where $n=\{n_1, n_2\}$ denotes the exterior unit vector normal to Γ_l. This B.V.P. may be considered as a special case[4] of the problem studied in Sec. 6.1.1 and thus Props. 6.1.1 and 6.1.5 of the minimum potential and com-

[4] Here $\Gamma_U\cap\Gamma_F\neq\emptyset$. This fact does not influence the derivation of the minimum propositions.

plementary energy hold. It follows that the displacement field u minimizes the potential energy

$$\Pi(u) = \int_\Omega w(\varepsilon(u))\, d\Omega \tag{10.2.6}$$

on the set U_{ad} of all kinematically admissible displacement fields (cf., 6.1.9) and that the stress field σ minimizes the complementary energy

$$\Pi^c(\sigma) = \int_\Omega w^c(\sigma)\, d\Omega - \alpha \int_{\Gamma_h} (x_1 \sigma_{23} - x_2 \sigma_{13})\, dx_1\, dx_2 \tag{10.2.7}$$

on the set Σ_{ad} of all statically admissible stress fields (cf., (6.1.27)). U_{ad} (resp. Σ_{ad}) is defined by the subsidiary conditions (10.2.4, 5) (resp. (10.2.1–3)) as well as by the inequality $w(\varepsilon(u)) < \infty$ (resp. $w^c(\sigma) < \infty$) (cf., the difference between (10.2.6, 7) and (6.1.10, 29)). The case of an elastic perfectly plastic material (cf. (3.3.83)) has already been extensively studied [118], [170], [316]. For homogeneous and isotropic material and for the v. Mises yield criterion the stress field σ which minimizes Π^c satisfies the relations (see [170] for the mathematical justification):

$$\sigma_{ij} = 0 \quad \text{in} \quad \Omega, \qquad \forall i, j = 1, 2, 3$$
$$\text{with the exception of } \sigma_{13} \text{ and } \sigma_{23} \tag{10.2.8}$$

and

$$\sigma_{ij,3} = 0 \quad \text{in} \quad \Omega. \tag{10.2.9}$$

This fact allows the introduction of a stress function $\psi \in \mathring{H}^1(\Sigma)$, which leads to a simple variational formulation of the problem. If the general material law (6.1.3) holds, then the proof of such a result is still an open problem, and hence in order to introduce a stress function we can reasonably assume that (10.2.8) is valid. Such an assumption suffices for an engineering-oriented theory.[5] From (10.2.1, 8) we obtain (10.2.9) and

$$\sigma_{13,1} + \sigma_{23,2} = 0, \tag{10.2.10}$$

and thus a stress function $\psi = \psi(x_1, x_2)$ can be introduced such that

$$\frac{\partial \psi}{\partial x_2} = \sigma_{13}, \qquad \frac{\partial \psi}{\partial x_1} = -\sigma_{23} \quad \text{in} \quad \Sigma. \tag{10.2.11}$$

Because the stress vector is tangent to Γ_Σ, $d\psi = 0$ on Γ_Σ, and due to (10.2.11) it is sufficient to take ψ as equal to 0 on Γ_Σ. Further, the functional $\tilde{w}^c$ is defined by setting $\tilde{w}^c(\psi) = w^c(\sigma(\psi))$. Thus the minimization problem of the complementary energy acquires the form

$$\tilde{\Pi}^c(\psi) = \min \{\tilde{\Pi}^c(\psi^0) \mid \psi^0 \in \tilde{\Sigma}_{ad}\}, \tag{10.2.12}$$

[5] Cf., [278] p. 241.

where

$$\tilde{\Pi}^c(\psi) = \int_\Omega \tilde{w}^c(\psi)\, d\Omega - 2\alpha \int_\Sigma \psi\, d\Sigma \tag{10.2.13}$$

and

$$\tilde{\Sigma}_{ad} = \{\psi^0 \mid \psi^0 \in \mathring{H}^1(\Sigma), \quad \tilde{w}^c(\psi^0) < \infty\}. \tag{10.2.14}$$

In the special case of an elastic perfectly plastic material obeying the von Mises yield condition, problem (10.2.12) takes the form

$$\tilde{\Pi}_1^c(\psi) = \min \{\tilde{\Pi}_1^c(\psi^0) \mid \psi^0 \in \mathring{H}^1(\Sigma), |\operatorname{grad} \psi^0| \le g\}, \tag{10.2.15}$$

with

$$\tilde{\Pi}_1^c(\psi) = h \int_\Sigma (\operatorname{grad} \psi)^2\, d\Sigma - \frac{2E\alpha}{(1+\nu)} \int_\Sigma \psi\, d\Sigma. \tag{10.2.16}$$

For this problem a free boundary is formed between the elastic ($|\operatorname{grad} \psi| < g$) and plastic ($|\operatorname{grad} \psi| = g$) regions in the cross section. When ψ is obtained from the minimization problem (10.2.12) for a given α, the corresponding torque M is given by

$$M = 2 \int_\Sigma \psi\, d\Sigma. \tag{10.2.17}$$

A generalization of the derived variational principles for the case of multiply-connected cross sections requires slight modifications to the procedure of this section (see [169]).

10.2.2 Discretization and Numerical Application

The numerical calculation is performed by minimizing the energy functionals after the discretization of the continuous problem by a finite element scheme. The cross section Σ is subdivided into Q elements E_q, $1 \le q \le Q$. Let N be the number of nodes P_n, $1 \le n \le N$. Of the total nodes N, N' belong to the interior of Σ and $N - N'$ to the boundary Γ_Σ (Fig. 10.2.2).

In the following example triangular elements are chosen and in every element the stress function is approximated by a linear polynomial. If ψ_q^λ, $\lambda = 1, 2, 3$, denote the values of ψ at the vertices of the element E_q, we may for this element write a relation

$$\psi_q^d(x_1, x_2) = \sum_{\lambda=1}^{3} f_q^\lambda(x_1, x_2)\psi_q^\lambda, \tag{10.2.18}$$

where $f_q^\lambda = 1$ at the λ-th node and $f_q^\lambda = 0$ at the remaining nodes of the element and outside it. Our main task is to determine the values of ψ_q^λ which minimize $\tilde{\Pi}^c$. After introducing in (10.2.12) the function ψ^d given by (10.2.18), we consider the vectors $\bar{\boldsymbol{\psi}} = [\psi_1^1, \ldots, \psi_Q^3]^T$ and $\boldsymbol{\psi} = [\psi_1, \ldots, \psi_{N'}]^T$

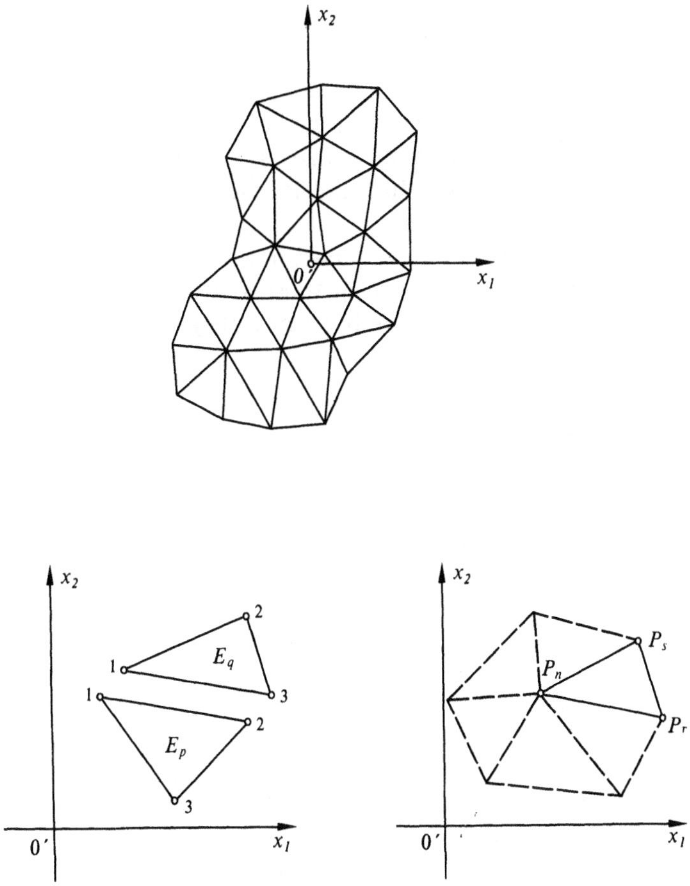

Fig. 10.2.2. Discretization of the cross section.

and the relation $\bar{\boldsymbol{\psi}} = \mathbf{S}\boldsymbol{\psi}$ between them. Here $\mathbf{S}$ is a $3Q \times N'$ matrix, composed of 0's and 1's which takes into account the element connection and the boundary conditions. If we write (10.2.18) for all elements in the form

$$\psi^d(x_1, x_2) = \mathbf{f}^T\bar{\boldsymbol{\psi}}, \tag{10.2.19}$$

where $\mathbf{f} = [f_1^1(x_1, x_2), \ldots, f_Q^3(x_1, x_2)]$, then the discretized form of (10.2.12) is

$$\tilde{\Pi}_d^c(\boldsymbol{\psi}) = \min\, \{\tilde{\Pi}_d^c(\boldsymbol{\psi}_0) \mid \boldsymbol{\psi}_0 \in \tilde{\Sigma}_{ad}^d\}, \tag{10.2.20}$$

where

$$\tilde{\Pi}_d^c(\boldsymbol{\psi}) = \tilde{\Pi}^c(\mathbf{f}^T\mathbf{S}\boldsymbol{\psi}) \tag{10.2.21}$$

and $\tilde{\Sigma}_{ad}^d$ is the discretized form of $\tilde{\Sigma}_{ad}$. The subsidiary condition $\tilde{w}^c(\psi) < \infty$ in Σ may be satisfied in the discretized problem at certain points of E_q, or at the center of gravity of the element, or in an integral form, e.g.,

$(\int_{E_q}|\tilde{w}^c(\psi^d(x_1, x_2))|dx_1 dx_2)|E_q|^{-1} < \infty$ where $|E_q|$ denotes the area of the element E_q.

The question of the convergence of the solution of the finite dimensional problem is discussed in [120], only for the holonomic, elastic perfectly plastic law. The numerical solution of (10.2.20) is obtained by a nonlinear optimization algorithm. As in Sec. 10.1, the flexible-tolerance algorithm of Paviani and Himmelblau is applied.

As a numerical example, let us consider a steel bar having as a cross section an equilateral triangle or a square (Fig. 10.2.3). The material of the bar is assumed to be elastic perfectly plastic, obeying the von Mises yield criterion (holonomic theory).

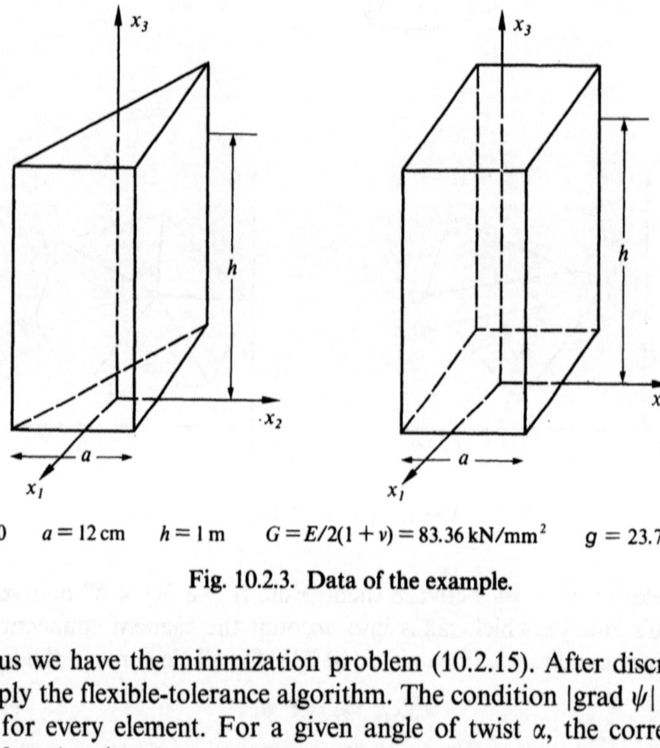

St 42/50 $a = 12$ cm $h = 1$ m $G = E/2(1+\nu) = 83.36$ kN/mm² $g = 23.78$ kN/cm²

Fig. 10.2.3. Data of the example.

Thus we have the minimization problem (10.2.15). After discretization, we apply the flexible-tolerance algorithm. The condition $|\text{grad}\,\psi| \leq g$ is satisfied for every element. For a given angle of twist α, the corresponding stress function ψ and torque M are calculated. Fig. 10.2.4 depicts the torque M versus the angle of twist per unit length $D = \alpha/h$ for both cross sections. The moment M^E (resp. M^P) for which at certain points of the cross section the yield limit is reached (resp. the whole cross section becomes plastic) is for the triangular cross section $M^E = 20.80$ kNm (resp. $M^P = 33.75$ kNm). These moments correspond to the angles of twist per unit length $D^E = 0.06$ rad/m and $D^P = 0.20$ rad/m. For the square cross section, $M^E = 88.25$ kNm for $D^E = 0.04$ rad/m and $M^P = 136.30$ kNm for $D^P = 0.22$ rad/m.

As a verification of the numerical results, the values of M^E and M^P are calculated by the formulas derived in the theory of applied elastoplasticity

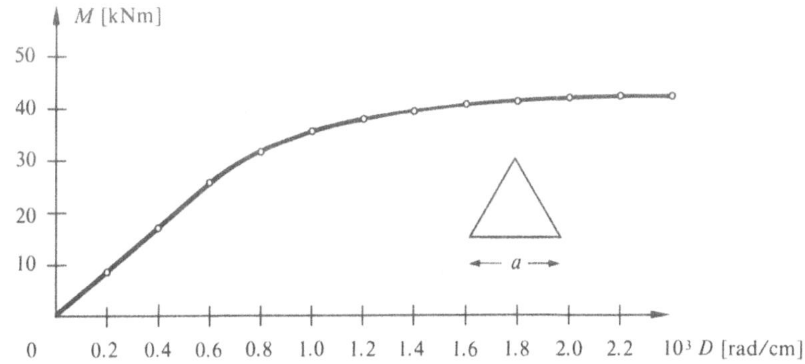

a)

D	M
0.0002	7.20
0.0004	13.90
0.0006	20.80
0.0008	25.70
0.0010	28.95
0.0012	30.51
0.0014	31.23
0.0016	31.90
0.0018	32.50
0.0020	33.75

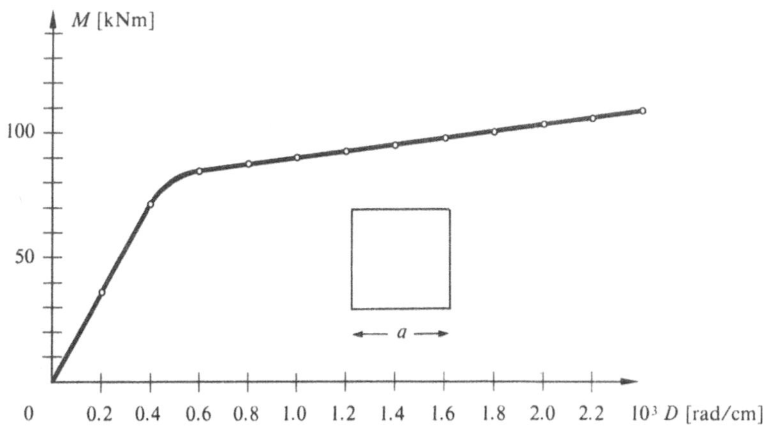

b)

D	M
0.0002	47.55
0.0004	88.25
0.0006	118.95
0.0008	126.35
0.0010	129.52
0.0012	131.05
0.0014	132.10
0.0016	133.55
0.0018	134.80
0.0020	135.40
0.0022	136.30

Fig. 10.2.4. *M-D* diagrams.

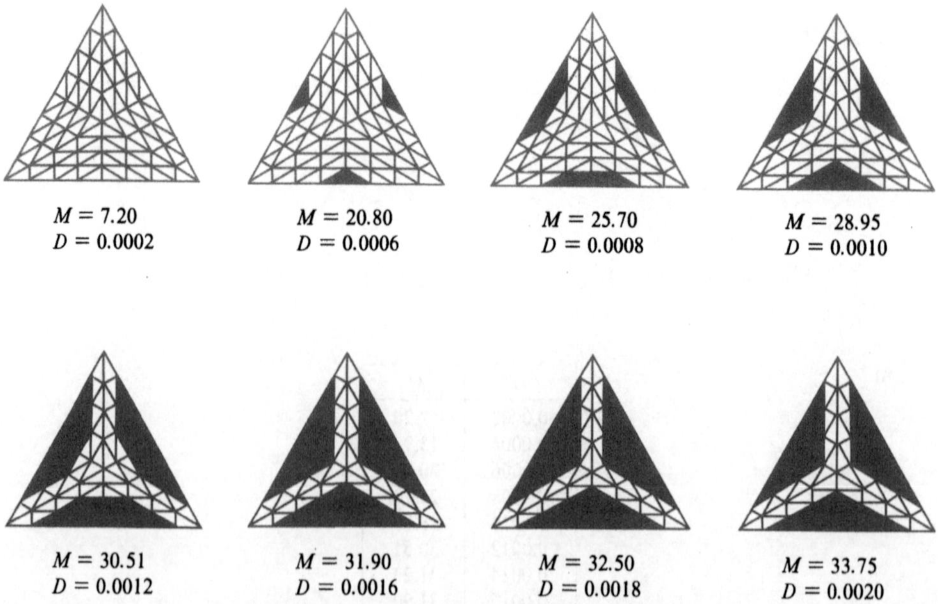

Fig. 10.2.5. Plastic and elastic zones.

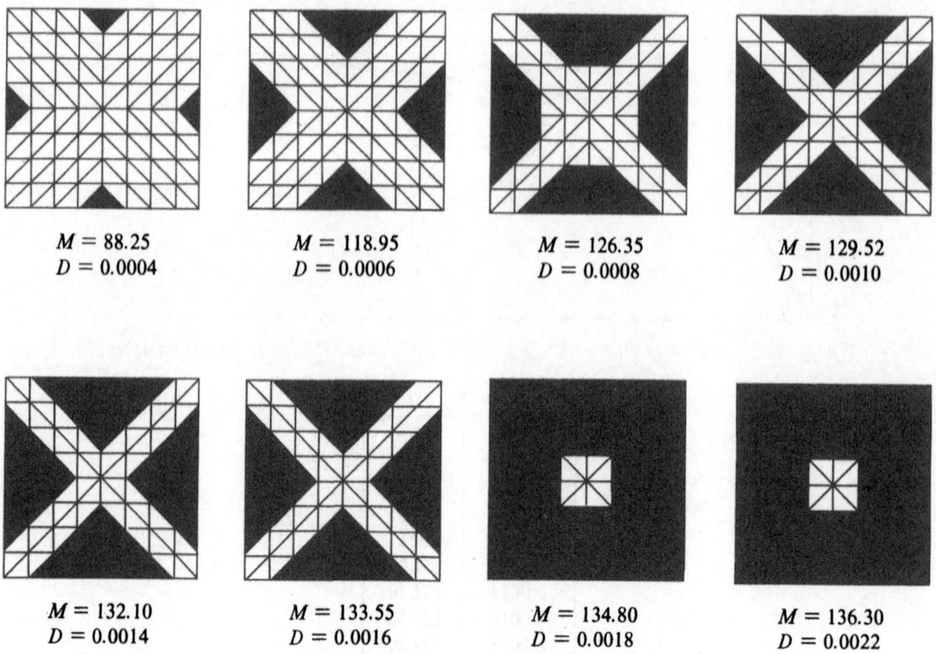

Fig. 10.2.6. Plastic and elastic zones.

[278]. For the triangle we find $M^E = ga^3/20 =$ 20.54 kNm, $M^P = ga^3/12 =$ 34.24 kNm, and for the square $M^E = (1/1.585)ga^3/3 =$ 86.42 kNm, $M^P = ga^3/3 =$ 136.98 kNm.

Figs. 10.2.5 and 10.2.6 depict the growth of the plastic zones for increasing D and M, and Figs. 10.2.7–10.2.10 the values of ψ (compared in certain cases with the values ψ_{th} obtained by the methods of the theory of elasticity) and the values of σ_{13}, σ_{23} for M^E and M^P.

$M^E = 20.80$
$D^E = 0.0006$

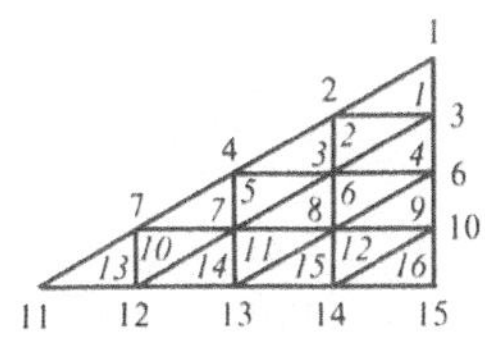

P_n	ψ	ψ_{th}	E_q	σ_{13} (MPa)	σ_{23} (MPa)
1	40.10	40.02	*1*	19.90	-41.36
2	32.61	33.84	*2*	47.95	-41.36
3	38.70	37.96	*3*	47.95	-55.82
4	19.65	19.91	*4*	73.86	-26.42
5	28.16	28.45	*5*	75.30	-55.82
6	31.80	31.20	*6*	123.10	-26.42
7	5.43	5.98	*7*	75.30	-50.85
8	13.15	13.24	*8*	123.10	-28.25
9	17.20	17.46	*9*	152.48	- 9.48
10	18.96	18.93	*10*	65.14	-50.85
11	0.00	0.00	*11*	152.68	-28.25
12	0.00	0.00	*12*	201.80	9.48
13	0.00	0.00	*13*	65.14	0.00
14	0.00	0.00	*14*	152.68	0.00
15	0.00	0.00	*15*	201.80	0.00
			16	233.48	0.00

Fig. 10.2.7. Triangular cross section: ψ, ψ_{th}, σ_{13}, σ_{23} for $M = M^E$.

$M^P = 33.75$
$D^P = 0.0020$

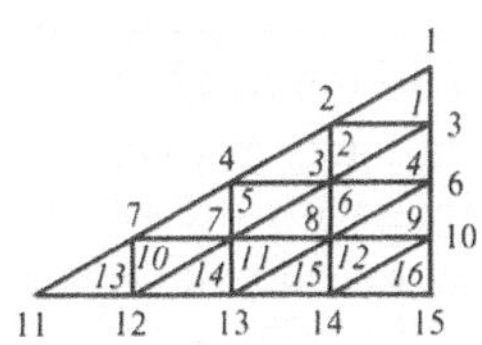

P_n	ψ	E_q	σ_{13} (MPa)	σ_{23} (MPa)
1	69.35	*1*	94.08	-39.26
2	55.30	*2*	167.36	-39.26
3	61.20	*3*	167.36	-39.26
4	34.85	*4*	235.46	0.00
5	40.80	*5*	167.75	-39.26
6	40.80	*6*	235.46	0.00
7	14.62	*7*	167.75	-39.26
8	20.38	*8*	235.46	0.00
9	20.38	*9*	235.46	0.00
10	20.38	*10*	167.75	-39.26
11	0.00	*11*	235.46	0.00
12	0.00	*12*	235.46	0.00
13	0.00	*13*	167.75	0.00
14	0.00	*14*	235.46	0.00
15	0.00	*15*	235.46	0.00
		16	235.46	0.00
Theoretical value: $D^P \to \infty$				

Fig. 10.2.8. Triangular cross section: ψ, σ_{13}, σ_{23} for $M = M^P$.

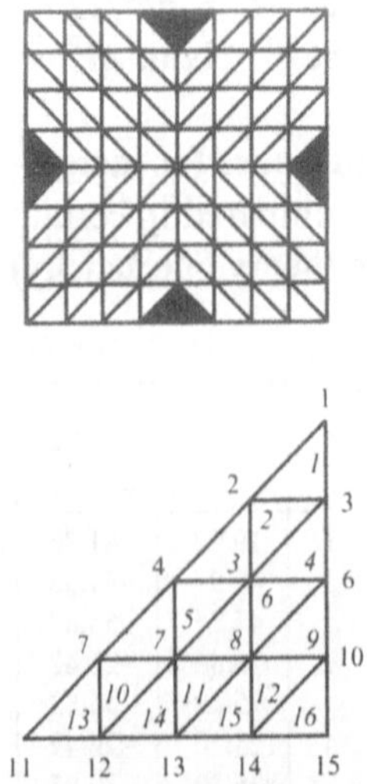

$M^E = 88.25$
$D^E = 0.0004$

P_n	ψ	E_q	σ_{13}	σ_{23}
			MPa	
1	72.31	*1*	36.00	-24.78
2	63.15	*2*	73.75	-24.78
3	66.80	*3*	73.75	-59.45
4	43.05	*4*	79.17	-19.40
5	52.00	*5*	10.01	-59.45
6	55.05	*6*	135.35	-19.40
7	17.25	*7*	108.11	-64.28
8	26.70	*8*	135.35	-32.27
9	31.62	*9*	144.01	-10.62
10	33.21	*10*	115.41	-64.28
11	0.00	*11*	179.72	-31.27
12	0.00	*12*	212.19	-10.62
13	0.00	*13*	115.41	0.00
14	0.00	*14*	179.72	0.00
15	0.00	*15*	212.19	0.00
		16	222.69	0.00

Fig. 10.2.9. Square cross section: ψ, σ_{13}, σ_{23} for $M = M^E$.

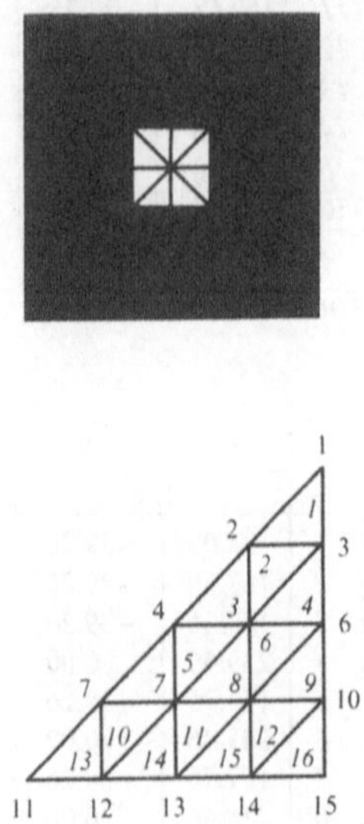

$M^P = 136.30$
$D^P = 0.0022$

P_n	ψ	E_q	σ_{13}	σ_{23}
			MPa	
1	130.90	*1*	166.87	0.00
2	105.98	*2*	235.40	0.00
3	105.98	*3*	235.40	0.00
4	70.78	*4*	235.40	0.00
5	70.78	*5*	235.40	0.00
6	70.78	*6*	235.40	0.00
7	35.62	*7*	235.40	0.00
8	35.62	*8*	235.40	0.00
9	35.62	*9*	235.40	0.00
10	35.62	*10*	235.40	0.00
11	0.00	*11*	235.40	0.00
12	0.00	*12*	235.40	0.00
13	0.00	*13*	235.40	0.00
14	0.00	*14*	235.40	0.00
15	0.00	*15*	235.40	0.00
		16	235.40	0.00
Theoretical value: $D^P \to \infty$				

Fig. 10.2.10. Square cross section: ψ, σ_{13}, σ_{23} for $M = M^P$.

Note that the application of an optimization algorithm in solving the elastoplastic torsion problem allows direct determination of the free boundary between the elastic and the plastic regions for a given angle of twist. In the classical theory of elastoplastic torsion, this is possible, with the exception of a small number of cross sections, only by means of trial-and-error methods or by the "sand-hill analogy" [275]. In [250], the same problem has been solved on the assumption that the geometrical and physical parameters of the bar are random variables. In order to solve the resulting stochastic inequality problem, the deterministic optimization algorithm is combined with the Monte Carlo simulation method (see [222] [247] [92], and for the underlying philosophy [223]). In our example, holonomic quantities have been used; similar variational expressions may also be obtained if (6.1.3) is replaced by (3.3.98) (see [170]).

10.3 A Linear Analysis Approach to Certain Classes of Inequality Problems

10.3.1 Description of the Method

In a variety of practical applications some unilateral problems appear which include a large number of unknowns, as with structures resting on tensionless elastic foundations, cylindrical tubes encased in soil, cable structures, etc. All these problems are nonlinear, even in the cases of linear elasticity and small deformations, and thus the classical linear analysis techniques cannot be used. However, efforts have been made to modify these techniques for use in unilateral problems as well; for example, many authors dealing with static analysis of contact problems assume the contact zone to be given or solve the problem by using iterative (trial-and-error) methods (see, e.g., [336] [294] [237] [149] [264]). However, convergence to the correct solution by these iterative procedures is not always ensured, since the contact area is not known *a priori* (see, e.g., [299] for a counter-example).

On the other hand, the incremental method for the solution of unilateral problems entails superfluous computational effort, and inaccurate solutions may be obtained if the load increments do not follow the changes of the static system of the structure.

The finite element method, applied to unilateral problems, leads, as we have seen, to discrete optimization problems with many unknowns, requiring the use of efficient optimization algorithms for the numerical calculations. However, most of these algorithms are efficient only in problems with a small number of unknowns. Moreover, general-purpose computer programs for linear elastic analysis cannot be used, which is a great disadvantage for the engineer.

In this section, a method is proposed which allows for the numerical treatment of certain classes of unilateral problems without the use of optimization algorithms. The method is suitable only for unilateral problems leading to a quadratic programming problem with linear-inequality subsidiary conditions [256]. Note that this class of unilateral problems is most important regarding practical applications (cf., Sec. 10.1). The method, which is based on propositions of Theil and van de Panne [166], replaces the unilateral problem by a number of classical bilateral problems. The aforementioned propositions allow control of the iterations, so that convergence to the solution of the initial unilateral problem is assured. Hence, available linear elastic analysis computer programs can be used without any modification and linear equation solvers can be applied to treat unilateral problems with many thousands of unknowns, this not being possible in the case of most optimization algorithms. We consider the following quadratic optimization problem[6]:

$$\min \{P(\mathbf{q}) = \tfrac{1}{2}\, \mathbf{q}^T\mathbf{A}\mathbf{q} + \mathbf{q}^T\mathbf{p} \,|\, \mathbf{a}_j^T \mathbf{q} \leq b_j, \qquad j \in \delta_0\}, \tag{10.3.1}$$

where δ_0 denotes the set of indices $1, 2, \ldots, m$. $\mathbf{A}$ is a positive-definite matrix (for the case "$\mathbf{A}$ positive semidefinite" see [256]), e.g., a stiffness or flexibility matrix, $\mathbf{p}$ is the load or initial strain vector, and $\mathbf{q}$ is the vector of the unknown variables. For our purposes we shall introduce the following notations. An inequality of problem (10.3.1) is denoted by u.c. (unilateral constraint). If only the corresponding equality is satisfied, this constraint is called an equality constraint. If $\mathbf{q}$ satisfies a constraint in the form $\mathbf{a}_j^T \mathbf{q} \gtreqless b_j$, then we say that the u.c. is replaced by a bilateral constraint, abbreviated b.c. The solution of (10.3.1) is denoted by $\hat{\mathbf{q}}$. Let $\hat{\delta}$ be the set of indices which $\hat{\mathbf{q}}$ satisfies in equality form, i.e., $\mathbf{a}_j^T \hat{\mathbf{q}} = b_j$. By $\mathring{\mathbf{q}}$ (resp. $\mathbf{q}^\delta$) we denote the solution of the problem

$$\min \{P(\mathbf{q})\} \tag{10.3.2}$$

resp.

$$\min \{P(\mathbf{q}) \,|\, \mathbf{a}_j^T \mathbf{q} = b_j, \qquad j \in \delta \subset \delta_0\}\,. \tag{10.3.3}$$

Problem (10.3.3) results if $|\delta|$ of the unilateral constraints are satisfied in equality form and $|\delta_0 - \delta|$ of the u.c.'s are replaced by b.c.'s.[7] Finally, let $V(\mathbf{q})$ be the set of indices corresponding to constraints which are violated by $\mathbf{q}$, i.e., $\mathbf{a}_j^T \mathbf{q} > b_j$ for $j \in V(\mathbf{q})$.

To keep the exposition brief, we shall give here without proofs the following propositions of Theil and van de Panne.[8]

Proposition 10.3.1. The equality

$$\hat{\mathbf{q}} = \mathbf{q}^{\hat{\delta}} \tag{10.3.4}$$

holds.

Proposition 10.3.2. If $\hat{\mathbf{q}} = \mathbf{q}^{\hat{\delta}}$ and $\hat{\delta} \neq \varnothing$, then for all subsets $\delta \subset \hat{\delta}$, including $\delta = \varnothing$,

$$h \in V(\mathbf{q}^\delta) \quad \text{for at least one} \quad h \in \hat{\delta} - \delta\,. \tag{10.3.5}$$

[6] We have chosen here the notation $\mathbf{q}$, $\mathbf{A}$, $\mathbf{p}$, etc. since $\mathbf{q}$ may be either a displacement vector (e.g., in the Signorini–Fichera problem of a linear elastic body when the potential energy is minimized) or a stress vector (e.g., in the friction problem of Sec. 5.1.1 when the complementary energy is minimized).

[7] $|\delta|$ denotes here the number of elements of the set δ.

[8] For the proofs see [166] p. 80.

Proposition 10.3.3. If $V(\mathbf{q}^\delta) = \varnothing$ and $h \in V(\mathbf{q}^{\delta - \{h\}})$, $\forall h \in \delta$, then $\mathbf{q}^\delta = \hat{\mathbf{q}}$.

Accordingly, if the solution $\mathbf{q}^0$ of the unrestricted problem (10.3.2) violates certain restrictions, the solution $\hat{\mathbf{q}}$ of (10.3.1) satisfies in equality form at least one of these restrictions.

If the solution $\mathbf{q}^\delta$ of (10.3.3) violates some restrictions, at least one of these restrictions must also be satisfied in equality form in order to yield the minimum $\hat{\mathbf{q}}$.

The solution $\mathbf{q}^\delta$ of (10.3.3), which satisfies all restrictions of the initial problem (10.3.1), is the solution $\hat{\mathbf{q}}$ we seek if and only if for every $h \in \delta$ the vector $\mathbf{q}^{\delta - \{h\}}$ violates the omitted restriction h.

These criteria give the first step (i.e., the solution of the unrestricted problem), the manner of proceeding to the next steps and finally an optimality test for the various solutions $\mathbf{q}^\delta$. Note that a vector $\mathbf{q}^\delta$ with $V(\mathbf{q}^\delta) = \varnothing$ (i.e., a vector which does not violate any restriction) is not necessarily the solution of the unilateral problem. This last remark distinguishes the proposed method from the usual trial-and-error methods.

The "replacement" of the initial unilateral problem by the finite sequence of the classical bilateral problems which correspond to the unconstrained, or equality-constrained, minimization problems (10.3.2, 3) permits the use of general-purpose computer programs, because (10.3.2, 3) give rise to linear equation systems. The behavior of the unilateral structure is the same as that of a bilateral structure. To apply this method, it is necessary in every problem to determine the u.c.'s, b.c.'s, and equality constraints—for instance, in the case of linear elastic structures containing cable elements (cf., (3.3.65) and Sec. 11.1), which cannot have compressive forces, i.e., $S_i \geq 0$. A u.c. is transformed into a b.c. by replacing the corresponding cable element with a rod element of the same cross-section. An equality constraint is realized by "cutting" the corresponding cable element. In the contact problem of a deformable body with a rigid support the u.c. $u_N \leq 0$ must be satisfied (cf., (3.3.22)). An equality constraint implies contact, i.e. $u_N = 0$, whereas a b.c. implies the removal of the support, i.e., $u_N \gtreqless 0$.

From the computational point of view, any finite element computer program may be used for the treatment of the bilateral structures. The consideration of the equality constraints in the unconstrained problem (10.3.2) leads to nonsymmetric matrices, which can be avoided by means of certain operations proposed in [64].

10.3.2 Applications

As a first example, let us consider the unilateral contact problem of a circular linear elastic Kirchhoff plate resting on a rigid support. The plate is loaded by a uniform surface load q and by bending moments M along its boundary (Fig. 10.3.1). These conditions appear in the bottom plate of a

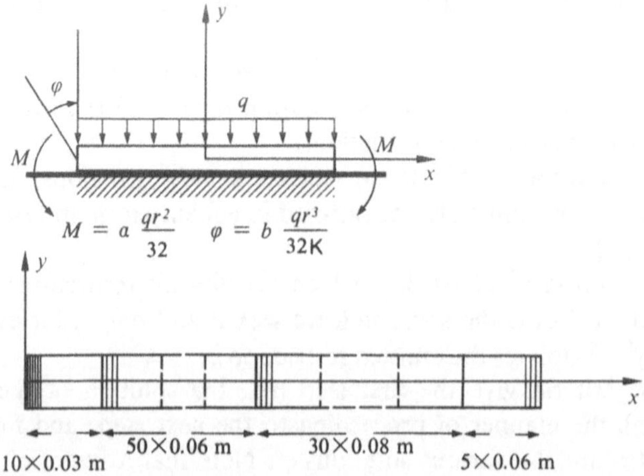

Fig. 10.3.1. The unilateral contact problem of a circular plate ($t = 0.06$ m, $r = 6$ m, $\nu = 0$, $E = 2.1 \times 10^8$ kN/m^2, $q = 0.6$ kN/m^2, $K = Eh^3/12(1-\nu^2)$).

cylindrical tank. The plate is discretized by means of axisymmetric two-dimensional elements, the thickness of the plate being divided into two layers. In order to compare the results of the calculation with the analytical solution given in [315], a small ratio of thickness t to radius r is chosen and a fine-mesh configuration is used, particularly around the plate center (Fig. 10.3.1). The solution of the simply supported circular plate constitutes the first step of the calculation for $a = 1$ and violates all restrictions. In the next iteration, equality constraints are considered simultaneously at all points

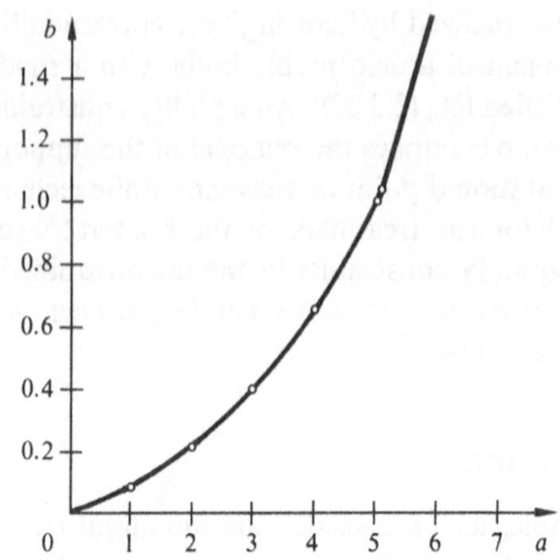

Fig. 10.3.2. Moment-rotation diagram (number of elements 190, number of constraints 95, maximum number of unknowns 571).

between $x = 0.0m$ and $x = 3.3m$. Since the solution does not satisfy the optimality criteria, we repeat this procedure, until for $x = 3.7$ a solution is obtained which passes the optimality test and is therefore the solution of the unilateral problem. This procedure is continued for $a = 2, 3, 4, 5$, and the resulting moment-rotation diagram is depicted in Fig. 10.3.2. For $a = 5$ the plate loses contact with the plane. The moment-rotation diagram is linear only for $a \geq 5$. For this value of a, the plate loses contact with the support and the problem becomes linear. In this application much computational effort is avoided by applying the optimality test only where required by physical circumstances. The numerical results obtained agree satisfactorily with the analytical solution of [315].

Fig. 10.3.3 gives the deformed shapes of the plate for $a = 1, 2, 3, 4, 5$ and $a = 5.1$ and some details of the deformation near the lifted parts of the plate. In Fig. 10.3.4 the numerical results of the method are compared with the analytical solution.

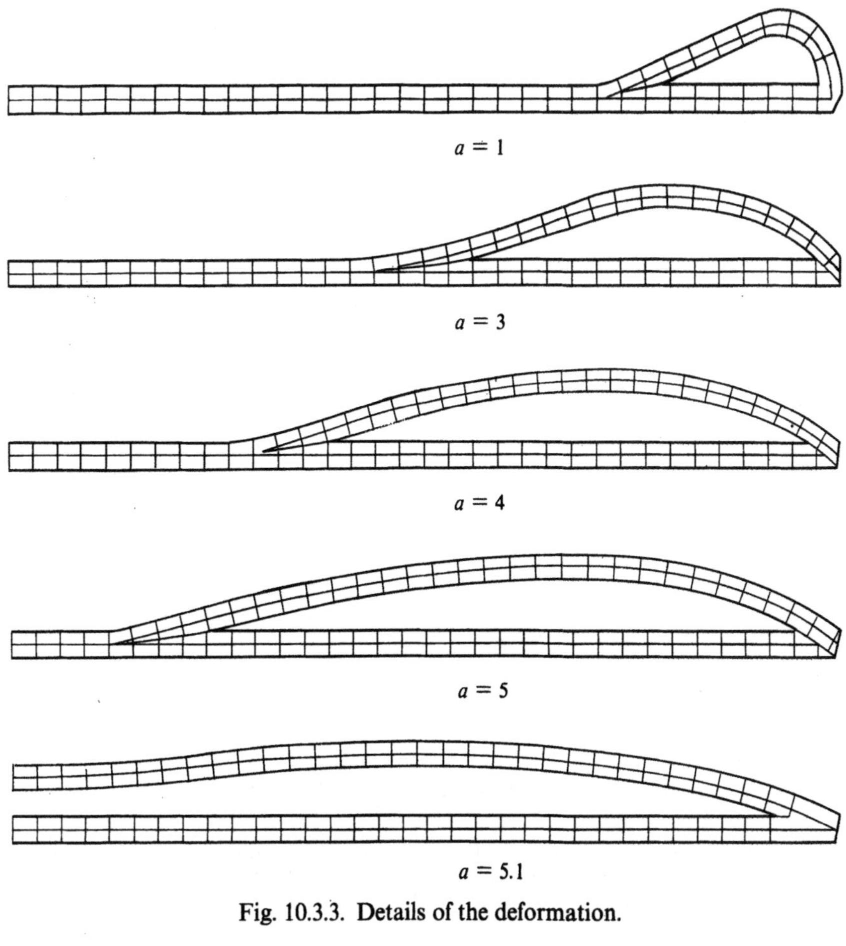

Fig. 10.3.3. Details of the deformation.

a	M kNm/m	$10^{-5}\varphi$ rad	$10^{-5}\varphi$* rad	b	*contact area* m	*contact area* m**
1	0.675	7.290	~ 7.5	0.0680	3.70	3.70/3.78
2	1.35	22.236	~ 23.0	0.2075	2.58	2.58/2.64
3	2.025	43.8	~ 43.9	0.4088	1.56	1.56/1.62
4	2.7	72.379	~ 71.5	0.6755	0.54	0.54/0.60
5	3.375	107.176	~107.14	1.0003	0.0	0.0
5.1	3.44	126.117	~126.5	1.1770	*no contact*	

Fig. 10.3.4. Comparison of the computed results with known analytical results: *) is obtained from [315] Fig. 155 and **) from the analytical method of [315].

Further we shall study the contact problem of the cylindrical tube of Fig. 10.3.5. The tube is encased in concrete (assumed to be rigid). Frictional effects and geometric nonlinearities are neglected.

To discretize the cylinder we use curved shell elements (a mixed model in terms of stresses and displacements [308]), and we proceed to the solution of every bilateral structure resulting from the method proposed.[9]

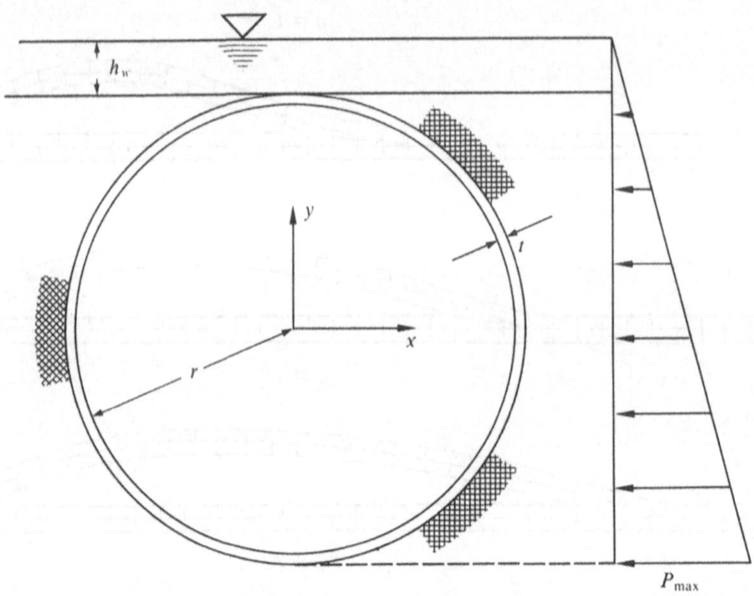

Fig. 10.3.5. The contact problem of a cylindrical tube ($h_w = 1.18$ m, $t = 4$ cm, $E = 60$ kN/cm^2, $\nu = 0.3$, $r = 1.315$ m, $P_{max} = (h_w + 2r)\ 5.8 \times 10^{-2}$ kN/m^2).

[9] This is an indirect use of mixed models for the numerical treatment of variational inequalities. The direct use of mixed models for variational inequalities is based on the min max formulation (wherever possible) of the inequalities.

The radial displacements are constrained against outward movement, whereas the tangential displacements are unrestricted. In Fig. 10.3.6, the deformed shape of the shell, the contact area and the distribution of the bending moments are depicted.

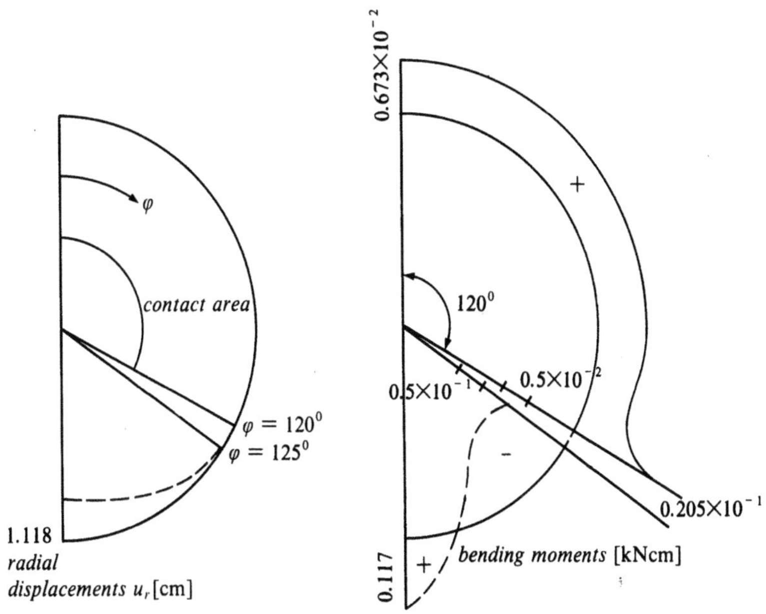

Fig. 10.3.6. Diagrams of contact area, deformed shape and bending moments.

To summarize, the method proposed in this section combines the basic advantages of the optimization algorithm approach to unilateral problems, i.e., the rational determination of free boundaries, with the basic advantage of linear elastic analysis computer programs, which is their capacity to treat a large number of unknowns.

Chapter 11

Incremental and Dynamic Inequality Problems

This chapter is devoted to the numerical treatment of incremental and dynamic inequality problems. The first Section deals with the elastoplastic analysis of cable structures, where we assume that the cables may become slack. With respect to this problem, we illustrate the application of multi-level decomposition techniques to the analysis of inequality problems with many unknowns. The second Section contains the incremental elastoplastic analysis of structures presenting geometric nonlinearities and physical destabilizing effects. Within every load increment a variational inequality is formulated which gives rise to a minimization problem. The last Section concerns the dynamic unilateral contact problem. The resulting variational inequality is discretized with respect to time, and within every time step a minimization problem arises which is solved by an appropriate optimization alogrithm.

11.1 The Elastoplastic Calculation of Cable Structures

11.1.1 Formulation of the Problem as a Linear Complementarity Problem and Related Expressions

This section deals with the elastoplastic calculation of cable structures, assuming geometric nonlinearity and unilateral behavior of the cables. The term unilateral means here that the cables cannot sustain compressive, i.e., negative, stresses, and that the cables may become slack during the loading process. This kind of unilateral behavior is also called stress-unilateral behavior [243]. The geometric nonlinearity due to the large displacements [1] and the physical nonlinearity due to the elastoplastic behavior of the cables are treated by means of an incremental procedure, i.e., the loads and the

[1] The strains are assumed to be small.

initial strains are simulated by a sequence of sufficiently small increments. This permits a linearization of the problem with respect to its geometrical and physical nonlinearity. But this is not the case with the phenomenological nonlinearity caused by the unilateral behavior of the cables. Because this latter nonlinearity introduces an inequality, we solve at every step an inequality problem.

The structure is divided into n elements, m of which correspond to the cables (which are idealized by pin-jointed bar elements). The stress and strain states of the individual elements are defined by means of natural generalized stress and strain vectors (cf., Sec. 10.1.1). An increment of a quantity is denoted by a dot over the symbol of the quantity. With respect to a reference configuration, the incremental stress-strain relationship for the i-th cable element has the form (see Fig. 11.1.1)

$$\dot{e}_i = F_{0,i}^{\cdot}\dot{s}_i + \dot{e}_{0i} - \dot{v}_i, \tag{11.1.1}$$

$$\sigma_i = \dot{s}_i + a_i \geq 0, \tag{11.1.2}$$

$$\gamma_i = \dot{v}_i + b_i \geq 0, \tag{11.1.3}$$

$$\sigma_i \gamma_i = 0. \tag{11.1.4}$$

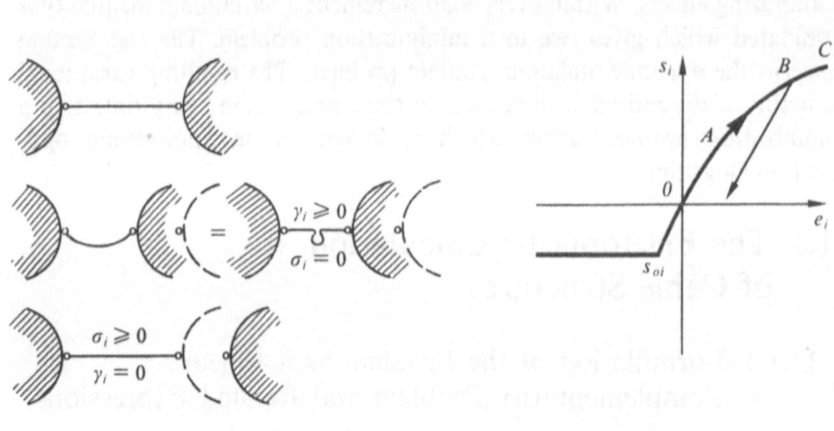

Fig. 11.1.1. On the stress-strain relation of a cable element.

Here s_i, e_i, $s_{0,i}$, $e_{0,i}$ are the stress, strain, initial stress and initial strain respectively, $F_{0,i}^{\cdot}$ is the tangential natural flexibility of the step under consideration, $-v_i$ is the slackness, a_i (resp. $-b_i$) denotes the total stress (resp. total slackness) at the beginning of the step, and finally σ_i (resp. $-\gamma_i$) denotes the total stress (resp. total slackness) at the end of the step. Relations (11.1.2–4) express the fact that at the end of each step a cable is either under tension or slack. (11.1.1) represents for $\dot{v}_i = 0$ the stress-strain incremental relations of each of the remaining *n-m* elements of the structure. Relations (11.1.1–4) take, for all the elements, the form

$$\dot{\mathbf{e}} = \mathbf{F}_0^{\cdot}\dot{\mathbf{s}} + \dot{\mathbf{e}}_0 - \dot{\mathbf{v}}, \tag{11.1.5}$$

$$\bar{\boldsymbol{\sigma}} = \dot{\bar{\mathbf{s}}} + \bar{\mathbf{a}} \geq \mathbf{0}, \tag{11.1.6}$$

$$\bar{\boldsymbol{\gamma}} = \dot{\bar{\mathbf{v}}} + \bar{\mathbf{b}} \geq \mathbf{0}, \tag{11.1.7}$$

$$\bar{\boldsymbol{\sigma}}^T\bar{\boldsymbol{\gamma}} = 0, \tag{11.1.8}$$

where $\dot{\mathbf{e}} = [\dot{\mathbf{e}}_1^T, \ldots, \dot{\mathbf{e}}_n^T]^T$, $\dot{\mathbf{s}} = [\dot{\mathbf{s}}_1^T, \ldots, \dot{\mathbf{s}}_n^T]^T$, $\dot{\mathbf{v}} = [\dot{v}_1, \ldots, \dot{v}_m, \ 0, \ldots, 0]^T = [\dot{\bar{\mathbf{v}}}^T | 0]^T$, $\bar{\boldsymbol{\sigma}} = [\sigma_1, \ldots, \sigma_m]^T$. Analogously, $\bar{\boldsymbol{\gamma}}$, $\bar{\mathbf{a}}$, $\bar{\mathbf{b}}$ are defined. $\mathbf{F}_0^{\cdot}$ is the tangential flexibility matrix of the unassembled elements. The part of a matrix or vector extended only over the cable elements is henceforth denoted by a bar.

At every step of the calculation the configuration of the structure is described by the vector $\mathbf{u}$ of nodal displacements, with respect to a fixed Cartesian coordinate system. The equations of equilibrium are linearized in every step by means of the geometric stiffness matrix $\mathbf{K}_G$ [9] and have the form

$$\mathbf{G}\dot{\mathbf{s}} + \mathbf{K}_G\dot{\mathbf{u}} = \dot{\mathbf{p}}. \tag{11.1.9}$$

Here $\mathbf{G}$ is the equilibrium matrix, $\mathbf{K}_G$ is the geometric stiffness matrix, which depends on the geometry and the stresses at the end of the preceding step, and $\dot{\mathbf{p}}$ is the load vector increment. In every step we assume that the structure behaves in a geometrically linear manner, and thus within this step the strain-displacement equations may be written in the form

$$\dot{\mathbf{e}} = \mathbf{G}^T\dot{\mathbf{u}}. \tag{11.1.10}$$

All incremental quantities in (11.1.9, 10) are referred to the configuration of the structure at the end of the preceding step. Relations (11.1.5–10) define a discrete unilateral problem. In terms of $\dot{\mathbf{u}}$ and $\dot{\mathbf{v}}$ we have the relations

$$\mathbf{K}\dot{\mathbf{u}} + \mathbf{G}\mathbf{K}_0\dot{\mathbf{v}} = \dot{\mathbf{p}} + \mathbf{G}\mathbf{K}_0\dot{\mathbf{e}}_0, \tag{11.1.11}$$

$$\bar{\boldsymbol{\sigma}} = \dot{\bar{\mathbf{s}}} + \bar{\mathbf{a}} = \bar{\mathbf{K}}_0\bar{\mathbf{G}}^T\dot{\mathbf{u}} - \bar{\mathbf{K}}_0\dot{\bar{\mathbf{e}}}_0 + \bar{\mathbf{K}}_0\dot{\bar{\mathbf{v}}} + \bar{\mathbf{a}} \geq \mathbf{0}, \tag{11.1.12}$$

$$\bar{\boldsymbol{\gamma}} = \dot{\bar{\mathbf{v}}} + \bar{\mathbf{b}} \geq \mathbf{0} \tag{11.1.13}$$

and

$$\bar{\boldsymbol{\sigma}}^T\bar{\boldsymbol{\gamma}} = 0. \tag{11.1.14}$$

Here $\mathbf{K} = \mathbf{K}_E + \mathbf{K}_G = \mathbf{G}\mathbf{K}_0\mathbf{G}^T + \mathbf{K}_G$, with $\mathbf{K}_0 = \mathbf{F}_0^{\cdot -1}$, is the total stiffness of the structure. Equation (11.1.11) is replaced by the two inequalities

$$\mathbf{K}\dot{\mathbf{u}} + \mathbf{G}\mathbf{K}_0\dot{\mathbf{v}} - \dot{\mathbf{p}} - \mathbf{G}\mathbf{K}_0\dot{\mathbf{e}}_0 \geq \mathbf{0}, \tag{11.1.15}$$

$$-\mathbf{K}\dot{\mathbf{u}} - \mathbf{G}\mathbf{K}_0\dot{\mathbf{v}} + \dot{\mathbf{p}} + \mathbf{G}\mathbf{K}_0\dot{\mathbf{e}}_0 \geq \mathbf{0}. \tag{11.1.16}$$

We write $\dot{u}_i$ in the form $\dot{u}_i = \dot{u}_{i+} - \dot{u}_{i-}$, where

$$\dot{u}_{i+} = \frac{\dot{u}_i + |\dot{u}_i|}{2} \geq 0 \quad \text{and} \quad \dot{u}_{i-} = \frac{-\dot{u}_i + |\dot{u}_i|}{2} \geq 0.$$

Let $\dot{\mathbf{u}}_+$ and $\dot{\mathbf{u}}_-$ be the vectors of $\dot{u}_{i+}$ and $\dot{u}_{i-}$ respectively, and let

$$\mathbf{U} = [\dot{\mathbf{u}}_+^T \,|\, \dot{\mathbf{u}}_-^T \,|\, \bar{\boldsymbol{\gamma}}^T]^T,\ \boldsymbol{\gamma}^T = [\bar{\boldsymbol{\gamma}}^T | \mathbf{0}],\ \mathbf{a}^T = [\bar{\mathbf{a}}^T | \mathbf{0}],\ \mathbf{b}^T = [\bar{\mathbf{b}}^T | \mathbf{0}] \tag{11.1.17}$$

$$\boldsymbol{\mu} = [(-\mathbf{G}\mathbf{K}_0\mathbf{b} - \dot{\mathbf{p}} - \mathbf{G}\mathbf{K}_0\dot{\mathbf{e}})^T \,|\, (\mathbf{G}\mathbf{K}_0\mathbf{b} + \dot{\mathbf{p}} + \mathbf{G}\mathbf{K}_0\dot{\mathbf{e}}_0)^T \,|\, (-\bar{\mathbf{K}}_0\bar{\mathbf{b}} - \bar{\mathbf{K}}_0\bar{\dot{\mathbf{e}}}_0 + \bar{\mathbf{a}})^T]^T. \tag{11.1.18}$$

Now, by introducing the symmetric matrix

$$\mathbf{R} = \begin{bmatrix} \mathbf{K} & -\mathbf{K} & \bar{\mathbf{G}}\bar{\mathbf{K}}_0 \\ -\mathbf{K} & \mathbf{K} & -\bar{\mathbf{G}}\bar{\mathbf{K}}_0 \\ \bar{\mathbf{K}}_0\bar{\mathbf{G}}^T & -\bar{\mathbf{K}}_0\bar{\mathbf{G}}^T & \bar{\mathbf{K}}_0 \end{bmatrix} \tag{11.1.19}$$

we can write (11.1.11–14) in the form

$$\mathbf{R}\mathbf{U} + \boldsymbol{\mu} \geq \mathbf{0}, \tag{11.1.20}$$

$$\mathbf{U} \geq \mathbf{0}, \tag{11.1.21}$$

$$\mathbf{U}^T(\mathbf{R}\mathbf{U} + \boldsymbol{\mu}) = 0. \tag{11.1.22}$$

The problem of determining a $\mathbf{U}$ to satisfy (11.1.20–22) is called a linear complementarity problem (L.C.P.). Let us give some attention to this interesting mathematical problem.

Assume first that $\mathbf{R}$ is positive definite. The L.C.P. (11.1.20–22) is equivalent to the variational inequalities

$$\{\mathbf{y}^T(\mathbf{U}^* - \mathbf{U}) \geq 0,\ \forall \mathbf{U}^* \geq \mathbf{0},\ \mathbf{U} \geq \mathbf{0},\ \mathbf{y} = \mathbf{R}\mathbf{U} + \boldsymbol{\mu}\} \tag{11.1.23}$$

and

$$\{\mathbf{U}^T(\mathbf{y}^* - \mathbf{y}) \geq 0,\ \forall \mathbf{y}^* \geq \mathbf{0},\ \mathbf{y} = \mathbf{R}\mathbf{U} + \boldsymbol{\mu} \geq \mathbf{0}\}. \tag{11.1.24}$$

Indeed, (11.1.20–22) are equivalent to the inclusions

$$-\mathbf{y} \in \partial I_K(\mathbf{U}) \quad \text{with} \quad K = \{\mathbf{U} \,|\, \mathbf{U} \geq \mathbf{0}\}, \qquad \mathbf{y} = \mathbf{R}\mathbf{U} + \boldsymbol{\mu} \tag{11.1.25}$$

and

$$-\mathbf{U} \in \partial I_{K^c}(\mathbf{y}) \quad \text{with} \quad K^c = \{\mathbf{y} \,|\, \mathbf{y} \geq \mathbf{0}\}, \qquad \mathbf{y} = \mathbf{R}\mathbf{U} + \boldsymbol{\mu}, \tag{11.1.26}$$

as easily results by taking into account the geometrical meaning of ∂I_K and ∂I_{K^c} (cf., Sec. 2.3.1). Applying Prop. 2.2.1 to (11.1.23) implies that a unique solution exists. Note that (11.1.2–4) are equivalent to the "local" variational inequalities

$$\{\sigma_i(\gamma_i^* - \gamma_i) \geq 0,\ \forall \gamma_i^* \geq 0,\ \gamma_i \geq 0\} \tag{11.1.27}$$

and

$$\{\gamma_i(\sigma_i^* - \sigma_i) \geq 0, \forall \sigma_i^* \geq 0, \sigma_i \geq 0\}, \tag{11.1.28}$$

which actually generate the "global" variational inequalities(11.1.23, 24).

By Prop. 2.2.2, the inequality (11.1.23) is equivalent to the minimization problem

$$\Pi(\mathbf{U}) = \min \{\Pi(\mathbf{U}^*) \mid \mathbf{U}^* \in K\}, \tag{11.1.29}$$

where

$$\Pi(\mathbf{U}^*) = \tfrac{1}{2}\mathbf{U}^{*T}\mathbf{R}\mathbf{U}^* + \boldsymbol{\mu}^T\mathbf{U}^* \quad \text{and} \quad K = \{\mathbf{U}^* \mid \mathbf{U}^* \geq \mathbf{0}\}. \tag{11.1.30}$$

The uniqueness of the solution implies that (11.1.24) is equivalent to the minimization problem

$$\Pi^c(\mathbf{U}) = \min \{\Pi^c(\mathbf{U}^*) \mid \mathbf{U}^* \in \bar{K}^c\}, \tag{11.1.31}$$

where

$$\Pi^c(\mathbf{U}^*) = \tfrac{1}{2}\mathbf{U}^{*T}\mathbf{R}\mathbf{U}^* \quad \text{and} \quad \bar{K}^c = \{\mathbf{U}^* \mid \mathbf{R}\mathbf{U}^* + \boldsymbol{\mu} \geq \mathbf{0}\}. \tag{11.1.32}$$

The duality theory of convex functionals shows that the dual of the minimization problem (11.1.29) is (11.1.31), and conversely. Due to the special structure of Π and Π^c, we can avoid the general duality theory of Sec. 3.2.2. Indeed, (11.1.29), or equivalently (11.1.25), may be written in the form (cf., Prop. 2.5.5)

$$\mathbf{U} \in \partial I_K^c(-\mathbf{y}), \qquad \mathbf{y} = \mathbf{R}\mathbf{U} + \boldsymbol{\mu}. \tag{11.1.33}$$

This inclusion, as we may easily verify, is equivalent to (11.1.26) and to (11.1.31).

It is apparent that the solution $\mathbf{U}$ of the problem satisfies the equality

$$\Pi(\mathbf{U}) + \Pi^c(\mathbf{U}) = 0, \tag{11.1.34}$$

and therefore

$$\min \{\Pi(\mathbf{U}^*) \mid \mathbf{U}^* \in K\} + \min \{\Pi^c(\mathbf{U}^*) \mid \mathbf{U}^* \in \bar{K}^c\} = 0. \tag{11.1.35}$$

If $\mathbf{R}$ is symmetric and positive semidefinite, for instance if $\mathbf{K}_G$ is positive semidefinite, we may verify, by the same methods used before, the equivalence of the L.C.P. to each of (11.1.23), (11.1.25), (11.1.26) and (11.1.29). The set of solutions of (11.1.29) (if any exists) is convex [2] (Sec. 2.2.1). We shall prove that the general form of the solutions of (11.1.29) is $\mathbf{U}^0 + \mathbf{r}$, with $\mathbf{U}^0 + \mathbf{r} \geq \mathbf{0}$, $\mathbf{R}\mathbf{r} = 0$, $\boldsymbol{\mu}^T\mathbf{r} = 0$, where $\mathbf{U}^0$ is a solution of (11.1.29). Indeed, let $\mathbf{U}_1$ and $\mathbf{U}_2$ be two solutions of (11.1.29), and thus of (11.1.23) as well. By setting $\mathbf{U}^* = \mathbf{U}_1$, $\mathbf{U} = \mathbf{U}_2$ and $\mathbf{U}^* = \mathbf{U}_2$, $\mathbf{U} = \mathbf{U}_1$ and adding, we obtain

$$(\mathbf{U}_1 - \mathbf{U}_2)^T\mathbf{R}(\mathbf{U}_1 - \mathbf{U}_2) \leq 0, \quad \text{which implies that} \quad \mathbf{R}(\mathbf{U}_1 - \mathbf{U}_2) = 0.$$

[2] For further discussion on this subject, see [256].

Then from (11.1.23) we find that $\boldsymbol{\mu}^T(\mathbf{U}_1 - \mathbf{U}_2) \geq 0$ and $\boldsymbol{\mu}^T(\mathbf{U}_2 - \mathbf{U}_1) \geq 0$, which yields $\boldsymbol{\mu}^T(\mathbf{U}_2 - \mathbf{U}_1) = 0$, completing the proof. Arguing similarly, we can prove that the general form of the solutions of the dual problem (11.1.31) is $\bar{\mathbf{U}}^0 + \mathbf{r}$, with $\mathbf{Rr} = 0$, where $\bar{\mathbf{U}}^0$ is a solution of (11.1.31). Further, we may verify by Prop. 2.2.2 that (11.1.31) is equivalent to the variational inequality (11.1.24) holding for every $\mathbf{U}^*$ with $\mathbf{y}^* = \mathbf{RU}^* + \boldsymbol{\mu} \geq \mathbf{0}$. On the other hand, (11.1.31) is equivalent (cf., (2.4.22)) to the inclusion

$$-\mathbf{RU} \in \partial I_{\bar{K}_c}(\mathbf{U}), \tag{11.1.36}$$

which is equivalent to the L.C.P.

$$\mathbf{R}(\mathbf{U} - \mathbf{V}) = 0, \qquad \mathbf{V} \geq \mathbf{0}, \qquad \mathbf{V}^T(\mathbf{RU} + \boldsymbol{\mu}) = 0, \qquad \mathbf{RU} + \boldsymbol{\mu} \geq \mathbf{0}, \tag{11.1.37}$$

as results directly from the fact that $\partial I_{\bar{K}_c}(\mathbf{U})$ is the outward normal cone to $\bar{K}_c$ at $\mathbf{U}$. By the substitution $\mathbf{V} - \mathbf{U} = \mathbf{r}$, we obtain

$$\mathbf{Rr} = 0, \mathbf{U} + \mathbf{r} \geq \mathbf{0}, \qquad (\mathbf{U} + \mathbf{r})^T(\mathbf{RU} + \boldsymbol{\mu}) = 0, \qquad \mathbf{RU} + \boldsymbol{\mu} \geq \mathbf{0}. \tag{11.1.38}$$

Accordingly, the solution of the L.C.P. (11.1.20–22)—i.e., the solution of the primal problem (11.1.29)—satisfies the dual problem, but not conversely unless $\mathbf{R}$ is positive definite. From the above, we may conclude also that the primal problem is solvable if and only if the dual problem is solvable. If $\mathbf{U}_1$ is a solution of the primal problem and $\mathbf{U}_2$ of the dual problem, then $\mathbf{U}_1$ is a solution of the dual problem too, and therefore $\mathbf{U}_2 - \mathbf{U}_1 = \mathbf{r}$. Accordingly, $\Pi^c(\mathbf{U}_2) = \Pi^c(\mathbf{U}_1)$, and thus

$$\Pi(\mathbf{U}_1) + \Pi^c(\mathbf{U}_2) = \mathbf{U}_1^T(\mathbf{RU}_1 + \boldsymbol{\mu}) = 0, \tag{11.1.39}$$

i.e., again (11.1.35) holds. If $\mathbf{U}$ is an admissible vector for both the primal and the dual problem, i.e., $\mathbf{U} \geq \mathbf{0}$ and $\mathbf{RU} + \boldsymbol{\mu} \geq \mathbf{0}$, and if (11.1.34) (or equivalently (11.1.22)) holds, then $\mathbf{U}$ is a solution of both the primal and the dual problems.

It is noteworthy that because of the discrete nature of the problem we could use the Kuhn–Tucker Theorem and the duality theorem of quadratic programming [166] to derive (11.1.29) and (11.1.31). Problems (11.1.29) and (11.1.31) can be written in terms of $\dot{\mathbf{u}}, \dot{\mathbf{v}}$ as

$$\begin{aligned} \min \{\Pi(\dot{\mathbf{u}}, \dot{\mathbf{v}}) = \tfrac{1}{2}\dot{\mathbf{u}}^T\mathbf{K}\dot{\mathbf{u}} + \tfrac{1}{2}\dot{\mathbf{v}}^T\mathbf{K}_0\dot{\mathbf{v}} + \dot{\mathbf{u}}^T\mathbf{G}\mathbf{K}_0\dot{\mathbf{v}} \\ -\dot{\mathbf{u}}^T(\mathbf{G}\mathbf{K}_0\dot{\mathbf{e}}_0 + \dot{\mathbf{p}}) + \dot{\mathbf{v}}^T(\mathbf{a} - \mathbf{K}_0\dot{\mathbf{e}}_0) \,|\, \dot{\mathbf{v}} + \mathbf{b} \geq \mathbf{0}\} \end{aligned} \tag{11.1.40}$$

and

$$\begin{aligned} \min \{\Pi^c(\dot{\mathbf{u}}, \dot{\mathbf{v}}) = \tfrac{1}{2}\dot{\mathbf{u}}^T\mathbf{K}\dot{\mathbf{u}} + \dot{\mathbf{u}}^T\mathbf{G}\mathbf{K}_0\dot{\mathbf{v}} + \dot{\mathbf{u}}^T\mathbf{G}\mathbf{K}_0\mathbf{b} + \tfrac{1}{2}\dot{\mathbf{v}}^T\mathbf{K}_0\dot{\mathbf{v}} + \dot{\mathbf{v}}^T\mathbf{K}_0\mathbf{b} \,|\, \dot{\mathbf{u}}, \dot{\mathbf{v}} \\ \text{satisfy (11.1.11) and the inequality (11.1.12)}\}. \end{aligned} \tag{11.1.41}$$

Functions Π and Π^c represent respectively the potential energy and the complementary energy of the structure for the step being considered. The present theory takes into account the possibility that a cable may become slack, and therefore the term $\dot{\mathbf{v}}$ appears in the expressions of Π and Π^c. If $\dot{\mathbf{v}}$ is zero, Π and Π^c are reduced to the classical expressions of potential and complementary energy of a linear elastic structure.

For the numerical calculation, the minimization problems (11.1.29) and (11.1.31) are used, written either in terms of the vector $\mathbf{U}$ or of the other unknowns of the problem. For the numerical implementation, quadratic programming algorithms may be used.

11.1.2 Multilevel Decomposition Techniques

In the majority of cable structures the number of cables and nodes is large, and so an optimization problem with a large number of unknowns and constraints must be solved. Here, a multilevel optimization technique suited to this kind of optimization problem is proposed. The initial optimization problem is decomposed into a number of subproblems. In the "first level" of the calculation, each subproblem is optimized separately, and in the "second level" the solutions of these subproblems are combined to yield the overall optimum.

It is interesting to note that some of these subproblems constitute minimization problems without inequality constraints.[3] They correspond to classical bilateral structures, and accordingly they can be treated numerically by any linear structural analysis computer program. Here the initial problem is decomposed into two subproblems: the first involves only the displacement terms and corresponds to a structure resulting from the given one by considering all the cables as bars (capable of having compressive forces), and the second, including only the slackness terms, corresponds to a hypothetical structure. To perform the decomposition, (11.1.40) is written as

$$\Pi(\dot{\mathbf{u}}, \dot{\mathbf{v}}) = \Pi_1'(\dot{\mathbf{u}}) + \Pi_1''(\dot{\mathbf{v}}) + \dot{\mathbf{u}}^T\mathbf{G}\mathbf{K}_0\dot{\mathbf{v}}, \tag{11.1.42}$$

where

$$\Pi_1'(\dot{\mathbf{u}}) = \tfrac{1}{2}\dot{\mathbf{u}}^T\mathbf{K}\dot{\mathbf{u}} - \dot{\mathbf{u}}^T(\mathbf{G}\mathbf{K}_0\dot{\mathbf{e}}_0 + \dot{\mathbf{p}})$$

and

$$\Pi_1''(\dot{\mathbf{v}}) = \tfrac{1}{2}\dot{\mathbf{v}}^T\mathbf{K}_0\dot{\mathbf{v}}^T + \dot{\mathbf{v}}^T(\mathbf{a} - \mathbf{K}_0\dot{\mathbf{e}}_0).$$

A new variable $\mathbf{w}$ is introduced and the minimization problem (11.1.42) takes the form

$$\min\{\Pi(\dot{\mathbf{u}}, \dot{\mathbf{v}}, \mathbf{w}) = \Pi_1'(\dot{\mathbf{u}}) + \Pi_1''(\dot{\mathbf{v}}) + \dot{\mathbf{u}}^T\mathbf{G}\mathbf{K}_0\mathbf{w} \mid \dot{\mathbf{v}} = \mathbf{w},\ \dot{\mathbf{v}} + \mathbf{b} \geq \mathbf{0}\}. \tag{11.1.43}$$

The Lagrangian of this problem is

$$\Pi_1(\dot{\mathbf{u}}, \dot{\mathbf{v}}, \mathbf{w}) = \Pi(\dot{\mathbf{u}}, \dot{\mathbf{v}}, \mathbf{w}) + \boldsymbol{\rho}^T(\dot{\mathbf{v}} - \mathbf{w}), \tag{11.1.44}$$

[3] In this respect, the method proposed in Sec. 10.3 could also be characterized as a decomposition technique.

where $\boldsymbol{\rho}$ is the vector of the Lagrange multipliers. The decomposition can be performed by means of two methods: the nonfeasible gradient controller method of Lasdon and Schoeffler and the feasible gradient controller method of Brosilow, Lasdon and Pearson [19]. In the nonfeasible gradient controller method, the value of $\boldsymbol{\rho}$ is supposed to be constant in the first level, say $\boldsymbol{\rho}_1$, and the minimization problem decomposes into the two subproblems (Fig. 11.1.2)

$$\min_{\dot{\mathbf{u}},\,\mathbf{w}} \{\Pi_1'(\dot{\mathbf{u}}) + \dot{\mathbf{u}}^T\mathbf{G}\mathbf{K}_0\,\mathbf{w} - \boldsymbol{\rho}_{(1)}^T\mathbf{w}\} \tag{11.1.45}$$

and

$$\min_{\dot{\mathbf{v}}} \{\Pi_1''(\dot{\mathbf{v}}) + \boldsymbol{\rho}_{(1)}^T\dot{\mathbf{v}} \mid \dot{\mathbf{v}} + \mathbf{b} \geq \mathbf{0}\}. \tag{11.1.46}$$

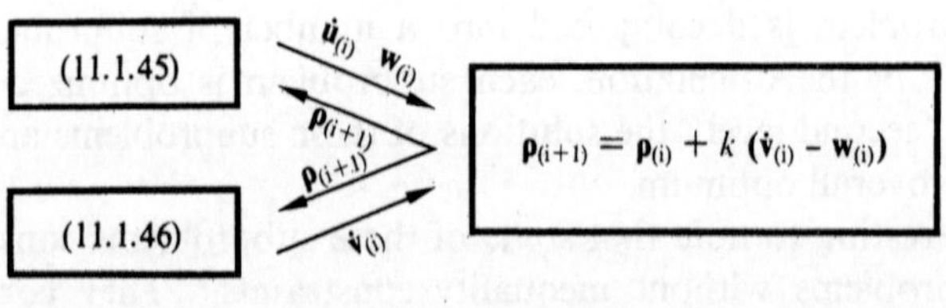

Fig. 11.1.2. The method of the nonfeasible gradient controller.

After performing the optimization, the values of $\dot{\mathbf{u}}$, $\dot{\mathbf{v}}$ and $\mathbf{w}$, say $\dot{\mathbf{u}}_{(1)}$, $\dot{\mathbf{v}}_{(1)}$ and $\mathbf{w}_{(1)}$, result. It is obvious that $\dot{\mathbf{v}}_{(1)} \neq \mathbf{w}_{(1)}$. The task of the second level is to estimate a new value of $\boldsymbol{\rho}$, say $\boldsymbol{\rho}_{(2)}$ by means of the equation

$$\boldsymbol{\rho}_{(2)} = \boldsymbol{\rho}_{(1)} + k(\dot{\mathbf{v}}_{(1)} - \mathbf{w}_{(1)}), \qquad k > 0, \tag{11.1.47}$$

where k is a properly chosen constant (see, e.g., [19]), and to transmit this new value to the first level. The optimization is performed again, new values $\dot{\mathbf{u}}_{(2)}$, $\dot{\mathbf{v}}_{(2)}$, and $\mathbf{w}_{(2)}$ result, and so on, until the differences $\dot{\mathbf{v}}_{(i)} - \mathbf{w}_{(i)}$ are made negligible. The algorithm converges in a finite number of steps, provided that the minima exist [19].

In the feasible gradient controller method, the value of $\mathbf{w}$ is taken as constant in the first level, say $\mathbf{w}_1$, and thus the initial problem decomposes into the two subproblems (Fig. 11.1.3)

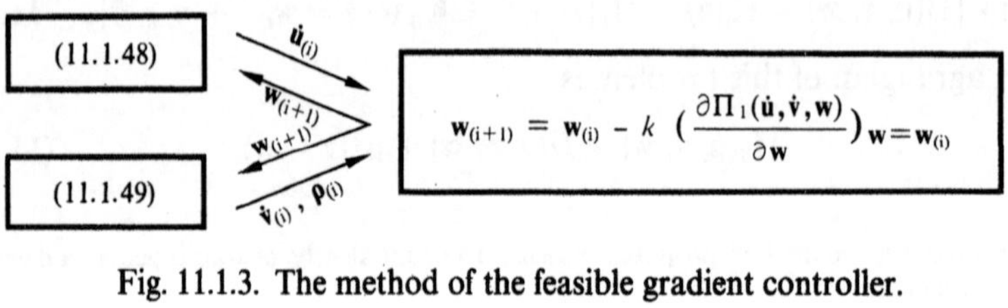

Fig. 11.1.3. The method of the feasible gradient controller.

$$\min_{\dot{\mathbf{u}}} \{\Pi_1'(\dot{\mathbf{u}}) + \dot{\mathbf{u}}^T \mathbf{G}\mathbf{K}_0 \mathbf{w}_{(1)}\} \tag{11.1.48}$$

and

$$\min_{\dot{\mathbf{v}}, \boldsymbol{\rho}} \{\Pi_1''(\dot{\mathbf{v}}) + \boldsymbol{\rho}^T(\dot{\mathbf{v}} - \mathbf{w}_{(1)}) \mid \dot{\mathbf{v}} + \mathbf{b} \geq \mathbf{0}\}. \tag{11.1.49}$$

As a result of the optimization, the values of $\dot{\mathbf{u}}$, $\dot{\mathbf{v}}$, and $\boldsymbol{\rho}$, say $\dot{\mathbf{u}}_{(1)}$, $\dot{\mathbf{v}}_{(1)}$ and $\boldsymbol{\rho}_{(1)}$, are calculated. By means of the second level a new value of $\mathbf{w}$, say $\mathbf{w}_{(2)}$, is estimated and transmitted to the first level. This value is given by the equation

$$\mathbf{w}_{(2)} = \mathbf{w}_{(1)} - k\left(\frac{\partial \Pi_1(\dot{\mathbf{u}}, \dot{\mathbf{v}}, \mathbf{w})}{\partial \mathbf{w}}\right)_{\mathbf{w}=\mathbf{w}_{(1)}}, \quad k > 0, \tag{11.1.50}$$

where k is a properly chosen constant (see, e.g., [19]). The optimization yields a new set of values $\dot{\mathbf{u}}_{(2)}$, $\dot{\mathbf{v}}_{(2)}$ and $\boldsymbol{\rho}_{(2)}$ and the procedure is continued until the difference between the consecutive values of vector $\mathbf{w}$ becomes sufficiently small.

11.1.3 Application

By means of this theory the cable structure of Fig. 11.1.4 is calculated. The upper and lower cables are parabolic and have a common horizontal pretensioning of 80 kN. Two kinds of proportional loading are considered:

(a) symmetric loading (Fig. 11.1.5), ranging from 0.1 kN/m to 1 kN/m per unit length of the horizontal axis; and
(b) nonsymmetric loading (Fig. 11.1.6), ranging from 0.1 kN/m to 1 kN/m per unit length of the horizontal axis.

In each step the load is increased by 0.1 kN/m. Only the vertical displacements of the nodes are taken into account in the calculations. The solution is obtained by solving the minimization problem (11.1.31) in each step of the calculation. We have used the Davidon–Fletcher–Powell algorithm, modified through the created response surface technique (c.r.s.t.) [143] in order to accomodate the constraints. This is a large-step gradient algorithm combined with a generalized projection method for the treatment of the constraints.

In Figs. 11.1.5–11.1.8 certain numerical results of the calculation are given.

The stress-unilateral theory of cable structures permits the calculation of such structures taking into account that the cables may become slack. Accordingly, the use of large prestressing forces, causing positive stresses in the cables (as happens mostly in practice), can be avoided. The stress field which results from the present method leads to a more economical design of the cable structure due to the reduction of the prestressing forces.

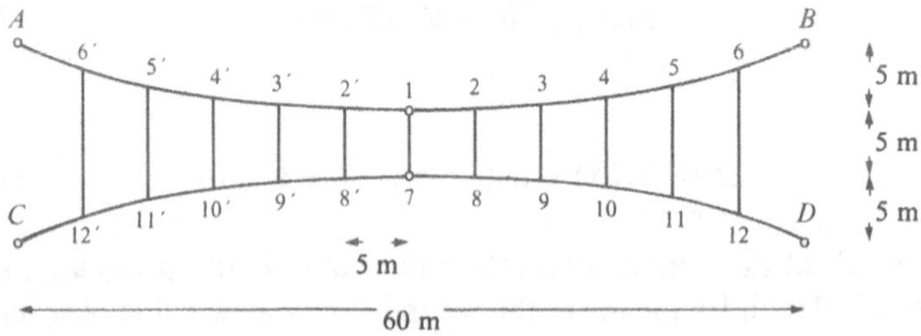

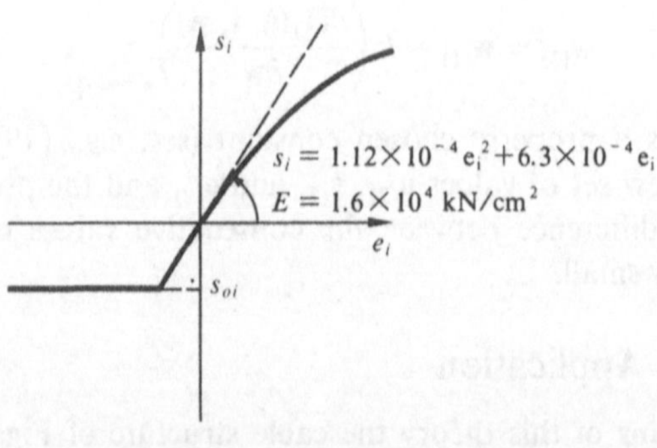

Fig. 11.1.4. Dimensions of the structure and stress-strain diagrams of the cables (cross sections of upper, lower, and vertical cables $F_u = 4.00$ cm^2, $F_l = 1.40$ cm^2 and $F_v = 1.20$ cm^2 respectively).

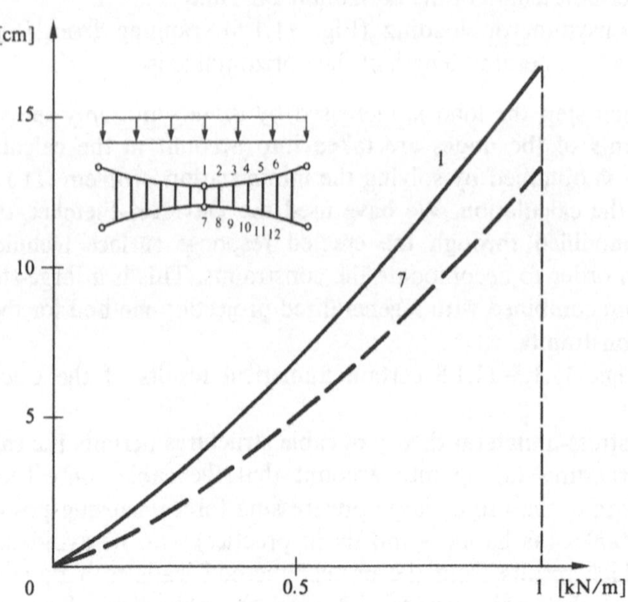

Fig. 11.1.5. Vertical displacements of nodes 1 and 7.

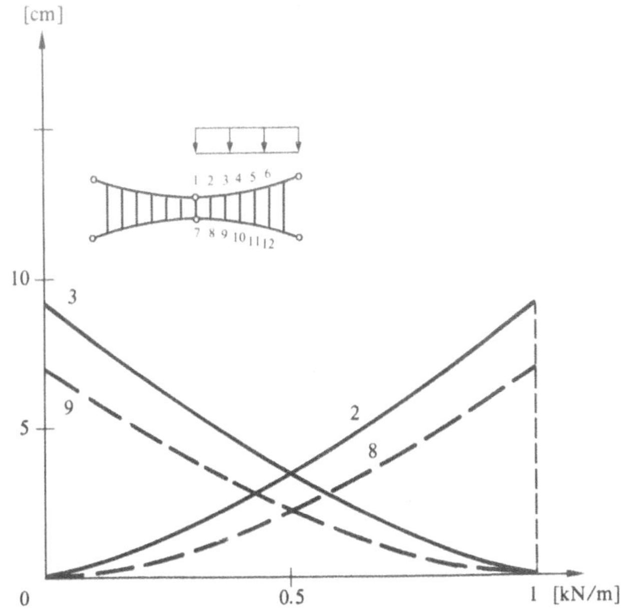

Fig. 11.1.6. Vertical displacements of nodes 2 and 8, and 3 and 9. (For 3 and 9 consider the "zero" (resp. the "one") of the horizontal axis at 1 (resp. 0))

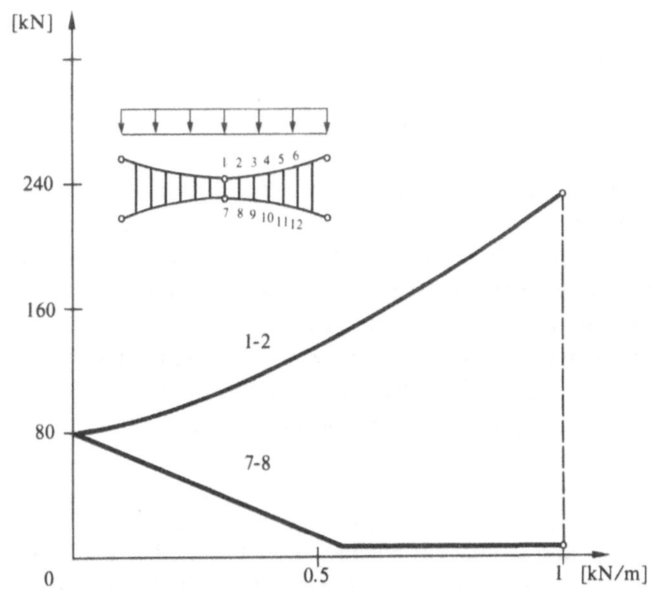

Fig. 11.1.7. Tension in cables 1–2 and 7–8 under symmetric loading.

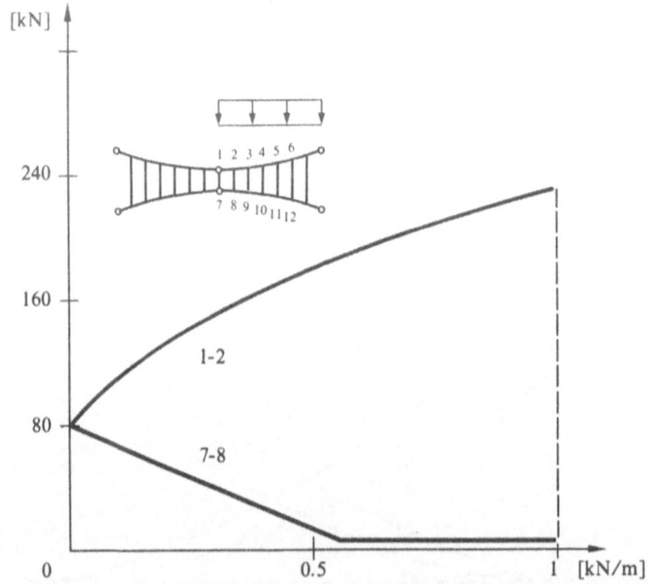

Fig. 11.1.8. Tension in cables 1–2 and 7–8 under nonsymmetric loading.

11.2 Incremental Elastoplastic Analysis. L.C.P.s, Variational Inequalities and Minimum Propositions

One of the most important areas for the application to structural analysis of variational inequalities and corresponding optimization techniques is elastoplasticity. Since the beginning of the 1950's, when the linear programming approach was applied to limit analysis and design, optimization has systematically entered the elastoplastic analysis of structures (see [190] [191] [52] [40]).

Here we shall study a representative general problem concerning the incremental elastoplastic analysis of discretized continua, assuming large displacements and physical destabilizing effects (see (3.3.120)).

The minimum propositions derived here through variational inequalities have already been obtained by means of the theory of L.C.P.s by Maier [190], who first applied quadratic programming methods to the theory of plasticity.

The structure is subdivided into n finite elements. For the description of the stress and strain fields within each element, the method of natural generalized stress and strain is applied (cf., Sec. 11.1.1). The loads are applied to the nodal points, and the structure is described by the nodal displacement vector $\mathbf{u}$ with respect to a fixed Cartesian coordinate system and to a given reference configuration. With the notation of the previous Section, and in the more general framework adopted here, the linearized equations of equilibrium and compatibility within every load increment take the form (11.1.9) and (11.1.10) respectively. The incremental elastoplastic linear flow law (3.3.118, 119) is considered for the i-th element. We only assume that the vectors $\mathbf{N}_i^j$ and $\mathbf{V}_i^j$ are linearly independent and that $\mathbf{V}_i^{jT}\mathbf{N}_i^j > 0$ for $j = 1, \ldots, m$.

In order to write (3.3.118, 119) for the entire structure, we denote by q the number of elements which are at the yield limit at the beginning of the present load step, and we then form the vectors $\dot{\boldsymbol{\lambda}} = [\dot{\boldsymbol{\lambda}}_1^T, \ldots, \dot{\boldsymbol{\lambda}}_q^T]^T$ and $\dot{\mathbf{F}} = [\dot{\mathbf{F}}_1^T, \ldots, \dot{\mathbf{F}}_q^T]^T$, and the matrices $\mathbf{H} = \text{diag}\ (\mathbf{H}_1, \ldots, \mathbf{H}_q)$, $\mathbf{F}_0 = \text{diag}\ (\mathbf{F}_{0,1}, \ldots, \mathbf{F}_{0,n})$, $\mathbf{N} = (\text{diag}\ (\mathbf{N}_1, \ldots, \mathbf{N}_q)^T \,\vdots\, \mathbf{0}^T)^T$ and $\mathbf{V} = (\text{diag}\ (\mathbf{V}_1, \ldots, \mathbf{V}_q)^T \,\vdots\, \mathbf{0}^T)^T$. Then (3.3.118, 119) become

$$\dot{\mathbf{e}} = \dot{\mathbf{e}}_0 + \mathbf{F}_0\dot{\mathbf{s}} + \mathbf{V}\dot{\boldsymbol{\lambda}}, \tag{11.2.1}$$

$$\dot{\mathbf{F}} = \mathbf{N}^T\dot{\mathbf{s}} - \mathbf{H}\dot{\boldsymbol{\lambda}}, \tag{11.2.2}$$

$$\dot{\boldsymbol{\lambda}} \geq \mathbf{0}, \qquad \dot{\mathbf{F}} \leq \mathbf{0}, \qquad \dot{\mathbf{F}}^T\dot{\boldsymbol{\lambda}} = 0, \tag{11.2.3}$$

if, additionally, initial strains $\dot{\mathbf{e}}_0$ exist. The linear complementarity relations (11.2.3) can be written equivalently (cf., Sec. 11.1.1) as the variational inequalities

$$-\dot{\mathbf{F}}^T(\dot{\boldsymbol{\lambda}}^* - \dot{\boldsymbol{\lambda}}) \geq 0, \qquad \forall\dot{\boldsymbol{\lambda}}^* \geq \mathbf{0}, \qquad \dot{\boldsymbol{\lambda}} \geq \mathbf{0} \tag{11.2.4}$$

and

$$-\dot{\boldsymbol{\lambda}}^T(\dot{\mathbf{F}}^* - \dot{\mathbf{F}}) \geq 0, \qquad \forall\dot{\mathbf{F}}^* \leq \mathbf{0}, \qquad \dot{\mathbf{F}} \leq \mathbf{0}. \tag{11.2.5}$$

From (11.1.9, 10) and (11.2.1, 2), by introducing the notations

$$\mathbf{K}_0 = \mathbf{F}_0^{-1}, \qquad \mathbf{G}\mathbf{K}_0\mathbf{G}^T = \mathbf{K}_E, \qquad \mathbf{G}\mathbf{K}_0\mathbf{V} = \mathbf{C}_V,$$

$$\mathbf{G}\mathbf{K}_0\mathbf{N} = \mathbf{C}_N, \qquad \mathbf{B} = \mathbf{H} + \mathbf{N}^T\mathbf{K}_0\mathbf{V},$$

we obtain the relations (dots denote rates)

$$\mathbf{K}\dot{\mathbf{u}} - \mathbf{C}_V\dot{\boldsymbol{\lambda}} - (\dot{\mathbf{p}} + \mathbf{G}\mathbf{K}_0\dot{\mathbf{e}}_0) = \mathbf{0} \tag{11.2.6}$$

and

$$\dot{\mathbf{F}} = \mathbf{C}_N^T\dot{\mathbf{u}} - \mathbf{B}\dot{\boldsymbol{\lambda}} - \mathbf{N}^T\mathbf{K}_0\dot{\mathbf{e}}_0, \tag{11.2.7}$$

which, together with (11.2.3), (11.2.4) or (11.2.5), describe the problem completely. Let us abbreviate by $\mathbf{A}$ the left-hand side of (11.2.6). As is obvious, (11.2.3, 6) are for $\dot{\mathbf{u}} \in \mathbb{R}^\nu$ and $\dot{\boldsymbol{\lambda}} \geq \mathbf{0}$ equivalent to the variational inequality

$$\mathbf{A}^T(\dot{\mathbf{u}}^* - \dot{\mathbf{u}}) - \dot{\mathbf{F}}^T(\dot{\boldsymbol{\lambda}}^* - \dot{\boldsymbol{\lambda}}) \geq 0, \qquad \forall\dot{\mathbf{u}}^* \in \mathbb{R}^\nu, \qquad \forall\dot{\boldsymbol{\lambda}}^* \geq \mathbf{0}. \tag{11.2.8}$$

Here ν denotes the dimension of $\dot{\mathbf{u}}$. (11.2.7, 8) may be written in the form

$$(\mathbf{Mx} + \mathbf{q})^T(\mathbf{x}^* - \mathbf{x}) \geq 0, \qquad \forall \mathbf{x}^* \in K, \qquad \mathbf{x} \in K, \tag{11.2.9}$$

where

$$\mathbf{x} = \left[\begin{array}{c}\dot{\mathbf{u}}\\ \hline \dot{\boldsymbol{\lambda}}\end{array}\right], \qquad K = \left\{\mathbf{x}^* = \left[\begin{array}{c}\dot{\mathbf{u}}^*\\ \hline \dot{\boldsymbol{\lambda}}^*\end{array}\right] \in \mathbb{R}^p \,|\, \dot{\mathbf{u}}^* \in \mathbb{R}^\nu, \quad \dot{\boldsymbol{\lambda}}^* \geq \mathbf{0}\right\} \tag{11.2.10}$$

$$\mathbf{q} = \begin{bmatrix} -\dot{\mathbf{p}} - \mathbf{GK}_0\dot{\mathbf{e}}_0 \\ \mathbf{N}^T\mathbf{K}_0\dot{\mathbf{e}}_0 \end{bmatrix} \text{ and } \mathbf{M} = \begin{bmatrix} \mathbf{K} & -\mathbf{C}_V \\ -\mathbf{C}_N^T & \mathbf{B} \end{bmatrix}. \tag{11.2.11}$$

Assume that $\mathbf{M}$ is symmetric and positive semidefinite. The former condition is equivalent to $\mathbf{V} = \mathbf{N}$ (normality) and $\mathbf{H}^T = \mathbf{H}$ (reciprocity of the interaction of yield modes). The positive semidefiniteness of the work-hardening matrix $\mathbf{H}$ (nonsoftening material behavior) and of the geometric stiffness matrix $\mathbf{K}_G$ ensures the latter condition. Then $\mathbf{x}^T\mathbf{Mx}$ is convex, and thus Prop. 2.2.2 applies. It results that (11.2.8) is equivalent to the minimization problem (primal problem)

$$\Pi(\mathbf{x}) = \min\{\Pi(\mathbf{x}^*) \,|\, \mathbf{x}^* \in K\}, \tag{11.2.12}$$

with

$$\Pi(\mathbf{x}^*) = \tfrac{1}{2}\mathbf{x}^{*T}\mathbf{Mx}^* + \mathbf{q}^T\mathbf{x}^*. \tag{11.2.13}$$

The foregoing problem may also be written in terms of $\dot{\mathbf{u}}$, $\dot{\boldsymbol{\lambda}}$ as

$$\begin{aligned}\min\{\Pi(\dot{\mathbf{u}}, \dot{\boldsymbol{\lambda}}) = {}& \tfrac{1}{2}\dot{\mathbf{u}}^T\mathbf{K}\dot{\mathbf{u}} - \dot{\mathbf{u}}^T\mathbf{GK}_0\mathbf{N}\dot{\boldsymbol{\lambda}} + \tfrac{1}{2}\dot{\boldsymbol{\lambda}}^T\mathbf{H}\dot{\boldsymbol{\lambda}} \\ &+ \tfrac{1}{2}\dot{\boldsymbol{\lambda}}^T\mathbf{N}^T\mathbf{K}_0\mathbf{N}\dot{\boldsymbol{\lambda}} + \dot{\mathbf{e}}_0^T\mathbf{K}_0(\mathbf{N}\dot{\boldsymbol{\lambda}} - \mathbf{G}^T\dot{\mathbf{u}}) - \dot{\mathbf{p}}^T\dot{\mathbf{u}} \,|\, \dot{\boldsymbol{\lambda}} \geq \mathbf{0}\}.\end{aligned} \tag{11.2.14}$$

Let us consider, as in Sec. 11.1.1, the dual problem

$$\Pi^c(\mathbf{x}) = \min\{\Pi^c(\mathbf{x}^*) \,|\, \mathbf{x}^* \in K^c\}, \tag{11.2.15}$$

where

$$\Pi^c(\mathbf{x}^*) = \tfrac{1}{2}\mathbf{x}^{*T}\mathbf{Mx}^* \tag{11.2.16}$$

and

$$K^c = \left\{\mathbf{x}^* \,|\, \mathbf{Mx}^* + \mathbf{q} = \left[\begin{array}{c}\mathbf{A}^*\\ \hline -\dot{\mathbf{F}}^*\end{array}\right], \quad \begin{array}{l}\mathbf{A}^* = \mathbf{K}\dot{\mathbf{u}}^* - \mathbf{C}_V\dot{\boldsymbol{\lambda}}^* - (\dot{\mathbf{p}} + \mathbf{GK}_0\dot{\mathbf{e}}_0) = \mathbf{0} \\ \dot{\mathbf{F}}^* = \mathbf{C}_N^T\dot{\mathbf{u}}^* - \mathbf{B}\dot{\boldsymbol{\lambda}}^* - \mathbf{N}^T\mathbf{K}_0\dot{\mathbf{e}}_0 \leq \mathbf{0}\end{array}\right\}. \tag{11.2.17}$$

By Prop. 2.2.2, for $\mathbf{x} \in K^c$ (11.2.15) is equivalent to the variational inequality

$$\mathbf{x}^T(\mathbf{Mx}^* - \mathbf{Mx}) \geq 0, \qquad \forall \mathbf{x}^* \in K^c. \tag{11.2.18}$$

In terms of $\dot{\mathbf{u}}$, $\dot{\boldsymbol{\lambda}}$, (11.2.16) becomes

$$\begin{aligned}\min\{\Pi^c(\dot{\mathbf{u}}, \dot{\boldsymbol{\lambda}}) = \tfrac{1}{2}\dot{\mathbf{u}}^T\mathbf{K}\dot{\mathbf{u}} - \dot{\mathbf{u}}^T\mathbf{G}\mathbf{K}_0\mathbf{N}\dot{\boldsymbol{\lambda}} + \tfrac{1}{2}\dot{\boldsymbol{\lambda}}^T\mathbf{H}\dot{\boldsymbol{\lambda}} \\ + \tfrac{1}{2}\dot{\boldsymbol{\lambda}}^T\mathbf{N}^T\mathbf{K}_0\mathbf{N}\dot{\boldsymbol{\lambda}} \mid \mathbf{K}\dot{\mathbf{u}} - \mathbf{G}\mathbf{K}_0\mathbf{N}\dot{\boldsymbol{\lambda}} - (\dot{\mathbf{p}} + \mathbf{G}\mathbf{K}_0\dot{\mathbf{e}}_0) = \mathbf{0}, \\ \mathbf{N}^T\mathbf{K}_0\mathbf{G}^T\dot{\mathbf{u}} - (\mathbf{H} + \mathbf{N}^T\mathbf{K}_0\mathbf{N})\dot{\boldsymbol{\lambda}} - \mathbf{N}^T\mathbf{K}_0\dot{\mathbf{e}}_0 \le 0\}.\end{aligned} \qquad (11.2.19)$$

The variational inequalities (11.2.9) and (11.2.18) are equivalent to the inclusions

$$-(\mathbf{M}\mathbf{x} + \mathbf{q}) \in \partial I_K(\mathbf{x}) \qquad (11.2.20)$$

and

$$-\mathbf{M}\mathbf{x} \in \partial I_{K^c}(\mathbf{x}) \qquad (11.2.21)$$

respectively. Note that (11.2.9) and (11.2.20) are equivalent to (11.2.3, 6, 7) for every $\mathbf{M}$ (without assuming symmetry and positivity). In this case, every solution of the initial problem satisfies (11.2.18) and (11.2.21), but not conversely. Indeed, from (11.2.5) and (11.2.6) we derive for $\dot{\mathbf{F}} \le \mathbf{0}$ and $\mathbf{A} = \mathbf{0}$ the inequality

$$\dot{\mathbf{u}}^T(\mathbf{A}^* - \mathbf{A}) - \dot{\boldsymbol{\lambda}}^T(\dot{\mathbf{F}}^* - \mathbf{F}) \ge 0, \qquad \forall \dot{\mathbf{F}}^* \le \mathbf{0}, \qquad \forall \mathbf{A}^* = \mathbf{0}, \quad (11.2.22)$$

which implies (11.2.18) or (11.2.21). The converse is true only if $\mathbf{M}$ is positive definite.

Concerning the relation between the primal and the dual problems for $\mathbf{M}$ symmetric and positive semidefinite, the same properties are valid as those obtained in Sec. 11.1.1 for the minimization problems (11.1.29) and (11.1.31).[4] If $\mathbf{M}$ is positive definite and symmetric, we have complete equivalence between the primal and dual problems, and between their equivalent forms (cf., Sec. 11.1.1).

The minimization problems (11.2.12) and (11.2.15) have been derived by Maier in [190]. By splitting the displacement rate vector $\dot{\mathbf{u}}$ into its positive and negative parts $\dot{\mathbf{u}}_+$ and $\dot{\mathbf{u}}_-$, a L.C.P. of the form (11.1.20–22) can be formulated, and the method of the previous section can be followed here as well, with

$$\mathbf{U} = \begin{bmatrix} \dot{\mathbf{u}}_- \\ \dot{\mathbf{u}}_+ \\ \dot{\boldsymbol{\lambda}} \end{bmatrix} \quad \boldsymbol{\mu} = \begin{bmatrix} +\dot{\mathbf{p}} + \mathbf{G}\mathbf{K}_0\dot{\mathbf{e}}_0 \\ -\dot{\mathbf{p}} - \mathbf{G}\mathbf{K}_0\dot{\mathbf{e}}_0 \\ \mathbf{N}^T\mathbf{K}_0\dot{\mathbf{e}}_0 \end{bmatrix} \quad \mathbf{R} = \begin{bmatrix} +\mathbf{K} & -\mathbf{K} & +\mathbf{C}_V \\ -\mathbf{K} & +\mathbf{K} & -\mathbf{C}_V \\ \mathbf{C}_N^T & -\mathbf{C}_N^T & \mathbf{B} \end{bmatrix}. \qquad (11.2.23)$$

Note that the L.C.P. holds for $\mathbf{M}$ nonsymmetric as well.

Next, we shall present a proposition on the properties of the solution of the inequality (11.2.9), where we assume that $\mathbf{M}$ is symmetric and positive

[4] The slightly different form of the sets K and K^c does not influence the reasoning.

semidefinite. We shall apply Prop. 4.1.2. Denote by ker $\mathbf{M}$ the kernel of the matrix $\mathbf{M}$, i.e.,

$$\ker \mathbf{M} = \{\mathbf{x} \mid \mathbf{x} \in \mathbb{R}^p, \quad \mathbf{M}\mathbf{x} = \mathbf{0}\}, \tag{11.2.24}$$

which is a linear subspace of $\mathbb{R}^p$, and let $\mathbf{Q}$ be the orthogonal projector of $\mathbb{R}^p$ onto ker $\mathbf{M}$. ker $\mathbf{M}$ coincides with the kernel of the symmetric form $\mathbf{x}^T\mathbf{M}\mathbf{x}$. If $\mathbf{I}$ is the identity matrix and if $\mathbf{P} = \mathbf{I} - \mathbf{Q}$, the semicoercivity relation

$$\mathbf{x}^T\mathbf{M}\mathbf{x} \geq c \,\| \mathbf{P}\mathbf{x} \|^2, \qquad \forall \mathbf{x} \in \mathbb{R}^p \tag{11.2.25}$$

holds[5] (c is a positive constant). Denote by T^* the linear subspace of $\mathbb{R}^p$ $\{\mathbf{x} | \mathbf{x} \in \ker \mathbf{M} \cap K \text{ implies } -\mathbf{x} \in \ker \mathbf{M} \cap K\}$, and by ker $\mathbf{q}$ the set

$$\ker \mathbf{q} = \{\mathbf{x} \mid \mathbf{q}^T\mathbf{x} = 0\}. \tag{11.2.26}$$

The following proposition holds.

Proposition 11.2.1. If for every $\mathbf{x}_0 \in \ker \mathbf{M} \cap K$

$$\mathbf{q}^T\mathbf{x}_0 \leq 0 \tag{11.2.27}$$

and if the equality sign holds if and only if $\mathbf{x}_0 \in T^*$, then problem (11.2.9) or (11.2.12) has a solution, say $\mathbf{x}$. Any other solution, say $\mathbf{x}_1$, is expressed as $\mathbf{x} + \mathbf{x}_0$, where $\mathbf{x}_0 \in \ker \mathbf{M} \cap \ker \mathbf{q}$ and $\mathbf{x} + \mathbf{x}_0 \in K$.

Proof. Follows immediately by applying Props. 4.1.2 and 4.1.4, q.e.d.

As is obvious from Prop. 11.2.1, if $\mathbf{M}$ is positive definite, a unique solution exists.[6] If $\delta\mathbf{x}$ is the difference between two solutions, then the relation $\mathbf{M}\delta\mathbf{x} = 0$ (cf., also Sec. 11.1.1) implies

$$\mathbf{K}_G(\delta\dot{\mathbf{u}}) + \mathbf{G}(\delta\dot{\mathbf{s}}) = \mathbf{0} \tag{11.2.28}$$

and

$$\mathbf{N}^T(\delta\dot{\mathbf{s}}) - \mathbf{H}(\delta\dot{\boldsymbol{\lambda}}) = \delta\dot{\mathbf{F}} = \mathbf{0}. \tag{11.2.29}$$

Thus the total stress-rate difference is equilibrated by the stress-rate difference which results from the geometric changes of the structure, and the yield-function rates $\dot{\mathbf{F}}$ are equal for all solutions. In the case of a geometrically linear theory, $\mathbf{K}_G = \mathbf{0}$ and thus $\delta\dot{\mathbf{s}} = \mathbf{0}$, i.e., the stress rates are uniquely determined.

[5] In a finite dimensional space the proof of (11.2.25) is immediate (cf., also the proof of (4.1.36)).

[6] If $\mathbf{M}$ is nonsymmetric but positive semidefinite, we can give a sufficient condition for the existence of a solution of (11.2.9) by applying a general proposition by Fichera (cf., [99], Prop. 2.I). (11.2.25) holds for this case as well, with $\mathbf{P} = \mathbf{I} - \mathbf{Q}$, $\mathbf{Q}$ being the orthogonal projector of $\mathbb{R}^p$ onto $\ker \mathbf{x}^T\mathbf{M}\mathbf{x} = \ker(\mathbf{M} + \mathbf{M}^T) \supset \ker \mathbf{M}$.

We shall now derive another class of variational principles with respect to the plastic multiplier rates $\dot{\boldsymbol{\lambda}}$, on the assumption that $\mathbf{K}$ is nonsingular. From (11.2.6, 7) we obtain the relation

$$\dot{\mathbf{F}} = -(\mathbf{H} - \mathbf{N}^T(\mathbf{K}_0\mathbf{G}^T\mathbf{K}^{-1}\mathbf{G}\mathbf{K}_0 - \mathbf{K}_0)\mathbf{V})\dot{\boldsymbol{\lambda}} + \mathbf{N}^T\mathbf{K}_0\mathbf{G}^T\mathbf{K}^{-1}\dot{\mathbf{p}} + \mathbf{N}^T(\mathbf{K}_0\mathbf{G}^T\mathbf{K}^{-1}\mathbf{G}\mathbf{K}_0 - \mathbf{K}_0)\dot{\mathbf{e}}_0. \quad (11.2.30)$$

Following [190], we introduce the notations

$$\mathbf{Z} = \mathbf{K}_0\mathbf{G}^T\mathbf{K}^{-1}\mathbf{G}\mathbf{K}_0 - \mathbf{K}_0 \quad (11.2.31)$$

and

$$\dot{\mathbf{s}}^E = \mathbf{K}_0\mathbf{G}^T\mathbf{K}^{-1}\dot{\mathbf{p}} + \mathbf{Z}\dot{\mathbf{e}}_0. \quad (11.2.32)$$

Matrix $\mathbf{Z}$ transforms any initial strain rate vector into the corresponding stress rate vector when only elastic deformations take place. Similarly, $\dot{\mathbf{s}}^E$ denotes the stress rate vector due only to the elastic deformation and caused by both $\dot{\mathbf{p}}$ and $\dot{\mathbf{e}}_0$. Thus we can write the relation

$$\dot{\mathbf{F}} = \mathbf{N}^T\dot{\mathbf{s}}^E - (\mathbf{H} - \mathbf{N}^T\mathbf{Z}\mathbf{V})\dot{\boldsymbol{\lambda}}, \quad (11.2.33)$$

which, together with (11.2.3), constitutes a L.C.P., taking, by means of (11.2.4), the equivalent variational inequality form

$$\dot{\boldsymbol{\lambda}}^T(\mathbf{H} - \mathbf{N}^T\mathbf{Z}\mathbf{V})^T(\dot{\boldsymbol{\lambda}}^* - \dot{\boldsymbol{\lambda}}) - (\mathbf{N}^T\dot{\mathbf{s}}^E)^T(\dot{\boldsymbol{\lambda}}^* - \dot{\boldsymbol{\lambda}}) \geq 0, \quad \forall\dot{\boldsymbol{\lambda}}^* \geq \mathbf{0}, \quad \dot{\boldsymbol{\lambda}} \geq \mathbf{0}. \quad (11.2.34)$$

Suppose now that $\mathbf{H} - \mathbf{N}^T\mathbf{Z}\mathbf{V}$ is symmetric (e.g., if $\mathbf{N} = \mathbf{V}$ and $\mathbf{H} = \mathbf{H}^T$) and positive semidefinite. Then (11.2.34) is equivalent to the minimization problem

$$\bar{\Pi}(\dot{\boldsymbol{\lambda}}) = \min\{\bar{\Pi}(\dot{\boldsymbol{\lambda}}^*) \mid \dot{\boldsymbol{\lambda}}^* \geq \mathbf{0}\}, \quad (11.2.35)$$

with

$$\bar{\Pi}(\dot{\boldsymbol{\lambda}}) = \tfrac{1}{2}\dot{\boldsymbol{\lambda}}^T(\mathbf{H} - \mathbf{N}^T\mathbf{Z}\mathbf{N})\dot{\boldsymbol{\lambda}} - (\mathbf{N}^T\dot{\mathbf{s}}^E)^T\dot{\boldsymbol{\lambda}}. \quad (11.2.36)$$

The dual problem of (11.2.35) reads

$$\bar{\Pi}^c(\dot{\boldsymbol{\lambda}}) = \min\{\bar{\Pi}^c(\dot{\boldsymbol{\lambda}}^*) \mid \mathbf{N}^T\dot{\mathbf{s}}^E - (\mathbf{H} - \mathbf{N}^T\mathbf{Z}\mathbf{N})\dot{\boldsymbol{\lambda}}^* \leq \mathbf{0}\}, \quad (11.2.37)$$

with

$$\bar{\Pi}^c(\dot{\boldsymbol{\lambda}}) = \tfrac{1}{2}\dot{\boldsymbol{\lambda}}^T(\mathbf{H} - \mathbf{N}^T\mathbf{Z}\mathbf{N})\dot{\boldsymbol{\lambda}}. \quad (11.2.38)$$

Concerning numerical techniques and the relation between the solutions of the primal and dual problems, we refer to Sec.11.1.1 and to [190] (see also [158]).

Note that minimization problems analogous to those presented here can also be formulated in the case of nonassociated friction laws [202].

11.3 Dynamic Unilateral Contact Problems

A representative class of dynamic inequality problems is that of dynamic unilateral contact problems, which we shall study in this Section from the standpoint of their numerical treatment. These problems are special cases of those we have already formulated in Sec. 6.4. If we assume that the boundary conditions (3.3.7) and (10.1.1) hold on Γ_S, then for a linear elastic body the solution of the problem fulfills the variational inequality (6.4.3). Φ is given by a relation of the form (6.2.6), but with j replaced by j_N. For a body obeying the material law (6.1.3), the variational inequality (6.4.4) holds. Suppose now that the linear elastic body is discretized (cf., Sec. 10.1.1). Then (6.4.3) acquires the form

$$\tfrac{1}{2}\mathbf{u}^T\mathbf{K}(\mathbf{v}-\mathbf{u}) + \Phi(\mathbf{v}_N) - \Phi(\mathbf{u}) - \mathbf{C}_T^T(\mathbf{v}_T-\mathbf{u}_T) \geq (\mathbf{f} - \mathbf{M}\mathbf{u}'' - \mathbf{C}\mathbf{u}')^T(\mathbf{v}-\mathbf{u}), \qquad \forall \mathbf{v} \in X^*, \quad (11.3.1)$$

where $\mathbf{K}$ is the stiffness matrix of the structure (see (10.1.11)), Φ is given by (10.1.12), X^* is the kinematically admissible set for the displacements, i.e.

$$X^* = \{\mathbf{v} \mid j_{N_i}(v_{N_i}) < \infty, \forall i \in \Gamma_S, \mathbf{v}_N = \mathbf{N}\mathbf{v}, \mathbf{v}_T = \mathbf{T}\mathbf{v}\}, \qquad (11.3.2)$$

$\mathbf{M}$ is the mass matrix and $\mathbf{C}$ is the damping matrix of the structure. For a structure with convex strain-energy density, instead of (11.3.1) the variational inequality (discrete version of (6.4.4))

$$W(\mathbf{v}) - W(\mathbf{u}) + \Phi(\mathbf{v}) - \Phi(\mathbf{u}) - \mathbf{C}_T^T(\mathbf{v}_T-\mathbf{u}_T) \geq (\mathbf{f} - \mathbf{M}\mathbf{u}'' - \mathbf{C}\mathbf{u}')^T(\mathbf{v}-\mathbf{u}), \qquad \forall \mathbf{v} \in \bar{X}^* \quad (11.3.3)$$

holds, where

$$\begin{aligned}\bar{X}^* = \{\mathbf{v} \mid w_i(\mathbf{e}_i(\mathbf{v}) - \mathbf{e}_{0i}) < \infty, \forall i \in \Omega,\\ j_{N_i}(v_{N_i}) < \infty, \forall i \in \Gamma_S, \mathbf{v}_N = \mathbf{N}\,\mathbf{v},\ \mathbf{v}_T = \mathbf{T}\mathbf{v}\}\end{aligned} \qquad (11.3.4)$$

is the kinematically admissible set, and W is given by (10.1.24). For both dynamic variational inequalities, the initial conditions

$$\mathbf{u} = \mathbf{u}_0 \quad \text{and} \quad \mathbf{u}' = \mathbf{u}_1 \quad \text{for} \quad t \doteq 0 \qquad (11.3.5)$$

must also be taken into account.

The numerical treatment of a dynamic inequality problem is performed by discretizing the corresponding variational inequality with respect to time (cf., [120] [251] [309]). Then we solve, for each time step δt, a static variational inequality. This gives rise to an optimization problem, whose solution is obtained by a suitable optimization algorithm [120] [251] [309]. Note that of the existing time-integration methods, only the direct integration schemes are applicable to the dynamic analysis of inequality problems. Response analysis by mode superposition requires calculation of eigenvalues and eigenvectors of the system, which do not exist here in the classical sense, because of the variational inequality (cf., Sec. 3.1.4). For the complete numerical analysis of certain other types of dynamic variational inequalities, refer to [120]. For the discretization of the problem with respect to time, the method of m-step linear difference operators (cf., e.g., [94]) is applied. At time instant $t^{(p)}$ we assume that

$$\sum_{r=0}^{q} \alpha^{(r)} \mathbf{u}^{(p-r)} = \delta t \sum_{r=0}^{q} \beta^{(r)} \dot{\mathbf{u}}^{(p-r)} \tag{11.3.6}$$

and

$$\sum_{r=0}^{q} \gamma^{(r)} \mathbf{u}^{(p-r)} = \mu(\delta t)^2 \sum_{r=0}^{q} \beta^{(r)} \ddot{\mathbf{u}}^{(p-r)}, \quad p \geq q, \quad p > 1. \tag{11.3.7}$$

The coefficients $\alpha^{(r)}$, $\beta^{(r)}$, $\gamma^{(r)}$ and μ depend on the finite-difference scheme[7] chosen. We assume that the step size δt remains constant. Moreover, in order to have an implicit integration scheme we take $\beta^{(0)}$ to be nonzero. From (11.3.6, 7) we obtain the relations

$$\alpha^{(0)} \mathbf{u}^{(p)} - \delta t \beta^{(0)} \dot{\mathbf{u}}^{(p)} = -\mathbf{U}\boldsymbol{\alpha} + \delta t \dot{\mathbf{U}} \boldsymbol{\beta} \tag{11.3.8}$$

and

$$\mu(\delta t)^2 \beta^{(0)} \ddot{\mathbf{u}}^{(p)} = \gamma^{(0)} \mathbf{u}^{(p)} + \mathbf{U}\boldsymbol{\gamma} - \mu(\delta t)^2 \ddot{\mathbf{U}} \boldsymbol{\beta} = \gamma^{(0)} \mathbf{u}^{(p)} + \tilde{\mathbf{U}}, \tag{11.3.9}$$

where

$$\boldsymbol{\alpha} = [\alpha^{(1)}, \ldots, \alpha^{(q)}]^T, \; \boldsymbol{\beta} = [\beta^{(1)}, \ldots, \beta^{(q)}]^T,$$
$$\boldsymbol{\gamma} = [\gamma^{(1)}, \ldots, \gamma^{(q)}]^T \quad \text{and} \quad \mathbf{U} = [\mathbf{u}^{(p-1)}, \ldots, \mathbf{u}^{(p-q)}] \; .$$

By substituting (11.3.8) and (11.3.9) into the variational inequality (11.3.3), we obtain for $t = t^{(p)}$

$$\begin{aligned} &W(\boldsymbol{\varepsilon}(\mathbf{v})) - W(\boldsymbol{\varepsilon}(\mathbf{u}^{(p)})) + \Phi(\mathbf{v}_N) - \Phi(\mathbf{u}_N^{(p)}) \\ &- \mathbf{C}_T^{(p)T}(\mathbf{v}_T - \mathbf{u}_T^{(p)}) \geq (\mathbf{f}^{(p)} - \mathbf{q}^{(p)})^T(\mathbf{v} - \mathbf{u}^{(p)}) \\ &- \frac{\gamma^{(0)} \mathbf{u}^{(p)T}}{\mu(\delta t)^2 \beta^{(0)}} \mathbf{M}^T(\mathbf{v} - \mathbf{u}^{(p)}) - \frac{\alpha^{(0)} \mathbf{u}^{(p)T}}{(\delta t)\beta^{(0)}} \mathbf{C}^T(\mathbf{v} - \mathbf{u}^{(p)}), \; \forall \mathbf{v} \in \bar{X}^*_{t=t(p)} \end{aligned} \tag{11.3.10}$$

where

$$\mathbf{q}^{(p)} = \frac{\mathbf{M}\tilde{\mathbf{U}}}{\mu(\delta t)^2 \beta^{(0)}} - \frac{\mathbf{C}[\delta t \dot{\mathbf{U}} \boldsymbol{\beta} - \mathbf{U}\boldsymbol{\alpha}]}{\delta t \beta^{(0)}}. \tag{11.3.11}$$

For every symmetric positive semidefinite matrix $\mathbf{A}$, the inequality

$$\tfrac{1}{2}\mathbf{v}^T\mathbf{A}\mathbf{v} - \tfrac{1}{2}\mathbf{u}^T\mathbf{A}\mathbf{u} \geq \mathbf{u}^T\mathbf{A}^T(\mathbf{v} - \mathbf{u}) \tag{11.3.12}$$

[7] Here $\dot{u}^{(p)}$ and $\ddot{u}^{(p)}$ denote simply the velocity and the acceleration at time $t^{(p)}$ respectively for the time-discretization scheme in use.

holds, as is easily verified. By applying (11.3.12) to the bilinear forms $\mathbf{v}^T\mathbf{M}\mathbf{v}$ and $\mathbf{v}^T\mathbf{C}\mathbf{v}$ ($\mathbf{C}$ is assumed to be symmetric positive semidefinite), we show that the solution of the inequality (11.3.10) satisfies the minimization problem

$$\Pi_{(p)}(\mathbf{u}^{(p)}) = \min \{\Pi_{(p)}(\mathbf{v}) \mid \mathbf{v} \in \bar{X}^*_{t=t(p)}\}, \tag{11.3.13}$$

where $(\alpha^{(0)} = \gamma^{(0)} = 1)$

$$\Pi_{(p)}(\mathbf{v}) = W(\varepsilon(\mathbf{v})) + \Phi(\mathbf{v}_N) - \mathbf{v}_T^T\mathbf{C}_T^{(p)} + \frac{1}{2}\frac{\mathbf{v}^T\mathbf{M}^T\mathbf{v}}{\mu(\delta t)^2\beta^{(0)}} + \frac{1}{2}\frac{\mathbf{v}^T\mathbf{C}^T\mathbf{v}}{(\delta t)\beta^{(0)}} - (\mathbf{f}^{(p)} - \mathbf{q}^{(p)})^T\mathbf{v}. \tag{11.3.14}$$

Next the dynamic unilateral contact problem of a linear elastic body is considered. For this case, a minimization problem of the form (11.3.13) must be solved in every time step, with

$$\Pi_{(p)}(\mathbf{v}) = \tfrac{1}{2}\mathbf{v}^T\mathbf{K}\mathbf{v} + \frac{1}{2}\frac{\mathbf{v}^T\mathbf{M}^T\mathbf{v}}{\mu(\delta t)^2\beta^{(0)}} + \frac{1}{2}\frac{\mathbf{v}^T\mathbf{C}^T\mathbf{v}}{\delta t\ \beta^{(0)}} - \mathbf{v}_T^T\mathbf{C}_T^{(p)} + \Phi(\mathbf{v}_N) - (\mathbf{f}^{(p)} - \mathbf{q}^{(p)})^T\mathbf{v}. \tag{11.3.15}$$

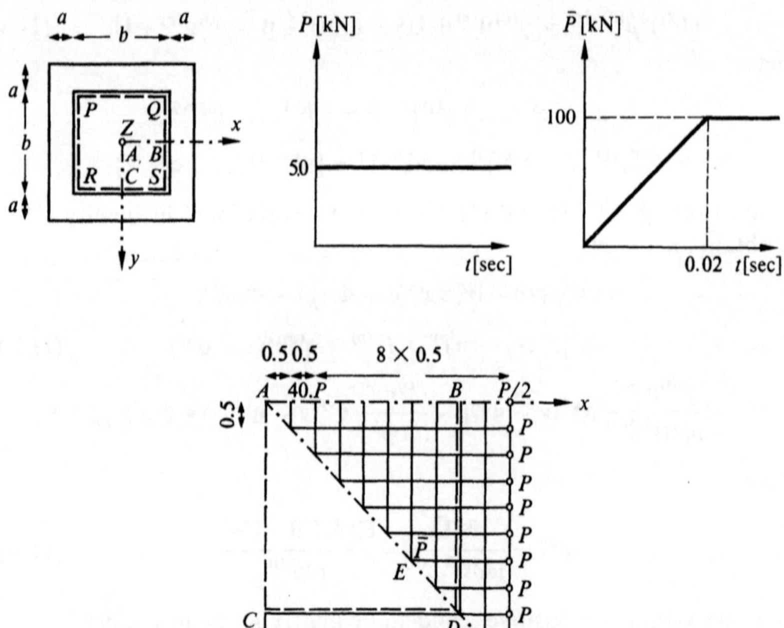

Fig. 11.3.1. The dynamic unilateral contact problem for a plate ($E = 4 \times 10^5$ kN/m^2, $\nu = 0.16$, $h = 5$ cm, $\rho = 2.5$ kN sec^2/m^4, $k = 3 \times 10^4$ kN/m^3).

The following numerical application illustrates the proposed method.

The linear elastic square plate of Fig. 11.3.1, with density ρ and thickness h rests, within the region PQRS, upon an elastic halfspace, which, for our purposes, is simulated by discrete unilateral Winkler springs having a spring constant k. The springs are considered to remain attached to the plate at their upper ends, whereas their lower ends are taken to be in contact with a rigid support. Thus for the discretized structure we must solve a Signorini–Fichera problem. Along PQ, QS, RS and RP the plate is simply supported. For simplicity, we assume that no tangential reaction exists between plate

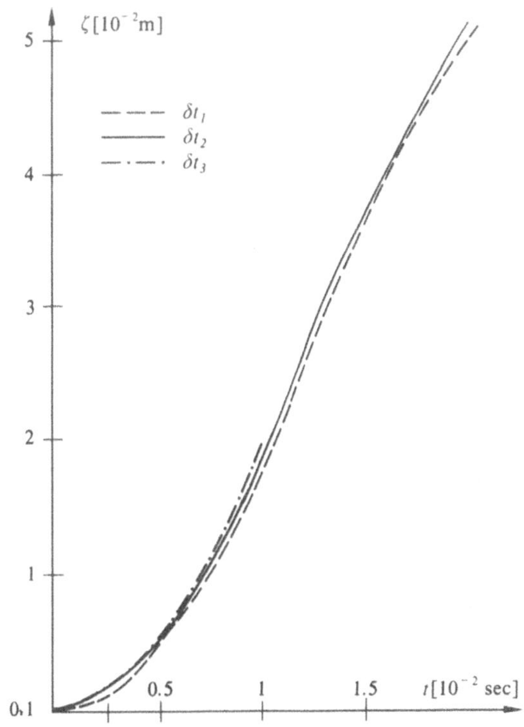

Fig. 11.3.2. Plate deflection ζ at $x = y = 5$ m.

and support, i.e., $\mathbf{C}_T = \mathbf{0}$ in (11.3.15). The plate is initially at rest; then constant loads P are instantaneously applied on the edges of the plate and its center. Another load $\bar{P}$ is applied to node E (Fig. 11.3.1). These load and support conditions permit us to distinguish three parts of the plate, each having different behavior: a free part (part I) lying outside the area ABCD (Fig. 11.3.1) and not affected by the contact conditions; a region (part II) in the close vicinity of the plate center which, due to the load acting upon A, remains in contact with the elastic subspace; and the remaining interior part (part III) of the plate in which transitions of the unilateral springs from active to inactive states take place.

Time integration is performed by Houbolt's method[8] (see, e.g., [18]). For every time step a quadratic optimization problem results, since $\Phi(\mathbf{v}_N)$ in (11.3.15) is incorporated into the elastic energy term $\frac{1}{2}\mathbf{u}^T\mathbf{K}\mathbf{u}$. For the numerical calculation, the linear analysis approach proposed in Sec. 10.3.1 is used.[9] After a small number of iterations, the contact zone is obtained for every increment.

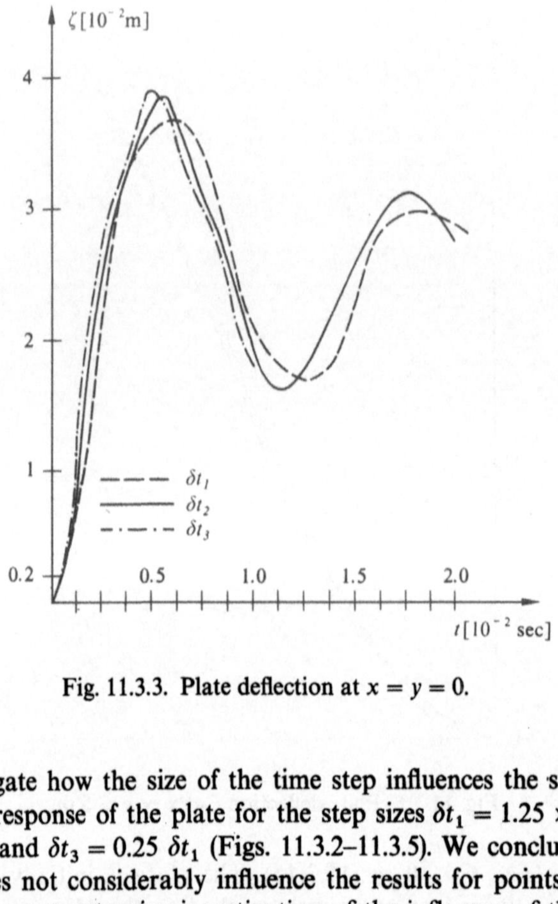

Fig. 11.3.3. Plate deflection at $x = y = 0$.

To investigate how the size of the time step influences the solution, we examine the response of the plate for the step sizes $\delta t_1 = 1.25 \times 10^{-3}$ sec, $\delta t_2 = 0.5\ \delta t_1$ and $\delta t_3 = 0.25\ \delta t_1$ (Figs. 11.3.2–11.3.5). We conclude that the time step does not considerably influence the results for points of Parts I and II. For a more extensive investigation of the influence of the step size on the results, see [309].

[8] We take in (11.3.6, 7) $q = 3$, $\alpha^{(0)} = \frac{11}{6}$, $\alpha^{(1)} = -3$, $\alpha^{(2)} = \frac{3}{2}$, $\alpha^{(3)} = -\frac{1}{3}$, $\beta^{(0)} = 1$, $\beta^{(1)} = \beta^{(2)} = \beta^{(3)} = 0$, $\gamma^{(0)} = 1$, $\gamma^{(1)} = -2.5$, $\gamma^{(2)} = 2$, $\gamma^{(3)} = -0.5$ and $\mu = 0.5$.

[9] Each quarter of the plate has 308 degrees of freedom and 64 constraints.

The foregoing calculations are performed with the simplifying assumption of totally plastic impact. It is possible to extend the calculation method presented herein to include other impact assumptions when necessary. This can be achieved by considering, e.g., a partial or total reversal of the velocity at the point of impact [207].

The impact phenomenon can be incorporated into the variational formulation of the problem by assuming appropriate functional spaces. In the numerical procedure, however, the impact must be considered separately during the time integration. Note that in dynamic unilateral contact problems it is not possible to determine, for a given value of the external action (e.g. by a direct optimization algorithm), the contact and noncontact regions as in static problems. In addition to the previous method in which, first we perform a time-discretization and then we solve the resulting minimum problems at every time step, the following method can also be proposed. The variational inequality is "replaced" through regularization by a sequence of nonlinear differential equations depending on the regularization parameter, and then the resulting differential equations have to be solved.

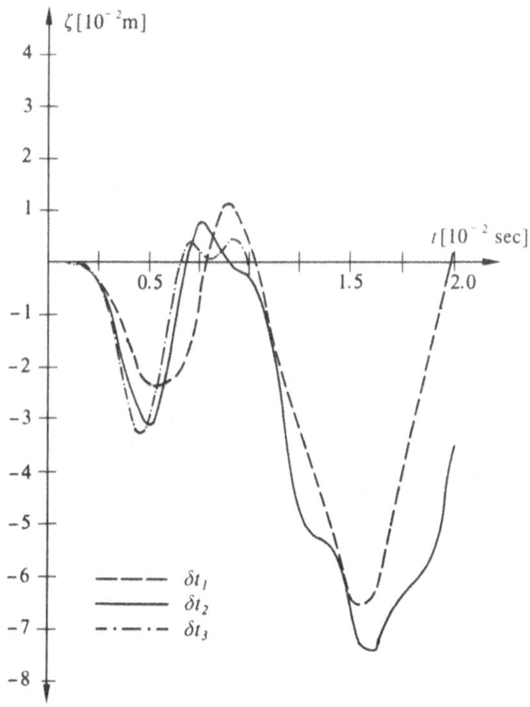

Fig. 11.3.4. Plate deflection at $x = y = 1$.

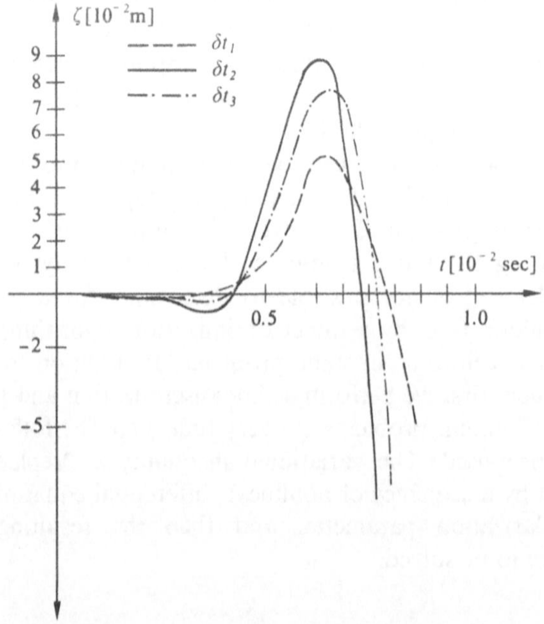

Fig. 11.3.5. Plate deflection at $x = 2.5$ m, $y = 1.0$ m.

In the case of hemivariational inequalities, both of the preceding procedures can be applied after appropriate modification. The regularization yields nonlinear differential equations and the time discretization yields static hemivariational inequalities. The latter ones through regularization yield sequences of nonlinear algebraic equations. In many cases these equations are semilinear and have only sparse nonlinear terms depending on the regularization parameter. Note also that within each time step every local minimum of the nonconvex potential energy of the time discretized problem (e.g. consider (11.3.14) with W and Φ nonconvex) is a substationarity point, and if (4.3.22) holds then every local minimum satisfies the static hemivariational inequality. Up to this point there is not much experience concerning the numerical treatment of hemivariational inequalities (see e.g. [348]).

Epilogue

In this book we have presented and studied certain classes of inequality problems in mechanics using as the basic tool the notion of subgradient in the convex case and the notion of generalized gradient in the nonconvex case.

The field of inequality problems in mechanics overlaps certain newly developed areas in mathematics, such as convex analysis, optimal control and optimization, and the theory of multivalued operators. The problems studied here are representative, but do not cover all possible applications; thus, for instance, in order to keep the length of the text within reasonable limits, we have not touched upon free-boundary problems in fluid mechanics (we refer the reader to [161] and to the literature given there).

Research on inequality problems has not yet been completed. Further study is needed, e.g., on the regularity of solutions of unilateral B.V.P.s, on eigenvalue problems for variational inequalities, on hemivariational inequalities, etc. Also, many questions on the numerical analysis of unilateral B.V.P.s remain open. Similarly, there are rich possibilities for research in various branches of mechanics. As an example, in analytical mechanics subsidiary inequality conditions of holonomic or nonholonomic nature can be studied and one can then determine the extent to which they affect the propositions of classical analytical mechanics (see, e.g., [303]). In this context, we could mention Hamiltonian equations expressed by means of subdifferentials [349] [350]. To give another example, a new area in elastomechanics, is that of the multivalued integral equation approach to unilateral B.V.P.s. For these topics see [348]-[361].

This book has treated variational and hemivariational inequalities in various fields of continuum mechanics; in other areas of physics as well, relativistic and nonrelativistic, inequality problems can be formulated—in essence, wherever generalized forces are derived by a convex or nonconvex superpotential. As we have seen, superpotentials permit the study of more complicated phenomena than do the methods of classical—"smooth"—

mechanics. The notions of subgradient, generalized gradient and derivate container are also seen to play a major role in one of the most important problems of physics, the coexistence of phases. It is obvious that every such generalization of the notion of classical derivative leads to a new type of variational problem and permits a formulation of variational principles for mechanical problems which could not be achieved using the classical calculus of variations. Thus the continuous development of new generalizations of the notion of classical derivative extends the frontiers of "nonsmooth" mechanics. The resulting variational problems have generally a concrete physical meaning: they express the principle of virtual work (or power), as indicated in the previous chapters in the case of variational and hemivariational inequalities. The "fans" of Ioffe [359] could offer another example in this context. Thus the principle of virtual work (or power) can be seen as a tool for the evaluation of such generalizations of the classical derivative. Let us take this idea a step further. Instead of defining new notions analogous to the directional differential of. F.H. Clarke or the upper subdifferential [293] in order to achieve a new generalization of the notion of classical derivative, we can define expressions-even of nonlocal nature-which have the properties of virtual work. Then we may use the inequality form of the principle of virtual work to define a new generalization of the classical derivative. The resulting notion would be a set consisting of all x' 's satisfying an inequality similar to (2.3.1) or to (4.3.13) but now having in its left-hand side the aforementioned virtual work.

Closely connected with the B.V.P.s studied herein is the theory of optimal control and parameter identification in systems governed by variational inequalities. As the study of such problems lies outside the scope of this book, we must restrict ourselves here to referring the reader to the existing literature on this topic (see [339] [204] [255] [246] [193] [242] [173] [254] [195] [259] and [16]).

Finally, one should not overlook various possibilities of applying inequality problems in the engineering sciences, especially in aeronautical, civil, and mechanical engineering. The last two chapters of this book have presented an essential survey on this subject.

It is hoped that the present study will encourage further research, both theoretical and applied, on the inequality problems in mechanics.

Appendices

The aim of the following four appendices is to recall certain notions of mechanics and to present certain others in the form in which they are employed in this book.

Appendix I. Some Basic Notions [20] [112] [321] [322]

Let Ω be an open bounded subset of $\mathbb{R}^3$ with boundary Γ. Ω is occupied by a deformable body and is referred to a fixed orthogonal Cartesian coordinate system $0X_1X_2X_3$ which establishes a one-to-one correspondence between the material particles X of the body and the point (X_1, X_2, X_3) (material coordinates of X). Henceforth we shall refer to the deformable body simply as (body) Ω. With respect to another coordinate system $\bar{0}x_1x_2x_3$ we may consider the coordinate equations

$$x_k = x_k(X_1, X_2, X_3), \qquad k = 1, 2, 3. \tag{AI.1}$$

Let $\bar{0}x_1x_2x_3$ be orthogonal Cartesian. Any deformation process may be generally described by means of the trajectory of each material particle X, i.e., by

$$x_i = \chi_i(X_k, t), \qquad i, k = 1, 2, 3, \tag{AI.2}$$

where $t \in [0, T]$ is the time variable. The point $x = \chi(X, t)$ is the place occupied by X at time t and let $x \equiv X$ for $t = 0$. The coordinates x_k, $k = 1, 2, 3$, are called the spatial coordinates of X. A mapping $\chi = \{\chi_i\} : \Omega \times [0, T] \to \Omega_t \subset \mathbb{R}^3$ is called a motion of Ω. Ω_t denotes the subset of $\mathbb{R}^3$ occupied by Ω at time t. We shall assume that χ, χ^{-1} are C^m-functions, where $m \in \mathbb{N}$ is sufficiently large. Let

$$g = g(X, t) \tag{AI.3}$$

be a C^1-function describing a quantity g. We call (AI.3) the material description of g, whereas

$$g = g(x, t) \tag{AI.4}$$

is the spatial description. Also, as is common in classical continuum mechanics, we call the use of x_i (resp. X_i) as independent variable the Eulerian (resp. Lagrangian) description of motion. We define the local or spatial derivative $\partial g(x, t)/\partial t$ of g, and its material derivative $\partial g(X, t)/\partial t$, which we will denote simply by dg/dt. Between material and local derivatives there holds the relation

$$\frac{dg}{dt}=\frac{\partial g(x, t)}{\partial t}+\frac{\partial g}{\partial x_i}\,\frac{\partial x_i(X, t)}{\partial t}, i = 1, 2, 3. \tag{AI.5}$$

The velocity $v = v(X, t)$ is obtained by differentiating (AI.2) with respect to t (keeping X fixed), i.e.,

$$v_i(X, t) = \frac{\partial \chi_i(X, t)}{\partial t}. \tag{AI.6}$$

The acceleration $\gamma = \gamma(X, t)$ is given by

$$\gamma_i = \frac{\partial v_i(X, t)}{\partial t} = \frac{dv_i}{dt} = \frac{\partial v_i(x, t)}{\partial t} + \frac{\partial v_i}{\partial x_j}\, v_j\,. \tag{AI.7}$$

By inverting (AI.2) we have $v = v(x, t)$. In the Eulerian description we consider the velocity gradient $L = \{L_{ij}\} = \{v_{i,j}\}$. Its symmetric and antisymmetric parts are called the rate of deformation and spin; thus

$$\begin{aligned} D &= \tfrac{1}{2}(L + L^T) = \text{sym (grad } v),\\ W &= \tfrac{1}{2}(L - L^T) = \text{asym (grad } v). \end{aligned} \tag{AI.8}$$

For a material particle we relate the instantaneous position x to the initial position X by means of the displacement vector u, i.e., $x = X + u$. We call $F = \{F_{i,\bar{j}}\} = \{\partial x_i/\partial X_j\}$ the deformation gradient. Then $F^T F$ is the right Cauchy-Green tensor (T denotes the transpose of a matrix) and

$$\mathsf{E} = \frac{F^T F - I}{2} \tag{AI.9}$$

is the Green strain tensor. Hereafter, for a vector a_i we denote the derivative $\partial a_i/\partial X_j$ by $a_{i,\bar{j}}$. The deformation gradient rate is expressed as

$$\frac{\partial F_{ik}(X, t)}{\partial t} = v_{i,j}\, x_{j,\bar{k}} \quad \text{or} \quad \frac{dF}{dt} = LF. \tag{AI.10}$$

Thus the rate $\dot{\mathsf{E}}$ of the Green strain tensor takes the form

$$\begin{aligned} \dot{\mathsf{E}} &= \frac{d\mathsf{E}}{dt} = \frac{1}{2}\left(\frac{dF^T}{dt}F + F^T\frac{dF}{dt}\right)\\ &= \tfrac{1}{2}(F^T L^T F + F^T L F) = F^T D F\,. \end{aligned} \tag{AI.11}$$

We obtain

$$\mathsf{E}_{ij} = \tfrac{1}{2}(u_{i,\bar{j}} + u_{j,\bar{i}} + u_{k,\bar{i}}u_{k,\bar{j}}) \tag{AI.12}$$

and

$$\frac{d\mathsf{E}_{ij}}{dt} = \tfrac{1}{2}(v_{i,\bar{j}} + v_{j,\bar{i}} + v_{k,\bar{i}}\, u_{k,\bar{j}} + u_{k,\bar{i}}\, v_{k,\bar{j}}). \tag{AI.13}$$

The Cauchy stress tensor $\sigma = \sigma(x, t)$ is employed in the case of spatial description. From the tensor σ the (second) Piola–Kirchhoff stress tensor $\mathbf{\Sigma} = \mathbf{\Sigma}(X, t)$ is defined by

$$\mathbf{\Sigma} = (\det F)F^{-1}\sigma(F^T)^{-1}. \tag{AI.14}$$

If the displacement gradients are sufficiently small, then

$$\mathsf{E}_{ij} \simeq \varepsilon_{ij} = \tfrac{1}{2}(u_{i,j} + u_{j,i}). \tag{AI.15}$$

We call the tensor $\varepsilon = \{\varepsilon_{ij}\}$ the (small) strain tensor, and the procedure of replacing E by ε geometric or kinematic linearization.

Appendix II. Rigidifying Velocity Fields. Objectivity [112] [197] [322]

Let $\Omega_t \subset \mathbb{R}^3$ be an open, bounded subset of $\mathbb{R}^3$ occupied by a body Ω at time t. Ω_t is referred to an orthogonal Cartesian coordinate system attached to the body. Let P be a point of Ω_t having coordinates $(X_{p_1}, X_{p_2}, X_{p_3})$. A velocity field such that

$$v(X_p, t) = v_0(t) + \omega(t) \wedge X_p, \ \forall X_p \in \Omega_t \tag{AII.1}$$

where $X_p = \{X_{pi}\}$, $i = 1, 2, 3$, is called a rigidifying velocity field. By v_0 we denote the velocity of the origin of the coordinate system. It is readily seen that such a velocity field results from a motion described by the equation

$$x = Q(t)X + x_0(t) \tag{AII.2}$$

where $Q^T Q = QQ^T = I$ and $\det Q = 1$. Let x_0 and Q be sufficiently smooth functions of t. From (AII.1,2), there results the relation

$$\omega_k = \tfrac{1}{2}\varepsilon_{kij}\Omega_{ji} \tag{AII.3}$$

where Ω is a skew-symmetric tensor given by $\Omega = (dQ/dt)Q^T$, and where $\varepsilon_{ijk} = 0$ if any two indices are alike, $\varepsilon_{ijk} = \varepsilon_{123} = 1$ if (i, j, k) is an even permutation of (1, 2, 3) and $\varepsilon_{ijk} = -1$ if (i, j, k) is an odd permutation.

Let x_i and $\bar{x}_i$, $i = 1, 2, 3$, be the coordinates of a material point with respect to two orthogonal Cartesian coordinate systems. If these systems rotate and translate each with respect to the other in an arbitrary manner, a relation of the form

$$\bar{x} = \tilde{Q}(t)\, x + \tilde{x}_0(t) \tag{AII.4}$$

holds, where $\tilde{Q}$ has the same properties as Q and $\tilde{x}_0$ is an arbitrary vector.

A tensor field is said to be frame-indifferent, or objective, if it transforms in the well-known tensorial manner, whenever (AII.4) is considered as a classical coordinate transformation. Similarly, a mapping between tensor fields is said to be objective or frame-indifferent if dependent and independent variables transform in the previous manner.

Appendix III. Dissipation [112]

Let Ω be a body having density ρ in the configuration under consideration. Let ε, η, $\psi = \varepsilon - \theta\eta$ be the specific (i.e., per unit mass) internal energy, specific entropy, and specific free energy, θ the absolute temperature, $q = \{q_i\}$ the heat flux vector and $\mathfrak{Q}$ the heat supply per unit volume[1] inside Ω (through radiation or chemical reactions). The First and Second Principles of Thermodynamics, in local form, read

$$\rho \frac{d\varepsilon}{dt} = \sigma_{ij} D_{ij} + \mathfrak{Q} - \operatorname{div} q \tag{AIII.1}$$

and

$$\frac{d\eta}{dt} + \frac{1}{\rho} \operatorname{div}\left(\frac{q}{\theta}\right) - \frac{\mathfrak{Q}}{\rho\theta} \geq 0. \tag{AIII.2}$$

Assuming that $\mathfrak{Q} = 0$, (AIII.2) along with (AIII.1) and the definition of ψ yields the inequality

$$\sigma_{ij} D_{ij} - \rho\left(\frac{d\psi}{dt} + \eta \frac{d\theta}{dt}\right) - \frac{q_i \theta_{,i}}{\theta} \geq 0. \tag{AIII.3}$$

(AIII.3) is written in the form

$$\mathscr{A} = \rho \mathscr{A} - \frac{q_i \theta_{,i}}{\theta} \geq 0, \tag{AIII.4}$$

where

$$\mathscr{A} = \frac{\sigma_{ij} D_{ij}}{\rho} - \left(\frac{d\psi}{dt} + \eta \frac{d\theta}{dt}\right). \tag{AIII.5}$$

$\mathscr{A}$ is called the intrinsic specific dissipation and $q_i\theta_{,i}/\theta$ the thermal dissipation. On the assumption that the thermal and intrinsic dissipations are "uncoupled", we may write the inequalities

$$\mathscr{A} \geq 0 \quad \text{and} \quad -\frac{q_i \theta_{,i}}{\theta} \geq 0. \tag{AIII.6}$$

Expressing ψ as a function of the state variables implies that $\mathscr{A}$ may be put in the form

$$\mathscr{A} = f_i v_i. \tag{AIII.7}$$

The terms v_i denote fluxes, and the terms f_i forces depending on the state variables.

Appendix IV. Plasticity and Thermodynamics [75] [196]

The thermodynamic state of a material element at time $t = t_0$ exhibiting plastic and/or viscoplastic behavior cannot be described completely by assuming as state variables the Piola–Kirchhoff stress tensor $\boldsymbol{\Sigma} = \boldsymbol{\Sigma}(X, t)$ and the absolute temperature $\theta = \theta(X, t)$, as in the case of an elastic material. Therefore, either the history of these variables for $-\infty < t \leq t_0$ should be given, or certain supplementary state variables

[1] $\mathfrak{Q}$ denotes the heat supply per unit mass only in Sec. 4.2.1.

(called hidden, i.e., not observable, variables) should be considered. The first method is not adequate for classical elastoplastic bodies because of the different material behavior during loading and unloading. Hidden variables (e.g., the dislocation distribution in a crystal, the residual stresses between the microelements of a polycrystal or of a granular medium, certain physical and chemical properties, etc.) are not necessarily internal variables (which always have a tensorial character) as is true with hidden variables depending on the orientation.

We denote by (I) the initial configuration of the body, in which we assume that the stresses are zero and the absolute temperature has a fixed value θ_0. The configuration at time t with stress σ and absolute temperature θ is denoted by (III). Let (II) denote another configuration, called a released configuration, which results from (III) by rapid, purely elastic unloading while the temperature is brought back to the initial value θ_0. If F denotes the deformation gradient of the "transformation" $(I)\to(III)$ and F_e and F_p those of the "transformations" $(II)\to(III)$ and $(I)\to(II)$ respectively, then the relation

$$F = F_e F_p \tag{AIV.1}$$

holds. With respect to the configuration (III), the rate of deformation tensor has the form

$$\begin{aligned} D &= \{\tfrac{1}{2}(v_{i,j} + v_{j,i})\} = \text{sym (grad } v) = \text{sym}\left(\frac{dF}{dt} F^{-1}\right) \\ &= \text{sym}\left(\frac{dF_e}{dt} F_e^{-1}\right) + \text{sym}\,(F_e V_p F_e^{-1}). \end{aligned} \tag{AIV.2}$$

Here $V_p = (\tilde{d}F_p/dt)\, F_p^{-1}$ is the plastic transformation rate with respect to the configuration (II). Each term of the last equality in (AIV.2) depends on changes, with time, of the orientation of the released configuration. Therefore Mandel has employed (cf., [196]) released configurations such that the director triad chosen always has the same orientation with respect to a fixed coordinate system. The resulting configurations (II) are then called isoclinic released configurations.

We take as state variables the Green strain tensor E_e of the elastic transformation $(II)\to(III)$, the absolute temperature θ and some hidden variables $a_j, j=1,\ldots,l$ of a tensorial nature. The specific power of the stresses is given by the expression

$$\frac{\sigma_{ij} D_{ij}}{\rho} = \text{tr}\left(\frac{\Sigma}{\rho_{II}} F_e^T D F_e\right) = \text{tr}\left(\frac{\Sigma}{\rho_{II}} \frac{\tilde{d}\mathsf{E}_e}{dt}\right) + \text{tr}\left(\frac{\Sigma}{\rho_{II}} F_e^T F_e V_p\right), \tag{AIV.3}$$

where tr (A) denotes the trace of the tensor A, $\tilde{d}/dt$ is the time derivative of a tensor in motion referred to the director triad of II and ρ_{II} is the density in the reference configuration (II). If $\psi = \psi(\mathsf{E}_e, a_j, \theta)$, the first of (AIII.6) acquires the form

$$\begin{aligned} \mathscr{A} = \text{tr}\left(\left(\frac{\Sigma}{\rho_{II}} - \frac{\partial\psi}{\partial \mathsf{E}_e}\right)\frac{\tilde{d}\mathsf{E}_e}{dt}\right) - \left(\frac{\partial\psi}{\partial\theta} + \eta\right)\frac{\tilde{d}\theta}{dt} \\ + \text{tr}\left(\frac{\Sigma}{\rho_{II}} F_e^T F_e V_p\right) - \frac{\partial\psi}{\partial a_j}\frac{\tilde{d}a_j}{dt} \geq 0. \end{aligned} \tag{AIV.4}$$

Next, we assume that $\tilde{d}\mathsf{E}_e/dt$ and $\tilde{d}\theta/dt$ may take arbitrary values, while both permanent deformations and $\tilde{d}a_j/dt$ remain bounded (e.g., in the case of a viscoplastic material). It results from (AIV.4) that

$$\frac{\boldsymbol{\Sigma}}{\rho_{II}} = \frac{\partial\psi}{\partial\mathsf{E}_e}, \tag{AIV.5}$$

$$\eta = -\frac{\partial\psi}{\partial\theta} \tag{AIV.6}$$

and thus

$$\mathscr{A} = \operatorname{tr}\left(\frac{\boldsymbol{\Sigma}}{\rho_{II}} \boldsymbol{F}_e^T \boldsymbol{F}_e \boldsymbol{V}_p\right) - \frac{\partial\psi}{\partial a_j}\frac{\tilde{d}a_j}{dt} \geq 0. \tag{AIV.7}$$

However, if instantaneous plastic deformations take place, i.e., in the case of ideal plasticity, $\tilde{d}\mathsf{E}_e/dt$, $\tilde{d}\theta/dt$, $\tilde{d}a_j/dt$, etc. are interrelated, thus not permitting similar reasoning. We may assume that (AIV.5, 6) remain valid because ideal plasticity is a limiting case of viscoplasticity. From (AIV.7) we associate with the flux $-\tilde{d}a_j/dt$ the force $A_j = \partial\psi/\partial a_j$, $j = 1, 2, \ldots, l$, and with the flux V_p the force

$$\boldsymbol{R} = \boldsymbol{F}_e^T \boldsymbol{F}_e \frac{\boldsymbol{\Sigma}}{\rho_{II}}. \tag{AIV.8}$$

Thus

$$\mathscr{A} = \operatorname{tr}(R^T V_p) - A_j \frac{\tilde{d}a_j}{dt}. \tag{AIV.9}$$

List of Notations

We have listed here the abbreviations and the main notations which are used throughout the text. We also refer the reader where necessary, to the section, wherein the definition is given. Throughout this book the summation convention with respect to a repeated index is employed, unless otherwise stated, i.e., we simply write $x_i y_{ij}$ instead of $\Sigma_i x_i y_{ij}$. Bold-face letters denote vectors and matrices of discretized continua. Since it is normally the case in this book, because of the function spaces used, that relations hold a.e. or μ-a.e., these two abbreviations are used in the interest of clarity only when necessary to remind the reader that a relation holds in this sense.

$\emptyset$	the empty set
$\forall$	for every
$\mathbb{N}$	the set of natural numbers including 0
$\mathbb{R}$	the set of real numbers
$\mathbb{R}_+$	the set of positive reals
$\bar{\mathbb{R}}$	the set of real numbers including $\pm\infty$
$\mathbb{R}^n$	Euclidean n-dimensional space
$\delta_{ij} = \begin{cases} 0 & \text{if } i \neq j \\ 1 & \text{if } i = j \end{cases}$	Kronecker's delta
$x = (x_1, \ldots, x_n) = \{x_i\}\ i = 1, \ldots, n$	coordinates of the point $x \in \mathbb{R}^n$
$\lvert x \rvert = (\sum_{i=1}^n x_i^2)^{1/2}$	length of $x \in \mathbb{R}^n$
$(x, y) = \sum_{i=1}^n x_i y_i = x_i y_i$	scalar product in $\mathbb{R}^n$
$x \wedge y$	vector product in $\mathbb{R}^n$
Ω	open, generally bounded and connected subset of $\mathbb{R}^n$

Γ	boundary of Ω
$\{\alpha_1, \alpha_2, \ldots\}$	set with elements $\alpha_1, \alpha_2, \ldots,$ etc. $\{\alpha_1\}$ is also written as α_1.
$A \cup B, A \cap B, A - B$	union, intersection and difference of sets A and B
$\mathscr{P}(A)$	set of all subsets of A (power set of A)
$A \times B$	product set of A, B: the set of all ordered pairs (α, β), $\alpha \in A$, $\beta \in B$. A subset R of $A \times A$ is termed an equivalence relation if $(\alpha, \alpha) \in R$; $(\alpha, \beta) \in \mathbb{R} \Rightarrow (\beta, \alpha) \in R$; $(\alpha, \beta) \in R, (\beta, \gamma) \in R \Rightarrow (\alpha, \gamma) \in R, \forall \alpha, \beta, \gamma \in A$
$[\alpha] = \alpha/R = \{\beta \mid (\alpha, \beta) \in R\}$	equivalence class *modulo R*
$A/R = \{[a] \mid a \in A\}$	quotient of A by R
$\operatorname{int} A$	interior of a set A
$\bar{A}$	closure of A
relint A	the relative interior of A (Sec. 2.1.1)
V.S.	vector space
T.V.S.	topological V.S.
L.C.T.V.S.	locally convex T.V.S.
L.C.H.T.V.S.	locally convex Hausdorff (or separated) T.V.S.
$f\colon A \to B$, $f(\cdot)\colon A \to B$, $\alpha \to f(\alpha)\colon A \to B$	function or mapping or transformation from A into B. If A and B have been specified, we write f, or $f(\cdot)$ or $\alpha \to f(\alpha)$. In order to indicate the independent variables, we also write $f = f(\alpha)$. A function f is one-to-one or injective if $f(\alpha_1) = f(\alpha_2) \Rightarrow \alpha_1 = \alpha_2$; it is onto or surjective if $f(A) = B$. A function which is injective and surjective is termed bijective. If $B = \mathbb{R}$, f is said to be a functional
$I\colon A \to A$	identity mapping; $I\colon A \to B$, $A \subset B$, $Ia = a$, is called natural or canonical injection (or simply injection if no ambiguity occurs) of A into B.
$\mathcal{G}(f)$	graph of f (Sec. 1.4.2)
$\Pi_{\alpha \in \Gamma} X_\alpha$	Cartesian product of the sets X_α; α belongs to a given set Γ. If $\Gamma = \{1, 2, \ldots, n\}$, then we write $\prod_{\alpha \in \Gamma} X_\alpha = X_1 \times X_2 \times \cdots \times X_n$
$f\|_{A_1}$	restriction $f_1\colon A_1 \to B$ of $f\colon A \to B$ to $A_1 \subset A$
$\sup_{\alpha \in A} f(a)$ or $\sup_A f(a)$	supremum of $f\colon A \to \mathbb{R}$
$\inf_{\alpha \in A} f(a)$ or $\inf_A f(a)$	infimum of $f\colon A \to \mathbb{R}$

$dx = dx_1 \cdots dx_n$, $d\Omega$, $d\Gamma$	Lebesgue measures on $\mathbb{R}^n$, Ω and Γ respectively
a.e.	almost every (everywhere)
μ – a.e.	almost every (everywhere) with respect to a measure μ
mes $\Omega = \lvert\Omega\rvert$, mes $\Gamma = \lvert\Gamma\rvert$	measure of Ω (the volume of Ω) and of Γ (the area of Γ)
$C^m(\Omega)$, $C^m(\bar{\Omega})$, $C^{m,\alpha}(\bar{\Omega})$	see Sec. 1.3.1
$C_c^\infty(\Omega)$, $D(\Omega)$, $D(\bar{\Omega})$, $L^p(\Omega)$, $L^\infty(\Omega)$	see Sec. 1.3.1
$\mathscr{D}'(\Omega)$, $\mathscr{M}^1(\Omega)$	see Sec. 1.3.2
$W^{l,p}(\Omega)$, $H^l(\Omega)$, $\mathring{H}^l(\Omega)$, $H^s(\Gamma)$	see Secs. 1.3.3 and 1.3.4
$BD(\Omega)$	Sec. 1.3.5
c, c_1, c_2	constant >0, unless otherwise specified (used in connection with norm inequalities, e.g., $\lvert\cdot\rvert \le c\lVert\cdot\rVert$ or $\lVert\cdot\rVert \le c$, etc.)
$(\cdot,\cdot)$, $\lvert\cdot\rvert_2$	inner product and norm of both $L^2(\Omega)$ and $[L^2(\Omega)]^n = L^2(\Omega) \times \cdots \times L^2(\Omega)$. Note that if $v = \{v_i\}$, $i = 1, \ldots, n$ then $\lvert v\rvert_{[L^2(\Omega)]^n} = (\sum_{i=1}^n \lvert v_i\rvert^2_{L^2(\Omega)})^{1/2}$
$\lvert\cdot\rvert_p$	norm of both $L^p(\Omega)$ and $[L^p(\Omega)]^n$
$((\cdot,\cdot))_l$, $\lVert\cdot\rVert_l$	inner product and norm of $H^l(\Omega)$ and $[H^l(\Omega)]^n$
$\lVert\cdot\rVert_{m,p}$	norm of $W^{m,p}(\Omega)$ and $[W^{m,p}(\Omega)]^n$
$\langle\cdot,\cdot\rangle_{X\times X'}$ (or $\langle\cdot,\cdot\rangle_X$), $\langle\cdot,\cdot\rangle_{p,q}$, $\langle\cdot,\cdot\rangle_s$	duality pairing between X and X', between $W^{p,q}$ and $[W^{p,q}]'$ and between H^s and $(H^s)'$ (or between $(H^s)^n$ and $[H^s]^{n\prime}$), $s \in \mathbb{R}$
$\langle\cdot,\cdot\rangle$	used for $\langle\cdot,\cdot\rangle_s$, $s = \frac{1}{2}$
$L^p(0T, X)$, $L_w^{p\prime}(0T, X)$, $W^{l,p}(0T, X)$, $C^l(0T, X)$	see Sec. 1.4.1
df/dt	strong derivative; see Sec. 1.4.1
dF/dx	Fréchet-derivative; see Sec. 1.4.2
F'	Gâteaux-derivative; see Sec. 1.4.2. In dynamic B.V.Ps. the prime also denotes the partial time derivative

$D^\alpha f = D^{\alpha_1} D^{\alpha_2} \ldots D^{\alpha_n} f = \partial^{\lvert\alpha\rvert} f/\partial x_1^{\alpha_1} \cdots \partial x_n^{\alpha_n}$, $\lvert\alpha\rvert = \sum_{i=1}^n \alpha_i$ $f_{,i} = \partial f/\partial x_i$, $\sigma_{ij,j} = \partial\sigma_{ij}/\partial x_j$

$\Delta = \sum_{i=1}^{n} \partial^2/\partial x_i^2$	Laplace operator, $\Delta\Delta$ biharmonic operator
$\ker T$	kernel of T
$\operatorname{epi} f$	epigraph of a functional f
$\operatorname{co} K$	convex hull of a set K
$D(f)$	effective domain of a functional f
I_K	indicator of a set K
P_K	projection operator onto a convex set K
l.s.c., u.s.c.	lower semicontinuous, upper semicontinuous
$\partial f(x)$	subgradient of f at x
$\bar{\partial} f(x)$	generalized gradient of f at x
$\Lambda f(x)$	derivate container of f at x
$D(\partial f)$	domain of ∂f
$R(\partial f)$	range of ∂f
f^c	conjugate functional
$\Gamma_0(X)$	the set of convex, l.s.c. and proper functionals on X (see Sec. 2.5.1)
d.a.u.d.	direction of approximately uniform descent
$N_C(x)$, $T_C(x)$	normal and tangent cones of C at x
Γ_U	part of the boundary Γ having given displacements (or velocities)
Γ_F	part of the boundary Γ having given forces F
Γ_S	part of the boundary Γ on which subdifferential boundary conditions hold
t	time, $t \in [0, T]$
E	modulus of elasticity
ν	Poisson's ratio
γ	acceleration vector
$\mathscr{U}$	space of generalized displacements or velocities
$\mathscr{F}$	space of generalized forces
Φ, G, j, j_N, j_T, J_μ, w, W	superpotentials
Φ^c, G^c, j^c, j_N^c, j_T^c, J_μ^c, w^c, W^c	conjugate superpotentials
β, β_N, β_i	maximal monotone operators from $\mathbb{R}$ into $\mathscr{P}(\mathbb{R})$
Φ_ε	regularized superpotential
$D = \{D_{ij}\}$	rate of deformation tensor or stretching tensor

$\sigma = \{\sigma_{ij}\}$	Cauchy stress tensor
$f = \{f_i\}$	volume force vector, body force vector (in Sec. 4.2)
$n = \{n_i\}$	outward unit normal to Γ
$S = \{S_i\}$	stress vector on Γ
ρ	density
$q = \{q_i\}$	heat flux vector
θ	temperature (absolute or not)
ε	specific internal energy
η	specific internal entropy
d/dt	material derivative (Sec. 3.1.1, 3.3.3, 4.2.1, App. I, III)
S_N, S_T	normal and tangential components of S with respect to Γ
u_N, u_T	normal and tangential components of u with respect to Γ
M, Q	bending moment, shearing force
ζ, z	deflection of a plate
$\partial\zeta/\partial n, \partial z/\partial n$	rotation of a plate
k	Winkler's coefficient
$\mathscr{k} = \{\mathscr{k}_{ij}\}$	thermal conductivity tensor
$\mathscr{m} = \{\mathscr{m}_{ij}\}$	thermal expansion tensor
$C = \{C_{ijhk}\}$	Hooke's tensor
$c = \{c_{ijhk}\}$	inverse Hooke's tensor
$\mathscr{c}_D$	specific heat at zero strain
E	Green strain tensor
$\boldsymbol{\Sigma} = \{\boldsymbol{\Sigma}_{ij}\}$	Piola–Kirchhoff stress tensor
$\varepsilon = \{\varepsilon_{ij}\}$	strain tensor
$\dot{\mathsf{E}} = d\mathsf{E}/dt$	time rate of the Green strain tensor
$\dot{\varepsilon} = \partial\varepsilon/\partial t$	strain rate tensor (kinematic linearization)
$\dot{\sigma} = \partial\sigma/\partial t$	
σ^D, ε^D	stress and strain deviators
$\sigma_{II} = \frac{1}{2}\sigma_{ij}^D\sigma_{ij}^D$	
$\varepsilon_{II} = \frac{1}{2}\varepsilon_{ij}^D\varepsilon_{ij}^D$	
$D_{II} = \frac{1}{2}D_{ij}D_{ij}$	
g	yield limit
Π	process
$\mathfrak{Q}$	heat supply through radiation or chemical reactions

χ	motion
v	specific volume (only in Sec. 4.2)
ϑ', p'	ambient temperature, ambient pressure
$\bar{\vartheta}, \bar{p}$	thermostatic temperature, thermostatic pressure
$\partial(\mathscr{G})$	see (4.2.13, 14)
Π	potential energy
Π^c	complementary energy
$\mu, \mu_\alpha, \mu_\beta$	coefficients of friction
$\tilde{d}/dt$	time derivative; see App. IV
$\{a_j\}$	scalar or tensor hidden variables
$\{A_j\}$	generalized forces associated with $\{-\tilde{d}a_j/dt\}$
$\mathscr{G}_1$ (resp. $\tilde{\mathscr{G}}_1$)	superpotential of dissipations in the sense of Clarke (resp. of Warga)
$\mathfrak{D}_S, \mathfrak{D}_u$	integral operators
$\mathbf{e}_i$ (resp. $\dot{\mathbf{e}}_i$)	strain (resp. strain increment or rate) vector of the i-th element
$\mathbf{s}_i$ (resp. $\dot{\mathbf{s}}_i$)	stress (resp. stress increment or rate) vector of the i-th element
$\mathbf{F}_0, \mathbf{G}$	flexibility matrix, equilibrium matrix
$\dot{\mathbf{F}}_0$	tangential flexibility matrix
$\mathbf{K}_E, \mathbf{K}_G$ and $\mathbf{K}$	elastic, geometric and total stiffness matrices
$\dot{\boldsymbol{\lambda}}_i$	vector of plastic multiplier rates
$\dot{\mathbf{F}}_i$	vector of yield function rates
$\mathbf{V}_i, \mathbf{H}_i, \mathbf{N}_i$	see Sec. 3.3.3
$\mathbf{v}_i$	slackness of the i-th cable element
$\mathbf{e}_0$	initial strain vector
$\mathbf{p}$ (resp. $\dot{\mathbf{p}}$)	load (resp. load increment or rate) vector
$\mathbf{u}$ (resp. $\dot{\mathbf{u}}$)	displacement (resp. displacement increment or rate) vector
$I, \mathbf{I}$	unit tensor, unit matrix

References

[1] A. F. Abeasis, J. P. Dias and A. Lopes-Pinto, Sur les valeurs-propres du sous-différentiel d' une fonction convexe avec un noyau non borné. *C. R. Acad. Sc. Paris.* **278** A (1974) 1197–1199.

[2] R. A. Adams, *Sobolev Spaces.* Academic Press, New York 1975.

[3] D' Alembert, J. Le Rond, *Traité de Dynamique,* Paris 1743.

[4] R. Ansorge and K. Taubert, Set-valued Collectively Compact Operators and Applications to Set-valued Differential Equations. *Computing* **23** (1979) 333–343.

[5] G. Anzellotti and M. Giaquinta, Existence of the Displacements Field for an Elasto-plastic Body Subject to Hencky's Law and von Mises Yield Condition. *Manuscripta Mathematica* **32** (1980) 101–136.

[6] P. Appell, *Traité de Mécanique rationelle, Tome I: Statique-Dynamique du Point.* Gauthier-Villars. Paris 1893.

[7] J. H. Argyris, Continua and Discontinua. Proc. 1st Conf. Matrix Meth. Struct. Mech. Wright Patterson Air Force Base, Dayton, Ohio 1965, AFFDL TR 66–80, 1966.

[8] J. H. Argyris, Three-Dimensional Anisotropic and Inhomogeneous Elastic Media. Matrix Analysis for Small and Large Displacements. *Ing. Archiv* **34** (1965) 33–55.

[9] J. H. Argyris and D. W. Scharpf, Some General Considerations on the Natural Mode Technique. *Aeron. J. Royal Aeron. Soc.* **73** (1969) 218–226 and 361–368.

[10] Aristoteles, *Metaphysica.*

[11] J. P. Aubin, *Approximation of Elliptic Boundary Value Problems.* Wiley-Interscience, New York 1972.

[12] J. P. Aubin and F. H. Clarke, Multiplicateurs de Lagrange en optimization non-convexe et applications. *C. R. Acad. Sc. Paris* **285** A (1977) 451–454.

[13] J. P. Aubin, *Applied Functional Analysis.* J. Wiley and Sons, New York 1979.

[14] J. P. Aubin and F. H. Clarke, Shadow Prices and Duality for a Class of Optimal Control Problems. *SIAM J. Control Optimization* **17** (1979) 567–586.

[15] A. Auslender, *Problèmes de Minimax via l' Analyse Convexe et les Inégalités Variationnelles: Théorie et Algorithmes.* Springer-Verlag, Berlin 1972.

[16] V. Barbu, Necessary Conditions for Non-Convex Distributed Control Problems Governed by Elliptic Variational Inequalities. *J. Math. Anal. Appl.* **80** (1981) 566–597.

[17] V. Barbu and Th. Precupanu, *Convexity and Optimization in Banach Spaces.* Editura Academiei, Bucuresti, Sijthoff and Noordhoff 1978.

[18] K. J. Bathe and E. L. Wilson, Stability and Accuracy Analysis of Direct Integration Methods. *Int. J. Earthq. Eng. Struct. Dyn.* **1** (1973) 283–291.

[19] E. J. Bauman, Multilevel Optimization Techniques and Application to Trajectory Decomposition. In: *Advances in Control Systems* (ed. by C. T. Leondes). Academic Press, New York 1968 (vol. 6).

[20] E. Becker and W. Bürger, *Kontinuumsmechanik*, B. G. Teubner Stuttgart 1975.

[21] J. F. Bell, The Experimental Foundations of Solid Mechanics. In: *Encyclopedia of Physics* Vol. VI a/1 (ed. by S. Flügge) Springer-Verlag, Berlin 1973.

[22] M. S. Berger, On Von Kármán's Equations and the Buckling of a Thin Elastic Plate, I. The Clamped Plate. *Comm. Pure Appl. Math.* **XX** (1967) 687–719.

[23] M. S. Berger and P. C. Fife, Von Kármán's Equations and the Buckling of a Thin Elastic Plate, II. Plate with General Edge Conditions. *Comm. Pure Appl. Math.* **XXI** (1968) 227–241.

[24] N. Bourbaki, *Eléments de Mathématique, Livre V. Espaces vectoriels topologiques.* Hermann, Paris 1953.

[25] N. Bourbaki, *Eléments de Mathématique, Livre III. Topologie générale.* Hermann, Paris 1958.

[26] N. Bourbaki, *Eléments de Mathématique, Livre VI. Intégration.* Hermann, Paris 1965 (2nd edition).

[27] H. Brézis, Equations et inéquations non linéaires dans les espaces vectoriels en dualité. *Ann. Inst. Fourier* **18** (1968) 115–175.

[28] H. Brézis, Problèmes unilatéraux. *J. Math. pures et appl.* **51** (1972) 1–168.

[29] H. Brézis, *Opérateurs maximaux monotones et semigroupes de contractions dans les espaces de Hilbert.* North-Holland Publ. Co., Amsterdam and American Elsevier Publ. Co., New York 1973.

[30] F. Browder, Variational methods for nonlinear elliptic eigenvalue problems. *Bull. Amer. Math. Soc.* **7** (1965) 176–183.

[31] F. Browder, On the Unification of the Calculus of Variations and the Theory of Monotone Non-linear Operators in Banach Spaces. *Proc. Nat. Acad. Sci. U.S.A.* **56** (1966) 419–425.

[32] P. L. Butzer and H. J. Wagner, On Duadic Analysis Based in the Pointwise Derivative. *Analysis Mathematica* **1** (1975) 171–196.

[33] M. Capurso and G. Maier, Incremental Elastoplastic Analysis and Quadratic Optimization. *Meccanica* **5** (1970) 1–10.

[34] C. Castaign, Intégrales convexes duales. In: *New Variational Techniques in Mathematical Physics*, C.I.M.E., Edizioni Cremonese, Roma 1974.

[35] C. Castaign and M. Valadier, *Convex Analysis and Measurable Multifunctions.* Springer-Verlag, Berlin 1977.

[36] C.E.T.I.M., Proc. Int. Symposium on Innovative Numerical Analysis in Applied Engineering Science, Versailles 1977.

[37] K. C. Chang, Variational Methods for Non-differentiable Functionals and their Applications to Partial Differential Equations. *J. Math. Anal. Appl.* **80** (1981) 102–129.

[38] Phan van Chuong, Solutions continues à droite d' une équation intégrale multivoque. Séminaire d' Analyse Convexe, Exposé No 3, Montpellier 1979.

[39] S. Christiansen, On Kupradze's Functional Equations for Plane Harmonic Problems. In: *Function Theoretic Methods in Differential Equations* (ed. by R. P. Gilbert and R. J. Weinacht). Pitman, London 1976.

[40] A. A. Chyras, *Linear Programming Methods in the Analysis of Elastic-Plastic Systems.* Strojizdat, Leningrad 1969. (in Russian).

[41] P. G. Ciarlet, *The Finite Element Method for Elliptic Problems.* North-Holland Pub. Co., Amsterdam 1978.

[42] P. G. Ciarlet, A Justification of the von Kármán equations. *Arch. Rat. Mech. Anal.* **73** (1980) 349–389.

[43] A. Cimetière, Un problème de flambement unilatéral en théorie des plaques. *J. Mécanique* **19** (1980) 183–202.

[44] A. Cimetière, Flambement unilatéral d' une plaque reposante sans frottement sur un support élastique tridimensionel. *C. R. Acad. Sc. Paris* **290** B(1980) 337–340.

[45] F. H. Clarke, Necessary Conditions for Nonsmooth Problems in Optimal Control and the Calculus of Variations. Ph. D. Thesis, University of Washington, Seattle 1973.

[46] F. H. Clarke, Generalized Gradients and Applications. *Trans. A.M.S.* **205** (1975) 247–262.

[47] F. H. Clarke, Generalized Gradients of Lipschitz Functionals. *Advances in Math.* **40** (1981) 52–67 (Tech. Report 1687, Math. Res. Center. Univ. of Wisconsin, Madison 1976).

[48] F. H. Clarke, On the Inverse Function Theorem. *Pacific J. Math.* **64** (1976) 97–102.

[49] F. H. Clarke, A New Approach to Lagrange Multipliers. *Math of Op. Research* **1** (1976) 165–174.

[50] F. H. Clarke, Inequality Constraints in the Calculus of Variations. *Can. J. Math.* **39** (1977) 528–540.

[51] F. H. Clarke, Multiple Integrals of Lipschitz Functions in the Calculus of Variations. *Proc. A.M.S.* **64** (1977) 260–264.

[52] M. Z. Cohn and G. Maier, *Engineering Plasticity by Mathematical Programming, Proc. NATO Advanced Study Inst., Waterloo Canada* 1977. Pergamon Press, New York 1979.

[53] B. D. Coleman, Thermodynamics of Materials with Memory. *Arch. Rat. Mech. Anal.* **17** (1964) 1–46.

[54] B. D. Coleman and J. M. Greenberg, Thermodynamics and the Stability of Fluid Motion, *Arch. Rat. Mech. Anal.* **25** (1967) 321–341.

[55] B. D. Coleman, On the Stability of Equilibrium States of General Fluids. *Arch. Rat. Mech. Anal.* **36** (1970) 1–32.

[56] T. F. Conry and A. Seireg, A Mathematical Programming Method for Design of Elastic Bodies in Contact. *J. Appl. Mech.* **52** (1971) 387–392.

[57] R. Contro, G. Maier and A. Zavelani, Inelastic Analysis of Suspension Structures by Nonlinear Programming. *Comp. Meth. App. Mech. Eng.* **5** (1975) 127–143.

[58] R. W. Cottle, F. Giannessi and J. L. Lions (eds), *Variational Inequalities and Complementarity Problems, Theory and Applications.* J. Wiley and Sons, New York 1980.

[59] N. Cristescu, *Dynamic Plasticity.* North-Holland Pub. Co., Amsterdam 1967.

[60] J. P. Crouzeix, Conjugacy in Quasiconvex Analysis. In: *Convex Analysis and its Applications* (ed. by A. Auslender). Springer-Verlag, Berlin 1977.

[61] T. A. Cruse, Numerical Solutions in Three-dimensional Elastostatics. *Int. J. Solids Struct.* **5** (1969) 1259–1274.

[62] T. A. Cruse, An Improved Boundary-Integral Equation Method for Three Dimensional Elastic Stress Analysis. *Comp. Struct.* **4** (1974) 741–754.

[63] T. A. Cruse and F. J. Rizzo, (eds.) *Boundary-Integral Equation Method: Computational Applications in Applied Mechanics.* ASME, Applied Mechanics Division Vol. 11, 1975.

[64] J. I. Curiskis and S. Valliappan, A Solution Algorithm for Linear Constraint Equations in Finite Element Analysis. *Comp. Struct.* **8** (1978) 117.

[65] C. Dafermos, On the Existence and the Asymptotic Stability of Solutions to the Equations of Linear Thermoelasticity. *Arch. Rat. Mech. Anal.* **29** (1968) 241–271.

[66] J. W. Daniel, *The Approximate Minimization of Functionals.* Prentice-Hall, Englewood Cliffs 1971.

[67] C. S. Desai and J. F. Abel, *Introduction to the Finite Element Method.* Van Nostrand Reinhold Co., New York 1972.

[68] J. P. Dias, Variational Inequalities and Eigenvalue Problem for Nonlinear Maximal Monotone Operators in Hilbert Spaces. *Am. J. Math.* **97** (1976) 905–914.

[69] U. Dieter, Optimierungsaufgaben in topologischen Vektorräumen. *Zeitschr. für Wahrscheinlichkeitstheorie* **5** (1966) 89–117.

[70] J. Dieudonné, *Eléments d' Analyse, Tome I, II Fondements de l' Analyse moderne.* Gauthier-Villars, Paris 1969.

[71] C. Do, The Buckling of a Thin Elastic Plate Subjected to Unilateral Conditions. In: *Applications of Methods of Functional Analysis to Problem of Mechanics* (ed. by P. Germain and B. Nayroles) Springer-Verlag, Berlin 1976.

[72] C. Do, Bifurcation Theory of Thin Elastic Plates Subjected to Unilateral Conditions. *J. Math. Anal. Appl.* **60** (1977) 435–448.

[73] C. Do, Nonlinear Bifurcation Problem and Buckling of an Elastic Plate Subjected to Unilateral Conditions in its Plane. In: *Contemporary Developments in Continuum Mechanics and Partial Differential Equations* (ed. by G. M. de la Penha and L. A. J. Medeiros). North-Holland Pub. Co., 1978.

[74] C. Do, Flambement élastoplastique d' une plaque mince. *C. R. Acad. Sc. Paris* **290** B (1980) 143–146.

[75] D. Domingon, M. N. R. Nina and J. H. Whitelaw (eds.), *Foundation of Continuum Thermodynamics.* McMillan, London 1974.

[76] O. De Donato and G. Maier, Local Unloading in Piecewise-Linear Plasticity. *ASCE*, EM3, **102** (1976) 383–393.

[77] D. C. Drucker, A more Fundamental Approach to Plastic Stress-Strain Relations. Proc. First U.S. Nat. Congr. on Appl. Mech. Chicago 1951.

[78] J. E. Dunn and R. L. Fosdick, The Morphology and Stability of Material Phases. *Arch. Rat. Mech. Anal.* **74** (1980) 1–99.

[79] N. Dunford and J. T. Schwartz, *Linear Operators Part I: General Theory.* Interscience Publishers, New York 1966.

[80] G. Dupuis and A. Probst, Etude d'une structure élastique soumise à des conditions unilatérales. *J. de Mécanique* **6** (1967) 1–41.

[81] A. Duschek and A. Hochrainer, *Tensorrechnung in analytischer Darstellung,* III. Springer-Verlag, Wien 1965.

[82] G. Duvaut and J. L. Lions, Ecoulement d'un fluide rigide viscoplastique incompressible. *C. R. Acad. Sc. Paris* **270** A (1970) 58–61.

[83] G. Duvaut and J. L. Lions, Un problème d'élasticité avec frottement. *J. de Mécanique* **10** (1971) 409–420.

[84] G. Duvaut and J. L. Lions, *Les inéquations en Mécanique et en Physique.* Dunod, Paris 1972.

[85] G. Duvaut and J. L. Lions, Inéquations en Thermoélasticité et Magnétohydrodynamique. *Arch. Rat. Mech. Anal.* **46** (1972) 241–279.

[86] G. Duvaut and J. L. Lions, Transfert de chaleur dans un fluide de Bingham dont la viscosité dépend de la température. *J. of Funct. Anal.* **11** (1972) 93–110.

[87] G. Duvaut and J. L. Lions, Problèmes unilatéraux dans la théorie de la flexion forte des plaques (I). Le cas stationnaire. *J. de Mécanique* **13** (1974) 51–74.

[88] G. Duvaut and J. L. Lions, Problèmes unilatéraux dans la théorie de la flexion forte des plaques (II). Le cas d'évolution. *J. de Mécanique* **13** (1974) 245–266.

[89] G. Duvaut, Problèmes dynamiques en elasto-viscoplasticité et plasticité parfaite avec conditions de frottement à la frontière. In: *New Variational Techniques in Mathematical Physics.* C.I.M.E. Ed. Cremonese, Roma 1974.

[90] G. Duvaut, Equilibre d'un solide élastique avec contact unilatéral et frottement de Coulomb. *C. R. Acad. Sc. Paris* **290** A (1980) 263–265.

[91] I. Ekeland and R. Temam, *Convex Analysis and Variational Problems.* North-Holland, Amsterdam and American Elsevier, New York 1976.

[92] S. M. Ermakov, *Die Monte-Carlo-Methode und verwandte Fragen.* Oldenburg Verlag, München 1975.

[93] R. S. Falk, Error Estimates for the Approximation of a Class of Variational Inequalities. *Math. of Computation* **28** (1974) 963–971.

[94] C. A. Felippa and K. C. Park, Direct Time Integration Methods in Nonlinear Structural Dynamics. *Comp. Meth. Appl. Mech. Eng.* **17/18** (1979) 277–313.

[95] W. Fenchel, On Conjugate Convex Functions. *Canad. J. Math.* **1** (1949) 73–77.

[96] G. Fichera, The Signorini Elastostatics Problem with Ambiguous Boundary Conditions. Proc. Int. Conf. Application of the Theory of Functions in Continuum Mechanics Vol. I, Tbilisi 1963.

[97] G. Fichera, Problemi elastostatici con vincoli unilaterali: il problema di Signorini con ambigue condizioni al contorno. *Mem. Accad. Naz. Lincei,* **VIII** 7 (1964) 91–140.

[98] G. Fichera, Existence Theorems in Elasticity. In: *Encyclopedia of Physics* (ed. by S. Flügge) Vol. VI a/2. Springer-Verlag, Berlin 1972.

[99] G. Fichera, Boundary Value Problems in Elasticity with Unilateral Constraints. In: *Encyclopedia of Physics* (ed. by S. Flügge) Vol. VI a/2. Springer-Verlag, Berlin 1972.

[100] F. D. Fischer, Zur Lösung des Kontaktproblems elastischer Körper mit ausgedehnter Kontaktfläche durch quadratische Programmierung. *Computing* **13** (1974) 353–384.

[101] G. Fix and G. J. Strang, *An Analysis of the Finite Element Method.* Prentice Hall, Englewood Cliffs 1973.

[102] J. Franču, On Signorini Problem for von Kármán Equations. The Case of Angular Domain. *Apl. Matematiky,* **24** (1979) 355–371.

[103] M. Frémond, Etude de structures viscoélastiques stratifiées soumises à des charges harmoniques et de solides élastiques reposant sur ces structures. Thèse de doctorat d'Etat, Univ. Paris VI, 1971.

[104] M. Frémond and M. Mucci, Comportement des chaussées rigides. Application à leur dimensionnement. *Bull. Liaison Labo. P. et Ch.* **65** (1973) 125–140.

[105] M. Frémond and A. Friâa, Analyse Limite. Comparaison des Méthodes Statique et Cinématique. *C. R. Acad. Sc. Paris* **286A** (1978) 107–110.

[106] A. M. Freudenthal, The Mathematical Theories of the Inelastic Continuum. In: *Encyclopedia of Physics.* Vol. VI (ed. by S. Flügge). Springer-Verlag, Berlin 1958.

[107] A. Friâa, La loi de Norton-Hoff généralisée en plasticité et viscoplasticité. Thèse de Doctorat d'Etat, Univ. Paris VI, 1979.

[108] E. Gagliardo, Caratterizzazioni delle tracce sulla frontiera relative ad alcune classi di funzioni in n variabili. *Rend. Semin. Mat. Padova* **27** (1957) 284–305.

[109] I. M. Gelfand and G. E. Schilow, *Verallgemeinerte Funktionen (Distributionen)* Vol. I. Deutscher Verlag der Wissenschaften, Berlin 1967.

[110] P. Germain, La méthode de puissances virtuelles en mécanique des milieux continus, 1ère partie. Théorie du second gradient. *J. de Mécanique* **12** (1973) 235–274.

[111] P. Germain, The Method of Virtual Power in Continuum Mechanics. Part 2: Microstructure. *SIAM J. Appl. Math.* **25** (1973) 556–575.

[112] P. Germain, *Cours de Mécanique des milieux continus I.* Masson, Paris 1973.

[113] P. Germain, The Role of Thermodynamics in Continuum Mechanics. In: *Foundations of Continuum Thermodynamics* (ed. by D. Domingon, M. N. R. Nina and J. H. Whitelaw). McMillan, London 1974.

[114] J. Ghaboussi, E. L. Wilson and J. Isenberg, Finite Element for Rock Joints and Interfaces. *ASCE, SM* 10, **99** (1973) 833–848.

[115] J. W. Gibbs, On the Equilibrium of Heterogeneous Substances (Trans. Conn. Acad. 1875–1878) The Scientific Papers I (1907) 55–353. (Yale Univ. Press, New Haven).

[116] K. Girkmann, *Flächentragwerke.* Springer-Verlag, Wien 1963.

[117] K. Glashoff and J. Sprekels, An Application of Glicksberg's Theorem to Set-Valued Integral Equations Arising in the Theory of Thermostats. *SIAM J. Math. Anal.* **12** (1981) 447–486.

[118] R. Glowinski and H. Lanchon, Torsion élastoplastique d'une barre cylindrique de section multiconnexe. *J. de Mécanique* **12** (1973) 151–171.

[119] R. Glowinski, Sur l'écoulement d'un fluide de Bingham dans une conduite cylindrique. *J. de Mécanique* **13** (1974) 601–621.

[120] R. Glowinski, J. L. Lions and R. Trémolières, *Analyse numérique des inéquations variationnelles.* Dunod, Paris 1976.

[121] R. Glowinski, Finite Elements and Variational Inequalities. In: *The Mathematics of Finite Elements and Applications* III 1978 (ed. by J. R. Whiteman) Academic Press, New York 1979.

[122] A. Göpfert, *Mathematische Optimierung in allgemeinen Vektorräumen.* B. G. Teubner, Leipzig 1973.

[123] A. E. Green and P. M. Naghdi, A General Theory of an Elastic-Plastic Continuum. *Arch. Rat. Mech. Anal.* **18** (1965) 251–281.

[124] S. R. De Groot, *Thermodynamik irreversibler Prozesse.* Bibliographisches Institut, Mannheim 1960.

[125] H. J. Grootenboer, S. F. C. H. Leijten and J. Blauuwendraad, Numerical Models for Reinforced Concrete Structures in Plane Stress. *Heron* **26** (1981) 1–83.

[126] M. Grun-Rehomme, Caractérisation du Sous-différentiel d'intégrandes convexes dans les espaces de Sobolev. *J. Math. Pures et Appl.* **56** (1977) 149–156.

[127] D. W. Haines and W. D. Wilson, Strain-energy Density Function for Rubberlike Materials. *J. Mech. Phys. Solids* **27** (1979) 345–360.

[128] B. Halphen and N. Q. Son, Plastic and Viscoplastic Materials with Generalized Potential. *Mech. Res. Comm.* **1** (1974) 43–47.

[129] B. Halphen and N. Q. Son, Sur les Matériaux Standards Généralisés. *J. de Mécanique* **14** (1975) 39–63.

[130] G. Hamel, *Theoretische Mechanik.* Springer-Verlag, Berlin 1967.

[131] J. Haslinger, Dual Finite Element Analysis for an Inequality of the 2nd Order. *Apl. Matematiky* **24** (1979) 118–132.

[132] J. Haslinger and J. Hlavaček, Approximation of the Signorini Problem with Friction by a Mixed Finite Element Method. *J. Math. Anal. Appl.* **86** (1982) 99–122.

[133] C. Heinz, Das d'Alembertsche Prinzip in der Kontinuumsmechanik. *Acta Mechanica* **10** (1970) 110–129.

[134] U. Heise, The Calculation of Cauchy Principal Values in Integral Equations for Boundary Value Problems of the Plane and Three-Dimensional Theory of Elasticity. *J. of Elasticity* **5** (1975) 99–110.

[135] U. Heise, Formulierung und Ordnung einiger Integralverfahren für Probleme der ebenen und räumlichen Elastostatik unter besonderer Berücksichtigung mechanischer Gesichtspunkte, Habilitation thesis, Aachen 1975.

[136] U. Heise, Application of the Singularity Method for the Formulation of Plane Elastostatical Boundary Value Problems as Integral Equations. *Acta Mechanica* **31** (1978), 33–69.

[137] U. Heise, Systematic Compilation of Integral Equations of the Rizzo Type and of Kupradze's Functional Equations for Boundary Value Problems of Plane Elastostatics. *J. of Elasticity* **10** (1980) 23–56.

[138] H. Hencky, Zur Theorie plastischer Deformationen und der hierdurch im Material hervorgerufenen Nachspannungen. *ZAMM* **4**(1924) 323–334.

[139] L. Herrmann, Finite Element Analysis of Contact Problems. *ASCE*, EM 5, **104**(1978) 1043–1057.

[140] P. Hess, On Semi-coercive Nonlinear Problems. *Ind. Univ. Math. J.* **23**(1974) 645–654.

[141] R. Hill, A Variational Principle of Maximum Plastic Work in Classical Plasticity. *Quar. J. Mech. Appl. Math.* **1**(1948) 18–28.

[142] R. Hill, *Mathematical Theory of Plasticity*. Univ. Press, Oxford 1950.

[143] D. M. Himmelblau, *Applied Nonlinear Programming*. McGraw-Hill Book Co., New York 1972.

[144] J. B. Hiriart-Urruty, Conditions nécessaires d'Optimalité en Programmation non différentiable. *C. R. Acad. Sc. Paris* **238**A(1976) 843–845.

[145] J. B. Hiriart-Urruty, Contributions à la Programmation Mathématique: Cas Déterministe et Stochastique, Thèse, Université de Clermont-Ferrand II, 1977.

[146] I. Hlavaček and J. Nečas, On Inequalities of Korn's Type I, II. *Arch. Rat. Mech. Anal.* **36**(1969) 305–311 and 312–334.

[147] I. Hlavaček and J. Naumann, Inhomogeneous Boundary Value Problems for the v. Kármán Equations I. *Apl. Matematiky* **19**(1974) 253–269.

[148] P. G. Hodge Jr., Numerical Applications of Minimum Principles in Plasticity. In: *Engineering Plasticity* (ed. by J. Heyman and F. A. Leckie). Cambridge Univ. Press, Cambridge 1968.

[149] T. J. R. Hughes, R. L. Taylor and W. Kanoknukulchai, A Finite Element Method for Large Displacement Contact and Impact Problems. In: *Formulations and Computational Algorithms in Finite Element Analysis* (ed. by K. F. Bathe, F. T. Oden and W. Wunderlich). MIT Press, Cambridge 1977.

[150] R. Hünlich and J. Naumann, On General Boundary Value Problems and Duality in Linear Elasticity, I, II. *Apl. Matematiky* **23**(1978) 208–229 and **25**(1980) 11–32.

[151] A. A. Ilyushin, On the Increment of Plastic Deformation and Yield Function. *Prikl. Math. Mech.* **24**(1960) 663–666. (in Russian).

[152] A. D. Ioffe and V. M. Tikhomirov, *Theory of Extremal Problems.* North Holland. Pub. Co., Amsterdam 1975.

[153] H. Jeggle, *Nichtlineare Funktionalanalysis.* B. G. Teubner, Stuttgart 1979.

[154] O. John, On Signorini Problem for von Kármán Equations. *Apl. Matematiky* **22**(1977) 52–68.

[155] C. Johnson, Existence Theorems for Plasticity Problems. *J. Math. pures et appl.* **55**(1976) 431–444.

[156] L. M. Kachanov, *Fundamentals of the Theory of Plasticity.* MIR Publishers, Moscow 1974 (Engl. translation).

[157] J. J. Kalker and Y. Van Randen, A Minimum Principle for Frictionless Elastic Contact with Application to Non-Hertzian Half-Space Contact Problems. *J. Eng. Math.* **6**(1972) 193–206.

[158] I. Kaneko, Complete solutions for a Class of Elastic-plastic Structures. *Comp. Meth. Appl. Mech. Eng.* **21**(1980) 193–209.

[159] B. Kawohl, Über nichtlineare gemischte Randwertprobleme für elliptische Differentialgleichungen zweiter Ordnung auf Gebieten mit Ecken, Doct. thesis, Techn. Universität Darmstadt 1978.

[160] Th. Kermanidis, Kupradze's Functional Equation for the Torsion Problem of Prismatic Bars Part 1, 2. *Comp. Meth. Appl. Mech. Eng.* **7**(1976) 39–46 and 249–259.

[161] D. Kinderlehrer and G. Stampacchia, *An Introduction to Variational Inequalities and their Applications*, Academic Press, New York 1980.

[162] G. H. Knightly and D. Sather, On Nonuniqueness of Solutions of the von Kármán Equations. *Arch. Rat. Mech. Anal.* **36**(1970) 65–78.

[163] W. T. Koiter, General Theorems for Elastic-Plastic Solids. In: *Progress in Solid Mechanics I.* North Holland Pub. Co. Amsterdam 1960.

[164] G. Köthe, *Topologische lineare Räume*, Vol. I. Springer-Verlag, Berlin 1960.

[165] M. Kučera, J. Nečas and J. Souček, The Eigenvalue Problem for Variational Inequalities and a New Version of the Ljusternik-Schnirelmann Theory. In: *Nonlinear Analysis* (ed. by L. Cesari, R. Kannan and H. F. Weinberger). Academic Press, New York 1978.

[166] H. Künzi and W. Krelle, *Nichtlineare Programmierung.* Springer-Verlag, Berlin 1962.

[167] V. D. Kupradze, *Potential Methods in the Theory of Elasticity.* Oldbourne Press, London 1965.

[168] V. D. Kupradze, T. G. Gegelia, M. O. Baschelejschwili and T. V. Burtschuladze, *Three-dimensional Problems of the Mathematical Theory of Elasticity and Thermoelasticity.* Nauka, Moscow 1976 (in Russian).

[169] H. Lanchon, Sur la solution du problème de torsion élastoplastique d'une barre cylindrique de section multiconnexe. *C. R. Acad. Sc. Paris* **271A**(1970) 1137–1140.

[170] H. Lanchon, Torsion élastoplastique d'un arbre cylindrique de section simplement ou multiplement connexe. *J. de Mécanique* **13**(1974) 267–320.

[171] C. Lanczos, *The Variational Principles of Mechanics*. University of Toronto Press, Toronto 1966.

[172] L. D. Landau and E. M. Lifschitz, *Lehrbuch der theoretischen Physik Vol. VII Elastizitätstheorie*. Akademie Verlag, Berlin 1966.

[173] H. H. E. Leipholz (ed), *Proc. IUTAM Conference on Structural Control, Waterloo 1979*. Univ. of Waterloo Press and Noordhoff-Sijthoff Pub. Co. 1980.

[174] H. H. E. Leipholz and M. A. Rohman, General Approach to Active Structural Control. *ASCE, EM* 6, **105**(1979) 1007–1023.

[175] C. Lemarechal, *Bundle Methods in Nonsmooth Optimization* (ed. by C. Lemarechal and R. Mifflin). Pergamon Press, Oxford 1978.

[176] F. Léné, Sur les matériaux élastiques à énergie de déformation non quadratique. Thèse de 3ème cycle, Université Paris VI, 1973.

[177] F. Léné, Sur les matériaux élastiques à énergie de déformation non quadratique. *J. de Mécanique* **13**(1974) 499–534.

[178] J. B. McLeod and R. E. L. Turner, Bifurcation of Non-differentiable Operators with an Application to Elasticity. *Arch. Rat. Mech. Anal.* **63**(1977) 1–45.

[179] J. L. Lions, *Problèmes aux limites dans les équations aux dérivées partielles*. Séminaire de Mathématiques supérieures. Presses de l'Université de Montréal, Montréal 1962.

[180] J. L. Lions and G. Stampacchia, Variational Inequalities. *Comm. pure and applied Math.* **XX** (1967) 493–519.

[181] J. L. Lions and E. Magenes, *Problèmes aux limites non homogènes et applications*, Vol. 1, 2, 3. Dunod, Paris 1968.

[182] J. L. Lions, *Quelques méthodes de résolution des problèmes aux limites non linéaires*. Dunod/Gauthier-Villars, Paris 1969.

[183] J. L. Lions, *Optimal Control of Systems Governed by Partial Differential Equations*. Springer-Verlag, Berlin 1971.

[184] G. Maier, A Quadratic Programming Approach for Certain Classes of Nonlinear Structural Problems. *Meccanica* **3**(1968) 121–130.

[185] G. Maier, Some Theorems for Plastic Strain Rates and Plastic Strains. *J. de Mécanique* **8**(1969) 5–19.

[186] G. Maier, A Method for Approximate Solutions of Stationary Creep Problems. *Meccanica* **4**(1969) 36–47.

[187] G. Maier, Complementary Plastic Work Theorems in Piecewise Linear Elastoplasticity. *Int. J. Solids Struct.* **5**(1969) 261–270.

[188] G. Maier, Linear Flow Laws of Elastoplasticity. A Unified General Approach. *Rend. Acc. Naz. Lincei* **VIII, XLVIII** (1969) 1–11.

[189] G. Maier, A Matrix Structural Theory of Piecewise Linear Elastoplasticity with Interacting Yield Planes. *Meccanica* **8**(1970) 54–66.

[190] G. Maier, Incremental Plastic Analysis in the Presence of Large Displacements and Physical Instabilizing Effects. *Int. J. Solids Struct.* **7**(1971) 345–372.

[191] G. Maier, Mathematical Programming Methods in Structural Analysis. In: *Variational Methods in Engineering* (ed. by C. A. Brebbia and H. Tottenham) Vol. II Southampton Univ. Press, Southampton 1973.

[192] G. Maier, Upper Bounds on Deformations of Elastic-Workhardening Structures in the Presence of Dynamic and Second-Order Geometric Effects. *J. Struct. Mech.* **2**(1973) 265–280.

[193] G. Maier, Future Directions in Engineering Plasticity. In: *Engineering Plasticity by Mathematical Programming* (ed. by M. Z. Cohn, G. Maier and D. E. Grierson) Pergamon Press, New York 1978.

[194] G. Maier, F. Andreuzzi, F. Giannessi, L. Jurina and F. Taddei, Unilateral Contact, Elastoplasticity and Complementarity with Reference to Offshore Pipeline Design. *Comp. Meth. App. Mech. Eng.* **17**/**18**(1979) 469–495.

[195] G. Maier, F. Giannessi and A. Nappi, Indirect Identification of Yield Limits by Mathematical Programming. *Eng. Structures* **4**(1982) 86–98.

[196] J. Mandel, *Plasticité classique et viscoplasticité.* CISM Course No. 97, Udine 1971. Springer-Verlag Wien 1972.

[197] G. A. Maugin, The Method of Virtual Power in Continuum Mechanics. Application to Coupled Fields. *Acta Meccanica* **35** (1980) 1–70.

[198] Ch. Massonnet, Résolution graphomécanique de problèmes généraux de l'élasticité plane. *Bull. CERES* **4**(1949) 3–183.

[199] Ch. Massonnet, Numerical Use of Integral Procedures. In: *Stress Analysis*, (ed. by O. C. Zienkiewicz and G. S. Holister) J. Wiley and Sons, London 1965.

[200] H. Matthies, G. Strang and E. Christiansen, The Saddle-point of a Differential Program. In: *Energy Methods in Finite Element Analysis.* (ed. by R. Glowinski, E. Y. Rodin, O. C. Zienkiewicz). John Wiley and Sons, New York 1979.

[201] H. Mecking, Deformation of Single Crystal. *Proc. 5th Int. Conf. on Textures of Materials* (ed. by E. Gottstein and G. Lücke) Springer-Verlag, Berlin 1978.

[202] R. Michalowski and Z. Mroz, Associated and Non-associated Sliding Rules in Contact Friction Problems. *Arch. of Mech.* (*Arch. Mech. Stosowanej*) **30**(1978) 259–276.

[203] E. Miersemann, Verzweigungsprobleme für Variationsungleichungen. *Math. Nachr.* **65**(1975) 187–209.

[204] F. Mignôt, Contrôle dans les inéquations variationnelles elliptiques. *J. Funct. Anal.* **22**(1976) 130–185.

[205] F. Mignôt and J. P. Puel, Sur un modèle de flambage pour une plaque mince élastoplastique. *C. R. Acad. Sc. Paris* **290A**(1980) 519–521.

[206] S. G. Mikhlin, *The Numerical Performance of Variational Methods.* Wolters-Noordhoff Pub., Groningen 1971.

[207] E. Mitsopoulou, Unilateral Contact, Dynamic Analysis of Beams by a Time-stepping, Quadratic Programming Procedure. *Meccanica* **18**(1983) 254–265.

[208] E. Mitsopoulou and P. D. Panagiotopoulos, On the Elastoplastic Torsion: Mathematical Theory of the Approximation of the Solution and Numerical Calculation. *Technica Chronica* **46**(1976) 29–37 (in Greek).

[209] J. J. Moreau, Fonctionnelles sous-différentiables. *C. R. Acad. Sc. Paris* **257A**(1963) 4117–4119.

[210] J. J. Moreau, Fonctionnelles Convexes. Séminaire sur les équations aux dérivées partielles. Collège de France 1967.

[211] J. J. Moreau, La notion de sur-potentiel et les liaisons unilatérales en élastostatique. *C. R. Acad. Sc. Paris* **267A**(1968) 954–957.

[212] J. J. Moreau, Sur les lois de frottement, de plasticité et de viscosité. *C. R. Acad. Sc. Paris* **271A**(1970) 608–611.

[213] J. J. Moreau, On Unilateral Constraints, Friction and Plasticity. In: *New Variational Techniques in Mathematical Physics*, C.I.M.E. Edizioni Cremonese, Roma 1974.

[214] J. J. Moreau, Evolution Problem Associated with a Moving Convex Set in a Hilbert Space. *J. Diff. Eqs.* **26**(1977) 347–374.

[215] U. Mosco, Approximation of the Solutions of Some Variational Inequalities. *Ann. Sc. Norm. Sup. Pisa* **21**(1967) 373–394.

[216] U. Mosco, On the Continuity of the Young-Fenchel Transform. *J. Math. Anal. Appl.* **35**(1972) 518–535.

[217] P. P. Mosolov and V. P. Miasnikov, Variational Methods in the Theory of the Fluidity of a Viscousplastic Medium. *P.M.M.* **29**(1965) 545–577.

[218] W. Moszyński, Frottement des corps solides dans le cas de l' anisotropie naturelle et artificielle. *Bull. Acad. Polon. Sci. Lettres Série* **7** (1951) 447–450.

[219] Z. Mroz, *Mathematical Models of Inelastic Material Behaviour*. Univ. of Waterloo Press, Waterloo 1973.

[220] M. Müller and J. Naumann, On Evolution Inequalities of a Modified Navier-Stokes type, I. *Apl. Matematiky* **23**(1978) 174–184.

[221] N. I. Muskhelishvili, *Some Basic Problems of the Mathematical Theory of Elasticity*. Noordhoff Int. Pub. Leyden 1975.

[222] M. A. Muspratt, Stochastic Plastic Analysis. *Arch. of Mech.* (*Arch. Mech. Stosowanej*) **25**(1973) 165–177.

[223] H. Mutschmann, Die Stufen der Wahrscheinlichkeit by Carneades. *Rheinishes Museum* **66**(1911) 79–85.

[224] A. Nadai, *Theory of Flow and Fracture of Solids*, Vol. II. McGraw-Hill, New York 1963.

[225] J. Naumann, On Some Unilateral Boundary Value Problems for the von Kármán Equations Part I. The coercive case. *Apl. Matematiky* **20**(1975) 96–125.

[226] J. Naumann and H. U. Wenk, On Eigenvalue Problems for Variational Inequalities. *Rend. Di Matematica, Univ. Roma* **9** VI(1976) 439–463.

[227] J. Naumann and I. Hlavaček, Inhomogeneous Boundary Value Problem for the von Kármán Equations II. *Apl. Matematiky* **20**(1975) 280–297.

[228] B. Nayroles, Opérations algébriques en Mécanique de structures. *C. R. Acad. Sc. Paris* **273A**(1971) 1075–1078.

[229] B. Nayroles, Point de vue algébrique. Convexité et intégrandes convexes en Mécanique des solides. In: *New Variational Techniques in Mathematical Physics.* C.I.M.E. Edizioni Cremonese, Roma 1974.

[230] B. Nayroles, J. J. Moreau's approach to a fundamental problem: the quasistatic evolution of a perfectly elastoplastic body. *Arch. of Mech.* (*Arch. Mech. Stosowanej*) **28**(1976) 115–132.

[231] J. Nečas, *Les méthodes directes en théorie des équations elliptiques.* Masson, Paris and Academia Prague 1967.

[232] J. Nečas, J. Jarušek and J. Haslinger, On the Solution of the Variational Inequality to the Signorini Problem with Small Friction. *Bulletino U.M.I.* **17B**(1980) 796–811.

[233] Nguyen Quoc Son, Méthodes énergétiques en Mécanique de la rupture. *J. de Mécanique* **19**(1980) 363–386.

[234] G. Nitsiotas, Zur Bildung und Berechnung statisch bestimmter Tragwerke mit Verbindungen aus Seilen. *Bauingenieur* **42**(1967) 412–414.

[235] G. Nitsiotas, Die Berechnung statisch unbestimmter Tragwerke mit einseitigen Bindungen. *Ing. Archiv* **41**(1971) 46–60.

[236] G. Nitsiotas, Elastostatics. Linear Theory Vol. I, II Thessaloniki 1978 (in Greek).

[237] N. Okamoto and M. Nakazawa, Finite Element Incremental Contact Analysis with Various Frictional Conditions. *Int. J. Num. Meth. Eng.* **14**(1979) 337–349.

[238] E. R. Oliveira, Plane Stress Analysis by a General Integral Method. *ASCE, EM* 1, **94**(1968) 79–101.

[239] W. Olszak, Z. Mróz and P. Perzyna, *Recent Trends in the Developments of the Theory of Plasticity.* Pergamon Press Oxford and PWN-Polish Sci. Publ. Warszawa 1963.

[240] A. C. Palmer, G. Maier and D. C. Drucker, Normality Relations and Convexity of Yield Surfaces for Unstable Materials or Structural Elements. *Trans. ASME* **24**(1967) 464–470.

[241] P. D. Panagiotopoulos, A Nonlinear Programming Approach to the Unilateral Contact- and Friction-Boundary Value Problem in the Theory of Elasticity. *Ing. Archiv* **44**(1975) 421–432.

[242] P. D. Panagiotopoulos, Subdifferentials and Optimal Control in Unilateral Elasticity. *Mech. Res. Comm.* **3**(1976) 91–96.

[243] P. D. Panagiotopoulos, A Variational Inequality Approach to the Inelastic Stress-Unilateral Analysis of Cable-Structures. *Comp. and Struct.* **6**(1976) 133–139.

[244] P. D. Panagiotopoulos, Convex Analysis and Unilateral Static Problems. *Ing. Archiv* **45**(1976) 55–68.

[245] P. D. Panagiotopoulos, A Systematic Approach to the Search for Variational Principles of Bilateral and Unilateral Problems. *ZAMM* **57**(1977) T246–T247.

[246] P. D. Panagiotopoulos, Optimal Control in the Unilateral Thin Plate Theory. *Arch. of Mech.* (*Arch. Mech. Stosowanej*) **29**(1977) 25–39.

[247] P. D. Panagiotopoulos, Analysis of the Stochastic "Unilateral Contact" Problem. *Comp. and Struct.* **7**(1977) 581–586.

[248] P. D. Panagiotopoulos, Ungleichungsprobleme in der Mechanik, Habilitation thesis, Aachen 1977.

[249] P. D. Panagiotopoulos, A Variational Inequality Approach to the Friction Problem of Structures with Convex Strain Energy Density and Application to the Frictional Unilateral Contact Problem. *J. Struct. Mech.* **6**(1978) 303–318.

[250] P. D. Panagiotopoulos, Stochastische holonome elastoplastische Torsion von zylindrischen oder prismatischen Stäben. *Ing. Archiv* **47**(1978) 421–431.

[251] P. D. Panagiotopoulos, A Variational Inequality Approach to the Dynamic Unilateral Contact Problem of Elastoplastic Foundations. *Proc. 3rd Int. Conf. Num. Meth. in Geomechanics* 1979 *Aachen*, Vol. I. A. A. Balkema, Rotterdam 1979.

[252] P. D. Panagiotopoulos and U. Zastrow, An Integral Equation Approach to the Unilateral Contact B.V.P. *ZAMM* **54**(1979) 392–393.

[253] P. D. Panagiotopoulos, Time-Space Unilateral Variations and Variational Inequalities in Relativistic Mechanics. *ZAMM* **60**(1980) T264–T265.

[254] P. D. Panagiotopoulos, Optimal Control of Unilateral Structural Analysis Problems. *Proc. IUTAM Conf.* 1979 *on Struct. Control, Waterloo*. Univ. of Waterloo Press and Noordhoff-Sijthoff Pub. Co. 1980.

[255] P. D. Panagiotopoulos, Optimal Control in the Theory of the Unilateral von-Kármán Plates. Proc. *IUTAM Conf.* 1978 *on Variational Methods in Mechanics of Solids, Evanston*. Pergamon Press, London 1980.

[256] P. D. Panagiotopoulos and D. Talaslidis, A Linear Analysis Approach to the Solution of Certain Classes of Variational Inequality Problems in Structural Analysis. *Int. J. Solids Struct.* **16**(1980) 991–1005.

[257] P. D. Panagiotopoulos, Dynamic and Incremental Variational Inequality Principles, Differential Inclusions and their Applications to Co-Existent Phases Problems. *Acta Mechanica* **40**(1981) 85–107.

[258] P. D. Panagiotopoulos, Non-convex Superpotentials in the Sence of F. H. Clarke and Applications. *Mech. Res. Comm.* **8**(1981) 335–340.

[259] P. D. Panagiotopoulos, Optimal Control and Parameter Identification of Structures with Convex and Non-Convex Strain Energy Density. Applications to Elastoplasticity and to Contact Problems. *S. M. Archives* **8**(1983) 363–411.

[260] P. D. Panagiotopoulos, Nonconvex Energy Functions. Hemivariational Inequalities and Substationarity Principles. *Acta Mechanica* **42**(1983) 160–183.

[261] P. D. Panagiotopoulos, Non-convex Energy Functionals. Application to Non-convex Elastoplasticity. *Mech. Res. Comm.* **9**(1982) 23–29.

[262] P. D. Panagiotopoulos, A Boundary Integral Inclusion Approach to Unilateral B.V.Ps.in Elastostatics. *Mech. Res. Comm.* **10**(1983) 91–96.

[263] J. Peetre, Mixed Problems for higher order Elliptic Equations in two variables, I, II. *Ann. Sc. Norm. Sup. Pisa* **15**(1961) 337–353 and **17** (1963) 1–12.

[264] H. Peterson, Application of the Finite Element Method in the Analysis of Contact Problems. In: *Finite Elements in Nonlinear Mechanics* (Ed. by P. Bergan et al.), Vol. 2, Tapir, New York 1978.

[265] A. Phillips, A Theory of ideal Locking Materials. *Trans. Soc. Rheol.* **3**(1959) 13–26.

[266] A. Phillips and M. Eisenberg, Observations on certain Inequality Conditions in Plasticity. *Int. J. Nonlinear Mech.* **1**(1966) 247–256.

[267] E. Pflaumann and H. Unger, *Funktionalanalysis Vol. I, II.* Bibliographisches Institut, Mannheim 1974.

[268] M. Potier-Ferry, Problèmes semi-coercifs. Applications aux plaques de von Kármán. *J. Math. pures et appl.* **53**(1974) 331–346.

[269] M. Potier-Ferry, Problèmes unilatéraux en théorie des plaques non linéaires. Thèse de 3ème cycle, Université Paris VI, 1974.

[270] M. Potier-Ferry, Sur un système d' équations rencontré en théorie des plaques. *C. R. Acad. Sc. Paris* **280A** (1975) 1385–1387.

[271] M. Potier-Ferry, Charges limites et théorème d' existence en théorie des plaques élastiques. *C. R. Acad. Sc. Paris* **280A** (1975) 1317–1319.

[272] W. Prager, On Ideal-locking Materials. *Trans. Soc. Rheol.* **1**(1957) 169–175.

[273] W. Prager, Elastic Solids of Limited Compressibility. Proc. 9th Int. Congress Appl. Mech. Brussels 1958, Vol. 5.

[274] W. Prager, *Einführung in die Kontinuumsmechanik.* Birkhäuser Verlag, Basel 1961.

[275] W. Prager and P. G. Hodge, *Theory of Perfectly Plastic Solids.* Dover Publ., New York 1968.

[276] J. S. Przemieniecki, *Theory of Matrix Structural Analysis.* McGraw-Hill Book Co., New York 1968.

[277] B. V. Querenburg, *Mengentheoretische Topologie.* Springer-Verlag, Berlin 1973.

[278] K. A. Reckling, *Plastizitätstheorie und ihre Anwendung auf Festigkeitsprobleme.* Springer-Verlag, Berlin 1967.

[279] G. Rieder, Iterationsverfahren und Operatorgleichungen in der Elastizitätstheorie. *Abh. Braunschw. Wiss. Ges.* **14**(1962) 109–343.

[280] G. Rieder, Mechanische Deutung und Klassifizierung einiger Integralverfahren der ebenen Elastizitätstheorie I, II. *Bull. Acad. Pol. Sci. Sér. Sci. Techn.* **16**(1968) 101–114.

[281] G. Rieder, On the Physical Interpretation of Certain Inner Products as a Guide to the Application of Functional Analysis. In: *Applications of Methods of Functional Analysis to Problems in Mechanics.* (ed. by P. Germain and B. Nayroles) Springer-Verlag, Berlin 1976.

[282] G. Rieder, U. Heise, U. Pahnke, H. Antes, H. Glahn and V. Kompiš, *Berechnung von elastischen Spannungen in beliebig krummlinig berandeten Scheiben und Platten.* Westdeutscher Verlag, Opladen 1976.

[283] F. J. Rizzo, An Integral Equation Approach to Boundary Value Problems of Classical Elastostatics. *Quar. Appl. Math.* **25**(1967) 83–95.

[284] A. P. Robertson and W. Robertson, *Topologische Vektorräume*. Bibliographisches Institut, Mannheim 1967.

[285] R. T. Rockafellar, Extension of Fenchel's Duality Theorem for Convex Functions. *Duke Math. J.* **33**(1960) 81–90.

[286] R. T. Rockafellar, *Monotone Processes of Convex and Concave Type*, Memoirs of the A.M.S. No. 77. A.M.S. Providence 1967.

[287] R. T. Rockafellar, Integrals which are Convex Functionals. *Pacific J. Math.* **24**(1968) 525–539.

[288] R. T. Rockafellar, *Convex Analysis*. Princeton Univ. Press, Princeton 1970.

[289] R. T. Rockafellar, On the Maximal Monotonicity of Subdifferential Mappings. *Pacific J. Math.* **33**(1970) 209–216.

[290] R. T. Rockafellar, Integrals which are Convex Functionals II. *Pacific J. Math.* **39**(1971) 439–469.

[291] R. T. Rockafellar, Convex Integral Functionals and Duality. In: *Contribution to Nonlinear Functional Analysis*. (ed. by E. H. Zarantonello). Academic Press, New York 1971.

[292] R. T. Rockafellar. *La théorie des sous-gradients et ses applications à l' optimization. Fonctions convexes et non-convexes.* Les Presses de l' Université de Montréal, Montréal 1979.

[293] R. T. Rockafellar, Generalized Directional Derivatives and Subgradients of Non-convex Functions. *Can. J. Math.* **XXXII** (1980) 257–280.

[294] S. K. Shan and I. S. Tuba, A Finite Element Method of Contact Problems of Solid Bodies-Part I: Theory and Validation. *Int. J. Mech. Sci.* **13**(1971) 615–627. Part II: Application to Turbine Blade Fastening. *Int. J. Mech. Sci.* **13**(1971) 627–638.

[295] M. Schatzman, Le Système Differentiel $d^2u/dt^2 + \partial\phi(u) = f$ avec conditions initiales. *C. R. Acad. Sc. Paris*, **284A**(1972) 603–606.

[296] M. Schatzman, Problèmes aux limites non-linéaires noncoercives. *Ann. Sc. Norm. Sup. Pisa* **XXVII, III**(1973) 641–686.

[297] L. Schwartz, *Théorie des distributions*. Hermann, Paris 1973.

[298] J. Salençon and A. Tristán-Lopez, Analyse de la stabilité des talus en sols cohérents anisotropes. *C. R. Acad. Sc. Paris*, **290B**(1980) 493–496.

[299] A. F. Sayegh, Elastic Analysis with Indeterminate Boundary Conditions. *A.S.C.E., EM*1, **100**(1974) 49–62.

[300] M. Sibony, Sur l' approximation d' équations et inéquations aux dérivées partielles non linéaires de type monotone. *J. Math. Anal. Appl.* **34**(1971) 502–564.

[301] A. Signorini, *Sopra alcune questioni di elastostatica.* Atti Soc. Ital. per il progesso delle Scienze 1933.

[302] W. I. Smirnow, *Lehrgang der höheren Mathematik*, *Teil V*. Deutscher Verlag der Wissenschaften, Berlin 1973.

[303] N. E. Stavrakova, The Principle of Hamilton-Ostrogradskii for Systems with One-sided Constraints. *PMM* **29**(1965) 738–741.

[304] P. M. Suquet, Un espace fonctionnel pour les équations de la plasticité. *C. R. Acad. Sc. Paris* **286A**(1978) 1129–1132.

[305] P. M. Suquet, Un espace fonctionnel pour les équations de la plasticité. *Ann. Fac. Sc. Toulouse* **1**(1979) 77–87.

[306] P. M. Suquet, Plasticité et Homogénéisation. Thèse de doctorat d' Etat, *Univ. Paris* **VI**, 1982.

[307] P. M. Suquet, Evolution Problems for a Class of Dissipative Materials. *Quart. Appl. Math.* **8**(1981) 391–414.

[308] D. Talaslidis and W. Wunderlich, Static and Dynamic Analysis of Kirchhoff Shells Based on a Mixed Finite Element Formulation. *Comp. and Struct.* **10**(1979) 239–251.

[309] D. Talaslidis and P. D. Panagiotopoulos, Linear Finite Element Approach to Unilateral Contact Problem in Structural Dynamics. *Int. J. Num. Meth. in Eng.* **18**(1982) 1505–1520.

[310] T. P. Tassios and P. J. Yannopoulos, Analytical Studies of Reinforced Concrete Members Under Cyclic Loading Based on Bond Stress-Slip Relationships, *ACI J.* **78**(1981) 206–216.

[311] K. Taubert, Über die Existenz und Approximation von Lösungen von Volterra Funktional-Differentialgleichungen mit schwacher rechten Seite. Habilitation thesis, Hamburg 1977.

[312] R. Temam, *Navier-Stokes Equations.* North Holland Pub. Co., Amsterdam 1972.

[313] R. Temam and G. Strang, Duality and Relaxation in the Variational Problems of Plasticity. *J. de Mécanique* **19**(1980) 493–527.

[314] R. Temam and G. Strang, Functions of Bounded Deformation. *Arch. Rat. Mech. Anal.* **75**(1980) 7–21.

[315] S. Timoshenko and S. Woinowsky-Krieger, *Theory of Plates and Shells.* McGraw-Hill, New York 1969.

[316] T. W. Ting, Elastic-plastic Torsion of Simply Connected Cylindrical Bars. *Ind. Univ. Math. J.* **20**(1971) 1047–1076.

[317] J. F. Toland, A Duality Principle for Non-convex Optimization and the Calculus of Variations. *Arch. Rat. Mech. Anal.* **71**(1979) 41–61.

[318] E. Tonti, A Systematic Approach to the Search for Variational Principles. In: *Variational Methods in Engineering* (ed. by C. A. Brebbia and H. Tottenham) Vol. I. Southampton Univ. Press, Southampton 1973.

[319] E. Tonti, The Reason for Analogies between Physical Theories. *Appl. Math. Modelling* **1**(1976) 37–50.

[320] F. Treves, *Topological Vector Spaces, Distributions and Kernels.* Academic Press, New York 1967.

[321] C. Truesdell and R. A. Tupin, The Classical Field Theories. In: *Encyclopedia of Physics.* Vol. III/1 (ed. by S. Flügge) Springer-Verlag, Berlin 1960.

[322] C. Truesdell and W. Noll, The Non-linear Field Theories of Mechanics. In: *Encyclopedia of Physics.* Vol. III/3 (ed. by S. Flügge) Springer-Verlag, Berlin 1965.

[323] C. Truesdell, *Six Lectures on Modern Natural Philosophy*, Springer-Verlag, Berlin 1966.

[324] M. M. Vainberg, *Variational Method and Method of Monotone Operators in the Theory of Nonlinear Equations.* John Wiley and Sons, New York and Israel Program for Scientific Translations, Jerusalem 1973.

[325] N. Ya. Vilenkin eta al., *Functional Analysis.* Wolters-Noordhoff Pub., Groningen 1972.

[326] W. Walter, *Differential- und Integral-Ungleichungen.* Springer-Verlag, Berlin 1964.

[327] J. Warga, Necessary Conditions without Differentiability Assumptions in Optimal Control. *J. Dif. Equations* **18**(1975) 41–62.

[328] J. Warga, Necessary Conditions without Differentiability Assumptions in Unilateral Control Problems. *J. Dif. Equations* **21**(1976) 25–38.

[329] J. Warga, Controlability and Necessary Conditions in Unilateral Problems without Differentiability Assumptions. *SIAM J. Control and Optimization* **14**(1976) 546–573.

[330] J. Warga, *Optimal Control of Differential and Functional Equations.* Nauka, Moskow 1978 (in Russian).

[331] J. Warga, Derivate Containers, Inverse Functions and Controllability. In: *Calculus of Variations and Control Theory* (ed. by D. L. Russell) Acad. Press New York 1976.

[332] J. Warga, On Bounding, Interior and Covering Functions (to appear).

[333] K. Washizu, *Variational Methods in Elasticity and Plasticity.* Pergamon Press, Oxford 1968.

[334] E. Weinel, Die Integralgleichung des ebenen Spannungszustandes und der Plattentheorie. *ZAMM* **11**(1931) 349–360.

[335] W. L. Wendland, E. Stephan and G. C. Hsiao, Integral Equations for Plane Mixed Boundary Value Problems. *Math. Meth. in the Appl. Sciences* **1**(1979) 265–321.

[336] E. A. Wilson and B. Parsons, Finite Element Analysis of Elastic Contact Problems Using Differential Displacements. *Int. J. Num. Meth. Eng.* **2**(1970) 387–395.

[337] K. Yosida, *Functional Analysis.* Springer-Verlag, Berlin 1966.

[338] L. C. Young, *Lectures on the Calculus of Variations and Optimal Control Theory.* W. B. Saunders Co., Philadelphia 1969.

[339] J. P. Yvon, Étude des quelques problèmes de contrôle pour des systèmes distribués. Thèse de Doctorat d' Etat, Université Paris VI 1973.

[340] M. W. Zemansky and R. H. Dittman, *Heat and Thermodynamics.* McGraw-Hill, New York 1980 (6th Edition).

[341] A. Zhukov, Plastic Deformation of Isotropic Metals in Combined Loading. *Izv. Akad. Nauk. SSSR. OTN* **12**(1956) 72–87 (in Russian).

[342] H. Ziegler, A Possible Generalization of Onsager's Theory. *Advances in Solid Mechanics,* Acad. Press, New York 1962.

[343] H. Ziegler, *Some Extremum Principles in Irreversible Thermodynamics with Application to Continuum Mechanics.* Progress in Solid Mechanics IV, North Holland Pub. Co., Amsterdam 1963.

[344] H. Zicgler, Discussion of some Objections to Thermomechanical Orthogonality. *Ing. Archiv.* **50**(1981) 149–164.

[345] O. C. Zienckiewicz, *The Finite Element Method,* McGraw-Hill, London 1977 (3rd edition).

References added in proof:

[346] H. Floegl and H. A. Mang, Tension Stiffening Concept Based on Bond Slip. *ASCE, ST* 12, **108** (1982) 2681–2701.

[347] J. Rauch, Discontinuous Semilinear Differential Equations and Multiple Valued Maps. *Proc. A.M.S.* **64** (1977) 277–282.

[348] P. D. Panagiopoulos and A. Avdelas, A Hemivariational Inequality Approach to the Unilateral Contact Problem and Substationarity Principles (to appear in *Ing. Archiv.* **54,** 1984).

[349] H. O. May, Das Hamiltonsche Prinzip für einseitige Probleme in einer Raum-Zeit-invarianten Analytischen Mechanik, Doct. thesis, RWTH Aachen 1983.

[350] P. D. Panagiotopoulos, Ungleichungsprobleme und Differentialinklusionen in der Analytischen Mechanik. *Annual Publ. School of Tech. Aristotle Univ. Thessaloniki* **Θ**' (1982) 1–30.

[351] P. D. Panagiotopoulos, Une génèralization non-convexe de la notion du surpotentiel et ses applications. *C. R. Acad. Sc. Paris* **296 II** (1983) 1105–1108.

[352] J. Haslinger and P. D. Panagiotopoulos, The reciprocal variational approach to the Signorini problem with friction. Approximation results. *Proc. Royal Soc. of Edinburgh* **98** (1984) 250–265.

[353] E. Delassus, Mémoire sur la théorie des liaisons finies unilatérales. *Ann. Éc. Norm* (3), **XXXIV** (1917) 95–179.

[354] I. Hlavaček, J. Haslinger, J. Nečas and J. Lovišek, *Riešenie variačných nerovností v mechanike.* ALFA, Bratislava 1982.

[355] R. Glowinki, *Numerical Methods for Nonlinear Variational Problems.* Springer-Verlag, New York-Berlin 1984.

[356] F. H. Clarke, *Optimization and Nonsmooth Analysis.* Wiley, New York 1983.

[357] V. Barbu, *Optimal Control of Variational Inequalities.* Pitman, Research Notes in Math. Vol. 100, Boston-London 1984.

[358] H. O. May, Generalized Variational Principles and Nondifferentiable Potentials in Analytical Mechanics (to appear in *J. Math. Phys.* **25,** 1984).

[359] A. D. Ioffe, Nonsmooth Analysis: Differential Calculus of Nondifferentiable Mappings. *Trans. A.M.S.,* **266** (1981) 1–56.

[360] N. Papageorgiou, Stochastic Nonsmooth Analysis and Optimization I (to appear in *Trans. A.M.S.*).

[361] P. D. Panagiotopoulos, Hemivariational Inequalities and Substationarity in the Static Theory of v. Kármán plates (to appear in *ZAMM* **65,** 1985).

Subject Index

GPSR Compliance

The European Union's (EU) General Product Safety Regulation (GPSR) is a set of rules that requires consumer products to be safe and our obligations to ensure this.

If you have any concerns about our products, you can contact us on ProductSafety@springernature.com

In case Publisher is established outside the EU, the EU authorized representative is:

Springer Nature Customer Service Center GmbH
Europaplatz 3
69115 Heidelberg, Germany

Batch number: 10578397

Printed by Printforce, the Netherlands